Circular Economy and Sustainable Business Performance Management

Circular Economy and Sustainable Business Performance Management

Editors

Claudio Sassanelli
Sergio Terzi

MDPI • Basel • Beijing • Wuhan • Barcelona • Belgrade • Manchester • Tokyo • Cluj • Tianjin

Editors
Claudio Sassanelli
Department of Mechanics, Mathematics and Management
Politecnico di Bari
Bari
Italy

Sergio Terzi
Department of Management, Economics and Industrial Engineering
Politecnico di Milano
Milano
Italy

Editorial Office
MDPI
St. Alban-Anlage 66
4052 Basel, Switzerland

This is a reprint of articles from the Special Issue published online in the open access journal *Sustainability* (ISSN 2071-1050) (available at: www.mdpi.com/journal/sustainability/special_issues/Circular_Economy_Sustainable_Business).

For citation purposes, cite each article independently as indicated on the article page online and as indicated below:

LastName, A.A.; LastName, B.B.; LastName, C.C. Article Title. *Journal Name* **Year**, *Volume Number*, Page Range.

ISBN 978-3-0365-8543-7 (Hbk)
ISBN 978-3-0365-8542-0 (PDF)

Cover image courtesy of Claudio Sassanelli

Contents

About the Editors

Claudio Sassanelli

Claudio Sassanelli is Assistant Professor at Politecnico di Bari, Department of Mechanics, Mathematics, and Management, and Senior Research Fellow at the École des Ponts Business School of École des Ponts ParisTech. He received his two master's degrees in Management and Civil Engineering from Politecnico di Bari, respectively, in 2010 and 2013, and his PhD in Management, Economics, and Industrial Engineering from Politecnico di Milano in 2017. He also held visiting researcher positions at Tokyo Metropolitan University (TMU) and Universidade de São Paulo (USP). His main research interest is Product-Service System (PSS) design, specifically addressing Product Lifecycle Management (PLM), Design for X (DfX) approaches, and Circular Economy and Industry 4.0 paradigms. To advance these research domains, he manages Special Issues in international journals as a Guest Editor and disseminates his research by being the co-author of more than 85 publications in international journals, international conference proceedings, and book chapters in the field. He is a member of IFIP WG 5.1 and of the editorial board of the Journals *Sustainability MDPI, Frontiers in Sustainability, Frontiers in Environmental Science*, and *Academia Engineering*. He has a 10-year experience in research, industrial and European research, and innovation action projects. He has been carrying out teaching activities in courses on production systems (environmental management, design and management, industrial technologies, and quality design and management) at Politecnico di Bari, Politecnico di Milano, SUPSI, and LIUC Università Cattaneo.

Sergio Terzi

Sergio Terzi is Full Professor at Politecnico di Milano, Department of Economics, Management and Industrial Engineering, Manufacturing Group. He holds a PhD from Politecnico di Milano and University of Lorraine (CRAN Laboratories). He is the author of four books and more than 200 papers at national and international level. He is member of the Editorial Board of the International Journal on Product Lifecycle Management , member of the IFIP WG 5.7 and vice-chair of the IFIP WG 5.1. He used to work on international and national research projects, generally with dissemination and exploitation responsibilities.

Preface to "Circular Economy and Sustainable Business Performance Management"

Today, more than ever, the world needs to be considered a finite and limited system, characterised by scarce resources and as a place where restocking is not possible in an infinite way. As such, careful resource management needs to be planned and set by the concurrent actions of heterogeneous stakeholders, from policymakers up to academics and practitioners, to effectively implement a circular economy. The different resources involved along the extended product lifecycle need to be adequately managed through innovative business models and design practises, coupled with reverse logistics and digital technology adoption, while methods and ways to measure and assess circularity performance are also needed.

This reprint highlights new opportunities and challenges for the Circular Economy and Sustainable Business Performance Management, focusing on technological advancements and management initiatives, including public-private partnerships between stakeholders.

The contributions gathered in this reprint are addressed to a wide spectrum of stakeholders. They are dedicated to academics who want to explore the research domain of performance measurements of Circular Economy systems, bolstering them in the development and application of new methods, models, and tools capable of easing and enhancing the activities needed to be performed to successfully measure circular and sustainable solutions supported by digital technologies. In addition, the reprint can also be useful for technology users and providers who would like to understand which technologies are most suitable to bolster Circular Economy performance measurement practises (given the tight bond between the Circular Economy and Industry 4.0 paradigms). This collection of contributions could constitute a guide for companies to lead their businesses towards a fair transition towards the Circular Economy, enabling them to measure circular operations under a data-driven circular manufacturing approach. Finally, it can be useful to policymakers, who should bolster companies and academics with practical actions and support the adoption of tailored solutions to compel them towards the total embracement, application, measurement, and control of the Circular Economy.

Editors would like to sincerely thank Mr. Samuel Li (Section Managing Editor of Sustainability MDPI) for his continuous, effective, and dedicated support both during the Special Issue time-lapse and the related editorial and reprint development.

Claudio Sassanelli and Sergio Terzi
Editors

sustainability

MDPI

Editorial

Circular Economy and Sustainable Business Performance Management

Claudio Sassanelli [1,2,*] and Sergio Terzi [3]

1 Department of Mechanics, Mathematics and Management, Politecnico di Bari, Via Orabona 4, 70125 Bari, Italy
2 Tech Center for Good, École des Ponts Business School of École des Ponts ParisTech (ENPC), 6 Place du Colonel Bourgoin, 75012 Paris, France
3 Department of Management, Economics and Industrial Engineering, Politecnico di Milano, Piazza L. da Vinci 32, 20133 Milan, Italy; sergio.terzi@polimi.it
* Correspondence: claudio.sassanelli@poliba.it

Today, more than ever, the world needs to be considered as a finite and limited system, characterized by scarce resources and as a place where restocking is not possible in an infinite way. As such, careful resource management needs to be planned and set by the concurrent actions of heterogeneous stakeholders, from policy makers up to academics and industrialists, in order to effectively implement the Circular Economy (CE) paradigm [1] and to be able to pursue sustainability in time. The different resources involved along the extended product lifecycle need to be adequately managed through innovative business models and design practices, coupled with reverse logistics and digital technology adoption. So far, several methods (e.g., Life Cycle Assessment (LCA) and Life Cycle Costing (LCC), Multi Criteria Decision Methods (MCDM), Material Flow Analysis (MFA), Design for X (DfX)) have been adopted and combined in different ways to measure and assess the circular performance of a system [2]. In addition, indicators able to measure the CE are not directly bonded to the firm's organizational functions involved in CE assessment [3]. With the aim of starting to fill this gap [4], conducted a systematic literature review to map them into 23 categories of CE micro-level metrics and compared them to Porter's Value Chain to detect the metrics' link with archetypal companies' organizational functions. Attempting to bridge methods and metrics in a unique methodology for CE performance assessment [5], proposed a novel methodology, the Circular Economy Performance Assessment (CEPA), mainly based on LCA and LCC and proposing a set of KPIs useful for the quantitative assessment of circular business models.

With the aim of addressing sustainable development, the CE can be adopted in manufacturing companies through the adoption of different Circular Manufacturing (CM) strategies (e.g., recycling, remanufacturing) [6]. Manufacturing companies are attempting to implement these strategies to limit their resource consumption and pollution generation. However, they are still not fully ready and mature enough to employ and deploy CE strategies and related practices in their processes. Indeed, the CE paradigm asks for multiple interventions in the organization (from business models and organizational ones through technological and competence-driven ones, up to data management ones). In this context, some qualitative models and methods have been proposed in the literature to help companies to realize which is their circular level and define roadmaps towards their circular level improvement. In addition, companies could benefit from a set of advantages led by fully embracing the CE paradigm. On the other side, throughout the circular transition, not only several hurdles can be encountered but also a rebound effect could result from the adoption of the CE [7,8]. Related to this, it is still not clear how to systematically measure both the negative and positive impacts occurring throughout the circular transition.

This Editorial refers to the Special Issue "*Circular Economy and Sustainable Business Performance Management*". The Special Issue highlights new opportunities and challenges for

Citation: Sassanelli, C.; Terzi, S. Circular Economy and Sustainable Business Performance Management. *Sustainability* **2023**, *15*, 8619. https://doi.org/10.3390/su15118619

Received: 17 May 2023
Accepted: 20 May 2023
Published: 25 May 2023

advancing the performance assessment of the CE, focusing on technological advancements and management initiatives, and including public–private partnerships between stakeholders.

Twenty-two manuscripts were submitted for consideration for the Special Issue, and all of them were subject to the rigorous *Sustainability* review process. In total, fourteen papers were finally accepted for publication and inclusion in this Special Issue (nine articles and five reviews). The contributions are listed below:

1. Acerbi, F., Sassanelli, C., Terzi, S., Taisch, M., 2021. A Systematic Literature Review on Data and Information Required for Circular Manufacturing Strategies Adoption. Sustainability 13, 1–27. https://doi.org/10.3390/su13042047
2. Negri, M., Neri, A., Cagno, E., Monfardini, G., 2021. Circular Economy Performance Measurement in Manufacturing Firms: A Systematic Literature Review with Insights for Small and Medium Enterprises and New Adopters. Sustain. 13, 1–27. https://doi.org/10.3390/su13169049
3. Tavera Romero, C.A., Castro, D.F., Ortiz, J.H., Khalaf, O.I., Vargas, M.A., 2021. Synergy between Circular Economy and Industry 4.0: A Literature Review Citation. Sustain. 13, 1–18. https://doi.org/10.3390/su13084331
4. Mangenda Tshiaba, S., Wang, N., Ashraf, S.F., Nazir, M., Syed, N., 2021. Measuring the Sustainable Entrepreneurial Performance of Textile-Based Small-Medium Enterprises: A Mediation-Moderation Model. Sustain. 13, 1–19. https://doi.org/10.3390/su131911050
5. Yang, L., Tan, J., Xia, W., Chi, Z., Qin, H., Gan, Q., Yang, Q., 2022. Corporate Performance, Market-Industry Competition and Enterprise Environmental-Protection Investment. Sustain. 14, 2–19. https://doi.org/10.3390/su14095459.
6. Nassani, A.A.; Hussain, H.; Rosak-szyrocka, J.; Yousaf, Z. Analyzing the Leading Role of High-Performance Work System towards Strategic Business Performance. Sustain. 2023, 15, 1–11. https://doi.org/10.3390/su15075697
7. Melo, I.C.; Junior, P.N.A.; Queiroz, G.A.; Yushimito, W.; Pereira, J. Do We Consider Sustainability When We Measure Small and Medium Enterprises' (SMEs') Performance Passing through Digital Transformation? Sustain. 2023, 15, 1–30. https://doi.org/10.3390/su15064917
8. Queiroz, G.A.; Delai, I.; Filho, A.G.A.; de Santa-Eulalia, L.A.; Torkomian, A.L.V. Synergies and Trade-Offs between Lean-Green Practices from the Perspective of Operations Strategy: A Systematic Literature Review. Sustain. 2023, 15, 1–27, https://doi.org/10.3390/su15065296.
9. Jankovic-Zugic, A.; Medic, N.; Pavlovic, M.; Todorovic, T.; Rakic, S. Servitization 4. 0 as a Trigger for Sustainable Business: Evidence from Automotive Digital Supply Chain. Sustain. 2023, 15, 1–13, https://doi.org/10.3390/su15032217.
10. Rehman, F.U.; Al-ghazali, B.M.; Farook, M.R.M. Interplay in Circular Economy Innovation, Business Model Innovation, SDGs, and Government Incentives: A Comparative Analysis of Pakistani, Malaysian, and Chinese SMEs. Sustain. 2022, 14, 1–31, https://doi.org/10.3390/su142315586.
11. Younas, A.; Kumar, L.; Deitch, M.J.; Qureshi, S.S.; Shafiq, J.; Naqvi, S.A.; Kumar, A.; Amjad, A.Q.; Nizamuddin, S. Treatment of Industrial Wastewater in a Floating Treatment Wetland: A Case Study of Sialkot Tannery. Sustain. 2022, 14, 1–20, https://doi.org/10.3390/su141912854.
12. Sá, M.M.; Oliveira-silva, C.; Cunha, M.P.; Gonçalves, A.; Diez, J.; Méndez-Tovar, I.; Izquierdo, E.C. Integration of the Circular Economy Paradigm in Companies from the Northwest of the Iberian Peninsula. Sustain. 2022, 14, 1–22, https://doi.org/10.3390/su14137940.
13. Alfarisi, S., Mitake, Y., Tsutsui, Y., Wang, H., Shimomura, Y., 2023. Nurture: A novel approach to PSS-rebound effect identification. Sustainability 15, 1–25, https://doi.org/10.3390/su15097359
14. Demko-Rihter, J., Sassanelli, C., Pantelic, M., Anisic, Z., 2023. A Framework to Assess Manufacturers' Circular Economy Readiness Level in Developing Countries: An Application Case in a Serbian Packaging Company. Sustain. 15, 1–25, https://doi.org/10.3390/su15086982

As shown in Table 1, the contributions covered large geographical areas, from specific country cases (e.g., Italy, Serbia, and China) to groups of countries (worldwide). The majority of the contributions (1, 2, 3, 4, 6, 7, 8, 9, 11, 12, 13, 14) relate to the field of business and management in the manufacturing sector. In detail, contributions 4 and 10 also pertain to macro aspects of CE performance assessment; contribution 5 argues about the financial aspects; and contribution 11 is more oriented to the hydrological factors related to the CE adoption (Table 1).

Table 1. Analysis of the published contributions in the Special Issue.

N# of Contribution	Research Area	Focus	Type of Research	Organization/Industry
1	Business and Management	Information and data management and sharing, circular manufacturing, digital technologies, data economy	Systematic Literature Review	Manufacturing
2	Business and Management	Performance indicators, small and medium enterprises	Systematic Literature Review	Manufacturing
3	Business and Management	Sustainability, Industry 4.0	Literature Review	Manufacturing
4	Business and Management, Entrepreneurship	Knowledge Management practices, sustainability entrepreneurship performance, SME, dynamic capabilities	Survey	Textile
5	Finance	Corporate performance; enterprise environmental protection investment; industry competition	Secondary data analysis and Multiple Regression Analysis	Manufacturing
6	Business and Management	Strategic business performance; organizational flexibility; high performance work system; manufacturing organizations	Quantitative research design (Structural Equation Modeling)	Manufacturing
7	Business and Management	Digitalization; small- and medium-sized enterprises (SMEs); Industry 4.0; triple bottom line of sustainability	Systematic literature review through the topic modeling method with a machine learning technique (Latent Dirichlet Allocation)	Manufacturing
8	Business and Management	Lean manufacturing; green manufacturing; competitive priorities; decision areas; sustainability	Systematic Literature Review	Manufacturing
9	Business and Management	Digital servitization; digital technologies; digital supply chain; automotive industry; Industry 4.0	Social Network Analysis (SNA) method	Automotive
10	Business and Management, Entrepreneurship	Circular economy innovation; business model innovation; government incentives; SMEs performance	Survey + Partial Least Squares Structural Equation Modeling (PLS-SEM)	Governments; Manufacturing

Table 1. *Cont.*

N# of Contribution	Research Area	Focus	Type of Research	Organization/Industry
11	Business and Management; Hydrology	Circular economy; circular bioeconomy; floating treatment wetland; phytoremediation; tropical wetlands; Typha latifolia	Model design and lab analysis	Tannery wastewater
12	Business and Management	Circular economy; industry/services strategies; manager/executive/technicians' perceptions	Survey + non-parametrical statistical tests	Manufacturing and service industry
13	Business and Management	Nurture; product-service system; rebound effect; sustainability; dematerialization; system dynamics	Feedback system thinking using system dynamics + case study	Manufacturing and service industry—Car sharing
14	Business and Management	Circular economy; readiness assessment; product lifecycle; manufacturing; KPI; developing country	Framework development and application case	Packaging

It is worth mentioning that contributions 9, 12, and 13 explored the relation between the CE and digital servitization of manufacturing [9] to advance the research on CE performance assessment through a consistent business model transition; in addition, relevance was given to the so-called rebound effect of the CE. Contribution 10 further explored the business model aspect, also evaluating the relationship of the level of business model innovation with the CE and the role of governments (through incentives). At a more micro-level, contribution 14 also raised the need for more support from the governments' side to improve CE performances and to provide a model for companies' CE readiness level assessment. On the other side, multiple researches (1, 4, 7, and 9) investigated the role of I4.0 technologies to support the gathering and evaluation of data and information with the aim of assessing CE performances. Contributions 4 and 8 focused on the adoption of practices (sustainable or lean/green) in manufacturing supporting CE performance improvement, and contribution 2 proposed a framework to ease decision-making processes from a CE perspective.

Finally, the main industry involved in these studies is manufacturing, with specific studies on the service industry (2), textile/tannery wastewater (2), automotive, and packaging.

Contribution 1 identified the pertinent data and information needed to assist the manufacturer's decision-making process in implementing and managing the various CM methods for pursuing the transition to CM using a comprehensive literature study. The research also suggests a theoretical framework based on the findings. It clarifies the four key areas that manufacturers must manage when adopting CM strategies and gives manufacturers an overview of what needs to be updated and enhanced inside the business.

Contribution 2 conducted a systematic literature analysis to better comprehend CE performance-measurement systems for manufacturing organizations from a general standpoint as well as to offer particular guidance for small and medium-sized enterprises and early adopters. The findings reveal a lack of an integrated, comprehensive, and scalable framework for monitoring the success of the circular economy, as well as a dearth of specialized guidance for small- to medium-sized businesses and early adopters.

Contribution 3 reviewed the most recent literature on the circular economy and the notion of Industry 4.0. This work's main goal was to outline the evolution of the CE and I4.0 as well as its multi-step approach of analysis. There have not been any studies up to this point that demonstrate how people are being prepared to deal with the transition from the linear economy, which is prevalent in most countries, towards a CE. It looked at the effects that technology advances have on the human person and on society.

Contribution 4 examined the role of knowledge management practices in sustainable entrepreneurship performance, also analyzing the connections between six concepts—knowledge sharing behavior, innovative capacity, absorptive capacity, dynamic capability, and opportunity recognition. The results demonstrate that knowledge management practices have a favorable and significant influence on the performance of sustainable entrepreneurship and the adaptability of SMEs. Additionally, the link between the dynamic skills of SMEs and sustained entrepreneurial success is strengthened through opportunity awareness. For scholars and practitioners interested in the topic of entrepreneurship, this study provides insightful information and useful recommendations.

Contribution 5 experimentally examined the link between corporate performance (CP) and the size of expenditure made by businesses in environmental protection (EI), starting with micro-enterprises, and looked into the moderating impact of industry rivalry on the relationship between CP and EI. The findings of the study indicate that performance has a significant impact on businesses' decisions about investments in environmental protection, and industry competitiveness can encourage businesses to make such expenditures.

Contribution 6 aimed to illustrate how high-performance work systems (HPWS) offer the foundation for strategic business performance (SBP) through the mediating function of organizational. This research, based on a quantitative approach, acquired information from top, middle, and operational management companies. The findings show that if organizational flexibility does not moderate the link between HPWS and SBP, HPWS will take a very long time to attain SBP. This research, which makes use of real data, shows useful methods for boosting manufacturing organizations' effectiveness in business growth.

Contribution 7 analyzed the literature about Digital Transformation in SMEs, focusing on performance measurement. The tools used by SMEs were analyzed under the triple bottom line perspective of sustainability (i.e., environmental, social, and economic aspects). A systematic literature review (SLR) was performed through the topic modeling method with a machine learning technique (Latent Dirichlet Allocation). The research shown that sustainability is treated as a separate topic in the literature, mostly neglecting the social and environmental aspects. This paper proposed a framework and research directions contributing to sustainable development goals (SDGs) 1, 5, 8, 9, 10, and 12 and able to guide policymakers and SMEs transitioning their production paradigm toward sustainability and digitalization.

Contribution 8 aimed to understand the relationships between Lean–Green practices from the point of view of the Operations Strategy. Synergies and potential trade-offs between competitive priorities and changes in decision areas were apparent when Lean–Green practices were investigated through a systematic literature review. The results found that Lean and Green are synergistic in most practices but must be managed according to the Operations Strategy.

Contribution 9 explored digital services in supply chains of the automotive industry. The research results indicated how suppliers affect car manufacturers to deliver digital services to their customers. Finally, this study shows that a closer interaction between manufacturers and suppliers in the manufacturing ecosystem is made possible by the integration of digital technology with product-related services. These connections let the production ecosystem withstand the impact of various conditions.

Contribution 10 investigated the impacts of the CE and business model innovation (BMI) on the economic, environmental, and social performance of Small and Medium Enterprises (SMEs) in Pakistan, Malaysia, and China, as well as the mediating function of governmental incentives. According to the findings, BMI and CE innovation have positive, noteworthy effects on the economic, environmental, and social performance of SMEs. The study also discovered that the link between CE innovation, BMI, and the economic, environmental, and social performance of SMEs can be mediated by government incentives.

Contribution 11 created a floating treatment wetland (FTW) to treat the effluent utilizing local plant species through phytoremediation in order to provide a cost-effective method for the treatment of tannery wastewater. Three distinct plant species were used to assess the FTW's effectiveness. The pilot model shows that FTWs are a cost-effective option in the installation of a costly treatment plant with high related running costs for treating effluent from tanneries. FTWs can assist in moving traditional wastewater treatment plans towards more sustainable ones in order to achieve the CE paradigm. Moreover, it is essential that the materials used for a wetland foundation have the ability to be recycled, are affordable, inexpensive, and available locally in order to adhere to the principles of the CE and ecologically friendly development.

Contribution 12 evaluated the integration of CE practices in both public and private organizations in the Iberian Peninsula's northwest. The perception of CE firms was evaluated through an online survey, containing information about the area(s) it was integrated in, why, the challenges it faced, or what was required to complete it, and how the effect of the adopted CE practices was quantified. According to the findings, businesses usually relate the CE to "resource optimisation". The primary strategic area where the CE was applied was the "Entity's vision and mission". "Environmental reasons" were the primary driving force behind entities' and organizations' adoption of the CE, while "lack of information and guidance" and "lack of financial resources" served as the primary hurdles.

Contribution 13 looked at whether nurturing should be a major concern in the product-service system, given that some features might have a rebound effect that has a big impact on meeting goals. This study showed that the business model system is intricate, with related problems and solutions. The results of this study show that the factor of nurture is a strong predictor of profit growth, but it also causes a decline in the environmental and social performance of the implementation of the product-service system, which has the impact of causing the system to rebound.

Contribution 14 developed a framework split on two levels (product and business model) for evaluating a company's CE readiness operating in underdeveloped nations. The framework helps businesses monitor a path for progress by defining subsequent activities and KPIs. Application of the methodology revealed areas for improvement, particularly in the policy environment, to encourage CE adoption in underdeveloped countries. In fact, the circular transformation process in businesses would be greatly aided by legislative incentives and tools of public authorities.

Several research gaps were detected by the set of contributions gathered in this Special Issue.

It is significant to note that, in relation to the usage of digital technologies, relatively few researchers have examined how moving to a CE backed by I4.0 might affect both people and society. The potential costs to society of the CE transition and the tools these players will have to prevent societal failure are also unknown. Adopting the many I4.0 technologies in developing economies may provide significant difficulties in addition to those relating to the system dynamics. In I4.0, the CE model enables the assessment of the revenue from chain production waste in a way that can boost ROI while lowering the environmental effects.

In addition, studies demonstrating the positive benefits of eco-conceptions, industrial and territorial ecology, the functional economy, second use, reuse, repair, recycling, and valuation from a social and political standpoint are still lacking in the field of CE performance assessment research.

Concerning data-driven circular manufacturing, to evaluate the potential effects that each piece of data and information could have on the pursuit of not only the specific circular manufacturing strategy to which it is connected but also the other strategies not theoretically intended to be bonded, a quantitative model might be created. Tracking and managing data and information might serve as the foundation for calculating the advantages of using circular manufacturing and gauging how well manufacturers use it. Long-term, this would lay the groundwork for the creation of a model that evaluates the amount of circular maturity in manufacturing firms through the formulation, computation, and monitoring of particular key performance metrics.

Additionally, a more thorough analysis of the discovered data and information should be carried out to better define their own qualities (such as accessibility and timeliness). The degree of granularity of the data might be one of the matters that receives special attention.

Future research should also focus on determining who is in charge of the collection and administration of the various types of data and information since the actors responsible for collecting them are connected to how those data and information are used. The manufacturing firm should be broken down into its individual functions as the unit of analysis, with relationships between internal managers and staff and external stakeholders from other organizations also being taken into account.

From an empirical standpoint, it could be possible to further explore the managerial and technological challenges that firms encounter when attempting to use data in circular manufacturing. In-depth research should be carried out to specifically identify and suggest to manufacturers the new processes required to capture the majority of the data and information that have emerged as being pertinent in circular manufacturing.

The integration of CE levels, theoretical development and empirical application, characteristics of the proposed indicators, considerations of sustainability, comprehensive perspectives on industrial systems, and scalability to adapt to firms' various characteristics are all areas where the current literature falls short.

An integrated, comprehensive, and scalable performance-measurement system for manufacturing enterprises is still lacking, according to the studied literature [10]. All of these qualities should be included in a successful performance-measuring system in order to decrease the measurement process' complexity.

Regarding integration, a successful performance-measurement system for the CE should give unmistakable instructions on how other paradigms inside manufacturing organizations, such as sustainability, would be covered concurrently. As a result, the CE should cover all CE levels while also comprehending how they relate to one another. Again, it is preferable to have a single, distinct method rather than many ones for gauging success at various levels and it is advised to take into account the viewpoints of various industrial decision-makers.

Concerning scalability, an efficient performance-measurement system for the CE should be adapted to various businesses, particularly SMEs and new adopters, in accordance with their unique characteristics and changing needs, in terms of the scope and depth of analysis, while also enabling internal performance measurement and benchmarking activities.

In particular, a scalable framework would permit the availability of many levels of analysis and, consequently, sets of indicators.

Grounded on the previous gaps reported above, there are several potential directions of study that could be implemented to establish a proficient performance assessment of the CE. This Special Issue, *"Circular Economy and Sustainable Business Performance Management"*, identifies the following directions:

- The exploration of how new digital technologies, gathered under the umbrella of Industry 4.0, can support the measurement and analysis of lifecycle data according to the industrial symbiosis level of the system. Since not only single companies but also industrial parks, global supply chains, urban territories, and municipal solid wastes can be taken into account as referring systems of a circular economy performance assessment, the analysis of data coming from an extended and circular supply chain gives the chance to approach very different scenarios in which circular business models have been adopted. This opens the way to also consider, in the measurement of the performance, specific building blocks such as reverse logistics and particular systems' conditions;
- The use of design practices to specifically empower product design and development according to specific measurement performances, such as DfX guidelines and checklists, and to give strategic directions to shift the linear lifecycle into a circular one;
- The development of methods and practices able to systematically and practically measure and assess the circularity degree of a given system and to take into account all the heterogeneous resources involved in its lifecycle;
- The development and adoption of methods and approaches, and of a set of related KPIs, suitable for the assessment of the circularity performance in different fields of application. These KPIs can deal with the circularity degree of the resources occurring

within the product life cycle and can also support the quantification of those that are the economic, environmental, and, most of all, social benefits of the CE. These KPIs, from a regulations and reporting perspective, can support the creation of a product certification system related to the circularity of resource flows, internal reporting and benchmarking in companies or support in the creation/enrichment of databases useful for LCA, etc. From a companies' portfolio circular innovation perspective, they can support not only the decision-making process along the design of new products but also the comparison of different versions of the same product based on their degree of circularity;

- The development of a CE maturity model, based on the definition of a set of KPIs for CE performance assessment, aimed at both defining companies' level of readiness in terms of circularity and proposing a roadmap to better address the CE.

Author Contributions: Conceptualization, C.S. and S.T.; methodology, C.S.; formal analysis, C.S.; investigation, C.S.; data curation, C.S.; writing—original draft preparation, C.S.; writing—review and editing, C.S. and S.T.; visualization, C.S.; supervision, S.T. All authors have read and agreed to the published version of the manuscript.

Conflicts of Interest: The authors declare no conflict of interest.

References

1. Geissdoerfer, M.; Savaget, P.; Bocken, N.M.P.; Hultink, E.J. The Circular Economy—A New Sustainability Paradigm? *J. Clean. Prod.* **2017**, *143*, 757–768. [CrossRef]
2. Sassanelli, C.; Rosa, P.; Rocca, R.; Terzi, S. Circular Economy Performance Assessment Methods: A Systematic Literature Review. *J. Clean. Prod.* **2019**, *229*, 440–453. [CrossRef]
3. Kristensen, H.S.; Mosgaard, M.A. A Review of Micro Level Indicators for a Circular Economy—Moving Away from the Three Dimensions of Sustainability? *J. Clean. Prod.* **2020**, *243*, 118531. [CrossRef]
4. Vinante, C.; Sacco, P.; Orzes, G.; Borgianni, Y. Circular Economy Metrics: Literature Review and Company-Level Classification Framework. *J. Clean. Prod.* **2021**, *288*, 125090. [CrossRef]
5. Rocca, R.; Sassanelli, C.; Rosa, P.; Terzi, S. Circular Economy Performance Assessment. In *New Business Models for the Reuse of Secondary Resources from WEEEs—The FENIX Project*; Rosa, P., Terzi, S., Eds.; Springer International Publishing: Berlin/Heidelberg, Germany, 2021; pp. 17–33, ISBN 9783030748869.
6. Acerbi, F.; Sassanelli, C.; Taisch, M. A Conceptual Data Model Promoting Data-Driven Circular Manufacturing. *Oper. Manag. Res.* **2022**, *15*, 838–857. [CrossRef]
7. Castro, C.G.; Trevisan, A.H.; Pigosso, D.C.A.; Mascarenhas, J. The Rebound Effect of Circular Economy: Definitions, Mechanisms and a Research Agenda. *J. Clean. Prod.* **2022**, *345*, 131136. [CrossRef]
8. Zink, T.; Geyer, R. Circular Economy Rebound. *J. Ind. Ecol.* **2017**, *21*, 593–602. [CrossRef]
9. Lamperti, S.; Cavallo, A.; Sassanelli, C. Digital Servitization and Business Model Innovation in SMEs: A Model to Escape from Market Disruption. *IEEE Trans. Eng. Manag.* **2023**, 1–15. [CrossRef]
10. Negri, M.; Neri, A.; Cagno, E.; Monfardini, G. Circular Economy Performance Measurement in Manufacturing Firms: A Systematic Literature Review with Insights for Small and Medium Enterprises and New Adopters. *Sustainability* **2021**, *13*, 9049. [CrossRef]

sustainability

Review

A Systematic Literature Review on Data and Information Required for Circular Manufacturing Strategies Adoption

Federica Acerbi *, Claudio Sassanelli, Sergio Terzi and Marco Taisch

Department of Management, Economics and Industrial Engineering, Politecnico di Milano, Via Lambruschini, 4/B, 20156 Milan, Italy; claudio.sassanelli@polimi.it (C.S.); sergio.terzi@polimi.it (S.T.); marco.taisch@polimi.it (M.T.)

* Correspondence: federica.acerbi@polimi.it

Abstract: In the extant literature, circular economy (CE) is considered a driver for sustainable development of the manufacturing sector, being it an industrial paradigm aiming at regenerating resources. CE is transferred to manufacturing companies through the adoption of different Circular Manufacturing (CM) strategies (e.g., recycling, remanufacturing, etc.). Nowadays, manufacturers are struggling to implement these strategies to limit their resource consumption and pollution generation. To enable their adoption, the extant literature unveiled the importance to control along the entire value chain different types of resource flows (i.e., material, energy, and information). Nevertheless, while for material and energy management some advancements were achieved, information management and sharing remains one of the major barriers in adopting these strategies. The present work, through a systematic literature review, aims to identify the relevant information and data required to support the manufacturer's decision process in adopting and managing the different CM strategies to pursue the transition towards CM. Furthermore, based on the results obtained, this research proposes a theoretical framework. It elucidates the four main areas to be managed by manufacturers in adopting CM strategies and it provides to the manufacturer an overview of what should be updated and upgraded inside the company to embrace CM strategies.

Keywords: circular economy; circular manufacturing; theoretical framework; information and data management; systematic literature review; industrial sustainability

Citation: Acerbi, F.; Sassanelli, C.; Terzi, S.; Taisch, M. A Systematic Literature Review on Data and Information Required for Circular Manufacturing Strategies Adoption. *Sustainability* **2021**, *13*, 2047. https://doi.org/10.3390/su13042047

Academic Editor: Antonella Petrillo

Received: 29 December 2020
Accepted: 5 February 2021
Published: 14 February 2021

1. Introduction

The last century has been characterised by an immoderate resource consumption trend followed by an uncontrollable increase of CO_2 emissions, which, if not stopped, might lead to planet collapse. Research dating back to the 1960s perceived this issue as a great problem to be addressed by society as a whole [1]. Therefore, while the concept behind sustainability is quite old, the term "sustainability" started to appear in the literature in the late 1970s [2]. Moreover, considering the importance of these issues, policymakers started promoting what is called "sustainable development" [3], grounded on environmental, economic, and social pillars [4]. Manufacturers are compelled to implement sustainable strategies to be taken on [5] to decouple their economic growth from resources consumption and maintain their high competitive advantage without creating negative externalities during their manufacturing activities. Among the possible solutions, the adoption of circular economy (CE), an industrial economy aiming at regenerating and restoring resources [6], started to be encouraged by policymakers worldwide, being it considered a great driver for manufacturing sustainability. The European Commission confirmed its position with the last recent update of the action plan [7]. In line with these actions, countries worldwide, as China [8], Australia [9] and USA [10] are promoting circular initiatives.

CE aims to extend the product lifecycle by slowing, narrowing, and closing resources loops [11] through biological, technical, and information cycles [12], and takes the name of Circular Manufacturing (CM) when applied in manufacturing companies [13]. CM is defined as *"the concurrent adoption of different CM strategies, which enable to reduce resources consumption, to extend resources lifecycles and to close the resources loops, by relying on manufacturers' internal and external activities that are shaped in order to meet stakeholders' needs" (p. 11)* [13]. More precisely, circular design, disassembly, remanufacture, reuse, recycle, servitization, cleaner production, resource efficiency, waste management, industrial symbiosis, and closed-loop supply chain and reverse logistics represent the CM strategies identified in the systematic literature review (SLR) performed by [13]. These strategies promote the alignment of the manufacturing sector with the sustainable development goals [14] and require a huge transformation from a socio-technical point of view [15]. This transformation needs support by intermediaries like policymakers and requires changes in the relations among actors involved, and their links with infrastructures, technologies, and contexts of application [16]. As reported in recently published reviews, the diffusion of CE in manufacturing is limited by several barriers. Taking as a representative sample those provided by Tura et al. (2019), and Ritzèn and Sandstrom (2017), these barriers could be economic and financial (due to a limited financial capability and support for companies), social (due to a lack of awareness among consumers), institutional (caused by the limited government support), technological and informational (caused by the reduced sharing of information and knowledge, organizational and attitudinal due to a silos approach and the fear of change), and last, along the supply chain, operational (due to a lack of support by the network and limited creations of appropriate partnerships) [17,18]. Most of the research agrees on these barriers in adopting CM, but among all, information management and sharing remains one of the major ones, causing a lack of support for the decision-making process requiring data and information standardisation [19]. Shortage of data flows generally represents a big issue for companies leading organizations to a silos setting [20]. Indeed, it would limit the potential of knowledge management (KM) practices in supporting a structured and aligned internal organization. To report an example specifically to CM, lack of data is detrimental for the selection of the right partner with whom to exchange resources in an industrial symbiosis network [19]. In addition, lack of data could undermine the right choice regarding the most appropriate strategy required to extend the lifecycle of the resources under analysis. This choice can be facilitated with the collection of data especially from end-of-life phases of the product lifecycle with the aim of improving the product design of next product generations [21]. Therefore, the sharing and usage of data, both internally and externally, will empower companies in enhancing their organizational and structural capabilities in extending the resources lifecycle. A preliminary and essential step to overcome this challenge is to identify which are the necessary data and information to be gathered, shared and managed [22]. Indeed, the complexity of this transition, determined also by the involvement of different stakeholders impacting of the decision-making process of manufacturers [23], requires us to gather data and information along each stage of the product lifecycle to pursue circular paths [24]. The gathering of data and information implies their management, by employing KM principles which consist of systematically discovering, acquiring, capturing, sharing and using productive knowledge in a cost-effective way to improve firms' and organizations' performances [25]. As proposed in the literature, KM can be based on a three-level framework: (i) knowledge requirements, (ii) knowledge reuse, and (ii) knowledge sharing [26].

Through an SLR, this work aims to put the basis to facilitate manufacturers' decision-making process in CM strategies adoption, by identifying and classifying all the relevant information and data required to pursue the transition in an efficient and structured way, concerning each CM strategy. In addition, to concretely provide exploitable support to manufacturers, the supporting technologies and tools, through which data and information are gathered and managed to create exploitable knowledge, are investigated. Therefore, although the extant literature presents a plethora of SLR investigating for instance how the technological advancements can support the establishment of circular systems [27], or what characterises the adoption of CE in manufacturing [13], or how a specific strategy is structured [28], until now a review aimed at clarifying in a structured and extensive way how to overcome the information management barrier issue is still missing; thus, it is essential to investigate the required data and information to be collected to adopt each CM strategy and the related supporting technologies.

Therefore, the objective of the present contribution is to overcome this scientific gap and to provide to practitioners a tool highlighting the main data and information required to design and manage circular systems through the adoption of different CM strategies based on data exploitation. To achieve the paper objectives, the research questions (RQ) addressed are the following: RQ1: "What data and information need to be gathered to support the decision-making process of manufacturers in CM adoption?" RQ2: "What are the tools (technologies or traditional tools) necessary to exploit data and information in CM?"

The article is structured as follows. Section 2 provides the description of the research methodology adopted. In Section 3, the literature review results show the state of the art regarding data, information and technologies supporting manufacturing companies' CM transition focusing the attention of each CM strategy. In Section 4, a theoretical framework is proposed discussing the literature review results. In Section 5, scientific, managerial and policymakers' implications are elucidated, opening the way to future research.

2. Materials and Methods

Considering information shortage as a barrier to adopt CM strategies, and the need to streamline information and data management under a circular perspective, this paper operates an SLR [29] to detect in a structured way the data and information required to enable manufacturing companies to embrace CM. Different contributions investigated in a systematic way the theory behind CE, e.g., to define the CE concept [30] or to explore the state of the art of academic research within the CE domain [31]. Despite the need to strengthen information flow, the systematic analysis of the data and information required in CM to support the decision-making process has not been performed yet. Indeed, so far, few studies supporting the decision-making process focused on one single CM strategy (e.g., on remanufacture [32]) have been conducted. Therefore, out of all the past reviews, the peculiarity of this contribution is the investigation of the data required to approach and manage each single CM strategy, considering those identified in Acerbi and Taisch (2020) (i.e., circular design, disassembly, remanufacture, recycle, reuse, servitization, cleaner production, resource efficiency, waste management, industrial symbiosis, and closed-loop supply chain) [13]. This analysis enables us to understand what relevant data, information and technologies are required for the adoption of each CM strategy, to create awareness and give a basic instrument, at least from an informational perspective, to the manufacturers in charge of the transition.

"Scopus" was used as scientific database for this review, being it the most diffused one for industrial engineering and having a broader coverage [33]. "Scopus" was queried using the following keywords: TITLE-ABS-KEY (("knowledge" OR "information" OR "data") AND "management") AND ("circular economy" OR "close-loop") AND ("Remanufacturing" OR "Recycling" OR "Reuse" OR "Reduce" OR "Redesign" OR "Recover") AND "manufacturing" AND ("digital technolog*" OR "Industry 4.0" OR "i4.0" OR "information technolog*" OR "platform*" OR "digital platform*" OR "authoring tool*" OR "PLM"). These keywords were defined after a random screening process in the extant scientific literature regarding contributions dealing with data and information in CM, selecting keywords coherent with the terminology adopted in the extant scientific literature. Nevertheless, the string of keywords enabled to partially limit the panel of papers within the boundaries of the research interest, and therefore other eligibility criteria were added. Only English written documents were selected, to include the latest global studies, without any restriction on timeframe and publication sources. A screening process was performed leveraging on three main criteria to discard out-of-topic papers, thus not dealing with data and information in CM:

- Papers not focused on CM (these account for the 47% of the papers discarded, that in absolute numbers correspond to a reduction of 35 papers from the sample of English-written documents);
- Papers not dealing with data and information for CM (these account for the 48% of the papers discarded, that in absolute numbers correspond to a reduction of 36 papers from the sample of English-written documents);
- Papers not focused on the manufacturing sector (these account for the 5% of the papers discarded, that in absolute numbers correspond to a reduction of 4 papers from the sample of English-written documents).

Indeed, starting from 178 documents identified through the keyword query, out of which 163 written in English, only 88 were selected for the review, as reported in Figure 1.

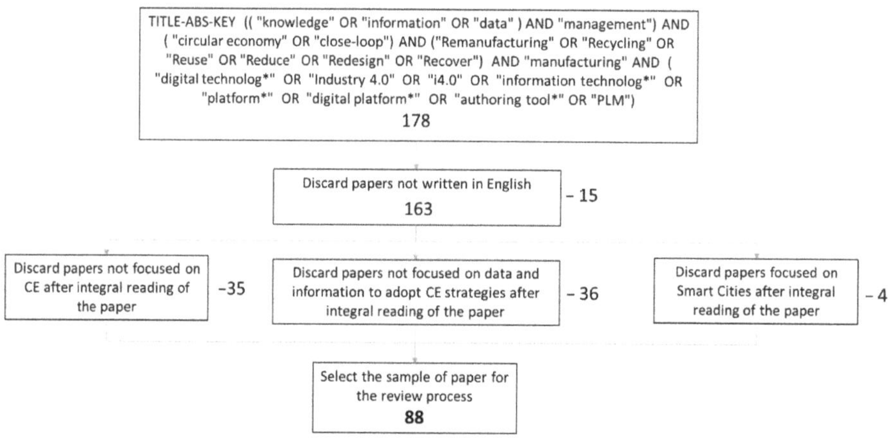

Figure 1. Paper screening process.

Using the SLIP method (which helps to Sort, Label, Integrate and Prioritize key concepts) [34], recently employed in a SLR in the CE context [35], the set of 88 documents has been analysed to detect and group the categories of the analysis. Specifically, for each CM strategy, the data and information were clustered into three categories: product, processes, and management. A fourth category, technologies and tools, was included in the analysis to show what tools can be adopted for data and information gathering and management.

3. Literature Review Results: Data and Information in CM

This section is divided into two main sub-sections: "Descriptive statistics" and "Data and Information required for CM". "Descriptive statistics" gives a general overview of quantitative results emerged from the sample of papers selected. "Data and Information required for CM" is focused on the narrative review of the papers selected for the analysis.

3.1. Descriptive Statistics

Observing the sample of selected papers, the top ten journals where these contributions have been published are reported in Figure 2. In particular, the number of publications in *Journal of Cleaner Production*, *Sustainability* (Switzerland) and *Resources, Conservation and Recycling* evidently overcome the rates of the others.

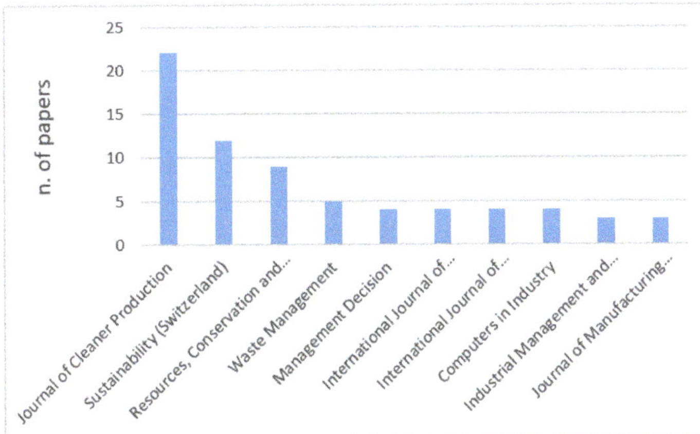

Figure 2. Top ten journal.

Some of the contributions were focused on specific manufacturing industries and the related statistics are reported in Figure 3. *Electronic* and *Construction* industries emerged to be the most advanced in CE, especially *Electronic* which is compelling to manage Waste Electrical and Electronic Equipment (WEEE) from years. Nevertheless, the majority of the papers did not provide indications regarding the specific manufacturing industry (see Figure 4).

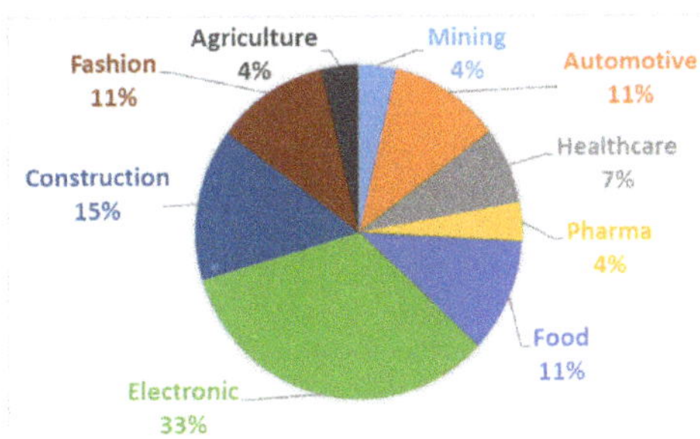

Figure 3. Paper statistics per industrial manufacturing industries.

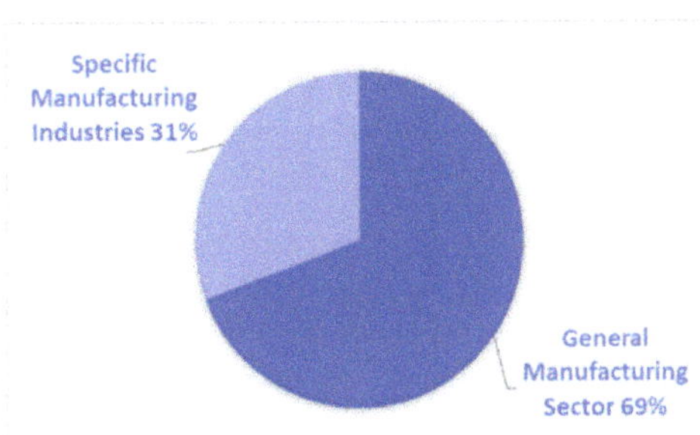

Figure 4. Paper statistics per manufacturing sector specificity.

Last, the contributions were clustered according to the CM strategy addressed, to evaluate the spectrum of strategies tackled and the number of papers focused on each strategy.

In some cases, more than one strategy has been considered in a single document analysed (see Figure 5).

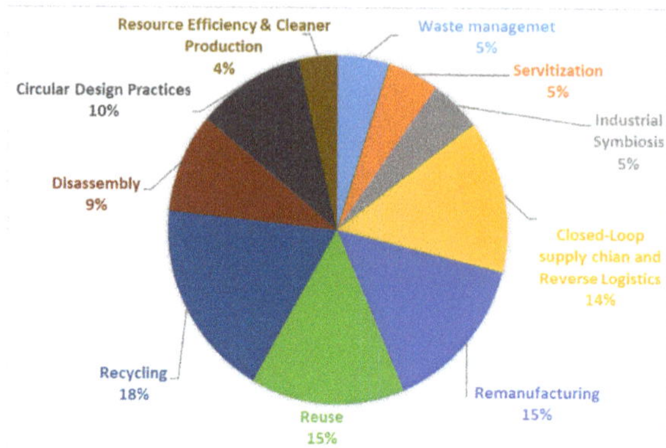

Figure 5. Paper statistics per CM strategy investigated in the paper.

3.2. *Data and Information Required for CM*

In this section, through the SLR, for each CM strategy, data and information are detected and clustered according to the three categories (i.e., product, process, management) and supportive technologies and tools are identified based on the SLIP method.

3.2.1. Circular Design

Circular design strategy is adopted for example by Fairphone to create modular and sustainable smartphones which can be achievable only if data regarding product characteristics and disassembly sequencing are gathered [36]. More precisely, Fairphone ensures the extension of their products lifecycle by gathering data during the design stage to start design the product to be easily repairable and flexible, to then improve these characteristics relying on consumers' behaviours data.

As also evident from this case, the decisions taken at the design stage, thus at the Beginning of Life (BoL), are the most important ones since they influence the potential environmental impacts that a product might generate during its lifecycle [37]. Decisions like the selection of the product material composition and the type of manufacturing processes, which strongly influence the rate of material or energy input per unit of product, gain a prominent position [37]. Moreover, the considerations regarding energy consumption cannot be limited only to product production, but required to be extended also over the product usage [38]. The materials selected must respect the sustainable requirements in terms of composition to avoid harmful materials, but also transportation modes to avoid pollution generation during the material delivery [39]. Besides the material, other information empower product circularity, among which the definition of product functionalities, estimation of disassembly tasks time and costs [40], product architecture, function, geometry, materials mix (weight and type), components specifications and assembly instruction [41]. To extend the product lifecycle, it is required to ensure, through product design, its maintainability, reparability, durability and the correct disposal [21]. The correct disposal is also influenced by customers' behaviours, which indeed need to be involved to ensure product circularity. Information such as product cost and customers' requirements complexity must be gathered [42]. This might require new managerial procedures and documentation which influence the traditional customers' management. Indeed, the influence of external stakeholders generates managerial implications during the transition. Other information to be gathered regards supply chain stakeholders' location, distribution model, retail prices, product demand, users' preferences. Moreover, the product must be designed to avoid its hibernation, by establishing recovery plans [43], and to keep high the possibility to extend its lifecycle, services to maintain product and production processes in use must be considered [44]. This can be done through maintenance or repairing activities that, if performed on internal physical assets, would foster the availability and the reliability of machinery and equipment, by also increasing the levels of safety [45] and product quality. Last, information costs regarding the planning (e.g., product requirements, finalization, materials), concept (e.g., reuse possibility), design (e.g., standardization, reusable parts), source (e.g., make or buy decisions, material procurement, supplier selection), manufacturing (e.g., process tooling, operational planning), launch (e.g., warranty analysis, predictive maintenance services), service (e.g., recycling, refurbishments) need to be estimated [46].

Regarding Circular Design strategy, the technologies and documentation usable to easily gather data are Bill Of Material (BOM), Product Life Cycle Management System (PLM), Sensors, Manufacturing Execution System (MES) and Enterprise Resource Planning (ERP) [47]. To use these data, visual analytics tools [41] and CAD 3D [38] can be implemented.

To summarise, linking these findings with RQ1 and RQ2, manufacturers embracing this strategy are recommended to gather data regarding product initial characteristics, functionalities and potentialities for their lifecycle extension. This information must be kept available until the product EoL to facilitate its management in terms of maintenance and recovery. It is important to select suppliers allowing that this can happen, and some peculiar relationships can be established with consumers thanks to the introduction for instance of warranty programs. Although external actors are relevant, technologies to be adopted for this strategy are necessary mainly for the internal sharing of data allowing first an internal alignment.

These results are summarised in Table 1.

Table 1. Circular design adoption data, information and technologies/tools.

[21,37–39,41–43,45–48]			
Product	**Process**	**Managerial**	**Technology/Tools**
- Product Functionalities, - Product Features, - Product Architecture, - Product Geometry, - Material Mix (weight and type), - Components Specifications, - Assembly Instruction, - Reuse Possibility, - Overall Costs, - Users preferences and requirements, - Reparability, - Durability, - Maintainability, - Modularity, - Joints	- Material and Energy used to produce and use product monitoring - Disassembly time and costs - Distribution: long/short/direct chain - Machinery and Equipment maintenance activities	- Warranty programs - Maintenance service - Material Procurement - Supplier selection - Leasing agreement - Take back service	- Visual Analytical Tools - CAD 3D - BOM - PLM - Sensors - MES - ERP

3.2.2. Disassembly

The disassembly strategy adoption highly facilitates the implementation of all the other CM strategies characterising the product EoL (e.g., remanufacture, waste management, etc.). A concrete example is given by the smartphones produced by Fairphone which are easily disassembled to ensure their circularity which is possible also thanks to an accurate data gathering at product EoL feeding the BoL with extensive data [36]. Regarding this strategy, Marconi et al. (2017) proposed quantitative measures to be considered during the disassembling process, that regard the disassembly depth, time, and costs. The first one is reflected in the number of operations to reach a target component; the second regards the time spent to reach it, and the third one regards the cost, in terms of both labour and tools, to disassemble it.

To summarise, linking these findings with RQ1 and RQ2, manufacturers embracing this strategy need to pay attention mainly on data and information regarding the type of tasks to disassemble a product and the time required, and need to invest in disassembling technologies to facilitate human work.

The results are summarised Table 2.

Table 2. Disassembly adoption data, information and technologies/tools.

[21,49,50]			
Product	**Process**	**Managerial**	**Technology/Tools**
- Product disassembly possibility - Product criticalities evaluation - Components substitution possibility	- Number of operations to reach a target component - Time spent to reach a target component - Labour and tools cost to reach a target component - Maintenance activities knowledge	-	- Disassembly technologies - Manual work

3.2.3. Recycle

Relight (2020), operating in the recycling of electronics, is a typical example of companies adopting the recycle strategy [51]. The company has to gather information regarding waste characteristics in order to ensure the right treatment path. In accordance with that, the recycling strategy requires product analysis to evaluate whether it is possible its recyclability [52]. In case it would not be possible to select an external "recycling partner", the manufacturer itself requires us to perform a self-assessment to evaluate the company's level of sustainability to perform a recycling process. This evaluation influences the decision-making process of the manufacturer needing specific information covering all the TBL principles [53]:

1. economic: need to evaluate the operation costs per unit, the product quality utility and value, the technical level and the profitability;
2. environmental: the evaluation is on resource consumption efficiency, pollution production, energy efficiency, environmental management system, environment equipment, and facilities;
3. social: the focus is on employee turnover rate, customer satisfaction, brand reputation, and local communities influence are evaluated.

For each sustainable principle, product factors (e.g., product value), processes factors (e.g., energy consumption), managerial factors (e.g., brand reputation), and technologies (e.g., environmental equipment) are taken into account.

Wrapping up, bonding these findings with RQ1 and RQ2, manufacturers embracing this strategy are recommended to gather data especially on product composition to evaluate whether, and in case, which materials can be recycled to then decide how to treat them for instance for a possible upcycling. Thus, an analytical analysis of the treatment has to be performed and process data must be collected to evaluate the social, economic and environmental impacts. Regarding the required technologies, they need to invest in specific advanced technologies, like recycling robots, to be efficient in embracing this strategy, and in environmental management system to monitor materials environmental impacts.

The results are reported in Table 3.

Table 3. Recycle adoption data, information and technologies/tools.

[50,52,53]			
Product	**Process**	**Managerial**	**Technology/Tools**
- Recyclable Materials and Components, - Quality utility value,	- Operation cost per unit - Technique level - Profitability - Resource consumption efficiency, - Pollution production - Energy efficiency	- Best recycling partner selection or self-assessment - Employee turnover rate - Brand reputation, - Local communities influence - Recycled product customer satisfaction	- Environmental management system - Environmental equipment and facilities - Recycling robots

3.2.4. Remanufacture

Remanufacture strategy is enacted for instance by America's remanufacturing co. (ARC) that bases its business model in ensuring to give the original quality level and characteristics to a turned back product [54]. ARC adopts advanced technologies to ensure to treat appropriately the product to be remanufactured according to the product conditions and they monitor the time and quality of turned back products. The remanufacturing strategy is characterized by a process based on different activities, which usually are: disassembly, cleaning, inspection, sorting, reconditioning or replacement, reassembly, and testing [55]. A remanufacturing process must satisfy some requirements to be introduced, for example, the components of the remanufactured product, including both collection and remanufacturing process costs, must cost less than the new ones. Moreover, to make valuable the introduction of the remanufacturing strategy, the manufacturer needs to introduce customer-oriented operations, by deploying adequate services, to take the benefits of an enlarging green market. This also implies that putting in place efficient core acquisition models to manage returned units and components, to appropriately plan remanufacturing processes, and to define the personnel involved would be beneficial. Therefore, regarding the returned products, the remanufacturer must gather specific information such as the product conditions, affected by consumers' usage, and the market demand; even though the level of uncertainties regarding the quality, quantity and time of returned products remains one of the major barriers [56]. On the product side, other information to be gathered regard the product type, model, original manufacturer, property (e.g., battery type) and the components; but also the localization, the amount of Work in Progress (WIP) to be remanufactured, the production plan, the resource status, the scheduling, the remanufacturing activities required for each product type (e.g., assembly, disassembly, inspection) and the related resources to perform these activities [57]. Indeed, returned products are characterized by different status and thus, during the inspection, according to the legislation and the conditions, specific processes are defined for each returned product. The MES on the shopfloor might support the gathering of these data. Indeed, this tool enables us to collect historical data to exploit the simulation benefits, extremely relevant in a context characterised by high uncertainty as returned products. Simulation supports the decision process by simulating real-time different scenarios and increasing the flexibility level of the company. Moreover, other technologies can be introduced, such as high levels of automation, advanced decision support tools based on data analytics, distributed data gathering, and cyber-physical systems to trace information to have them always available. Last, there must be a high level of safety and ergonomics conditions for humans working on the shopfloor [58].

To summarise, linking these findings with RQ1 and RQ2, manufacturers embracing this strategy are recommended to first gather data on product conditions which help in deciding which activities are required to remanufacture it. Then, an analysis regarding the costs and time to remanufacture the product must be performed to balance the activities scheduling, bearing in mind the technologies available to be adopted. Therefore, investments in advanced remanufacturing technologies linked to decision-support systems are suggested to be done. More in details, decision-support systems can be linked to monitoring systems allowing to keep track of product conditions and design characteristics to evaluate the best remanufacturing option.

The results are reported in Table 4.

Table 4. Remanufacture adoption data, information and technologies/tools.

[49,50,55–58]			
Product	**Process**	**Managerial**	**Technology/Tools**
- Product conditions - Market requirements - Product type - Product model - Propriety (e.g., battery type) - Components - Components manufacturer information	- Flexible processes according to market and product requirements - Monitor collection and remanufacturing processes costs - Production plan scheduling balancing remanufacturing activities - Remanufacturing activities scheduling (e.g., disassembly cleaning, inspection, sorting, reconditioning/replacement, reassembly, testing)	- Safety and ergonomics condition for workers - Planning of remanufacturing processes - Monitor time, quantity and quality of returned products - Monitor resources status	- Remanufacturing technologies with advance automation levels - Tracking technologies to trace information - Decision support tools based on data analytics tools, distributed data gathering and cyber-physical systems - Simulations tools - MES on the shopfloor

3.2.5. Reuse

The reuse strategy is becoming widely diffused to encounter circular principles, as the case of Lush that sells reusable packaging, which can be reused only if their characteristics are still functional for the object purpose [59]. Therefore, reuse, as CM strategy, is enabled only in case there is an adequate product EoL management. On one side, this must be supported by managerial factors such as the right marketing activities to empower customers demand and make them aware of the benefits of turning the products back, but also through the monitoring of the legislation respectfulness. Moreover, to put in place an appropriate reuse strategy, the company has to monitor collection and transportation activities and related costs [60].

To summarise, linking these findings with RQ1 and RQ2, manufacturers embracing this strategy are recommended to gather data especially regarding the product functionalities and their location to ensure the convenience to reuse the product. Moreover, the most important thing remains the possibility to make reusable products which is reflected in the available regulations and consumers' awareness. No specific technologies are required for this strategy.

The results are summarised in Table 5.

Table 5. Reuse adoption data, information and technologies/tools.

[60]			
Product	**Process**	**Managerial**	**Technology/Tools**
- Product functionalities	- Transportation costs - Collection costs - Supply costs	- Marketing - Legislation respectfulness	-

3.2.6. Waste Management

Another CM strategy adoptable by manufacturers is waste management. A concrete example of the adoption of this strategy is Greentronics (2020) which treats electronic waste after having checked for instance the type of waste, the possible hazardous substances present in the waste to evaluate how to treat them [61]. Before deciding whether to dispose or reprocess a product, there are different activities to be performed. More precisely, the activities are collection, sorting, disassembly, and data analytics regarding material type and weight, material flows quality, mass and volume [62] and the amount of waste reused [63]. In case the company deals with more peculiar and hazardous waste, such as batteries, besides the data concerning battery weight and waste collection centres location, more detailed data, like raw material composition per battery, are gathered [64]. A waste management strategy can be put in place also in case of product deletion which choice has huge impacts on operations, end-users, financials, and supply chain, due to the fact that the inventory of this product becomes obsolete. Therefore, other information is required such as: product type, product components, component materials type, components materials sourcing, product lifecycle stage, product quality, product circularity level, product location in the supply chain, internal storage capacity, end-user consumption and delivery modes [65]. Concerning the management category, from the Chinese experience, it has been suggested to create alliances among firms to establish unified collection channels modes, introduce platforms usable internally and with third parties [66].

From a technological point of view, all these activities can be supported by I4.0 technologies such as smart bins with sensors to detect and analyse material and waste, self-automated vehicles and containers, methods to automatically analyse images. For sorting and disassembly, different types of robots can be used [62]. For data analytics on waste management, sensors, blockchain, deep learning techniques, multivariate model equation and Building Information Modelling (BIM) are suggested to be used [65]. Last, predictive models can be introduced to facilitate waste management [67].

To summarise, linking these findings with RQ1 and RQ2, manufacturers embracing this strategy are recommended to gather data regarding waste type and quantity to evaluate if it can be partially reused after specific treatments and whether are present hazardous substances to evaluate how to manage them to avoid negative impacts. For the sorting of waste, sensorized robotics empowered by big data analytics are suggested to be considered for future investments. In addition, considering also the importance to track the localisation of waste, in the near future blockchain can become the right means towards trusted tracking of resources.

The results are reported in Table 6.

Table 6. Waste Management adoption data, information and technologies/tools.

[62–72]			
Product	**Process**	**Managerial**	**Technology/Tools**
- Product components - Product materials composition, - Product material quality - Raw material composition - Raw material regeneration, reuse and restoration - Hazardous substances - Product consumers' demand - Product storage amount - Waste weight - Waste type - Waste location - Waste reused amount	- Production process energy and material consumption, - Air emissions - Waste generation during production processes - Water consumption - Energy consumption	- Logistics - Labour force - Waste collection centres (location)	- Blockchain - Sensors - IoT - Big Data - Smart bins - Sorting robots - Recycling robots - Data analytics - Deep learning technique - Multivariate model equation - BIM - CAD/CAE 3D - Web technologies (to identify the localization of waste per category)

3.2.7. Industrial Symbiosis

To exchange valuable resources, Industrial Symbiosis might be adopted as CM strategy, as done by Nespresso that takes back the used capsules to exploit the remaining coffee as fertilizer to be given to rice fields [73]. To ensure that this partnership can be suitable for both sides, Nespresso has to create awareness in its consumers to stimulate the product turn back and it has to analyse the type of waste it has, to be aligned with the requests for the rise fields. Therefore, this strategy involves different stakeholders among which the waste producer, the waste user, the waste treatments companies and waste transportation companies [74]. Indeed, this strategy aims to make companies collaborate by exchanging by-products, resources, and scraps under the common goals of environmental and economic sustainability that might be enabled by collaborative platforms or existing web platforms to develop input–output matching tools [75]. Considering product and process categories, the necessary information refers to the types and quantities of resources consumed, the types and quantities of waste and by-products produced and the conversion processes, all backed by economic information concerning the exchange. This allows us to evaluate what resources can be recovered from waste or by-products, what waste or by-products can be used as raw material and last, what technologies and processes are necessary for the conversion [76]. In addition, information regarding availability and requirements of resources in terms of quantity, time and price should be gathered together with process factors like the temperature, the flow rate, the pressure, the enthalpy [75], the concentration of chemical species and their toxicity, the flow of energy, water used for the concrete usage of these resources [77]. All these issues influence the managerial decisions, for instance, it is important to evaluate the distance among the entities exchanging resources, and the profitability related to the exchanges [76].

To summarise, linking these findings with RQ1 and RQ2, manufacturers embracing this strategy are recommended to gather information either internally and externally to evaluate the possible matches with other actors. Therefore, they must collect data regarding type and quantity of waste and by-products produced internally. This information is necessary to be shared externally to allow the selling of these resources or to allow the exchange of the resources in case waste and by-products produced by others could be useful for their productive activities. For this strategy is important the investment in collaborative platforms to be able to share and obtain relevant data. Therefore, this also requires investing in data format standardisation and system integration to ensure an easier access to data for different actors.

The results are reported in Table 7.

Table 7. Industrial Symbiosis adoption data, information and technologies/tools.

[19,74–78]			
Product	**Process**	**Managerial**	**Technology/Tools**
- Type and quantity of by-products and waste produced - Type (e.g., material, energy, water) and quantity of resources consumed - Timespan of availability - Resource nutrient (e.g., the characterization of the resource regarding its reinstatement into biological or technical cycles) - Toxicity of resources - Cyclicity of resources	- Conversion processes (i.e., from waste and by-product to resource) information (e.g., skills, material, energy, water) - Monitoring of physical system data during the conversion, such as temperature, flow rate, pressure, enthalpy, the concentration of chemical species of streams - Emissions - Costs	- Location of the entities collaborating in the industrial symbiosis - Storage of information (e.g., name, address, industry, etc.) of the entities collaborating in the industrial symbiosis - Profitability estimation coming from the exchange	- Collaboration platforms - Existing web platforms (e.g., U. S. Materials Market Place, WasteIsNotWaste, Resource Efficient Scotland, Minnesota Materials Exchange, The Waste Exchange) - Ontology and Ontological frameworks - PLM systems - Waste exchange registry - Lifecycle inventory database

3.2.8. Closed-Loop Supply Chain and Reverse Logistics

A valid example of closed-loop supply chain or reverse logistics strategy adoption is the one of ARC, which creates partnerships with both manufacturers and distributors to ensure the maximisation of the returned products value while protecting the new products value [54].To allow that this can happen in an efficient way, ARC has to collect data regarding the type, the quantity, the time and the quality of returned products.

The reverse logistics is based on all the logistic activities required to transform used products into reusable products [79]. Being products classified according to the remaining useful life and their status, the information to be gathered on the product side are product condition [80], product lifecycle stage, product design, inventory of returned products, product perishability, product complexity and hazardous material composition [81]. Moreover, considering the uncertainties regarding the returned products, there should be the necessity to gather data regarding time, quality, quantity, and types of returned products, to then develop a plan and decide whether to recondition, reuse or sell the returned product in a secondary market [49]. Whenever a product is turned back, it is important to evaluate the reasons why it has been returned. It might be a defective product, or it needs maintenance or repair activities, or it has been returned due to excess of products [82]. This information influence disassembly activities and remanufacturing activities required by the product [49].

Moreover, to establish an efficient reverse logistic network, other information concerning managerial and processes issues needs to be gathered to direct investments such as managerial commitment, internal reorganization, institutional issues management in terms of regulation and taxation, technological investments to establish processes supporting CM strategies and last, informational challenges, considering the difficulties encountered for information exchanges [83]. In reverse logistics, supplier management practices should be revised. Environmental auditing of suppliers must be performed (e.g., on injury reduction in terms of deforestation and material recycling, or on emission reduction or cost savings [84] that might be done also through a questionnaire, the introduction of a compliance statement, a product testing report, the bill of material (BOM) of products and components purchased and the establishment of environmental requirements for purchasing [85]. The hygiene and safety conditions of suppliers must be checked, as well as their partnerships with green organizations, their adherence to green policies, their pollution control initiatives, the appropriate staff training, the environmental standards adoption, the quality and time of the service provided, the economic data, and the usage of environmentally friendly technology and materials [86]. Customer management should be revised too since it is necessary to stimulate new demand for reconditioned products and a remarketing strategy can be put in place [87].

The gathering of these data might be eased by the adoption of RFID or IoT sensors and actuators that can be furtherly integrated through the usage of PLM, Relational Database Management Systems or Database Handling Systems [87]. ICT systems are considered essential elements to support reverse logistics and to store the right information [88]. Among them, ERP is considered one of the most important instruments to streamline information flows and to enable the returned products monitoring [79].

To summarise, bonding these findings with RQ1 and RQ2, manufacturers embracing this strategy are recommended to ensure the tracking of the product along its lifecycle to monitor both the conditions and localisation allowing its efficient and economically convenient recovery. A network must be designed appropriately thus, data regarding the actors involved are extremely relevant. With this intent, supportive technologies can be considered valuable investments, among which the integration of information and communication systems allowing the right setting of reverse logistics activities and tracking technologies such as RFID.

The results are reported in Table 8.

Table 8. Closed-loop supply chain and Reverse Logistics adoption data, information and technologies/tools.

[44,49,79–103]			
Product	**Process**	**Managerial**	**Technology/Tools**
- Product condition (e.g., Quality of returned product), - Product value - Product life cycle stage, - Product design (e.g., weight and dimension) - Product material composition - Inventory of returned products (e.g., quantity of returned product) - Type of returned product - Time of returned product - Reason for returned products (e.g., defective product, need for maintenance and repair activities, excess of products) - Product demand - Product perishability - Product complexity - Hazardous substances - Product packaging	- Disassembly scheduling and cost - Green and circular design - Life cycle assessment for an eco-report - Recondition process cost - Repair process cost - Transportation requirements (e.g., tools to be used) - ISO 14001/ ISO 14002 requirements respectfulness - Delivery methods - Environmental policies respect - Collaboration with local recycling organization for product recycling - Energy and Water monitoring, - Production scrap monitoring - By-products monitoring - Solid or liquid waste - Emissions, - Transportations used	- Adequate functions introduction (e.g., receive, store, sort, remanufacture, repair, de-manufacture, dispose) - Design reverse logistic network - Sustainable supplier selection - Supplier environmental auditing - Supplier compliance statement. - New environmental requirements for purchasing - Supplier BOM and product testing report - Supplier partnership with green organization - New design practices for recovery products - Inventory management - Definition of new sales channels - New take-back programs, advanced recycling fee introduction - Supplier Hygiene and safety check - Supplier quality and time of the service deployed - Supplier costs - Human health and staff training, - Supplier pollution control initiatives and environmental protection standards - Supplier environmental technology and material usage - Top management commitment, - Partnerships with suppliers	- ICT - Information systems as ERP - Recovery BOM - Material Requirements Planning (MRP) - Recovery Requirements Planning - Sharing platforms - PLM, - Relational Database Management Systems - Database Handling Systems - Green information technologies systems - Eco-database for products

3.2.9. Servitization

Another CM strategy is the servitization and linked to that, the PSS is one of the most diffused business models in CM adoption. For instance, Philips Lighting is switching to PSS to control the entire product lifecycle [104]. Since Philips Lighting provides a service, data collection but be ensured along the product lifecycle to give the right support to consumers. These business models, on the process side, are monitored through data regarding carbon emissions, renewable energy consumption, climate change impact, waste production, resource depletion [50], production scheduling, work procedure for disassembly, energy and materials used from renewable resources, waste, and energy recovered by the system [105]. On the product side, information on the possibility to disassemble, remanufacture, recycle and reuse the product to provide the right service is required [50], as well as the product characteristics, the usage of biodegradable materials, and the level of product flexibility in terms of the number of users that potentially can use the product [105]. Last, another important aspect to be monitored regards the auxiliary services provided to the users, such as product maintenance, repair, parts replacement, recall, and scrap recycling. These services are required to be integrated with information throughout the product lifecycle [52] to update customer management practices. Indeed, a market analysis must be performed by monitoring customers demand, needs, requirements, and competitors' actions. Regarding the suppliers' management, data concerning the supplier innovation level of products, the transportation requirements and expertise on processes and technologies adopted need to be gathered [106].

One of the most diffused technologies supporting servitization is the Big Data Analytics technique leveraging on smart products [50]. Big Data Analytics can support companies in detecting hidden knowledge, to improve their competitive advantage, among which the relation between lifecycle decisions and process parameters [33].

Wrapping up, linking these findings with RQ1 and RQ2, manufacturers embracing this strategy are recommended to collect data regarding consumers' behaviours and product conditions, during the usage phase, to ensure the provision of tailored services. This allows the stable maintenance of product functionalities and it might support an improved design for the future. Therefore, big data analytics can be suitable to back smart products tracking and analysing data on both the service delivery and the usage phase.

The results are reported in Table 9.

Table 9. Servitization adoption data, information and technologies/tools.

[33,50,52,105–110]			
Product	**Process**	**Managerial**	**Technology/Tools**
- Product characteristics, - Product design specifications - Core quality assurance - Remanufactured quality assurance - Biodegradable materials used - Product flexibility (e.g., Product possibility to be reused) - Product possibility to be remanufactured - Product possibility to be recycled - Product possibility to be repaired - Product possibility to be shared - Returned product time, quantity and quality - Product demand - Customers' requirements - Technical requirements - Product innovation - Product parts replacement, - Product assembly - Product usage - Disposal methods	- Production scheduling - Manufacturing specification - Disassembly work procedure - Material and scrap recycling process - Energy and Material used from renewable resources - Waste and energy recovered by the system - Transportation requirements - Product operations - Product maintenance - Product repair	- Insurance policies - Competitors analysis - Collection centres - Training the employees to deploy the service - Sharing service - Product recall management	- Big data analytics - Smart products - IoT - Cyber Physical Systems - Data mining - Cloud Manufacturing - Artificial Intelligence - Intranet

3.2.10. Resource Efficiency and Cleaner Production

Resource efficiency (i.e., material and energy efficiency) and cleaner production are considered relevant CM strategies too. Both of these strategies are promoted for instance by Apple (2020) which on one side, regarding the resource efficiency, is forcing to design lighter products (and packaging) to ensure a reduced amount of material usage, on the other side, for cleaner production, establishes relationships only with suppliers using renewable energy sources and accepting their Zero Waste Program aiming to divert the 100% of waste from landfills [111]. Apple has to collect data regarding its suppliers, both on their activities and certifications, but it needs also to collect data regarding the product design to improve the product sustainable characteristics generation by generation.

Concerning the adoption of these strategies, during the manufacturing processes, energy usage should be monitored and this is affected by product type, machines used and state of the final product [112]. Technologies like photovoltaic panels to produce energy, sensorized glasses to adjust heat and light inside the building and water reuse technologies are used for the right implementation of these strategies [113].

To summarise, linking these findings with RQ1 and RQ2, manufacturers embracing these strategies are recommended to collect data regarding the sustainability of their productive activities or of those of the partners selected, such as of the suppliers. This allows us to keep under control the environmental impacts of the company and its network. For the adoption of these strategies, investments in advanced and eco-technologies are suggested on one side to monitor the assets conditions reducing the scrap generation, and on the other side to ensure high environmental standard of the activities.

The results are reported in Table 10.

Table 10. Resource Efficiency and Cleaner Production adoption data, information and technologies/tools.

[112,113]			
Product	**Process**	**Managerial**	**Technology/Tools**
- Product type and characteristics - Final status of the product	- Energy used during manufacturing activities - Machine characteristics used to produce the product	-	- Sensorized glasses to adjust heat and light inside the building - Photovoltaic panels - Water reuse technologies

Regardless of the strategy adopted, the top management commitment is important to drive the modification of internal managerial factors, to respect the adoption of new regulations, and to introduce new customers and suppliers' management strategies to embrace CM. All these factors required to be supported also by government financial resources and new technological investments [103].

4. Discussion

The transition towards CM involves the entire organization and implies an orchestration with the whole ecosystem [114]. Manufacturers, through the gathering of the right data and information, are facilitated in managing more effectively the circular requirements that might arise along the product lifecycle [115]. On one side, this SLR showed a high correlation among the data and information necessary to adopt each strategy, strengthening the possibility to exploit the same data for the concurrent adoption of different CM strategies in embracing CM. For instance, data regarding product characteristics and composition are the basement for each strategy to be adopted, and the information for the scheduling of the activities, with various purposes, is fundamental for the optimization of the efforts required to extend resources lifecycle. More precisely, to highlight these concepts, data regarding product composition can be extremely useful and common to be gathered if looking at circular design strategy. At the same time this data is extremely important for the correct adoption of all the other strategies, like recycling and remanufacturing for example, since without this information the resources are more difficult to be treated. In addition, a common ground for all the strategies is the necessary engagement with external actors and thus, the importance to introduce integrated platforms allowing the streamlining of the information sharing among different entities is underlined. This is true in general, no matter the scale of the manufacturer considered since the common required data can be available in case of a correct data gathering and sharing. On the other side, it emerged that each strategy is characterised by four main categories, highly integrated among them, out of which three correspond to the categories of data and information (i.e., product, process, management) to be gathered and managed by the manufacturer in charge of the transition, while the fourth category corresponds to the technologies and tools to gather and manage data and information to create exploitable knowledge.

The analysis of these categories enabled to define a theoretical framework (see Figure 6) which provides an overview of the relevant areas to be monitored in CM by a manufacturing company. Other frameworks already exist in the literature, such as the "Butterfly" framework developed by The Ellen MacArthur Foundation (2015) to investigate the possible restorative paths of materials and energy flows [6], and the one developed by OECD (2018) to investigate circular business models enhancing CE values [116]. The framework proposed in this contribution, evidently characterized by great commonalities with the previous ones, aims to revisit them from the perspective of data and information valorisation. Therefore, both the frameworks just mentioned, i.e., the "Butterfly framework" and the "OECD framework", are not specific for the manufacturing sector and they mainly depict the different solutions and strategies to be adoptable in CE by describing the possible outcomes in a qualitative manner. They both do not provide a structured analysis regarding the requirements for the distinctive solutions. Actually, the present framework is sector specific, being focused only on the manufacturing sector. It has been built to allow manufacturers to have a wide and complete spectrum of the information needed to adopt CM. Indeed, it provides in detail both the data and the technologies required to address each CM strategy. Moreover, this structured analysis allows manufactures to evaluate the synergic potentials among the different strategies and to take more informed decisions on the different adoptable CM strategies."

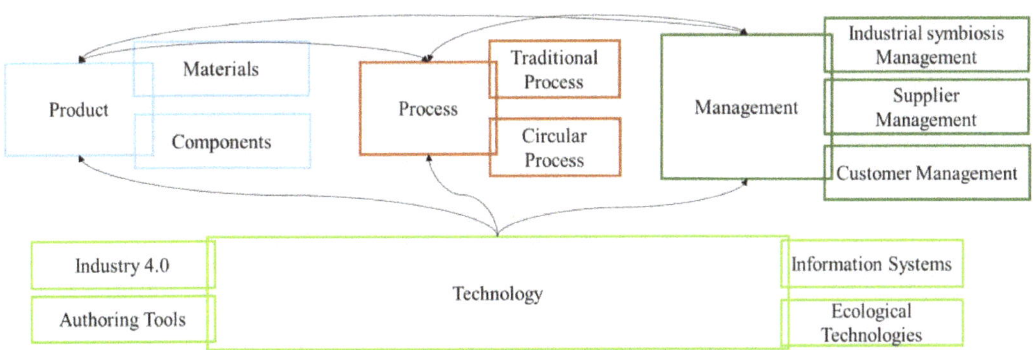

Figure 6. Theoretical framework.

The framework categories are discussed in the sub-sections below.

4.1. Product Category

The "product" category is the gear boosting the entire ecosystem towards CM and the findings of this review revealed the need to gather data and information regarding the physical composition of products, in terms of both components and materials (see the light blue block in Figure 6). The importance of this category has been also highlighted by the data detected for it which are required mainly by all the different CM strategies.

This category gains importance especially dealing with the circular design, representing the origin of the product lifecycle. Thus, considering that the main goal of designing "circular" products is to enable the extension of their lifecycle [39], and the reduction of resources used [11], some data such as assembly instruction, product modularity, features, and geometry are revealed to be important to understand the structure of the components.

This would be linked to the gathering of data necessary to evaluate the possibility to disassemble and repair the product and to ensure to address both technical requirements and customers' needs. An important piece of data, common to all the CM strategies, is the customers' demand-type since it enables us to understand if the company is able to both address the market needs and stimulate a demand aligned with the company circular values. Besides, the review showed the importance to assess material toxicity, material mix, and material weight to limit the negative social and environmental impacts. Data regarding material composition are important also for reverse logistics strategy to select the right material type, for industrial symbiosis strategy to appropriately exchange waste, and for waste management and recycling to properly manage the material to be wasted and recycled. To close the resources loops, information like product final status, product usage, product remaining useful life and product value, gain importance at product EoL. These data are common for remanufacturing, reuse, recycle, servitization and reverse logistics strategies. Peculiar information is required while dealing with reverse logistics like the product perishability and the time and quantity of turn-back products.

As final remark regarding future trends, the gathering of data and information on products will be facilitated more and more thanks to the introduction of sensors into products design characteristics. Thus, smart products development will encourage more circular behaviors and will ensure the real time collection of data along the product lifecycle enabling to facilitate the resources lifecycle. Therefore, product characteristics and conditions will be easily used in this scenario, and consumers will be supported in managing products in a more responsible manner. Moreover, the static perspective will be still representative for the product category and thus, the BOM and the recovery BOM will cover a promising role in ensuring the embracement of CM by manufacturers since they allow to have under control information regarding product characteristics starting from their initial design and material requests.

4.2. Process Category

The "product" category is strictly related to the "process" one (see the brown block in Figure 6) since, to enable the concurrent adoption of different CM strategies, it is not enough to act on the product, by adapting and changing it through a suitable design. New processes necessitate to be introduced in the company (i.e., circular processes) and, traditional processes are required to be modified, which is possible through the appropriate data and information gathering.

Circular processes usually take their names from the related strategy, and represent the core for the adoption of CM strategies like remanufacture, reuse, recycle, disassembly, servitization, industrial symbiosis, and reverse logistics, since they concretely make regenerate resources.

Some data are distinctive for a certain strategy. For instance, data characterising disassembly refer to the estimation of the number of operations to reach a target component and the time spent in doing it. Industrial symbiosis requires the evaluation of the necessary resources to convert the waste into a resource (e.g., skills required, amount and type of necessary material, the amount of energy and water) and to monitor the physical system during the conversion process by gathering data like pressure, temperature, and emissions. The distinctive data characterising the reverse logistics are those regarding the transportation methods and requirements (e.g., ad hoc tools necessary to transport different types of products).

Other data and information belonging to the process category are instead common to all, or to the majority, of the CM strategies. Actually, economic data to ensure economic sustainability are relevant for several strategies, among which reuse, remanufacturing, reverse logistics, and servitization that require to monitor transportation, collection, supply and reconditioning costs. Several strategies share also environmental-related data such as the evaluation of the amount and the typology of resources used, and the emissions generated in each process. Moreover, social-oriented data allowing to satisfy the entire demand, like the flexibility of internal processes, the scheduling of production plan and the scheduling of disassembling and remanufacturing activities, are common to different strategies. Concerning social aspects, data and information regarding workers' daily activities are gathered for all the strategies to monitor workers' ergonomic and safety conditions that sustain improvement of company efficiency by narrowing resources usage. All these shared data are required to be gathered also for the traditional manufacturing processes such as design, production, and logistics. Indeed, both new circular processes and updated traditional manufacturing processes necessitate to be implemented in a concurrent and coordinated way whenever CM strategies are adopted, requiring to enlarge the spectrum of data to be gathered.

All these processes influence one another. For instance, circular processes can be efficiently adopted only if it is delivered an appropriate service of maintenance and repair, and if it is possible the "turn back" of products. Therefore, the circular processes must be backed by the introduction of an adequate reverse logistic network [49] influencing the traditional logistics processes.

As final remark regarding the future trends on process category, manufacturers will need to ensure an appropriate internal organization encountering the needs from the traditional and circular processes. Therefore, the collection of data regarding these processes allows us to balance the resources needed and satisfy consumers demand. Data and information gathering and usage will allow the right scheduling of the activities to concurrently address different goals. This will be facilitated by the introduction of advanced technologies and information systems enabling both the internal alignment and the sharing of information with external actors whenever required.

4.3. Management Category

To ensure the orchestration of the entire ecosystem, also the management is influenced (see the dark green block in Figure 6). Managerial procedures might require to be modified and new ones need to be introduced, especially while dealing with industrial symbiosis and reverse logistics. They both require the involvement of stakeholders both internal and external to the supply chain, all needing to be appropriately managed. This review enlightens the relevance of data and information concerning three main aspects:

1. Supplier management: suppliers selection criteria need to be adjusted to produce products with the appropriate materials, such as non-toxic, biodegradable, and recyclable materials [84]. For instance, a new auditing procedure including ad hoc questionnaires could be introduced to gather data assessing supplier sustainability and circularity. In addition, to boost the competitiveness of the company and ensure product circularity, it is necessary to strengthen the relationships with the selected suppliers on the long-term, inevitably requiring data transparency;
2. Customer management: circular products production is justified only if there is an acceptable demand from customers which requires stimulation through marketing actions [60], among which are take-back programs, which allow customers to bring back, instead of discard, those products no more valuable for them. Therefore, customer management has to be updated to understand how to address customers' needs by taking advantage of an increasing "green market". Besides, it would help in training customers' behaviours towards the acceptance of circular products (e.g., recycled products);
3. Industrial symbiosis management: to retain value from the scrap, generated due to internal production inefficiencies, this can be either analysed to be reused or it can be exchanged with other external industrial companies. This second option requires an ad hoc management of these third parties to establish partnerships whenever possible.

As final remark regarding the future trends on management category, new solutions to establish trusted relationships will be developed. Therefore, partnerships with external stakeholders represent the right means embracing CM to fully exploit the benefits originated from this paradigm. To make this happen, data sharing is extremely important and thus, platforms or integrated systems implementation will be encouraged in the future. Moreover, the three types of stakeholders above mentioned (i.e., consumers, suppliers and partners in an industrial symbiosis network) cover the most relevant actors to engage with allowing the manufacturer to cover the entire product lifecycle having the necessary data to embrace CM.

4.4. Technology

Last, technology becomes the backbone of the CM transition through data valorisation (see the light green block in Figure 6). Both advanced I4.0 technologies, as recycling robots supporting recycle strategy [62], and information systems, like ERP supporting closed-loop supply chain [79], start covering relevant positions in CM. Moreover, information systems can be fed through different sources of data among which the authoring tools. Considering the rising need to appropriately manage products from their BoL, authoring tools emerged to be another important tool allowing to have lots of information regarding product characteristics, especially relevant for the circular design strategy [37]. In addition, ecological technologies emerged to be useful in supporting cleaner production and resource efficiency adoption embracing CE principles, such as the use of renewable sources of energy. For instance, photovoltaic panels and sensorized glasses are considered suitable for appropriate energy management [113].

To summarise, below are reported the four main classes of supportive technologies in CM:

1. Industry 4.0 technologies, such as Big Data analytics and robots;
2. Information Systems such as ERP, MRP, and PLM;
3. Authoring Tools such as CAD/CAEX and CAD 3D;
4. Ecological Technologies such as photovoltaic panels and sensorized glasses.

As final remark regarding the future trends of the technology category, the introduction of the four types of technologies just mentioned need to be considered in the strategic plans of manufacturers. Therefore, ecological technologies investments need to be specific according to the CM strategy willing to be adopted, and investments in ICT for instance will be required regardless the strategy adopted. This is especially true if looking at the CM strategies concurrent adoption, which requires the synergic implementation and integration of these technologies to allow a more comprehensive data gathering and usage.

5. Conclusions

This research operated an SLR intending to identify and classify all the relevant information, data, and supportive technologies and tools, required to aid manufacturers to approach to CM and manage the CM strategies. In detail, this research provides an overview of relevant areas to be monitored by manufacturers in CM, adopting different but correlated strategies, in an efficient and structured way. The review has been operated for each CM strategy identified in [13], analysed through four main categories.

From a scientific perspective, this research enables us to cover the identified gap facilitating data and information exploitation as necessary resources to adopt CM.

The first result of this SLR is the detection of the specific data and information needed by manufacturers in CM, classified per each strategy and category. From this review, it emerged that some data and information are peculiar for specific strategies (e.g., product maintenance in servitization), while others are common to different strategies (e.g., product characteristics) which might help and sustain their concurrent adoption. Therefore, the need to gather similar, or in some cases exactly the same data and information, strengthen the necessity to concurrently adopt CM strategies to embrace CM. Moreover, the data gathering and usage can be facilitated by the introduction of specific technologies which might be used in a synergic manner to support the concurrent adoption of the different strategies too. These findings allowed to address both of the RQs. In addition, relying on these findings, specific recommendation for manufacturers, regarding the most relevant data to be gathered and technologies to invest in, have been provided for each single CM strategy.

The second finding of this SLR is that manufacturers, for the appropriate embracement of the different CM strategies, necessitate to monitor and manage data and information concerning the "product", the "processes" and the "management", relying on the adoption of supporting "technologies/tools" to gather and use them. From this, the development of the theoretical framework (depicted in Figure 6). More in detail, for each of the four categories are elucidated the possible future trends which can be useful both for scholars, to undertake new studies in those directions, and for manufacturers to evaluate adequate future investments.

In line with that, the third result is reflected in the development of the theoretical framework (see Figure 6), embodying the review results. This gives a paramount perspective of classes of data required in CM and supporting technologies to gather and use data. It raises manufacturers' awareness regarding the type of internal and external data that should be gathered from internal activities, and from external stakeholders like consumers and suppliers to embrace CM strategies.

From a managerial perspective, this review provides an overview of the different information and data that manufacturers have to gather to adopt each CM strategy. This analysis allows us to create awareness in manufacturers regarding the need to exploit data gathered not only internally but also externally, by underlining the importance to share data in CM adoption. This requires us to promote data standardisation in terms of format too, in order to facilitate the exchange of data. Therefore, data exchange would favour the collaboration and the exchange of knowledge which benefits the exchange of all the other resources flows (e.g., materials). Actually, the theoretical framework developed elucidates the macro categories in which data should be gathered to be used for circular purposes, supporting manufacturers to clarify the areas to be monitored in CM. Therefore, this work represents the basement to allow a structured support for the decision-making process of manufacturers in embracing CM strategies. In addition, relying on this research, manufacturers can be stimulated in using specific technologies to gather and use data according to the CM strategy adopted. On one side, this can sustain tailored technological investments for the exploitation of data relying on the integration of different systems (such as the MES with the PLM and ERP). On the other side, the digitization path undertaken recently by manufacturing companies can be seen through new lenses. Thus, new usage of the technologies, sometimes already considered for investment to optimise companies' daily activities such as Industry 4.0 technologies, can be operated to enhance sustainable performances by supporting the concurrent adoption of CM strategies relying on the appropriate data and information.

From the policymakers' perspective, this review puts the basis to define adequate countermeasures and actions to promote the sharing of information management among industrial actors and with final consumers, to boost the sustainability of the manufacturing sector through the adoption of CM. Policymakers can take advantage for their future plans re-directing investments towards specific technologies that emerged to be the most promising to exploit data value under a CM perspective.

One of the main limitations of the present analysis is that the framework proposed is based only on the scientific literature and calls for further research to give practical and empirical evidence to the results obtained. In addition, all these findings opened the possibility for future studies:

- A conceptual data model structured on the data and information detected in this review needs to be developed to operationalise what has been so far theoretically obtained to bridge the managerial and operational levels of manufacturing companies throughout the transition towards CM. This model should enable to demonstrate how the concurrent adoption of different CM strategies can be eased through the exploitation of their data, information and technologies commonalities and it should be empirically verified;

- A quantitative model could be developed to assess the impact that each data and information might have to pursue not only the specific CM strategy to which it is related but also the other strategies not theoretically supposed to be bonded. This would validate the links detected among the different CM strategies and also unveil new connections among them through practical evidence;
- The tracking and management of data and information could represent the basement to quantify the benefits derived from the usage of CM [117] and to measure the circular performances of manufacturers [118]. In the long run, this would create the foundation for the development of a model assessing the circular maturity level of manufacturing companies through the definition, calculation, and monitoring of specific key performance indicators;
- A deeper investigation of the identified data and information should be performed to better define their own characteristics (e.g., accessibility, timeliness). Among them, particular attention could be dedicated to data's level of granularity;
- Since the usage of the different type of data and information are linked to the actors in charge of gathering them, future studies should be dedicated to investigate who is responsible for their gathering and management. The unit of analysis should be shifted from the manufacturing company as a whole to its single functions, also including the need to engage connections not only among internal managers and employees but also with external stakeholders belonging to other entities;
- Both managerial and technological barriers faced by manufacturers in exploiting data in CM might be further investigated also from an empirical point of view;
- Last, extensive studies should be performed to uniquely define and propose to manufacturers the new procedures necessary to gather most of the data and information emerged to be relevant in CM.

Author Contributions: Conceptualization, F.A.; methodology, F.A.; validation, C.S.; formal analysis, F.A.; investigation, F.A.; writing—original draft preparation, F.A.; writing—review and editing, C.S. and M.T.; visualization, C.S.; supervision, C.S., S.T. and M.T. All authors have read and agreed to the published version of the manuscript.

Funding: This research received no external funding.

Informed Consent Statement: Not applicable.

Conflicts of Interest: The authors declare no conflict of interest.

Acronyms

BOM	Bill Of Material	MES	Manufacturing Execution System
BoL	Begging of Life	MoL	Middle of Life
CE	Circular Economy	PLM	Product Lifecycle Management System
CM	Circular Manufacturing	PSS	Product-Service System
EoL	End of Life	SLR	Systematic Literature Review
ERP	Enterprise Resource Planning	TBL	Triple Bottom Line
ICT	Information and Communication Technology	WEEE	Waste Electrical and Electronic Equipment
KM	Knowledge Management	WIP	Work in Progress

References

1. Boulding, K.E. The economics of the coming spaceship earth. In *Environmental Quality in a Growing Economy: Essays from the Sixth RFF Forum*; Resources for the Future/Johns Hopkins University Press: Baltimore, MD, USA, 2013; Volume 3, pp. 1–20, ISBN 9781315064147.
2. Caradonna, J. *Sustainability: A history*; Oxford University Press: New York, NY, USA, 2014.
3. WCED. *Report of the World Commission on Environment and Development: Our Common Future*; Oxford University Press: Oxford, UK, 1987.
4. Elkington, J. *The Triple Bottom Line: Does It All Add up? Assessing the Sustainability of Business and CSR*; Richardson, J., Henriques, A., Eds.; Taylor and Francis: Abingdon, UK, 2013; ISBN 9781849773348.
5. Garetti, M.; Taisch, M. Sustainable manufacturing: Trends and research challenges. *Prod. Plan. Control* **2012**, *23*, 83–104. [CrossRef]

6. The Ellen MacArthur Foundation. *Towards a Circular Economy: Business Rationale for an Accelerated Transition*. 2015. Available online: https://www.ellenmacarthurfoundation.org/assets/downloads/publications/TCE_Ellen-MacArthur-Foundation_26-Nov-2015.pdf (accessed on 10 November 2019).
7. European Commission. *Circular Economy Action Plan*. 2020. Available online: https://ec.europa.eu/environment/circular-economy/pdf/new_circular_economy_action_plan.pdf (accessed on 11 October 2020).
8. Zhu, Q.; Geng, Y.; Lai, K. Circular economy practices among Chinese manufacturers varying in environmental-oriented supply chain cooperation and the performance implications. *J. Environ. Manag.* **2010**, *91*, 1324–1331. [CrossRef]
9. Pagotto, M.; Halog, A. Towards a Circular Economy in Australian Agri-food Industry: An Application of Input-Output Oriented Approaches for Analyzing Resource Efficiency and Competitiveness Potential. *J. Ind. Ecol.* **2016**, *20*, 1176–1186. [CrossRef]
10. Ranta, V.; Aarikka-Stenroos, L.; Ritala, P.; Mäkinen, S.J. Exploring institutional drivers and barriers of the circular economy: A cross-regional comparison of China, the US, and Europe. *Resour. Conserv. Recycl.* **2018**, *135*, 70–82. [CrossRef]
11. Bocken, N.; de Pauw, I.; Bakker, C.A.; van der Grinten, B. Product design and business model strategies for a circular economy. *J. Ind. Prod. Eng.* **2016**, *33*, 308–320. [CrossRef]
12. Valkokari, P.; Tura, N.; Ståhle, M.; Hanski, J.; Ahola, T. *Advancing Circular Business*; Tampere University: Tampere, Finland, 2019; ISBN 9789521543074.
13. Acerbi, F.; Taisch, M. A literature review on circular economy adoption in the manufacturing sector. *J. Clean. Prod.* **2020**, 123086. [CrossRef]
14. United Nations. Sustainable Development Goals. 2015. Available online: https://sustainabledevelopment.un.org/?menu=1300 (accessed on 16 September 2020).
15. Schot, J.; Steinmueller, W.E. Three frames for innovation policy: R&D, systems of innovation and transformative change. *Res. Policy* **2018**, *47*, 1554–1567. [CrossRef]
16. Kivimaa, P.; Boon, W.; Hyysalo, S.; Klerkx, L. Towards a typology of intermediaries in sustainability transitions: A systematic review and a research agenda. *Res. Policy* **2019**, *48*, 1062–1075. [CrossRef]
17. Tura, N.; Hanski, J.; Ahola, T.; Ståhle, M.; Piiparinen, S.; Valkokari, P. Unlocking circular business: A framework of barriers and drivers. *J. Clean. Prod.* **2019**, *212*, 90–98. [CrossRef]
18. Ritzén, S.; Sandström, G.Ö. Barriers to the Circular Economy—Integration of Perspectives and Domains. *Procedia CIRP* **2017**, *64*, 7–12. [CrossRef]
19. Halstenberg, F.A.; Lindow, K.; Stark, R. Utilization of Product Lifecycle Data from PLM Systems in Platforms for Industrial Symbiosis. *Procedia Manuf.* **2017**, *8*, 369–376. [CrossRef]
20. Grieves, M.W. Product lifecycle management: The new paradigm for enterprises. *Int. J. Prod. Dev.* **2005**, *2*, 71–84. [CrossRef]
21. Marconi, M.; Germani, M. An end of life oriented framework to support the transition toward circualr economy. In Proceedings of the 21st International Conference on Engineering Design, ICED17, Vancouver, BC, Canada, 21–25 August 2017; Volume 5.
22. Schmidt, N.; Lueder, A. The Flow and Reuse of Data: Capabilities of AutomationML in the Production System Life Cycle. *IEEE Ind. Electron. Mag.* **2018**, *12*, 59–63. [CrossRef]
23. Acerbi, F.; Taisch, M. Towards a data classification model for circular product life cycle management. In *Product Lifecycle Management Enabling Smart X, Proceedings of the 17th IFIP WG 5.1 International Conference, PLM 2020, Rapperswil, Switzerland, 5–8 July 2020*; Nyffenegger, F., Ríos, J., Rivest, L., Bouras, A., Eds.; Springer: Cham, Switzerland, 2020; pp. 473–486.
24. Sassanelli, C.; Rossi, M.; Pezzotta, G.; Pacheco, D.A.; Terzi, S. Defining Lean Product Service Systems (PSS) features and research trends through a systematic literature review. *Int. J. Prod. Lifecycle Manag.* **2019**. [CrossRef]
25. Olivier, S.; Pires, S.P.; Loures, E.R.F.; Santos, E.A.P.; Cestari, J.M.P.A. Knowledge management for sustainable performance in industrial maintenance. In Proceedings of the IIE Annual Conference and Expo 2015, Nashville, TN, USA, 30 May—2 June 2015.
26. Xin, Y.; Ojanen, V.; Huiskonen, J. Knowledge Management in Product-Service Systems—A Product Lifecycle Perspective. In Proceedings of the 10th CIRP Conference on Industrial Product-Service Systems, IPS2 2018, Linköping, Sweden, 29–31 May 2018; Volume 73, pp. 203–209.
27. Alcayaga, A.; Wiener, M.; Hansen, E.G. Towards a framework of smart-circular systems: An integrative literature review. *J. Clean. Prod.* **2019**, *221*, 622–634. [CrossRef]
28. Farooque, M.; Zhang, A.; Thürer, M.; Qu, T.; Huisingh, D. Circular supply chain management: A definition and structured literature review. *J. Clean. Prod.* **2019**, *228*, 882–900. [CrossRef]
29. Tranfield, D.; Denyer, D.; Smart, P. Towards a Methodology for Developing Evidence-Informed Management Knowledge by Means of Systematic Review. *Br. J. Manag.* **2003**, *14*, 207–222. [CrossRef]
30. Kirchherr, J.; Reike, D.; Hekkert, M. Conceptualizing the circular economy: An analysis of 114 definitions. *Resour. Conserv. Recycl.* **2017**, *127*, 221–232. [CrossRef]
31. Merli, R.; Preziosi, M.; Acampora, A. How do scholars approach the circular economy? A systematic literature review. *J. Clean. Prod.* **2018**, *178*, 703–722. [CrossRef]
32. Sitcharangsie, S.; Ijomah, W.; Wong, T.C. Decision makings in key remanufacturing activities to optimise remanufacturing outcomes: A review. *J. Clean. Prod.* **2019**, *232*, 1465–1481. [CrossRef]
33. Ren, S.; Zhang, Y.; Liu, Y.; Sakao, T.; Huisingh, D.; Almeida, C.M.V.B. A comprehensive review of big data analytics throughout product lifecycle to support sustainable smart manufacturing: A framework, challenges and future research directions. *J. Clean. Prod.* **2019**, *210*, 1343–1365. [CrossRef]

34. Maeda, J. *The Laws of Simplicity. Design, Technology, Business, Life*; MIT Press: Cambridge, MA, USA, 2006.
35. Sassanelli, C.; Urbinati, A.; Rosa, P.; Chiaroni, D.; Terzi, S. Addressing Circular Economy through Design for X approaches: A Systematic Literature Review. *Comput. Ind.* **2020**, *120*, 1–23. [CrossRef]
36. Fairphone. Fairphone as an Example of 'Circular Design'. 2019. Available online: https://forum.fairphone.com/t/fairphone-as-an-example-of-circular-design/50303 (accessed on 16 September 2020).
37. Laurenti, R.; Sinha, R.; Singh, J.; Frostell, B. Some pervasive challenges to sustainability by design of electronic products—A conceptual discussion. *J. Clean. Prod.* **2015**, *108*, 281–288. [CrossRef]
38. Khetriwal, D.S.; First, I. Enabling Closed Resource Loops in Electronics: Understanding Consumer Disposal Behaviour using insights from Diffusion models. *Econ. Res. Istraz.* **2012**, *25*, 47–68. [CrossRef]
39. Wang, L.; Shen, B. A product line analysis for eco-designed fashion products: Evidence from an outdoor sportswear brand. *Sustainability* **2017**, *9*, 1136. [CrossRef]
40. Yang, Y.; Chen, L.; Jia, F.; Xu, Z. Complementarity of circular economy practices: An empirical analysis of Chinese manufacturers. *Int. J. Prod. Res.* **2019**. [CrossRef]
41. Ramanujan, D.; Chandrasegaran, S.K.; Ramani, K. Visual Analytics Tools for Sustainable Lifecycle Design: Current Status, Challenges, and Future Opportunities. *J. Mech. Des. Trans. ASME* **2017**, *139*. [CrossRef] [PubMed]
42. Kumar, S.; Luthra, S.; Haleem, A. Customer involvement in greening the supply chain: An interpretive structural modeling methodology. *J. Ind. Eng. Int.* **2013**. [CrossRef]
43. Wilson, G.T.; Smalley, G.; Suckling, J.R.; Lilley, D.; Lee, J.; Mawle, R. The hibernating mobile phone: Dead storage as a barrier to efficient electronic waste recovery. *Waste Manag.* **2017**, *60*, 521–533. [CrossRef]
44. Wong, C.W.Y.; Lai, K.H.; Cheng, T.C.E.; Lun, Y.H.V. The roles of stakeholder support and procedure-oriented management on asset recovery. *Int. J. Prod. Econ.* **2012**, *135*, 584–594. [CrossRef]
45. Jasiulewicz-Kaczmarek, M. Identification of maintenance factors influencing the development of sustainable production processes—A pilot study. *IOP Conf. Ser. Mater. Sci. Eng.* **2018**, *400*, 062014. [CrossRef]
46. Nagiligari, B.K.; Shah, J.; Sha, Z.; Thirugnanam, S.; Jain, A.; Panchal, J. Integrated Part Classification for Product Cost and Complexity Reduction. In Proceedings of the ASME 2014 International Design Engineering Technical Conferences and Computers and Information in Engineering Conference, Buffalo, NY, USA, 17–20 August 2014.
47. Zhang, Z.; Liu, G.; Jiang, Z.; Chen, Y. A cloud-based framework for lean maintenance, repair, and overhaul of complex equipment. *J. Manuf. Sci. Eng. Trans. ASME* **2015**, *137*. [CrossRef]
48. Baran, J. Designing a Circular Product in the Light of the Semantic Design Areas. In proceeding of 29th International Business Information Management Association Conference, Vienna, Austria, 3–4 May 2017.
49. Shaharudin, M.R.; Govindan, K.; Zailani, S.; Tan, K.C.; Iranmanesh, M. Product return management: Linking product returns, closed-loop supply chain activities and the effectiveness of the reverse supply chains. *J. Clean. Prod.* **2017**, *149*, 1144–1156. [CrossRef]
50. Doni, F.; Corvino, A.; Bianchi Martini, S. Servitization and sustainability actions. Evidence from European manufacturing companies. *J. Environ. Manag.* **2019**, *234*, 367–378. [CrossRef] [PubMed]
51. Relight. 2020. Available online: https://www.relightitalia.it/en/company (accessed on 16 September 2020).
52. Cavalcante, J.; Gzara, L. Product-Service Systems lifecycle models: Literature review and new proposition. *Procedia CIRP* **2018**. [CrossRef]
53. Zhou, F.; Wang, X.; Lim, M.K.; He, Y.; Li, L. Sustainable recycling partner selection using fuzzy DEMATEL-AEW-FVIKOR: A case study in small-and-medium enterprises (SMEs). *J. Clean. Prod.* **2018**. [CrossRef]
54. America's Remanufacturing Company (ARC). 2020. Available online: http://www.remancouncil.org/companies/americas-remanufacturing-company (accessed on 16 September 2020).
55. Tan, Q.; Zeng, X.; Ijomah, W.L.; Zheng, L.; Li, J. Status of end-of-life electronic product remanufacturing in China. *J. Ind. Ecol.* **2014**. [CrossRef]
56. Sakao, T.; Sundin, E. How to Improve Remanufacturing? A Systematic Analysis of Practices and Theories. *J. Manuf. Sci. Eng. Trans. ASME* **2019**, *141*. [CrossRef]
57. Goodall, P.; Sharpe, R.; West, A. A data-driven simulation to support remanufacturing operations. *Comput. Ind.* **2019**, *105*, 48–60. [CrossRef]
58. Tolio, T.; Bernard, A.; Colledani, M.; Kara, S.; Seliger, G.; Duflou, J.; Battaia, O.; Takata, S. Design, management and control of demanufacturing and remanufacturing systems. *CIRP Ann. Manuf. Technol.* **2017**, *66*, 585–609. [CrossRef]
59. Lush. Reduce, Reuse, Recycle Lush Fresh Handmade Cosmetics. 2020. Available online: https://www.lushusa.com/stories/article_reduce-reuse-recycle.html (accessed on 16 September 2020).
60. Ongondo, F.O.; Williams, I.D.; Dietrich, J.; Carroll, C. ICT reuse in socio-economic enterprises. *Waste Manag.* **2013**, *33*, 2600–2606. [CrossRef]
61. Greentronics. 2020. Available online: http://www.greentronics.ro/ (accessed on 16 September 2020).
62. Sarc, R.; Curtis, A.; Kandlbauer, L.; Khodier, K.; Lorber, K.E.; Pomberger, R. Digitalisation and intelligent robotics in value chain of circular economy oriented waste management—A review. *Waste Manag.* **2019**, *95*, 476–492. [CrossRef] [PubMed]
63. Schilkowski, C.; Shukla, M.; Choudhary, S. Quantifying the circularity of regional industrial waste across multi-channel enterprises. *Ann. Oper. Res.* **2019**. [CrossRef]

64. Leba, M.; Ionica, A.; Dovleac, R.; Dobra, R. Waste management system for batteries. *Sustainability* **2018**, *10*, 332. [CrossRef]
65. Kouhizadeh, M.; Sarkis, J.; Zhu, Q. At the Nexus of Blockchain Technology, the Circular Economy, and Product Deletion. *Appl. Sci.* **2019**, *9*, 1712. [CrossRef]
66. Cao, J.; Xu, J.; Wang, H.; Zhang, X.; Chen, X.; Zhao, Y.; Yang, X.; Zhou, G.; Schnoor, J.L. Innovating collection modes for waste electrical and electronic equipment in China. *Sustainability* **2018**, *5*, 1446. [CrossRef]
67. Akinadé, O.O.; Oyedele, L.O. Integrating construction supply chains within a circular economy: An ANFIS-based waste analytics system (A-WAS). *J. Clean. Prod.* **2019**. [CrossRef]
68. Zhang, A.; Zhong, R.Y.; Farooque, M.; Kang, K.; Venkatesh, V.G. Blockchain-based life cycle assessment: An implementation framework and system architecture. *Resour. Conserv. Recycl.* **2020**, *152*, 104512. [CrossRef]
69. Belaud, J.-P.; Prioux, N.; Vialle, C.; Sablayrolles, C. Big data for agri-food 4.0: Application to sustainability management for by-products supply chain. *Comput. Ind.* **2019**, *111*, 41–50. [CrossRef]
70. Honic, M.; Kovacic, I.; Sibenik, G.; Rechberger, H. Data- and stakeholder management framework for the implementation of BIM-based Material Passports. *J. Build. Eng.* **2019**, *23*, 341–350. [CrossRef]
71. Bai, C.; Shah, P.; Zhu, Q.; Sarkis, J. Green product deletion decisions: An integrated sustainable production and consumption approach. *Ind. Manag. Data Syst.* **2018**. [CrossRef]
72. Nascimento, D.L.M.; Alencastro, V.; Quelhas, O.L.G.; Caiado, R.G.G.; Garza-Reyes, J.A.; Lona, L.R.; Tortorella, G. Exploring Industry 4.0 technologies to enable circular economy practices in a manufacturing context: A business model proposal. *J. Manuf. Technol. Manag.* **2019**, *30*, 607–627. [CrossRef]
73. Nespresso. Nespresso Commitments Doing Is Everything Nespresso. 2020. Available online: https://www.nespresso.com/it/it/commitments (accessed on 16 September 2020).
74. Fraccascia, L.; Yazdanpanah, V.; van Capelleveen, G.C.; Yazan, D.M. A framework for industrial symbiosis systems for agent-based simulation. In Proceedings of the 2019 IEEE 21st Conference on Business Informatics (CBI), Moscow, Russia, 15–17 July 2019.
75. Yeo, Z.; Masi, D.; Low, J.S.C.; Yen, N.T.; Tan, P.S.; Barnes, S. Tools for promoting industrial symbiosis: A systematic review. *J. Ind. Ecol.* **2019**. [CrossRef]
76. Raabe, B.; Low, J.S.C.; Juraschek, M.; Herrmann, C.; Tjandra, T.B.; Ng, Y.T.; Kurle, D.; Cerdas, F.; Lueckenga, J.; Yeo, Z.; et al. Collaboration Platform for Enabling Industrial Symbiosis: Application of the By-product Exchange Network Model. *Procedia CIRP* **2017**, *61*, 263–268. [CrossRef]
77. Gómez, A.M.M.; González, F.A.; Bárcena, M.M. Smart eco-industrial parks: A circular economy implementation based on industrial metabolism. *Resour. Conserv. Recycl.* **2018**, *135*, 58–69. [CrossRef]
78. Cutaia, L.; Antonella, L.; Barberio, G.; Sbaffoni, S.; Mancuso, E.; Scagliarino, C.; La Monica, M. The experience of the first industrial symbiosis platform in Italy. *Environ. Eng. Manag. J.* **2015**. [CrossRef]
79. Oltra-Badenes, R.; Gil-Gomez, H.; Guerola-Navarro, V.; Vicedo, P. Is It Possible to Manage the Product Recovery Processes in an ERP? Analysis of Functional Needs. *Sustainability* **2019**, *11*, 4380. [CrossRef]
80. Viegas, C.V.; Bond, A.; Vaz, C.R.; Bertolo, R.J. Reverse flows within the pharmaceutical supply chain: A classificatory review from the perspective of end-of-use and end-of-life medicines. *J. Clean. Prod.* **2019**, *238*, 117719. [CrossRef]
81. Chileshe, N.; Jayasinghe, R.S.; Rameezdeen, R. Information flow-centric approach for reverse logistics supply chains. *Autom. Constr.* **2019**, *106*, 102858. [CrossRef]
82. Bernon, M.; Tjahjono, B.; Ripanti, E.F. Aligning retail reverse logistics practice with circular economy values: An exploratory framework. *Prod. Plan. Control* **2018**, *29*, 483–497. [CrossRef]
83. Masi, D.; Day, S.; Godsell, J. Supply chain configurations in the circular economy: A systematic literature review. *Sustainability* **2017**, *9*, 1602. [CrossRef]
84. Rogers, Z.S.; Carter, C.R.; Kwan, V. Making tough choices: A policy capturing approach to evaluating the tradeoffs in sustainable supplier development initiatives. *J. Purch. Supply Manag.* **2019**, *25*, 100574. [CrossRef]
85. Aalirezaei, A.; Noorbakhsh, A.; Esfandi, N. Evaluation of relationships between GSCM practices and SCP using SEM approach: An empirical investigation on Iranian automobile industry. *J. Remanuf.* **2018**. [CrossRef]
86. Liou, J.J.H.; Tamošaitiene, J.; Zavadskas, E.K.; Tzeng, G.H. New hybrid COPRAS-G MADM Model for improving and selecting suppliers in green supply chain management. *Int. J. Prod. Res.* **2016**, *54*, 114–134. [CrossRef]
87. Gan, S.-S. The Conceptual Framework of Information Technology Adoption Decision-Making in a Closed-Loop Supply Chain. In Proceeding of the International Conference on Productivity and Sustainability in Science Engineering and Technology, Jakarta, Indonesia, 5–7 December 2017; Volume 8.
88. Garrido-Hidalgo, C.; Olivares, T.; Ramirez, F.J.; Roda-Sanchez, L. An end-to-end Internet of Things solution for Reverse Supply Chain Management in Industry 4.0. *Comput. Ind.* **2019**, *112*, 103127. [CrossRef]
89. Mahadevan, K. Collaboration in reverse: A conceptual framework for reverse logistics operations. *Int. J. Product. Perform. Manag.* **2019**. [CrossRef]
90. Akcil, A.; Agcasulu, I.; Swain, B. Valorization of waste LCD and recovery of critical raw material for circular economy: A review. *Resour. Conserv. Recycl.* **2019**, *149*, 622–637. [CrossRef]
91. Temur, G.T.; Bolat, B. Evaluating efforts to build sustainable WEEE reverse logistics network design: Comparison of regulatory and non-regulatory approaches. *Int. J. Sustain. Eng.* **2017**. [CrossRef]

92. Romero, D.; Molina, A. Reverse—Green Virtual Enterprises and their Breeding Environments: Closed-loop networks. In Proceedings of the IFIP Advances in Information and Communication Technology, Klagenfurt, Austria, 9–13 September 2013; Springer: New York, NY, USA, 2013; Volume 408, pp. 589–598.
93. Bahtia, M.S.; Kumar, R.S. Antecedents of implementation success in closed-loop supply chain: An empirical investigation. *Int. J. Prod. Res.* **2019**. [CrossRef]
94. Coenena, J.; der Heijdena, R.E.C.M.; Riel, A.C.R. Understanding approaches to complexity and uncertainty in closed-loop supply chain management: Past findings and future directions. *J. Clean. Prod.* **2018**. [CrossRef]
95. Bai, C.; Kusi-Sarpong, S.; Sarkis, J. An implementation path for green information technology systems in the Ghanaian mining industry. *J. Clean. Prod.* **2017**, *164*, 1105–1123. [CrossRef]
96. Islam, S.; Karia, N.; Fauzi, F.B.A.; Soliman, M.S.M. A review on green supply chain aspects and practices. *Manag. Mark.* **2017**, *12*, 12–36. [CrossRef]
97. Wong, C.W.Y.; Lai, K.; Lun, Y.H.V.; Cheng, T.C.E. Environmental management practices with supply chain efforts. In *SpringerBriefs in Applied Sciences and Technology*; Springer Nature: Berlin, Germany, 2015; pp. 29–72.
98. Agi, M. Analysis of the influence of organisational and inter-organisational factors on the implementation of Green Supply Chain Management practices. In Proceedings of the 2015 International Conference on Industrial Engineering and Systems Management (IESM), Seville, Spain, 21–23 October 2015.
99. Subramanian, N.; Gunasekaran, A. Cleaner supply-chain management practices for twenty-first-century organizational competitiveness: Practice-performance framework and research propositions. *Int. J. Prod. Econ.* **2015**, *164*, 216–233. [CrossRef]
100. Papettia, A.; Marconi, M.; Rossia, M.; Germani, M. Web-based platform for eco-sustainable supply chain management. *Sustain. Prod. Consum.* **2019**. [CrossRef]
101. Pal, R.; Sandberg, E. Sustainable value creation through new industrial supply chains in apparel and fashion. *IOP Conf. Ser. Mater. Sci. Eng.* **2017**, *254*. [CrossRef]
102. Kalverkamp, M.; Pehlken, A.; Wuest, T.; Young, S.B. Sustainability of cascading product lifecycles: The need for adaptive management to end-of-life supply chains. In Proceedings of the IFIP Advances in Information and Communication Technology, Turin, Italy, 2–4 July 2018; Springer: New York, NY, USA, 2018; Volume 540, pp. 159–168.
103. Tseng, M.L.; Islam, M.S.; Karia, N.; Fauzi, F.A.; Afrin, S. A literature review on green supply chain management: Trends and future challenges. *Resour. Conserv. Recycl.* **2019**, *141*, 145–162. [CrossRef]
104. Philips Lighting. Light as a Service Philips Lighting. 2020. Available online: https://www.lighting.philips.co.uk/campaigns/art-led-technology (accessed on 16 September 2020).
105. De Sousa Jabbour, A.B.L.; Rojas Luiz, J.V.; Rojas Luiz, O.; Jabbour, C.J.C.; Ndubisi, N.O.; Caldeira de Oliveira, J.H.; Junior, F.H. Circular economy business models and operations management. *J. Clean. Prod.* **2019**, *235*, 1525–1539. [CrossRef]
106. Xin, Y.; Ojanen, V.; Huiskonen, J. Dealing with Knowledge Management Practices in Different Product Lifecycle Phases within Product-service Systems. *Procedia CIRP* **2019**, *83*, 111–117. [CrossRef]
107. Wu, Y.; Zhu, D. Bicycle Sharing Based on PSS-EPR Coupling Model: Exemplified by Bicycle Sharing in China. *Procedia CIRP* **2017**, *64*, 423–428. [CrossRef]
108. Wiesner, S.; Freitag, M.; Westphal, I.; Thoben, K.D. Interactions between service and product lifecycle management. *Procedia CIRP* **2015**, *30*, 36–41. [CrossRef]
109. Farnioli, M.; Llshaj, A.; Lombardi, M.; Sciarretta, N.; Di Gravio, G. A BIM-based PSS approach for the management of maintenance operations of building equipment. *Buildings* **2019**, *9*, 139. [CrossRef]
110. Chen, Z.; Huang, L. Application review of LCA (Life Cycle Assessment) in circular economy: From the perspective of PSS (Product Service System). *Procedia CIRP* **2019**, *83*, 210–217. [CrossRef]
111. Apple. Environment—Apple. 2020. Available online: https://www.apple.com/lae/environment/ (accessed on 16 September 2020).
112. Hasan, S.M.M.; Rokonuzzaman, M.; Tuhin, R.A.; Salimullah, S.M.; Ullah, M.; Sakib, T.H.; Thollander, P. Drivers and barriers to industrial energy efficiency in textile industries of Bangladesh. *Energies* **2019**, *9*, 1775. [CrossRef]
113. Daú, G.; Scavarda, A.; Scavarda, L.F.; Taveira, V.J. The healthcare sustainable supply chain 4.0: The circular economy transition conceptual framework with the corporate social responsibility mirror. *Sustainability* **2019**, *11*, 3259. [CrossRef]
114. Parida, V.; Burström, T.; Visnjic, I.; Wincent, J. Orchestrating industrial ecosystem in circular economy: A two-stage transformation model for large manufacturing companies. *J. Bus. Res.* **2019**, *101*, 715–725. [CrossRef]
115. De Oliveira, S.F.; Soares, A.L. A PLM vision for circular economy. In Proceedings of the IFIP Advances in Information and Communication Technology. In Proceedings of the 18th Working Conference onVirtual Enterprises (PROVE), Vicenza, Italy, 18–20 September 2017; Springer: New York, NY, USA, 2017; Volume 506, pp. 591–602.
116. OECD. Business Models for the Circular Economy. *Bus. Model Circ. Econ.* **2018**. [CrossRef]
117. Rosa, P.; Sassanelli, C.; Terzi, S. Circular Business Models versus Circular Benefits: An Assessment in the Waste from Electrical and Electronic Equipments Sector. *J. Clean. Prod.* **2019**, *231*, 940–952. [CrossRef]
118. Sassanelli, C.; Rosa, P.; Rocca, R.; Terzi, S. Circular economy performance assessment methods: A systematic literature review. *J. Clean. Prod.* **2019**, *229*, 440–453. [CrossRef]

sustainability

MDPI

Review

Circular Economy Performance Measurement in Manufacturing Firms: A Systematic Literature Review with Insights for Small and Medium Enterprises and New Adopters

Marta Negri , Alessandra Neri *, Enrico Cagno and Gabriele Monfardini

Department of Management, Economics and Industrial Engineering, Politecnico di Milano, 20156 Milan, Italy; marta.negri@polimi.it (M.N.); enrico.cagno@polimi.it (E.C.); gabriele.monfardini@mail.polimi.it (G.M.)
* Correspondence: alessandra.neri@polimi.it

Abstract: The circular economy is a central and increasingly important topic within managerial and academic discourse. Although the circular economy could bring benefits to their performance, manufacturing firms still struggle with its adoption. As an effective adoption should pass for adequate performance measurement, the present study performs a systematic literature review to deepen the knowledge of circular economy performance-measurement systems for manufacturing firms, both from a general perspective and to provide specific insights for small–medium enterprises and new adopters. The results show the lack of an integrated, holistic, and scalable framework for measuring circular economy performance, and only a few and dispersed specific indications for small–medium enterprises and new adopters. Shortcomings of the extant literature are identified in terms of integration of the circular economy's level, theoretical development and empirical application, characteristics of the indicators proposed, considerations of sustainability, holistic perspectives on industrial systems, and scalability to adapt to firms' different characteristics. The study paves the way for further research while offering theoretical and practical implications.

Keywords: circular economy; performance indicators; small and medium enterprises; new adopters; manufacturing sector; systematic literature review

Citation: Negri, M.; Neri, A.; Cagno, E.; Monfardini, G. Circular Economy Performance Measurement in Manufacturing Firms: A Systematic Literature Review with Insights for Small and Medium Enterprises and New Adopters. *Sustainability* **2021**, *13*, 9049. https://doi.org/10.3390/su13169049

Academic Editor: Ioannis Nikolaou

Received: 7 July 2021
Accepted: 9 August 2021
Published: 12 August 2021

1. Introduction

The concept of a circular economy (CE) emerged in the 1990s, as the link between environmental preservation and industry's economic performance [1]. Nonetheless, only recently has the CE paradigm gained adequate significance [2], as the manufacturing sector perceived the need to reduce resource depletion, minimize waste, and lower its environmental impact [3,4]. According to Kirchherr et al.'s definition [2], CE can be applied at three different levels, namely the micro (single firm, from a single product to the advertisement), the meso (industrial systems and networks), and the macro (society or Country) level, and it entails the inclusion of the waste hierarchy (prevention, reduce, reuse, recycle, recovery, disposal). Additionally, CE is strongly linked with sustainability, and it is necessary to properly understand their relationship for a complete overview of CE as a concept [5,6].

Manufacturing firms can significantly benefit from the CE paradigm, both in terms of environmental preservation and economic gains [7,8]. Nonetheless, firms still struggle with the adoption of the CE paradigm and its related practices and interventions [9]. The measurement of performance is paramount to track progress and foster the implementation of the CE paradigm [10,11]. From this standpoint, the literature proposed several different frameworks and methodologies to measure CE-related performance [4]; the methods employed are rather varied, such that additional work to combine them has been called out for [12]. The extant efforts also differ in terms of context investigated, such as sector [13] or geographical area [14,15], or in terms of levels of application [16,17]. Although such diversity allows for a tailored performance measurement, it might (i) undermine interoperability; (ii) not be appropriate to support firms toward their incremental adoption of the CE paradigm; and (iii) restrict benchmarking activity [18]. From this perspective, it would be of great interest to synthesize and organize the ideas and contributions towards the measurement of CE-related performance in manufacturing firms [12], and, particularly, towards the features that an effective performance-measurement system should have [18]. The first question the present study aims at answering is:

RQ1. How can CE performance be measured in manufacturing firms?

The proper adoption of a specific performance-measurement system could nonetheless clash with some firms' characteristics. For example, the lack of resources and a blurred strategy might exert a negative influence on the development of a performance-measurement system [19]; resource and capability constraints, as well as the lack of operative instruments, could negatively impact the process of reporting [20]; besides limited resources, differences among firms depend also on strategy design and organizational setting [21], while the level of awareness results as fundamental for the diffusion of new concepts and paradigms within a firm [22,23]. The abovementioned characteristics distinguish firms to a different extent, but apply particularly to two types of firms: small-medium enterprises (SMEs) and the new adopters (NAs). SMEs refer to firms with less than 250 employees and an annual turnover not exceeding 50 million € [24]. Despite their prominent role, in the European context, in terms of economic, social, and environmental impacts [25,26], SMEs often lack appropriate know-how, resources, and support [27]. SMEs are usually resource-constrained, in terms of time, staff, and economic resources, compared to larger firms [26]. SMEs are also generally less prone than large enterprises (LEs) to undertake transformational changes [28] and appear limitedly conscious of their impact on the environment and society [29]. Their size could be, at first glance, considered a proxy of their competencies [30], although this might not always be the case [31]. CE adopters can be defined as firms born in a linear economy, making efforts to implement CE in their existing business models [32]. NAs can be defined as firms at the first stages of the circular transition [33,34] and at the initial stage of the adoption process [35]. NAs can face specific problems in relation to aspects of strategy, organizational structure, and performance management [21]. They might require additional support during the adoption process, as they may not be able to develop the required know-how [4,36]. Despite the relevance of the CE in the current debate, still, a large share of firms can be considered NAs, as the adoption of the CE paradigm occurs slowly Table 1 distinguishes SMEs and NAs according to four characteristics that emerged as pivotal from the above discussion, namely resource constraints, awareness, competences, and maturity level. Each of these characteristics might apply in different ways and to different extents to SMEs and NAs. As it can be inferred from Table 1, a firm could be both an SME and an NA; nonetheless, also LEs can be included within NAs, while native firms, i.e., firms founded on CE principles, might include SMEs [32].

Table 1. Categorization of SMEs and NAs according to pivotal characteristics. The table categorizes SMEs and NAs according to four characteristics that emerged as pivotal from the literature, namely resource constraints, awareness, competences and maturity level. The ✔ indicates that the feature is recognized as a main characteristic for the specific types of firm; the ✓ indicates that the characteristic could or could not apply for the specific firms. Supporting references are provided.

Characteristics	SMEs	NAs	Supporting References
Resources constraints. Availability of resources as time, staff, and economic resources	✔	✓	[37,38]
Awareness. Awareness over a specific aspect of interest and over the firm's impact on it	✓	✓	[35,39–41]
Competence. Adequate competences to tackle the specific aspect of interest	✓	✔	[30–32,35]
Maturity Level. Maturity level in terms of transition towards the specific aspect of interest	✓	✔	[27,32,35]

Such a categorization is relevant for properly understanding the needs of the different types of firms, and, as for the measurement of performance, reasoning on their specific characteristics [42,43]. An effective performance-measurement system should indeed allow for both a general and a tailored application, having flexibility as one of its main features, so as to adapt to a firm's continuous and dynamic evolution [18]. Despite its relevance, such an approach is still missing in the extant literature [44,45]. Focusing on SMEs and NAs, some contributions attempted to study CE implementation in SMEs [5], but evidence is scant and there is no indication of such effort for NAs. On the other hand, methods not specifically addressing SMEs or NAs might result in being too burdensome or difficult to adopt [42]. From this stand point, considering how CE performance can be measured in manufacturing firms, it is pivotal to understand the implications of the different characteristics of SMEs and NAs on the development of effective performance-measurement systems. Thus, the second question this study aims to answer is:

RQ2. What are the related implications for SMEs and NAs?

Leveraging the above and considering the specific targets of synthetizing and organizing the previous knowledge, the present study conducts a systematic literature review to answer the two research questions. The present study will address the measurement of CE performance from the perspective of the characteristics that an effective performance-measurement system should have, from a general viewpoint and particularly for being adequate and appropriate for application in SMEs and NAs.

The remainder of this paper is structured as follows. A detailed description of the systematic literature review methodology is provided by clearly outlining the followed steps (Section 2). After a descriptive analysis and evaluation of emerging themes (Section 3), the areas for which additional research is necessary are identified (Section 4). Finally (Section 5), pivotal implications of the study and possible further research are outlined.

2. Methodology

This study adopts a systematic literature review approach, identifying the contributions provided by the extant literature and possible emerging paths for future research [46]. The review is based on the procedure suggested by Howard et al. [47] and Tranfield and Denyer [48], adopting the following steps: (1) question formulation; (2) source identification; (3) study selection and evaluation; (4) analysis and synthesis; (5) reporting and using results.

2.1. Question Formulation

The objective of the review is to analyze in a comprehensive and informative manner the research questions illustrated in the previous section. According to the CIMO logic [49], we want to identify, considering manufacturing firms (context), frameworks and assessment methods (intervention) allowing the evaluation (mechanism) of CE-related performance (outcome).

2.2. Source Identification

A keywords-based search of the Scopus database was performed [50,51]. The search was performed on 4 May 2021 and updated on 2 July 2021. As per this study's goal, the keywords selected relate to performance measurement (indicator, KPI, performance, metric) or assessment procedure (assessment, measurement, analysis, evaluation), and terms related to the topic (circular economy, circularity). The selection of the keywords was based on previous publications, so as to ensure that the selected keywords were complete and appropriate for the scope of the present work: see [42,52] for keywords related to performance measurement and the assessment procedure; and [52–54] for keywords related to CE.

As for exclusion criteria, the analysis was limited to contributions published in English from the year 2000 onwards, while subject areas out of this study's interest, like medicine, the arts, and immunology were excluded. The procedure is reported in Table 2. A total of

5820 contributions were retrieved. The set was purged, eliminating 66 contributions for which relevant and necessary information was not provided (no author/title information), obtaining a set of 5754 contributions considered for source selection.

Table 2. Query used for the source identification.

Criteria Selection for the Literature Review			
Keywords	**Language**	**Publication Year**	**Subject Areas Excluded**
TITLE-ABS-KEY ("circular economy" OR circularity) AND TITLE-ABS-KEY (indicator OR KPI OR performance OR metric OR assessment OR measurement OR analysis OR evaluation)	English	>1999	MEDI; PHYS; AGRI; EART; ARTS; NEUR; PSYC; PHAR; IMMU; NURS; DENT; VETE

2.3. Source Selection and Evaluation

The selection of the contribution was performed based on the PRISMA methodology, which was deployed in the screening, eligibility, and inclusion phases. Each phase was conducted independently by each author and followed by a discussion within the research group, leading to a common decision on the inclusion or exclusion of a contribution, ensuring internal validity [46]. Additionally, no specific setting was addressed during the literature search and review other than the industrial context, strengthening the external validity [55]. The steps followed and assumptions made were explicit, so as to reduce bias. The PRISMA diagram, reported in Figure 1, shows the details for each phase.

The screening of the contributions was performed in two steps [56]. The first step employed a "title analysis". For the analysis, a manual coding was performed, wherein suggestions were made regarding the exclusion of contributions not considered relevant for the present work, or addressing a broader perspective on CE, such as the region or country level. As a result of the title analysis, 5573 contributions were excluded, while 181 contributions were considered eligible for an abstract analysis. The second step consisted of this "abstract analysis"; here, 87 contributions were excluded and 94 were identified as suitable for undergoing the eligibility phase. The evaluation of the contributions' eligibility was based on a "full-text analysis". From this, 20 contributions were discharged, mainly because they were not proposing indicators for measuring CE or had a high-level scope. Finally, 74 contributions were deemed as relevant for our analysis and thus included in the final set.

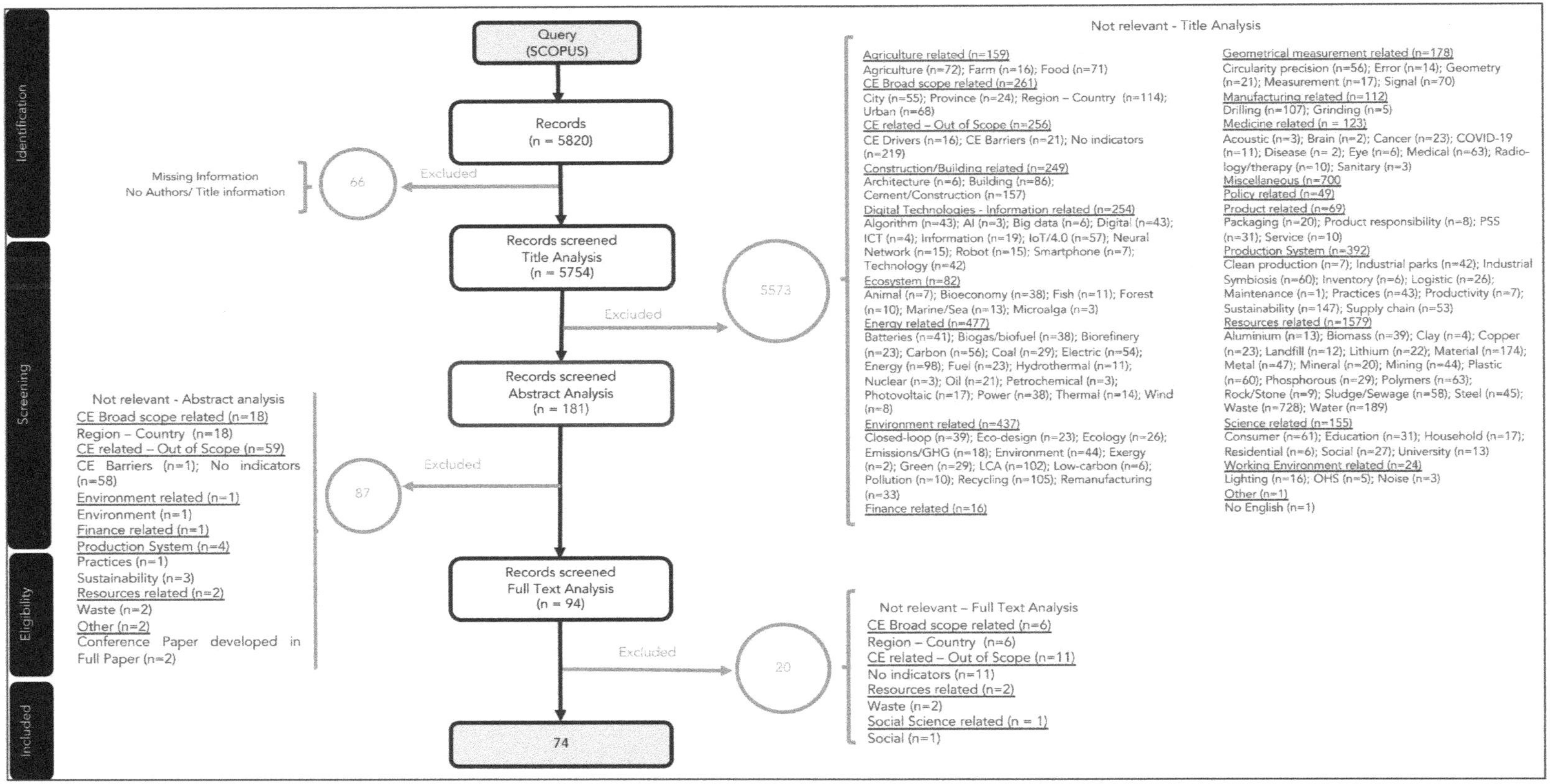

Figure 1. Details of the PRISMA procedure's phases.

For the evaluation phase, the retrieved contributions were classified according to the selected critical dimensions of analysis, appropriate to the scope of the review. A content analysis for a qualitative material evaluation was performed [57]. The contributions were divided into literature reviews and original studies. As for literature reviews, the following were highlighted: general information (authors, date, journal) and characteristics of the review (focus; timespan; database; number of papers included). The axes of analysis for literature reviews were selected so to better understand the focus of the selected contributions and identify areas that still need to be addressed with regards to the aim of the present work [43]. concerning original studies, the following were pinpointed: general information (authors, date, journal); theoretical development (context of development; methodology); indicators (number; categorization; sustainability pillars; prioritization; index); empirical application (context of application; methodology; sample/set). the axes of analysis for original studies were selected to (i) characterize the performance-measurement systems according to features that emerged as interesting from the literature [43,58]; (ii) understand the context of development and application of the frameworks [59,60].

The screening of the contributions and the content analysis were performed independently by three researchers (three of the four authors of the present paper). At every step, individual results were discussed, and a common result was defined and agreed upon. The results of the evaluation phase are presented and commented on in the next section.

2.4. Data Analysis

The data analysis was performed capturing relevant information through a critical investigation of the retrieved contributions [61], with a twofold perspective: (i) a descriptive evaluation, identifying quantitative trends, and (ii) a review of content in terms of emerging themes for the qualitative evaluation of research outcomes, identifying areas for which additional research is necessary.

3. Findings: Descriptive Evaluation and Emerging Themes

The section presents a descriptive evaluation and a discussion of emerging themes. The argument is divided between literature reviews and original research, and organized according to the axes of analysis introduced in Section 2.3.

3.1. Literature Reviews

The 13 retrieved reviews are recent and published after the year 2017. Most of them (n = 8) are published in the Journal of Cleaner Production; the others on Sustainability (Switzerland) (n = 2), Sustainable Production and Consumption (n = 2), and Resources, Conservation and Recycling (n = 1); therefore, the topic is addressed predominantly by journals combining environmental and managerial subject areas. Only one review explores methodologies the others analyze assessment frameworks for the micro level, either individually (n = 7), or combined with the meso and macro levels (n = 5).

Generally speaking, the literature reviews do not explicitly distinguish among the resource, material, product or firm levels of investigation [62]. The contributions still appear focused on the process or product level, and the possible actions integrated into industrial operations that can be undertaken at an industrial-plant level [63,64] are not considered. Such a perspective could limit the potential of interventions, as many practices for enhanced CE go beyond the boundaries of production processes [5]. Additionally, the literature reviews address the measurement of performance from a general perspective. Only Lindgreen et al. [51] provide insights for SMEs, while no implications for NAs were investigated. Considering our research questions, a further step in the current literature appears necessary, focusing attention on the measurement of CE performance in manufacturing firms, with a focus on specific implications for SMEs and NAs.

The overall evaluation of the reviewed Literature Reviews s reported in Table 3.

Table 3. Source Evaluation—Literature Reviews. The table reports general information of the considered literature reviews and provide insights on the characteristics of the review in terms of years, database and number of contributions considered, the circular economy level addressed, and the specific focus on SMEs and NAs.

Ref.	Journal	Years Considered	Databases Considered	Num. of Contributions	Level Considered	Focus on SMEs/NAs
[65]	Sustain. Prod. Consum	2013–2020	Scopus	58	micro; nano.	no
[58]	J. Clean. Prod	2000–2019	Scopus, WoS	61	micro; meso; macro	no
[45]	Sustain. Prod. Consum	2010–2019	Scopus, Google Scholar	135	micro; meso; macro	no
[66]	Sustainability	2009–2019	Springer; Scopus; MDPI, Whiley	60	micro; meso; macro	no
[67]	J. Clean. Prod	2007–2015	Scopus, WoS	107	micro	no
[51]	Sustainability	2007–2019	Scopus	74	micro (firm; product)	insights for SMEs
[44]	J. Clean. Prod	2006–2019	Scopus, WoS	31	micro	no
[68]	Resour. Conserv. Recycl	2008–2018	WoS	72	micro; meso; macro	no
[62]	J. Clean. Prod	n.a.	Scopus, Web of Science	52	micro; manufacturing	no
[34]	J. Clean. Prod	2010–2018	Academic; Non-academic	49	micro	no
[69]	J. Clean. Prod	2009–2018	Scopus	45	methodology	no
[1]	J. Clean. Prod	2004–2017	Scopus	601	micro; meso; macro	no
[70]	J. Clean. Prod	2003–2017	Scopus, Google Scholar	41	micro (reuse; recycle)	no

3.2. Original Studies

3.2.1. General Information

The temporal distribution of original studies (Figure 2) allows identifying two agglomerations. The first ($n = 13$) consists of contributions published before 2015, and it is related to the Chinese context, due to China's early focus on CE [71]. The second ($n = 48$) starts from the year 2015 and mirrors the increasing interest of Europe [72].

The reviewed contributions were published in both peer-reviewed journals ($n = 48$) and conference proceedings ($n = 13$). Focusing on the former set, the most represented journals are Journal of Cleaner Production ($n = 14$), Resources, Conservation and Recycling ($n = 6$) and Sustainability (Switzerland) ($n = 5$). Considering the subject area(s) of these journals (Figure 3), the topic is mainly considered from an environment and management related perspectives ($n = 25$).

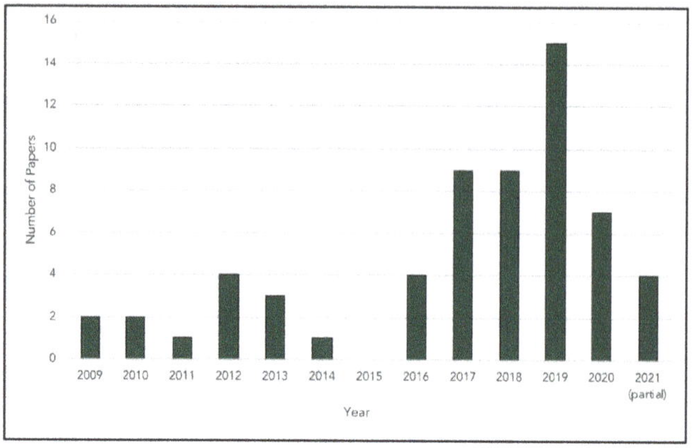

Figure 2. Original studies by year of publication.

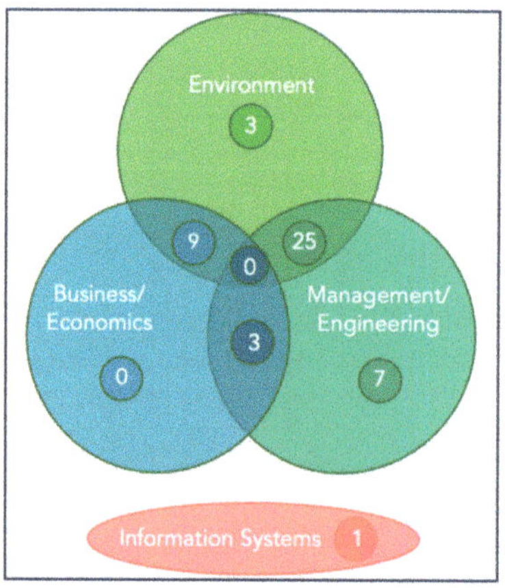

Figure 3. Original studies by area of journal.

3.2.2. Circular Economy Level

Despite all the reviewed contributions are focusing on issues of interest to the industrial decision-maker (IDM), they take different perspectives. The majority address frameworks at the single-product (n = 21) or the materials-and-resources level (n = 6), some consider the firm level (n = 19), others consider a system perspective (n = 8), while only a few studies (n = 6) analyze different levels simultaneously.

Micro Level. At the micro level, the largest share of the reviewed contributions focuses on a resource, material, or product level. Several methodologies for measuring the CE performance of products are proposed, and a common approach is not yet established. Besides proposing different approaches, the reviewed contributions also present diverse foci. For example, Di Maio et al. [73] and Linder et al. [74] developed indicators based on product economic and market value; Figge et al. [75] proposed longevity indicators to consider the closed loops of products; others focus only on one phase of the life cycle, such as reparability [76], recycling [77], or end-of-life [78,79].

Within the micro level, models addressing a product's circularity are largely diffused, but might present drawbacks for the measurement of the overall firm performance on CE, as it is not clear how to scale them up at the firm level. From this standing, their use for supporting the enhancement of firms' CE performance might be limited. A possible solution could be applying product-level indicators to all products manufactured by a firm, assigning weights based on quantities [80]. As a drawback, such a procedure would require a considerable amount of information for firms producing several different products, as it typically happens in manufacturing firms; besides, the procedure would be time-consuming, requiring high levels of awareness and competency [51]. Bearing that in mind, such a solution for the evaluation of a firm's performance would not be suitable for resource-constrained firms or for firms at the beginning of their CE transition [19,20]. It would be too burdensome, with the possible pitfall of discouraging firms or not providing proper support during their transition. Moreover, the scale-up process might not include plant-level practices [63], which represents a further limitation.

The reviewed contributions also propose methods to gauge the CE performance of the entire firm. For example, Koksharov et al. [81] propose the use of indicators' values trends, over time, as a proxy of circularity development; Garza-Reyes et al. [4] assess firm's circularity with a qualitative questionnaire based on the practices implemented; Rincón-Moreno et al. [82] adapted the indicators, already present at the macro level, at the micro level, also leveraging interviews with IDMs; Rossi et al. [13] link indicators with levels of applicability and usefulness in reaching desired performance, guiding firms along the implementation of a measurement system, and facilitating the CE transition. A limited number of contributions focusing on a firm's level address specific firms' sizes. Four contributions focus on LEs; for example, Rossi et al. [13] state that frameworks should employ an industrial-systems perspective and include multi-dimensional indicators; Yadav et al. [15] propose a framework that includes advanced indicators, such as managerial, organizational and policy indicators. Three contributions focus on SMEs, among them Aravossis et al. [83] underline the need for tailoring tools to the requirements of SMEs in terms of efficiency; Garza-Reyes et al. [4] propose a toolkit for SMEs with nine progressive levels of circularity. The literature stressed how the systems developed for LEs might not be suitable for SMEs [18,20], as LEs are supposed to have more resources available. Nonetheless, the literature also showed that LEs do not necessarily imply implementing more advanced practices and tools [31].

Integration of Levels. Competitiveness is increasingly played among industrial systems—as supply chains, industrial parks or industrial districts [18,84], rather than single firms [85,86]. Systems are pivotal for enhancing CE [2], so that a single firm's performance should be considered within the broader system in which it operates [4,43]. In the reviewed contributions, the micro and meso levels are separated, without an integrated micro–meso perspective. Mostly, the focus is generically only on meso level [87,88] supply networks [89], supply chains [90,91], and industrial parks [92]. Only Parchomenko et al. [93] consider both the micro and meso levels, but include the macro one, too. An integrated framework for the evaluation of performance that considers the micro level while also providing insights for the meso level would be of particular interest [6,45,90], but it has not been properly addressed thus far. Such a framework would help firms appropriately allocating their resources to prioritize those systems addressing performance on two levels, and thus is expected to outperform the single-perspective solutions. Such a characteristic would be of great support for resource-constrained firms [4], as it would allow having a single and straightforward system, rather than several with excessively detailed information. At the same time, it could also be of relevance for firms with a low level of competence and maturity, as the system would allow the focus on those interventions and actions able to foster the enhancement of their transitions from a multi-level perspective, eventually maximizing outcomes. The overall evaluation of the reviewed original studies, with respect to the circular-economy level considered, is reported in Table 4.

Table 4. Source Evaluation—Original Studies (CE Level and Theoretical Development). The table reports general information of the considered contributions and provides insights on: the circular economy level addressed; the theoretical development of the proposed framework in terms of context and methodology (AHP: analytic hierarchy process; BSC: balanced scorecard; CBA: cost-benefit analysis; EDIT: eco-innovation development and implementation tool; EF: ecological footprint; ETV: environmental technology verification; GMA: general morphological analysis; LCA: life-cycle assessment; LR: literature review; MCA: multiple correspondence analysis; MF: material flow; V: value-based; WIO: waste input-output; TOPSIS: technique for order preference by similarity to ideal solution).

	CE Level	Theoretical Development	
Ref.	**Level**	**Context of Development (Sector; Size; Area; Process)**	**Methodology**
[94]	micro	-	LR; experts; GMA; AHP; TOPSIS
[95]	firm	-	LR
[17]	material	-	LR
[82]	firm	-	LR
[96]	product	-	LR
[97]	product	-	LR; LCA
[98]	product	plastic	LR; experts' opinions; case studies
[99]	product	manufacturing	LR; experts' opinions
[13]	firm	electronic, textile, plastic	interviews; surveys; Focus group
[14]	firm	Europe	LR; experts' opinions
[15]	firm	emerging economies	LR; experts' opinions; Interviews; Survey
[100]	product	pharmaceutical	LR; MF
[79]	product	end of life	LR
[83]	firm—SME	SMEs	LR; LCA; MF
[101]	product	-	LR
[102]	firm	pulp and paper	LR
[4]	firm—SMEs	manufacturing; SMEs	LR; MF
[78]	product	tires production; end of life	LR
[103]	micro		LR; LCA
[54]	firm	LEs	LR; MCA
[104]	product	plastic (waste)	LR; experts' opinions
[81]	firm	-	LR; LCA
[16]	micro & macro	-	LR; experts' opinions
[105]	product	-	CBA; LCA
[93]	micro, meso, macro	-	LR
[106]	firm	food	LR; EDIT; experts' opinions
[89]	supply network	-	LR
[107]	product	food	LR
[75]	resource	-	LR; LCA
[77]	product	recycling	LR; MF
[90]	supply chain	-	LCA
[91]	supply chain	-	LR; companies' reports; interviews
[108]	product	-	MF; LCA; British standard
[109]	micro		LR; MF
[110]	product		LR; LCA; MF; ETV
[111]	product	-	LCA
[112]	firm	manufacturing	LR
[88]	meso	-	LR; survey

Table 4. *Cont.*

	CE Level	Theoretical Development	
Ref.	**Level**	**Context of Development (Sector; Size; Area; Process)**	**Methodology**
[53]	product	manufacturing	LR; V
[73]	resource	-	LR
[113]	product	-	V
[74]	product	-	LR
[114]	product	-	LR; BSC
[31]	firm	biotech and pharmaceutical	LR
[76]	product	repairing	LR; experts' opinions
[115]	material	-	LR; LCA; simulation
[116]	product	vehicles end of life	MF; V
[117]	firm	metallurgy	LR; companies' reports; survey
[118]	material	-	LR
[119]	firm	manufacturing	LR
[87]	meso	-	LR; LCA
[120]	firm	manufacturing	LR; MF
[121]	meso and macro	-	LR
[122]	firm	-	LR
[123]	meso	-	LR
[124]	firm	chemical	LR; WIO
[92]	industrial parks	-	Chinese regulations
[125]	firm	manufacturing	LR; Chinese regulations
[126]	firm	manufacturing	LR
[127]	supply chain	-	LR; MF; EF
[128]	firm	energy-intensive sectors	LR

3.2.3. Theoretical Development: Methods and Contexts

Following [129,130], contributions are grouped into theoretical (only introducing a new theoretical framework) and theoretical–empirical (theoretical framework coupled with an empirical application). Some of the reviewed contributions (n = 20) are only theoretical, while most of them (n = 41) are theoretical–empirical. The methodology employed for their theoretical development is, for almost the totality of cases (n = 54), a literature review, sometimes complemented by other considerations based, for example, on lifecycle assessment or material flow analysis. The previous literature is generally considered a solid base and a good proxy for the relevance of the specific aspects tackled [131,132]. Nonetheless, the literature alone is not sufficient, and should be integrated with other development backgrounds, such as IDMs and experts [42,43]. Only a few contributions included a panel of experts [14], case studies [98], surveys [88], or a mix of the above [15] in their theoretical development. Only a multi-perspectives approach could avoid bias due to the consideration of only a specific viewpoint [31]. The involvement of IDMs is fundamental for tailoring a general system to the specific characteristics of the reference firm.

As for the theoretical development, several contributions focus on a single context. In terms of sector, studies address manufacturing in general [99,120], but many are tailored to specific sectors such as pulp and paper [102], tire production [78,95], plastics [104], or chemicals [124]. Concerning geographical areas, some studies focus on Europe [14] or emerging economies [15]. In both cases, the development of the model could be biased by the selected contextual factors, so the applicability in different contexts should be further investigated [133], possibly leading to the development of indicators suitable for various contexts [42,53]. As for size, three studies address SMEs [4,83], one LEs [54], while the remaining do not focus their theoretical development on one or more specific firms' dimensions. If this, on the one hand, allows for a general system to be proposed, on the other hand, it might then represent a drawback to the adoption of the proposed frameworks by firms of different sizes, and, particularly, by SMEs. SMEs indeed might need a more straightforward tool [42,133], specifically, as they potentially have limited resources and capability to collect the large amount of data required for some of the frameworks proposed in the literature [20]. The overall evaluation of the reviewed original studies as for the theoretical development is reported in Table 4.

3.2.4. Circular Economy Performance Indicators

Number of indicators. The studies show a high variance in terms of the number of indicators proposed, with a range from 1 to 189 (Figure 4), for a total of 1066 performance indicators. Almost half of the studies (n = 26) propose less than 10 indicators, while very few more than 35 indicators (n = 3). The average number of indicators was 17.48; the average is far from the median of 13 and a high standard deviation of 26 indicators is observed. As the average appears biased by the contribution proposing 189 indicators [91], the exclusion of this study leads to an average of 14.62 indicators, with 12.5 indicators as median and 13.54 indicators as standard deviation.

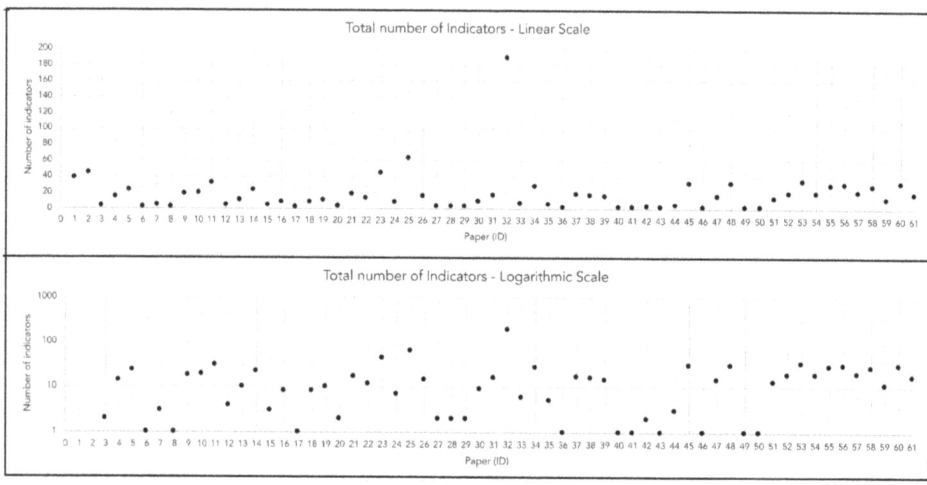

Figure 4. Original studies by the number of indicators.

The variety in the number of the proposed indicators is significant, with the risk of confusing firms and slow down CE adoption [82]. CE is a complex topic, requiring several indicators to be properly addressed [13]. A framework for measuring CE performance should be thus characterized by an appropriate breadth and depth. Such a level of detail would require a high number of indicators [134], but it is advisable to use a limited set of indicators, keeping them informative and not confusing [42,135,136]. An accepted threshold number of indicators has not yet been established in the literature, and suggestions range between 5 and 60 indicators [137,138], while empirical applications suggest a range from 20 to 70, depending on the context [133,139]. As some of the reviewed contributions can be placed in this range [31,117,124], the literature is still missing a proposal for adapting a system based on the specific context under evaluation.

Previous studies suggest that a framework for measuring performance should be manageable and easy to handle, while also guiding firms towards enhanced performance and a more structured approach towards CE [42]. Firms with low awareness, resources and competencies may require few indicators, providing an effective and efficient measurement of performance. With an increasing availability of resources and maturity, the system may be expanded to include more indicators [4]. Ideally, the frameworks should be scalable to support a firm at all the stages of resources, awareness, competence levels during their CE transition [43,134].

Categorization of indicators. Given the complex nature of the concept, systems for the measurement of CE-related performance should be addressed from a multi-dimensional perspective [82]. In general terms, authors provide frameworks of indicators that include economic- and productivity-related aspects [31,88,91], environmental aspects [117,121,124], and social and institutional aspects [91,123].

Almost half of the studies still do not propose a categorization of their indicators. The most common categorization is based on the traditional or slightly modified triple bottom line. All the remaining studies propose their categorizations, leading to a high level of heterogeneity and unclear indications for IDMs. Categories should be designed to allow ease of understanding, above all for those firms with limited competences. Nonetheless, it could be difficult to strictly assign a performance indicator to a single and unique category [42,132]. In this way, indicators able to cover multiple aspects (i.e., be part of multiple categories) could provide better indications to IDMs [43]. Focusing on those indicators able to maximize the content of information could be a great help for firms characterized by limited resources or at the beginning of their circular transitions [42]. Indeed, such a system would allow focusing on the most impactful indicators, obtaining the greatest amount of information with a minimum number of indicators.

Integration with the sustainability pillars. When the association was possible and not doubtful (734 out of 1066), we categorized indicators based on the three pillars of sustainability (Figure 5).

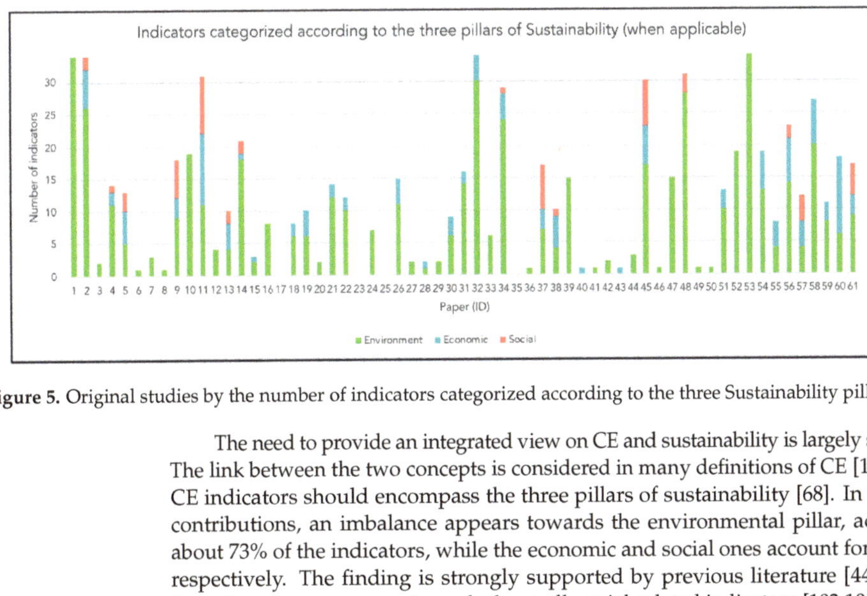

Figure 5. Original studies by the number of indicators categorized according to the three Sustainability pillars.

The need to provide an integrated view on CE and sustainability is largely stressed [6,62]. The link between the two concepts is considered in many definitions of CE [140], such that CE indicators should encompass the three pillars of sustainability [68]. In the reviewed contributions, an imbalance appears towards the environmental pillar, accounting for about 73% of the indicators, while the economic and social ones account for 18% and 9%, respectively. The finding is strongly supported by previous literature [44], urging the inclusion of more economic- and, above all, social-related indicators [102,105].

Environmental indicators often show a life-cycle perspective [96], including cumulative energy demand, global-warming potential, and water-stress index [98]. Some authors focus on the dimension of material utilization, as CE promises to minimize resource consumption: Rossi et al. [13] list reduction of raw material, renewability, recyclability, reduction of toxic substances, reuse, remanufacturing, refurbishment, product longevity, stakeholder structure, and diversity; Sánchez-Ortiz et al. [14] develop an environment-focused system, including indicators ranging from material consumption, toxicity of materials, generation of waste, and recycling rate and quality. Another main focus provided by the reviewed contributions is end-of-life solutions, like recycling, somewhat failing to address the other 'Rs' and the waste hierarchy [77], or clearly distinguish between strategies aimed at managing the end-of-life and the lifespan of products [17]. Since retaining the value of products and materials is one of the key features of CE, it makes sense to measure this aspect with particular care. However, CE is a more complex and multi-faceted paradigm, and complete frameworks of indicators should also include other environmental and social indicators. CE and social aspects are related in many ways, as concerns labor practices and decent work, human rights, society, and product responsibility. As employment, health and safety, and participation are most commonly included in CE [13,66], an additional step should be taken to include aspects other than internal ones [62]. Padilla-Rivera et al. [141] focus their attention on social indicators for CE, providing a valuable perspective that nonetheless needs integration with environmental and economic aspects.

To have an integrated model, able to reconcile sustainability and CE and their intersection, would be of great relevance [142,143]. Particularly, firms with limited resources or competences could benefit from the availability of an integrated performance-measurement systems able to include aspects related to two different yet interrelated concepts [13,74]. Such a system would indeed allow for a more straightforward measurement of performance, helping firms better organize resources towards those efforts indicating positive outcomes on both the concepts.

Prioritization of Indicators. A limited set of contributions (n = 4) provides a prioritization of indicators. The prioritization of indicators could be very useful for firms to identify and focus on those considered relevant [144]. Indeed, such prioritization could allow firms to begin their performance measurement with a limited number of indicators, only subsequently moving to a larger set [145]. Given the difficulty in measuring and improving all indicators at once, prioritization aims to identify the most urgent, providing beneficial insights to IDMs. The prioritization proposed by the reviewed studies follows different approaches: Cristóbal et al. [107] prioritize indicators according to the waste hierarchy; Wenbo [92] prioritizes indicators based on the 5R principle (rethinking, reduce, reuse, recycle, repair); Cayzer et al. [53] prioritize according to the relative importance of indicators, following IDMs' perspectives; Yadav et al. [15] prioritize and divide indicators into categories based on empirical evidence from a case study. Intuitively, the feature would be extremely useful for firms with limited resources and competences, or at the initial stages of their circular transition, as it would allow them to identify the key indicators to focus on [19].

A relevant aspect of prioritization concerns who performs it [18]. A proper prioritization should consider different perspectives, such as those of IDMs, experts, external stakeholders, or academics [146]. This would help overcome the subjectivity of a single perspective [83], resulting in, nonetheless, more complexity [43]. Particularly, the identification of the right IDMs can be rather challenging in firms with little coordination among different departments [19] or with problems assigning decisional power to project champions [147], given their size or immaturity.

Development of an index. About half of the studies (n = 26) consider the development of an index, i.e., a combination of indicators providing a snapshot of a given performance area, although the trend decreases from the year 2016 onward [122,126]. The use of an index presents several benefits: it is easy to understand, communicate, and benchmark efforts towards CE [107]. Among the most common indexes, it is possible to cite the Circular Economy Indicator Prototype [53], the Circular Economy Toolkit [148]; the Material Circularity Indicator [80]. All three are nonetheless related to the product level and focus mainly on environmental aspects, although business opportunities are described by the Circular Economy Toolkit. Albeit straightforward in their use, indexes present drawbacks in their application, as they neither distinguish between different loops (e.g., reuse, refurbish, recycle) nor provide guidance for circular product development [34]. Garza-Reyes et al. [4] tried to shift the evaluation of circularity from the product to the firm level, according to the CE practices implemented, providing practical suggestions to IDMs mainly from a qualitative perspective.

One main issue related to the development of an index lays in the possible subjectivity of evaluation [83] of how different indicators should be weighted in their contributions to the overall index [91]. Some studies apply the analytic hierarchy process [87,124], use fuzzy methods [125,126], or a combination of them [123]. Others propose the use of the multi-criteria evaluation method [96,105]. The majority of the studies, nonetheless, develop their index according to a ratio of quantities, considering the quantity of a particular material used over the total weight of a product [97] or the correspondent economic value [99]. These approaches can present drawbacks, as the development of the index is largely dependent on the perspective of whoever performs the selection [92,147], leading to possible inconsistency and subjectivity [149,150]. A way to overcome this impediment would be to include more than one perspective, such as those of internal and external stakeholders, experts, and academics [146]. For example, Cayzer et al. [53] rely on insights from the literature and IDMs, while Cristóbal et al. [107] test weighting factors retrieved from both grey and academic literature.

Indexes could represent valid help for firms with limited resources or a low level of maturity. Indeed, indexes allow quick assess of performance and provide benchmarks between different firms or different years, which also allows the tracking of progress. Nonetheless, a single index could entail a narrow scope [68], unable to consider all the multi-faced aspects of CE [16,53]. A proper starting point could entail two or three main indexes [151,152], with indications and guidelines towards enhanced CE and an integrated measurement of the related performance.

The overall evaluation of the reviewed original studies, as regards circular economy performance indicators, is reported in Table 5.

Table 5. Source Evaluation—Original studies (circular economy performance indicators). The table reports general information of the considered contributions and provides insights on the circular economy performance indicators in terms of number (° based on practices; retrieve from previous contributions), categorization (TBL: triple bottom line; BSC: balanced scorecard; SCOR: supply chain operations reference), sustainability pillars considered, prioritization and index development. The ✔ indicates whether the contribution performs a prioritization of the indicators and/or develops an index.

Circular Economy Performance Indicators							
Ref.	Num.	Categorization	TBL Pillars			Prioritization	Index
			Eco	Env	Soc		
[94]	38	9R	-	38	-	✔	-
[95]	44	categories—circular model; material circularity; economic model; environmental sustainability; social.	6	26	9	-	-
[17]	2	-		2		-	-
[82]	14	categories—production and consumption; waste management; secondary raw material; competitiveness and innovation	2	11	1	-	-
[96]	23	TBL + legislative; technical; business	5	5	3	-	✔
[97]	1	-	0	1	0	-	✔
[98]	3	-	0	3	0	-	-
[99]	1	-		1		-	✔
[13]	18	TBL	3	9	6	-	-
[14]	19	-	0	19	0	-	-
[15]	31	categories—managerial; organizational; supply chain; informational and technological; strategy and policy	11	11	9	✔	-

Table 5. *Cont.*

Circular Economy Performance Indicators							
Ref.	**Num.**	**Categorization**	**TBL Pillars**			**Prioritization**	**Index**
			Eco	**Env**	**Soc**		
[100]	4	-	0	4	0	-	-
[79]	10	TBL	4	4	2	-	-
[83]	22	categories—administrative; waste; energy; emissions; water	1	18	2	-	✓
[101]	3	categories—circular product design; servitised business models; supply chain management & reverse logistics; digital technologies; supply chain; product range; user; market; business cost; product structure; failure; end-of-use; technical; usage; usage cost; economic impact; environmental impact; social impact	1	2	0	-	-
[102]	8	categories—eco-efficiency; reuse	0	8	0	-	✓
[4]	1°	-	*	*	*	-	✓
[78]	8	-	2	6	0	-	-
[103]	10°	categories—production; use end of life; across life cycle	4	6	0	-	-
[54]	12	-	2	10	0	-	-
[104]	2	-	0	2	0	-	-
[81]	17	-	2	12	0	-	✓
[16]	44°	categories—micro; macro				-	-
[105]	7°	categories—material circularity; lifecycle	0	7	0	-	✓
[93]	63°	-				-	-
[106]	15	categories—lifecycle environmental impact; lifecycle cost	4	11	0	-	-
[89]	2	-	0	2	0	-	-
[107]	2	-	1	1	0	✓	✓
[75]	2	-	0	2	0	-	-
[77]	9	categories—technical; economic; sustainable	3	6	0	-	-
[90]	16	SCOR Processes	2	14	0	-	-
[91]	189	categories—environmental; economic; operational; logistics; organizational; marketing	12	30	0	-	-
[108]	6	-	0	6	0	-	✓
[109]	28	categories—circular economy; life cycle resource efficiency; climate energy & other; stocks and sufficiency	4	24	1	-	-
[110]	5°	-				-	-

Table 5. *Cont.*

Circular Economy Performance Indicators							
Ref.	Num.	Categorization	TBL Pillars			Prioritization	Index
			Eco	Env	Soc		
[111]	1	-	0	1	0	-	✓
[112]	17	TBL + Circularity	3	7	7	-	✓
[88]	16	BCS's perspectives	5	4	1	-	-
[53]	15	categories design; manufacturing; commercialization; in-use; end of use	0	15	0	✓	✓
[73]	1	-	1	0	0	-	✓
[113]	2	-	0	2	0	-	-
[74]	1	-	1	0	0	-	-
[114]	3	-	0	3	0	-	✓
[31]	31	-	6	17	7	-	-
[76]	1	-	0	1	0	-	✓
[115]	1	-	0	1	0	-	-
[116]	15	-	0	15	0	-	✓
[117]	31	categories—climate change; water; energy; land use; chemical risks; resource depletion; material efficiency; unrecovered materials; impacts from emissions; end use & end of life	0	28	3	-	-
[118]	1	-		1		-	✓
[119]	1	-		1		-	✓
[87]	13	categories—economic; resources; environment; recycling	3	10		-	✓
[120]	19	categories—energy consumption; material consumption; waste disposal and recycle; product and packing material recovery; green design; raw material production	0	19	0	-	✓
[121]	34	levels—macro; meso categories—resource output rate; resource consumption rate; resource utilization rate; waste disposal	0	34	0	-	-
[122]	19	categories—resource recycling; pollution and management; protection money	6	13	0	-	✓

Table 5. *Cont.*

Circular Economy Performance Indicators							
Ref.	Num.	Categorization	TBL Pillars			Prioritization	Index
			Eco	Env	Soc		
[123]	29	factors—production flexibility; product flexibility; delivery flexibility; marketing flexibility; external coordinated flexibility; organization flexibility; manufacturing of flexibility; research and development flexibility; fiscal policy system; legal system; public opinion influence; natural factors; build mechanisms; incentive mechanisms; trust mechanisms; interest and risk-sharing mechanisms; economic factors; technical factors; inter-enterprise links way	4	4	0	-	✔
[124]	30	positive vs negative	7	14	2	-	✔
[92]	20	TBL + element; management	4	4	4	✔	✔
[125]	27	categories—economic benefits; resources and energy utilization; material recycling; pollution control; production process; development potential	7	20	0	-	✔
[126]	11	categories—economy; resource used; resource reused; waste disposed	3	8	0	-	✔
[127]	31	categories—financial value; customer service; costs and benefits; business process; environmental performance	12	6	0	-	✔
[128]	17	TBL	3	9	5	-	-

3.2.5. Empirical Application: Methods Contexts

Several studies do not provide an empirical application of the proposed theoretical framework. This prevents an assessment of the proposed frameworks' capability to survive the test of real-case confrontation [153], leading to possible incomplete considerations [34]. The studies providing an empirical application mostly employ the case study methodology (n = 33). Few authors adopt the case study for a theory-building perspective, while most of them considered it for theory testing, to understand the usefulness and applicability of their proposed models and framework(s). Looking at the developed case studies, several shortcomings can be noticed. On the one hand, many contributions carried out a very limited number of case studies (one or two), so that a strong linkage with empirical application might be missing [5], and a higher number of case studies would be needed to extend the robustness of validation [4,112]. On the other hand, contributions providing a higher number of case studies focus on a narrow context, such as sector [31], geographical area [73,99], or type of firm, i.e., multi national enterprises [54]. Only a few contributions concentrate on specific sizes, such as SMEs or LEs ([82,83,154] and [13,31,125], respectively), while no attention is specifically dedicated to firms with different levels of maturity, resources, competences, or awareness. Additional research should aim to better understand the specific needs of these firms. To identify insights, distinct empirical investigations are precious and essential within the exact context of interest.

The overall evaluation of the reviewed original studies as for the empirical application is reported in Table 6.

Table 6. Source Evaluation—original studies (empirical application). The table reports general information of the considered contributions and provides insights on the empirical application in terms of context, methodology, sample employed.

	Empirical Application		
Ref.	Context of Application (Sector; Size; Area; Specific Process)	Methodology	Sample/Set
[94]	-	-	-
[95]	manufacturing of tires	case study (theory testing)	1
[17]	electronic and wood	case study (theory testing)	2
[82]	Spain, SMEs	case study (theory testing)	17
[96]	furniture; Sweden	case study (theory testing)	2
[97]	washing machine	case study (theory testing)	1
[98]	plastic: Sicily (Italy)	case study (theory testing)	1
[99]	manufacturing; Sweden	case study (theory testing)	18
[13]	electronic, textile, plastic; Brazil; LEs	case study (theory building)	3
[14]	-	-	-
[15]	heavy manufacturing; LEs; India	Best Worst Method; Decision-Making Trial; Evaluation Laboratory (theory testing)	1
[100]	-	-	-
[79]	engine manufacturing	case study (theory testing)	1
[83]	food; SMEs; Greece	case study (theory building)	1
[101]	washing machines; North Europe	case study (theory testing)	1
[102]	pulp and paper; Spain and Portugal	case study (theory testing)	2
[4]	manufacturing; SMEs; Mexico	case study (theory testing)	1
[78]	tires production; Italy, Switzerland; end of life	case study (theory testing)	1
[103]	-	-	-
[54]	multinational enterprises	case study (theory building)	13
[104]	plastic; Belgium	case study (theory testing)	1
[81]	-	-	-
[16]	-	-	-
[105]	beer packaging; UK and India	case study (theory testing)	1
[93]	-	-	-
[106]	food; UK	case study (theory testing)	4
[89]	plastics	case study (theory testing)	1
[107]	food; EU	case study (theory testing)	1
[75]	mobile phone manufacturer	case study (theory testing)	1
[77]	-	-	-
[90]	-	-	-
[91]	-	-	-
[108]	biomedical	case study (theory testing)	1
[109]	-	-	-
[110]	energy device manufacturer	case study (theory building)	1

Table 6. *Cont.*

Ref.	Empirical Application Context of Application (Sector; Size; Area; Specific Process)	Methodology	Sample/Set
[111]	biomedical	case study (theory testing)	1
[112]	-	-	-
[88]	-	-	-
[53]	leather manufacturer	case study (theory testing)	1
[73]	The Netherlands	case study (theory building)	40
[113]	plastic (waste); Belgium	case study (theory testing)	1
[74]	engine manufacturer	examples (theory testing)	2
[114]	catalytic-converter manufacturer	case study (theory testing)	1
[31]	biotech and pharmaceutical; LEs	case study (theory testing); survey (theory testing)	8
[76]	mobile phone repairer	survey (theory building)	400
[115]	mobile phones	case study (theory testing)	1
[116]	vehicles; China; end-of-life	case study (theory testing)	1
[117]	-	-	-
[118]	coal combustion by-products; USA	case study (theory testing)	1
[119]	-	-	-
[87]	-	-	-
[120]	-	-	-
[121]	-	-	-
[122]	-	-	-
[123]	coal; China	case study (theory testing)	1
[124]	chemical; China	case study (theory testing)	1
[92]	China	case study (theory testing)	5
[125]	brewery; LEs	case study (theory testing)	1
[126]	-	-	-
[127]	not specified	case study (theory testing)	1
[128]	metallurgy and energy; China	case study (theory testing)	2

4. Discussion

4.1. *How Can CE Performance Be Measured in Manufacturing Firms?*

Although valuable frameworks are present in the extant literature, some features still need to be properly tackled so as to provide an effective performance-measurement system for the evaluation of CE performance in manufacturing firms. In the following, specific features are outlined based on the results obtained from the review of the literature; the features are also discussed and confronted against the frameworks currently proposed by the extant literature.

4.1.1. The Need for an Integrated Performance-Measurement System

CE should not be a stand-alone concept within manufacturing firms [58,155], and a particularly strong relationship is identified with sustainability [156,157]. The literature started considering the three pillars of sustainability for the development of CE-measurement systems, but a proper integration is still missing [6,158], particularly concerning the social pillar, as also previously underlined [65].

The environmental pillar is traditionally more developed in the measurement of CE. Almost all the reviewed contributions include environmental indicators (see Figure 5), although the association is not always straightforward. In many cases, the environmental assessment is the only aspect monitored in the frameworks proposed, as in [98,105]. Although contributions typically stress the link between the economic and the environmental pillars in the CE paradigm, the social aspect should not be left behind [66]. The majority of contributions, nonetheless, do not include social indicators, and in general, their inclusion is not explicit, as in [15]. Although specific contributions have begun reconciling CE and sustainability, as in [31] and [112], full integration between the CE and sustainability paradigms needs to be properly addressed.

4.1.2. The Need for a Holistic Performance-Measurement System

To properly evaluate CE, performance should be assessed considering not only the single firm (micro level) but also the industrial systems in which it operates (meso level) [86]. Firms should thus be provided with a framework, able to cover the micro level, that also allows insight at the meso level [6]. The literature has nonetheless shown that, so far, manufacturing firms are still focusing their efforts on the adoption of internal practices [4], and also practices involving the industrial system are considered rather complex to implement [41]. Being able to simultaneously measure performance at different CE levels would require the adoption of a holistic system, as the different levels are intimately interconnected and explicable only by their reference in the overall picture. The reviewed contributions do not provide such a perspective, as they focus exclusively on either the product level [96,97], the firm level [15], or on the micro level, in general [103]. Only three contributions consider multiple levels: product and firm [109], micro and macro [16], and macro and meso [121], but a specific holistic integration of the micro and meso levels should be proposed.

4.1.3. The Need for an Appropriate Theoretical and Empirical Development

The measurement of performance should allow internal improvement [159,160], communication with external stakeholders [161,162], and benchmarking with peers [163]. In this way, a performance-measurement system should be general enough to be applied in different contexts, such as sector and geographical area, while also allowing a tailored approach to possible distinct needs [18]. Careful considerations need to be made in terms of theoretical development and empirical application. As for the former, a framework should be general and its included indicators should undergo an objective selection and prioritization, based on insights deriving from the different perspectives of IDMs, external stakeholders, academics, and experts [58]. Concerning the last, the capability of the framework to survive empirical tests of its data should be assessed [153]. It would thus be necessary to conduct case studies in heterogeneous contexts [164], effectively corroborating insights from different IDMs within firms operating in the same industrial system. The reviewed studies present shortcomings in this sense. Firstly, several studies are only theoretical and do not provide an empirical analysis [94]. Secondly, even when the proposed frameworks are tested empirically, most consider only one case study [98,118]. Many are also focused on a single sector and do not provide indications on how to generalize their results to different firms [83].

4.2. What Are the Related Implications for SMEs and NAs?

The reviewed contributions offer little insight for the applications of their developed frameworks in SMEs and NAs. Additionally, as highlighted in the Findings section, the proposed frameworks often clash with the specific needs of SMEs and NAs.

A scalable CE performance-measurement system could be of great help for SMEs and NAs: considering their characteristics, such a framework could represent valid help in fostering the enhancement of a firm's overall performance. To be specifically adopted by SMEs and NAs, a system should allow for ease of use and understanding. It should not be excessively complex or time-consuming, given the potentially limited availability of resources and limited level of awareness and competences of SMEs and NAs (see Table 1); rather, a performance-measurement system should be practical and manageable [42], while also providing a quick and comprehensive overview of its main aspects for evaluating CE performance in an integrated and holistic manner. Scalability would thus be of great importance [37,38], as it would allow a system to adapt easily to the changing needs of firms, as their awareness, competences, and maturity grow throughout their CE transition. Indeed, SMEs and NAs might initially face several difficulties in enhancing their CE performance by adopting practices; at this stage, a framework should be simple and effective at the same time. As firms become more aware and competent about the topic and can better organize their resources towards enhanced CE, their needs, in terms of a framework's features, could change. At this stage, different degrees of depth and breadth of performance measurement should be necessary. From this standing, the performance-measurement systems should allow for different levels of analysis, with a progressive inclusion of more advanced aspects [42].

Two of the reviewed contributions offer a specific perspective on SMEs, while no indications are provided for NAs. Shortcomings are nonetheless still present: Aravossis et al. [83] focus on SMEs, but their assessment is slightly shifted towards sustainability rather than CE and does not provide any indication on how to tailor a framework to the needs of specific SMEs. Garza-Reyes et al. [4] provide a measurement toolkit tailored for SMEs, with indications on how to assess the level of circularity of firms; nonetheless, their measurement is based on the number and depth of practices implemented, not on indicators.

4.3. Towards the Development of an Integrated, Holistic, and Scalable Performance-Measurement System for Manufacturing Firms

Considering the reviewed literature, a performance-measurement system able to address the above-mentioned features is still missing. To reduce the complexity of the measurement process [18], it is advisable for an effective performance-measurement system to meet all these features.

As for integration, an effective performance-measurement system for CE should provide clear indications regarding the simultaneous coverage of other paradigms within the manufacturing firms, such as sustainability. The indicators included in the system should be assessed in terms of the extent of simultaneous coverage of these two paradigms. Indeed, as the two paradigms are strongly interrelated, it would be sound to understand how the same indicator can provide information in both paradigms, rather than using separate and different indicators for each, as, for example, proposed by [112]. For developing such an integrated system, a great and deep understanding of the interrelations and overlaps between the two paradigms would be, of course, required, and additional value could derive from the simultaneous consideration of the perspectives of multiple IDMs within the same manufacturing firms and their industrial systems [147].

As for the holistic perspective, an effective performance-measurement system for CE should thus provide coverage of different CE levels, understanding the interrelations among them. Again, it is advisable to have a single, unique system for measuring performance at different levels, rather than separate ones. This, as well, should issue from a deep understanding of the information provided by the different indicators included in a framework and of the extent of their coverage of different levels of CE application. Once more, it is suggested to consider the perspectives of multiple IDMs [147].

As for scalability, an effective performance-measurement system for CE should be adapted to different firms, specifically SMEs and NAs, according to their characteristics and their evolving needs, in terms of breadth and depth of analysis, while also simultaneously allowing for internal performance measurement and benchmarking activities [18]. Particularly, a scalable framework would allow the presence of different levels of analysis and thus sets of indicators. The minimum set would include a limited number of indicators, so as to not distract from the pursuit of a focused strategy [135,136]. A minimum set should nonetheless be developed, so as to be able to maximize the content of information for an integrated and holistic evaluation of CE performance [42]. From such an initial set, additional sets should be derived, with an increased number of indicators and an increased content of information provided. The presence of different sets would allow a firm to move among them, according to its characteristics and stage of CE transition [42].

As for the development of such a performance-measurement system, it would be advisable to confront theoretical development with empirical evidence, deriving from its application in manufacturing firms. Particularly, to evaluate the scalability of such a system, longitudinal empirical analyses in SMEs and NAs would be of great relevance.

5. Concluding Remarks

The present study critically reviewed the literature proposing frameworks of indicators to measure CE, particularly, investigating how CE performance can be measured in manufacturing firms, and the related implications for SMEs and NAs.

The analysis shows a growing focus on the development of frameworks, but (i) scarce integration of the proposed frameworks with relevant aspects within manufacturing firms, such as sustainability; (ii) limited empirical validation and application of the developed frameworks; and (iii) limited attention given to the distinct needs of SMEs and NAs. Based on these identified shortcomings, relevant insights for further research are suggested, summarized in the need for an integrated, holistic, and scalable performance-measurement system for measuring CE.

The present study offers contributions from both a theoretical and managerial perspective, following the recommendations of Wickert et al. [165] for literature review. First, 74 literature contributions were analyzed, allowing a comprehensive list of axes for the evaluation of literature reviews and original studies: these axes could prove useful, for scholars and managers alike, as a reference guide to continue the exploration of the topic. Second, a detailed analysis of the previous literature, according to these axes of evaluation, was provided, underlining how each axis should be properly considered, spurring interest in future research. Third, an emerging need for the development of an integrated, holistic, and scalable framework of performance indicators for measuring CE is strongly called for. It would be of great interest to academia, fostering further research, and to practitioners, supporting them in understanding what features to look for in a framework for CE performance measurement. The theoretical development of such research should be then necessarily assessed against empirical applications in manufacturing firms, and particularly in SMEs and NAs. The present study has not considered or evaluated the specific indicators proposed by the reviewed contributions; having understood the main features that an effective performance-measurement system for CE should entail, future research should tackle analysis of indicators, focusing on the identification and selection of the right indicators to be included in integrated, holistic, and scalable frameworks.

The analysis was conducted following the principles of ethical research, quality, and accuracy. Nonetheless, some limitations should be highlighted. First, the study was conducted considering only the Scopus scientific research database; different findings may be obtained from other databases. Second, as the measurement of CE performance is currently a hot topic in the managerial and academic debate—the number of studies on the argument is constantly increasing—and the time frame used could have excluded some relevant recent contributions. Future research should be thus directed to consider the abovementioned limitations, while also investigating the evolution of the research topic.

Author Contributions: Conceptualization, M.N., A.N. and E.C.; methodology, M.N. and A.N.; validation, M.N., A.N. and E.C.; formal analysis, M.N., A.N. and G.M.; writing—original draft preparation, M.N. and A.N.; writing—review and editing, A.N.; visualization, A.N; supervision, A.N.; project administration, A.N. All authors have read and agreed to the published version of the manuscript.

Funding: This research received no external funding.

Institutional Review Board Statement: Not applicable.

Informed Consent Statement: Not applicable.

Data Availability Statement: Not applicable.

Conflicts of Interest: The authors declare no conflict of interest.

Abbreviations

CE	Circular Economy
IDM	Industrial Decision Maker
LEs	Large Enterprises
KPI	Key Performance Indicator
NAs	New Adopters
SMEs	Small and Medium Enterprises

References

1. Merli, R.; Preziosi, M.; Acampora, A. How do scholars approach the circular economy? A systematic literature review. *J. Clean. Prod.* **2018**, *178*, 703–722. [CrossRef]
2. Kirchherr, J.; Reike, D.; Hekkert, M. Conceptualizing the circular economy: An analysis of 114 definitions. *Resour. Conserv. Recycl.* **2017**, *127*, 221–232. [CrossRef]
3. Dey, P.; Malesios, C.; De, D.; Budhwar, P.; Chowdhury, S.; Cheffi, W. Circular Economy to Enhance Sustainability of Small and Medium sized Circular Economy to Enhance Sustainability of Small and Medium sized Enterprises. *Bus. Strateg. Environ.* **2020**, *29*, 2145–2169. [CrossRef]
4. Garza-Reyes, J.A.; Salomé Valls, A.; Peter Nadeem, S.; Anosike, A.; Kumar, V. A circularity measurement toolkit for manufacturing SMEs. *Int. J. Prod. Res.* **2019**, *57*, 7319–7343. [CrossRef]
5. Mura, M.; Longo, M.; Zanni, S. Circular economy in Italian SMEs: A multi-method study. *J. Clean. Prod.* **2020**, *245*, 118821. [CrossRef]
6. Fehrer, J.A.; Wieland, H. A systemic logic for circular business models. *J. Bus. Res.* **2021**, *125*, 609–620. [CrossRef]
7. Parida, V.; Burström, T.; Visnjic, I.; Wincent, J. Orchestrating industrial ecosystem in circular economy: A two-stage transformation model for large manufacturing companies. *J. Bus. Res.* **2019**, *101*, 715–725. [CrossRef]
8. Patwa, N.; Sivarajah, U.; Seetharaman, A.; Sarkar, S.; Maiti, K.; Hingorani, K. Towards a circular economy: An emerging economies context. *J. Bus. Res.* **2021**, *122*, 725–735. [CrossRef]
9. Kumar, V.; Sezersan, I.; Garza-Reyes, J.A.; Gonzalez, E.D.R.S.; AL-Shboul, M.A. Circular economy in the manufacturing sector: Benefits, opportunities and barriers. *Manag. Decis.* **2019**, *57*, 1067–1086. [CrossRef]
10. Engida, T.G.; Rao, X.; Berentsen, P.B.M.; Oude Lansink, A.G.J.M. Measuring corporate sustainability performance– the case of European food and beverage companies. *J. Clean. Prod.* **2018**, *195*, 734–743. [CrossRef]
11. European Commission. A Circular Economy for Plastics—Insights from Research and Innovation to Inform Policy and Funding Decisions. 2019. Available online: https://op.europa.eu/en/publication-detail/-/publication/33251cf9-3b0b-11e9-8d04-01aa75ed71a1/language-en/format-PDF/source-87705298 (accessed on 8 August 2021).
12. Walzberg, J.; Lonca, G.; Hanes, R.J.; Eberle, A.L.; Carpenter, A.; Heath, G.A. Do We Need a New Sustainability Assessment Method for the Circular Economy? A Critical Literature Review. *Front. Sustain.* **2021**, *1*, 620047. [CrossRef]
13. Rossi, E.; Bertassini, A.C.; dos Ferreira, C.S.; Neves do Amaral, W.A.; Ometto, A.R. Circular economy indicators for organizations considering sustainability and business models: Plastic, textile and electro-electronic cases. *J. Clean. Prod.* **2020**, *247*, 119137. [CrossRef]
14. Sánchez-Ortiz, J.; Rodríguez-Cornejo, V.; Del Río-Sánchez, R.; García-Valderrama, T. Indicators to Measure Efficiency in Circular Economies. *Sustainability* **2020**, *12*, 4483. [CrossRef]
15. Yadav, G.; Mangla, S.K.; Bhattacharya, A.; Luthra, S. Exploring indicators of circular economy adoption framework through a hybrid decision support approach. *J. Clean. Prod.* **2020**, *277*, 124186. [CrossRef]
16. Moraga, G.; Huysveld, S.; Mathieux, F.; Blengini, G.A.; Alaerts, L.; Van Acker, K.; de Meester, S.; Dewulf, J. Circular economy indicators: What do they measure? *Resour. Conserv. Recycl.* **2019**, *146*, 452–461. [CrossRef] [PubMed]
17. Moraga, G.; Huysveld, S.; De Meester, S.; Dewulf, J. Development of circularity indicators based on the in-use occupation of materials. *J. Clean. Prod.* **2021**, *279*, 123889. [CrossRef]
18. Neri, A. Industrial Sustainability Performance Measurement System—Challenges for the development. In *Methods in Sustainability Science*; In, press, Ren, J., Eds.; Elsevier: Amsterdam, The Netherlands, 2021; pp. 87–104, ISBN 9780128239872.
19. Falle, S.; Rauter, R.; Engert, S.; Baumgartner, R.J. Sustainability management with the Sustainability Balanced Scorecard in SMEs: Findings from an Austrian case study. *Sustainability* **2016**, *8*, 545. [CrossRef]

20. Arena, M.; Azzone, G. A process-based operational framework for sustainability reporting in SMEs. *J. Small Bus. Enterp. Dev.* **2012**, *19*, 669–686. [CrossRef]
21. Cosenz, F.; Bivona, E. Fostering growth patterns of SMEs through business model innovation. A tailored dynamic business modelling approach. *J. Bus. Res.* **2021**, *130*, 658–669. [CrossRef]
22. Johnson, M.P. Sustainability management and small and medium-sized enterprises: Managers' awareness and implementation of innovative tools. *Corp. Soc. Responsib. Environ. Manag.* **2015**, *22*, 271–285. [CrossRef]
23. Witjes, S.; Vermeulen, W.J.V.V.; Cramer, J.M. Exploring corporate sustainability integration into business activities. Experiences from 18 small and medium sized enterprises in The Netherlands. *J. Clean. Prod.* **2017**, *153*, 528–538. [CrossRef]
24. European Union. Commission Recommendation of 6 May 2003 Concerning the Definition of Micro, Small and Medium-Sized Enterprises. 2003. Available online: http://data.europa.eu/eli/reco/2003/361/oj (accessed on 13 May 2021).
25. Ghenţa, M.; Matei, A. Smes and the circular economy: From policy to difficulties encountered during implementation. *Amfiteatru Econ.* **2018**, *20*, 294–309. [CrossRef]
26. Micheli, G.J.L.; Cagno, E.; Neri, A.; Cieri, E. Non-safety costs: A novel methodology for an ex-ante evaluation. *Saf. Sci.* **2021**, *133*, 105025. [CrossRef]
27. Albats, E.; Alexander, A.; Mahdad, M.; Miller, K.; Post, G. Stakeholder management in SME open innovation: Interdependences and strategic actions. *J. Bus. Res.* **2019**, *119*, 291–301. [CrossRef]
28. Mitchell, S.; O'Dowd, P.; Dimache, A. Manufacturing SMEs doing it for themselves: Developing, testing and piloting an online sustainability and eco-innovation toolkit for SMEs. *Int. J. Sustain. Eng.* **2020**, *13*, 159–170. [CrossRef]
29. Feil, A.A.; De Quevedo, D.M.; Schreiber, D. An analysis of the sustainability index of micro- and small-sized furniture industries. *Clean Technol. Environ. Policy* **2017**, *19*, 1883–1896. [CrossRef]
30. Katz-Gerro, T.; López Sintas, J. Mapping circular economy activities in the European Union: Patterns of implementation and their correlates in small and medium-sized enterprises. *Bus. Strateg. Environ.* **2019**, *28*, 485–496. [CrossRef]
31. Veleva, V.; Bodkin, G.; Todorova, S. The need for better measurement and employee engagement to advance a circular economy: Lessons from Biogen's "zero waste" journey. *J. Clean. Prod.* **2017**, *154*, 517–529. [CrossRef]
32. Rovanto, I.K.; Bask, A. Systemic circular business model application at the company, supply chain and society levels—A view into circular economy native and adopter companies. *Bus. Strateg. Environ.* **2021**, *30*, 1153–1173. [CrossRef]
33. Gusmerotti, N.M.; Testa, F.; Corsini, F.; Pretner, G.; Iraldo, F. Drivers and approaches to the circular economy in manufacturing firms. *J. Clean. Prod.* **2019**, *230*, 314–327. [CrossRef]
34. Saidani, M.; Yannou, B.; Leroy, Y.; Cluzel, F.; Kendall, A. A taxonomy of circular economy indicators. *J. Clean. Prod.* **2019**, *207*, 542–559. [CrossRef]
35. Moric, I.; Jovanović, J.Š.; Đoković, R.; Peković, S.; Perović, Đ. The Effect of Phases of the Adoption of the Circular Economy on Firm Performance: Evidence from 28 EU Countries. *Sustainability* **2020**, *12*, 2557. [CrossRef]
36. Sehnem, S.; Campos, L.M.S.; Julkovski, D.J.; Cazella, C.F. Circular business models: Level of maturity. *Manag. Decis.* **2019**, *57*, 1043–1066. [CrossRef]
37. An, W.; Xu, Y.; Zhang, J. Resource constraints, innovation capability and corporate financial fraud in entrepreneurial firms. *Chin. Manag. Stud.* **2018**, *12*, 2–18. [CrossRef]
38. Musso, P.; Schiavo, S. The impact of financial constraints on firm survival and growth. *J. Evol. Econ.* **2008**, *18*, 135–149. [CrossRef]
39. Yongtao, W. SMEs in the Circular Economy Development Strategy. *Manag. Sci. Eng.* **2015**, *9*, 76–80. [CrossRef]
40. Ormazabal, M.; Prieto-Sandoval, V.; Puga-Leal, R.; Jaca, C. Circular Economy in Spanish SMEs: Challenges and opportunities. *J. Clean. Prod.* **2018**, *185*, 157–167. [CrossRef]
41. Masi, D.; Kumar, V.; Garza-Reyes, J.A.; Godsell, J. Towards a more circular economy: Exploring the awareness, practices, and barriers from a focal firm perspective. *Prod. Plan. Control* **2018**, *29*, 539–550. [CrossRef]
42. Cagno, E.; Neri, A.; Howard, M.; Brenna, G.; Trianni, A. Industrial sustainability performance measurement systems: A novel framework. *J. Clean. Prod.* **2019**, *230*, 1354–1375. [CrossRef]
43. Neri, A.; Cagno, E.; Lepri, M.; Trianni, A. A triple bottom line balanced set of Key Performance Indicators to measure the sustainability performance of industrial supply chains. *Sustain. Prod. Consum.* **2021**, *26*, 648–689. [CrossRef]
44. Kristensen, H.S.; Mosgaard, M.A. A review of micro level indicators for a circular economy—Moving away from the three dimensions of sustainability? *J. Clean. Prod.* **2020**, *243*, 118531. [CrossRef]
45. Harris, S.; Martin, M.; Diener, D. Circularity for circularity's sake? Scoping review of assessment methods for environmental performance in the circular economy. *Sustain. Prod. Consum.* **2021**, *26*, 172–186. [CrossRef]
46. Thomé, A.M.T.; Scavarda, L.F.; Scavarda, A.J. Conducting systematic literature review in operations management. *Prod. Plan. Control* **2016**, *27*, 408–420. [CrossRef]
47. Howard, J.; Piacentino, J.; MacMahon, K.; Schulte, P. Using systematic review in occupational safety and health. *Am. J. Ind. Med.* **2017**, *60*, 921–929. [CrossRef] [PubMed]
48. Tranfield, D.; Denyer, D. Producing a systematic review. In *The Sage Handbook of Organizational Research Methods*; SAGE: Thousand Oaks, CA, USA, 2009; pp. 672–688.
49. Denyer, D.; Tranfield, D.; Van Aken, J.E. Developing design propositions through research synthesis. *Organ. Stud.* **2008**, *29*, 393–413. [CrossRef]

50. Palmaccio, M.; Dicuonzo, G.; Belyaeva, Z.S. The internet of things and corporate business models: A systematic literature review. *J. Bus. Res.* **2021**, *131*, 610–618. [CrossRef]
51. Roos Lindgreen, E.; Salomone, R.; Reyes, T. A Critical Review of Academic Approaches, Methods and Tools to Assess Circular Economy at the Micro Level. *Sustainability* **2020**, *12*, 4973. [CrossRef]
52. Cagno, E.; Neri, A.; Negri, M.; Bassani, C.A.; Lampertico, T. The Role of Digital Technologies in Operationalizing the Circular Economy Transition: A Systematic Literature Review. *Appl. Sci.* **2021**, *11*, 3328. [CrossRef]
53. Cayzer, S.; Griffiths, P.; Beghetto, V. Design of indicators for measuring product performance in the circular economy. *Int. J. Sustain. Eng.* **2017**, *10*, 289–298. [CrossRef]
54. Howard, M.; Hopkinson, P.; Miemczyk, J. The regenerative supply chain: A framework for developing circular economy indicators. *Int. J. Prod. Res.* **2019**, *57*, 7300–7318. [CrossRef]
55. Avellar, S.A.; Thomas, J.; Kleinman, R.; Sama-Miller, E.; Woodruff, S.E.; Coughlin, R.; Westbrook, T.R. External Validity: The Next Step for Systematic Reviews? *Eval. Rev.* **2017**, *41*, 283–325. [CrossRef]
56. Xiao, Y.; Watson, M. Guidance on Conducting a Systematic Literature Review. *J. Plan. Educ. Res.* **2019**, *39*, 93–112. [CrossRef]
57. Engert, S.; Rauter, R.; Baumgartner, R.J. Exploring the integration of corporate sustainability into strategic management: A literature review. *J. Clean. Prod.* **2016**, *112*, 2833–2850. [CrossRef]
58. De Pascale, A.; Arbolino, R.; Szopik-Depczyńska, K.; Limosani, M.; Ioppolo, G. A systematic review for measuring circular economy: The 61 indicators. *J. Clean. Prod.* **2021**, *281*, 124942. [CrossRef]
59. Sousa, R.; Voss, C.A. Contingency research in operations management practices. *J. Oper. Manag.* **2008**, *26*, 697–713. [CrossRef]
60. Choudhury, R.G. Relationship between contextual factors, business performance, and strategy: A study of manufacturing and service industries in India. *Bus. Manag. Rev.* **2016**, *7*, 295–303.
61. Maestrini, V.; Luzzini, D.; Maccarrone, P.; Caniato, F. Supply chain performance measurement systems: A systematic review and research agenda. *Int. J. Prod. Econ.* **2017**, *183*, 299–315. [CrossRef]
62. Kravchenko, M.; Pigosso, D.C.; McAloone, T.C. Towards the ex-ante sustainability screening of circular economy initiatives in manufacturing companies: Consolidation of leading sustainability-related performance indicators. *J. Clean. Prod.* **2019**, *241*, 118318. [CrossRef]
63. Trianni, A.; Cagno, E.; Neri, A. Modelling barriers to the adoption of industrial sustainability measures. *J. Clean. Prod.* **2017**, *168*, 1482–1504. [CrossRef]
64. Tonelli, F.; Evans, S.; Taticchi, P. Industrial sustainability: Challenges, perspectives, actions. *Int. J. Bus. Inov. Res.* **2013**, *7*, 143–163. [CrossRef]
65. de Oliveira, C.T.; Dantas, T.E.T.; Soares, S.R. Nano and micro level circular economy indicators: Assisting decision-makers in circularity assessments. *Sustain. Prod. Consum.* **2021**, *26*, 455–468. [CrossRef]
66. Padilla-Rivera, A.; Russo-Garrido, S.; Merveille, N. Addressing the social aspects of a circular economy: A systematic literature review. *Sustainability* **2020**, *12*, 7912. [CrossRef]
67. Vinante, C.; Sacco, P.; Orzes, G.; Borgianni, Y. Circular economy metrics: Literature review and company-level classification framework. *J. Clean. Prod.* **2021**, *288*, 125090. [CrossRef]
68. Corona, B.; Shen, L.; Reike, D.; Rosales Carreón, J.; Worrell, E. Towards sustainable development through the circular economy—A review and critical assessment on current circularity metrics. *Resour. Conserv. Recycl.* **2019**, *151*, 104498. [CrossRef]
69. Sassanelli, C.; Rosa, P.; Rocca, R.; Terzi, S. Circular economy performance assessment methods: A systematic literature review. *J. Clean. Prod.* **2019**, *229*, 440–453. [CrossRef]
70. Sandin, G.; Peters, G.M. Environmental impact of textile reuse and recycling—A review. *J. Clean. Prod.* **2018**, *184*, 353–365. [CrossRef]
71. Su, B.; Heshmati, A.; Geng, Y.; Yu, X. A review of the circular economy in China: Moving from rhetoric to implementation. *J. Clean. Prod.* **2013**, *42*, 215–227. [CrossRef]
72. Fitch-Roy, O.; Benson, D.; Monciardini, D. Going around in circles? Conceptual recycling, patching and policy layering in the EU circular economy package. *Env. Polit.* **2020**, *29*, 983–1003. [CrossRef]
73. Di Maio, F.; Rem, P.C.; Baldé, K.; Polder, M. Measuring resource efficiency and circular economy: A market value approach. *Resour. Conserv. Recycl.* **2017**, *122*, 163–171. [CrossRef]
74. Linder, M.; Sarasini, S.; van Loon, P. A Metric for Quantifying Product-Level Circularity. *J. Ind. Ecol.* **2017**, *21*, 545–558. [CrossRef]
75. Figge, F.; Thorpe, A.S.; Givry, P.; Canning, L.; Franklin-Johnson, E. Longevity and Circularity as Indicators of Eco-Efficient Resource Use in the Circular Economy. *Ecol. Econ.* **2018**, *150*, 297–306. [CrossRef]
76. Flipsen, B.; Bakker, C.; Van Bohemen, G. Developing a reparability indicator for electronic products. In Proceedings of the 2016 Electronics Goes Green 2016+ (EGG), Berlin, Germany, 6–9 September 2016; pp. 1–9.
77. Grimaud, G.; Perry, N.; Laratte, B. Development of an Evaluation Tool for Engineering Sustainable Recycling Pathways. *Procedia CIRP* **2018**, *69*, 781–786. [CrossRef]
78. Gigli, S.; Landi, D.; Germani, M. Cost-benefit analysis of a circular economy project: A study on a recycling system for end-of-life tyres. *J. Clean. Prod.* **2019**, *229*, 680–694. [CrossRef]
79. Alamerew, Y.A.; Brissaud, D. Circular economy assessment tool for end of life product recovery strategies. *J. Remanuf.* **2019**, *9*, 169–185. [CrossRef]

80. Ellen MacArthur Foundation. Circularity Indicators: An Approach to Measuring Circularity. 2015. Available online: https://www.ellenmacarthurfoundation.org/assets/downloads/insight/Circularity-Indicators_Project-Overview_May2015.pdf (accessed on 8 August 2021).
81. Koksharov, V.; Starodubets, N.; Ponomareva, M. Assessment of an enterprise circular economy development. *WSEAS Trans. Bus. Econ.* **2019**, *16*, 559–567.
82. Rincón-Moreno, J.; Ormazábal, M.; Álvarez, M.J.; Jaca, C. Advancing circular economy performance indicators and their application in Spanish companies. *J. Clean. Prod.* **2021**, *279*, 123605. [CrossRef]
83. Aravossis, K.G.; Kapsalis, V.C.; Kyriakopoulos, G.L.; Xouleis, T.G. Development of a holistic assessment framework for industrial organizations. *Sustainability* **2019**, *11*, 3946. [CrossRef]
84. The Royal Academy of Engineering. Industrial Systems: Capturing Value through Manufacturing. 2012. Available online: https://www.raeng.org.uk/publications/reports/industrial_systems_capturing_value) (accessed on 1 August 2021).
85. Massaroni, E.; Cozzolino, A.; Wankowicz, E. Sustainability in supply chain management—A literature review. *Sinergie* **2015**, *33*, 331–355. [CrossRef]
86. Shibin, K.T.; Gunasekaran, A.; Dubey, R. Explaining sustainable supply chain performance using a total interpretive structural modeling approach. *Sustain. Prod. Consum.* **2017**, *12*, 104–118. [CrossRef]
87. Zhang, J.; Wang, N.; Hong, J. Comprehensive evaluation on the development of industry cluster circular economy. *Adv. Mater. Res.* **2013**, *779*, 1777–1780. [CrossRef]
88. Butzer, S.; Schötz, S.; Petroschke, M.; Steinhilper, R. Development of a Performance Measurement System for International Reverse Supply Chains. *Procedia CIRP* **2017**, *61*, 251–256. [CrossRef]
89. Brown, P.J.; Bajada, C. An economic model of circular supply network dynamics: Toward an understanding of performance measurement in the context of multiple stakeholders. *Bus. Strateg. Environ.* **2018**, *27*, 643–655. [CrossRef]
90. Jain, S.; Jain, N.K.; Metri, B. Strategic framework towards measuring a circular supply chain management. *Benchmarking* **2018**, *25*, 3238–3252. [CrossRef]
91. Kazancoglu, Y.; Kazancoglu, I.; Sagnak, M. A new holistic conceptual framework for green supply chain management performance assessment based on circular economy. *J. Clean. Prod.* **2018**, *195*, 1282–1299. [CrossRef]
92. Li, W. Comprehensive evaluation research on circular economic performance of eco-industrial parks. *Energy Procedia* **2011**, *5*, 1682–1688. [CrossRef]
93. Parchomenko, A.; Nelen, D.; Gillabel, J.; Rechberger, H. Measuring the circular economy—A Multiple Correspondence Analysis of 63 metrics. *J. Clean. Prod.* **2019**, *210*, 200–216. [CrossRef]
94. Franco, N.G.; Almeida, M.F.L.; Calili, R.F. A strategic measurement framework to monitor and evaluate circularity performance in organizations from a transition perspective. *Sustain. Prod. Consum.* **2021**, *27*, 1165–1182. [CrossRef]
95. Vimal, K.E.K.; Kulatunga, A.K.; Ravichandran, M.; Kandasamy, J. Application of multi grade fuzzy approach to compute the circularity index of manufacturing organizations. *Procedia CIRP* **2021**, *98*, 476–481. [CrossRef]
96. Alamerew, Y.A.; Kambanou, M.L.; Sakao, T.; Brissaud, D. A multi-criteria evaluation method of product-level circularity strategies. *Sustainability* **2020**, *12*, 5129. [CrossRef]
97. Bracquené, E.; Dewulf, W.; Duflou, J.R. Measuring the performance of more circular complex product supply chains. *Resour. Conserv. Recycl.* **2020**, *154*, 104608. [CrossRef]
98. Cascone, S.; Ingrao, C.; Valenti, F.; Porto, S.M.C. Energy and environmental assessment of plastic granule production from recycled greenhouse covering films in a circular economy perspective. *J. Environ. Manage.* **2020**, *254*, 109796. [CrossRef] [PubMed]
99. Linder, M.; Boyer, R.H.W.; Dahllöf, L.; Vanacore, E.; Hunka, A. Product-level inherent circularity and its relationship to environmental impact. *J. Clean. Prod.* **2020**, *260*, 121096. [CrossRef]
100. Akrivos, V.; Haines-Gadd, M.; Mativenga, P.; Charnley, F. Improved metrics for assessment of immortal materials and products. *Procedia CIRP* **2019**, *80*, 596–601. [CrossRef]
101. Bressanelli, G.; Perona, M.; Saccani, N. Assessing the impacts of circular economy: A framework and an application to the washing machine industry. *Int. J. Manag. Decis. Mak.* **2019**, *18*, 282. [CrossRef]
102. de Abreu Ferreira, I.; de Castro Fraga, M.; Godina, R.; Barreiros, M.S.; Carvalho, H. A proposed index of the implementation and maturity of circular economy practices-the case of the pulp and paper industries of Portugal and Spain. *Sustainability* **2019**, *11*, 1722. [CrossRef]
103. Helander, H.; Petit-Boix, A.; Leipold, S.; Bringezu, S. How to monitor environmental pressures of a circular economy: An assessment of indicators. *J. Ind. Ecol.* **2019**, *23*, 1278–1291. [CrossRef]
104. Huysveld, S.; Hubo, S.; Ragaert, K.; Dewulf, J. Advancing circular economy benefit indicators and application on open-loop recycling of mixed and contaminated plastic waste fractions. *J. Clean. Prod.* **2019**, *211*, 1–13. [CrossRef]
105. Niero, M.; Kalbar, P.P. Coupling material circularity indicators and life cycle based indicators: A proposal to advance the assessment of circular economy strategies at the product level. *Resour. Conserv. Recycl.* **2019**, *140*, 305–312. [CrossRef]
106. Slorach, P.C.; Jeswani, H.K.; Cuéllar-Franca, R.; Azapagic, A. Environmental and economic implications of recovering resources from food waste in a circular economy. *Sci. Total Environ.* **2019**, *693*, 133516. [CrossRef]
107. Cristóbal, J.; Castellani, V.; Manfredi, S.; Sala, S. Prioritizing and optimizing sustainable measures for food waste prevention and management. *Waste Manag.* **2018**, *72*, 3–16. [CrossRef]

108. Mesa, J.; Esparragoza, I.; Maury, H. Developing a set of sustainability indicators for product families based on the circular economy model. *J. Clean. Prod.* **2018**, *196*, 1429–1442. [CrossRef]
109. Pauliuk, S. Critical appraisal of the circular economy standard BS 8001:2017 and a dashboard of quantitative system indicators for its implementation in organizations. *Resour. Conserv. Recycl.* **2018**, *129*, 81–92. [CrossRef]
110. Walker, S.; Coleman, N.; Hodgson, P.; Collins, N.; Brimacombe, L. Evaluating the Environmental Dimension of Material Efficiency Strategies Relating to the Circular Economy. *Sustainability* **2018**, *10*, 666. [CrossRef]
111. Angioletti, C.M.; Despeisse, M.; Rocca, R. Product Circularity Assessment Methodology Cecilia. *IFIP Adv. Inf. Commun. Technol.* **2017**, *514*, 411–418. [CrossRef]
112. Azevedo, S.; Godina, R.; Matias, J. Proposal of a Sustainable Circular Index for Manufacturing Companies. *Resources* **2017**, *6*, 63. [CrossRef]
113. Huysman, S.; De Schaepmeester, J.; Ragaert, K.; Dewulf, J.; De Meester, S. Performance indicators for a circular economy: A case study on post-industrial plastic waste. *Resour. Conserv. Recycl.* **2017**, *120*, 46–54. [CrossRef]
114. Saidani, M.; Yannou, B.; Leroy, Y.; Cluzel, F. How to Assess Product Performance in the Circular Economy? Proposed Requirements for the Design of a Circularity Measurement Framework. *Recycling* **2017**, *2*, 6. [CrossRef]
115. Franklin-Johnson, E.; Figge, F.; Canning, L. Resource duration as a managerial indicator for Circular Economy performance. *J. Clean. Prod.* **2016**, *133*, 589–598. [CrossRef]
116. Pan, Y.; Li, H. Sustainability evaluation of end-of-life vehicle recycling based on emergy analysis: A case study of an end-of-life vehicle recycling enterprise in China. *J. Clean. Prod.* **2016**, *131*, 219–227. [CrossRef]
117. Rönnlund, I.; Reuter, M.; Horn, S.; Aho, J.; Aho, M.; Päällysaho, M.; Ylimäki, L.; Pursula, T. Eco-efficiency indicator framework implemented in the metallurgical industry: Part 1—A comprehensive view and benchmark. *Int. J. Life Cycle Assess.* **2016**, *21*, 1473–1500. [CrossRef]
118. Park, J.Y.; Chertow, M.R. Establishing and testing the "reuse potential" indicator for managing wastes as resources. *J. Environ. Manage.* **2014**, *137*, 45–53. [CrossRef]
119. Li, S.L. Study on the model of quantitative evaluation of circular economy development for industry manufacturing based on WIOA. *Appl. Mech. Mater.* **2013**, *345*, 384–387. [CrossRef]
120. Zheng, J.; Zheng, C. Research on a Circular Economy Index System Frame of Manufacturing Industrial Chain. In *International Asia Conference on Industrial Engineering and Management Innovation (IEMI2012) Proceedings*; Qi, E., Shen, J., Dou, R., Eds.; Springer: Berlin/Heidelberg, Germany, 2013; pp. 1389–1399. ISBN 978-3-642-38445-5.
121. Geng, Y.; Fu, J.; Sarkis, J.; Xue, B. Towards a national circular economy indicator system in China: An evaluation and critical analysis. *J. Clean. Prod.* **2012**, *23*, 216–224. [CrossRef]
122. Jiang, C.-H.; Jiang, P. Analysis on evaluation index system of enterprise environmental performance under circular economy. In Proceedings of the 2012 International Conference on Management Science & Engineering 19th Annual Conference Proceedings, Dallas, TX, USA, 20–22 September 2012; pp. 658–663. [CrossRef]
123. Tang, L.H.; Tian, Y.H. The fuzzy comprehensive evaluation of the flexible development of circular economy in enterprise. In Proceedings of the 2012 International Conference on Management Science & Engineering 19th Annual Conference Proceedings, Dallas, TX, USA, 20–22 September 2012; pp. 408–415. [CrossRef]
124. Zheng, J.; Huang, Y.; Wang, Z. Study on establishment and application of circular economy evaluation index system for the chemical industry. *Adv. Mater. Res.* **2012**, *524–527*, 3455–3458. [CrossRef]
125. An, X.H.; Cui, Y.M.; Qi, E.S. Study on eco-efficiency evaluation of manufacturing system based on Circular economy. In Proceedings of the 2010 IEEE 17th International Conference on Industrial Engineering and Engineering Management, Xiamen, China, 29–31 October 2010; pp. 591–593. [CrossRef]
126. Chen, L.J.; Li, J.; Chen, L.K. Evaluating performance of circular economy for manufacturing enterprise based on the theories of EFA and MFA. In Proceedings of the 2010 IEEE 17th International Conference on Industrial Engineering and Engineering Management, Xiamen, China, 29–31 October 2010; pp. 515–519. [CrossRef]
127. Xu, J. Model of cluster green supply chain performance evaluation based on circular economy. In Proceedings of the 2009 Second International Conference on Intelligent Computation Technology and Automation, Changsha, China, 10–11 October 2009; Volume 3, pp. 941–944. [CrossRef]
128. Zheng, J. Structural design and empirical analysis on the circular economy index system for energy-intensive industries. In Proceedings of the 2009 16th International Conference on Industrial Engineering and Engineering Management, Beijing, China, 21–23 October 2009; pp. 2141–2145. [CrossRef]
129. Martín-Peña, M.L.; Díaz-Garrido, E.; Sánchez-López, J.M. Analysis of benefits and difficulties associated with firms' Environmental Management Systems: The case of the Spanish automotive industry. *J. Clean. Prod.* **2014**, *70*, 220–230. [CrossRef]
130. Murillo-Luna, J.L.; Garcés-Ayerbe, C.; Rivera-Torres, P. Barriers to the adoption of proactive environmental strategies. *J. Clean. Prod.* **2011**, *19*, 1417–1425. [CrossRef]
131. Veleva, V.; Ellenbecker, M. Indicators of sustainable production: Framework and methodology. *J. Clean. Prod.* **2001**, *9*, 519–549. [CrossRef]
132. Ahi, P.; Searcy, C. An analysis of metrics used to measure performance in green and sustainable supply chains. *J. Clean. Prod.* **2015**, *86*, 360–377. [CrossRef]

133. Trianni, A.; Cagno, E.; Neri, A.; Howard, M. Measuring industrial sustainability performance: Empirical evidence from Italian and German manufacturing small and medium enterprises. *J. Clean. Prod.* **2019**, *229*, 1355–1376. [CrossRef]
134. Garengo, P.; Biazzo, S.; Bititci, U.S. Performance measurement systems in SMEs: A review for a research agenda. *Int. J. Manag. Rev.* **2005**, *7*, 25–47. [CrossRef]
135. Epstein, M.J.; Widener, S.K. Identification and use of sustainability performance measures in decision-making. *J. Corp. Citizsh.* **2010**, *40*, 43–73. [CrossRef]
136. Medini, K.; Da Cunha, C.; Bernard, A. Tailoring performance evaluation to specific industrial contexts—Application to sustainable mass customisation enterprises. *Int. J. Prod. Res.* **2015**, *53*, 2439–2456. [CrossRef]
137. Collins, A.J.; Hester, P.; Ezell, B.; Horst, J. An improvement selection methodology for key performance indicators. *Environ. Syst. Decis.* **2016**, *36*, 196–208. [CrossRef]
138. Globerson, S. Issues in developing a performance criteria system for an organization. *Int. J. Prod. Res.* **1985**, *23*, 639–646. [CrossRef]
139. Saeed, M.A.; Kersten, W. Sustainability performance assessment framework: A cross-industry multiple case study. *Int. J. Sustain. Dev. World Ecol.* **2020**, *27*, 496–514. [CrossRef]
140. Nobre, G.C.; Tavares, E. The quest for a circular economy final definition: A scientific perspective. *J. Clean. Prod.* **2021**, *314*, 127973. [CrossRef]
141. Padilla-Rivera, A.; do Carmo, B.B.T.; Arcese, G.; Merveille, N. Social circular economy indicators: Selection through fuzzy delphi method. *Sustain. Prod. Consum.* **2021**, *26*, 101–110. [CrossRef]
142. Millar, N.; McLaughlin, E.; Börger, T. The Circular Economy: Swings and Roundabouts? *Ecol. Econ.* **2019**, *158*, 11–19. [CrossRef]
143. Lieder, M.; Rashid, A. Towards circular economy implementation: A comprehensive review in context of manufacturing industry. *J. Clean. Prod.* **2016**, *115*, 36–51. [CrossRef]
144. Thakkar, J.; Kanda, A.; Deshmukh, S.G. Supply chain performance measurement framework for small and medium scale enterprises. *Benchmarking* **2009**, *16*, 702–723. [CrossRef]
145. Eckerson, W.W. Performance Management Strategies: How to Create and Deploy Effective Metrics. 2009. Available online: https://tdwi.org/research/2009/01/bpr-1q-performance-management-strategies.aspx (accessed on 13 February 2021).
146. Li, T.; Zhang, H.; Yuan, C.; Liu, Z.; Fan, C. A PCA-based method for construction of composite sustainability indicators. *Int. J. Life Cycle Assess.* **2012**, *17*, 593–603. [CrossRef]
147. Cagno, E.; Neri, A.; Trianni, A. Broadening to sustainability the perspective of industrial decision-makers on the energy efficiency measures adoption: Some empirical evidence. *Energy Effic.* **2018**, *11*, 1193–1210. [CrossRef]
148. Evans, J.; Bocken, N. Developing a Tool for Manufacturers to Find Opportunity in the Circular Economy. *KES Trans. Sustain. Des. Manuf.* **2014**, *2011*, 303–320.
149. Calabrese, A.; Costa, R.; Levialdi, N.; Menichini, T. A fuzzy analytic hierarchy process method to support materiality assessment in sustainability reporting. *J. Clean. Prod.* **2016**, *121*, 248–264. [CrossRef]
150. Feng, S.C.; Joung, C.B. A measurement infrastructure for sustainable manufacturing. *Int. J. Sustain. Manuf.* **2011**, *2*, 204–221. [CrossRef]
151. Gong, M.; Simpson, A.; Koh, L.; Tan, K.H. Inside out: The interrelationships of sustainable performance metrics and its effect on business decision making: Theory and practice. *Resour. Conserv. Recycl.* **2018**, *128*, 155–166. [CrossRef]
152. Joung, C.B.; Carrell, J.; Sarkar, P.; Feng, S.C. Categorization of indicators for sustainable manufacturing. *Ecol. Indic.* **2013**, *24*, 148–157. [CrossRef]
153. Voss, C.; Tsikriktsis, N.; Frohlich, M. Case research in operations management. *Int. J. Oper. Prod. Manag.* **2002**, *22*, 195–219. [CrossRef]
154. Garza-Reyes, J.A.; Villarreal, B.; Kumar, V.; Molina Ruiz, P. Lean and green in the transport and logistics sector—A case study of simultaneous deployment. *Prod. Plan. Control* **2016**, *27*, 1221–1232. [CrossRef]
155. Cristoni, N.; Tonelli, M. Perceptions of Firms Participating in a Circular Economy. *Eur. J. Sustain. Dev.* **2018**, *7*, 105–118. [CrossRef]
156. Geissdoerfer, M.; Savaget, P.; Bocken, N.M.P.; Hultink, E.J. The Circular Economy—A new sustainability paradigm? *J. Clean. Prod.* **2017**, *143*, 757–768. [CrossRef]
157. Panchal, R.; Singh, A.; Diwan, H. Does circular economy performance lead to sustainable development?—A systematic literature review. *J. Environ. Manag.* **2021**, *293*, 112811. [CrossRef]
158. Kravchenko, M.; McAloone, T.C.; Pigosso, D.C.A. To what extent do circular economy indicators capture sustainability? *Procedia CIRP* **2020**, *90*, 31–36. [CrossRef]
159. Bhanot, N.; Rao, P.V.; Deshmukh, S.G. An integrated sustainability assessment framework: A case of turning process. *Clean Technol. Environ. Policy* **2016**, *18*, 1475–1513. [CrossRef]
160. Winroth, M.; Almström, P.; Andersson, C. Sustainable production indicators at factory level. *J. Manuf. Technol. Manag.* **2016**, *27*, 842–873. [CrossRef]
161. Fuente, J.A.; García-Sánchez, I.M.; Lozano, M.B. The role of the board of directors in the adoption of GRI guidelines for the disclosure of CSR information. *J. Clean. Prod.* **2017**, *141*, 737–750. [CrossRef]
162. Székely, N.; Vom Brockem, J. What can we learn from corporate sustainability reporting? Deriving propositions for research and practice from over 9,500 corporate sustainability reports. *PLoS ONE* **2017**, *12*, e0174807. [CrossRef]

163. Ferrari, A.M.; Volpi, L.; Pini, M.; Cristina, S.; García-Muiña, F.E.; Settembre-Blundo, D. Building a sustainability benchmarking framework of ceramic tiles based on Life Cycle Sustainability Assessment (LCSA). *Resources* **2019**, *8*, 11. [CrossRef]
164. Hillebrand, B.; Kok, R.A.W.; Biemans, W.G. Theory-testing using case studies. *Ind. Mark. Manag.* **2001**, *30*, 651–657. [CrossRef]
165. Wickert, C.; Post, C.; Doh, J.P.; Prescott, J.E.; Prencipe, A. Management Research that Makes a Difference: Broadening the Meaning of Impact. *J. Manag. Stud.* **2021**, *58*, 297–320. [CrossRef]

sustainability

MDPI

Article

A Framework to Assess Manufacturers' Circular Economy Readiness Level in Developing Countries: An Application Case in a Serbian Packaging Company

Jelena Demko-Rihter [1], Claudio Sassanelli [2,*], Marija Pantelic [3] and Zoran Anisic [1]

[1] Department of Industrial Engineering and Management, Faculty of Technical Sciences, University of Novi Sad, Trg Dositeja Obradovica 6, 21000 Novi Sad, Serbia
[2] Department of Mechanics, Mathematics and Management, Politecnico di Bari, Via Orabona, 4, 70125 Bari, Italy
[3] Bosis, 14221 Valjevo, Serbia
* Correspondence: claudio.sassanelli@poliba.it

Abstract: Researchers highlighted the gap between the circular economy (CE) theory and real manufacturing practices. In developing countries, the background for CE development is quite different from developed countries, where there is an established waste management structure and a robust environmental policy. In addition, a shortage of best practices, guidelines, learning experiences, frameworks, and models capable of guiding manufacturers in measuring their circular level and track a roadmap towards an improvement of their circular readiness is raised in the literature. Therefore, this research develops and proposes a framework for assessing company's CE readiness and is tailored for companies operating in developing countries. In detail, the framework investigates the two main perspectives (product and business model) that companies should consider adopting and implementing CE in their operations and business. The framework also supports companies to track an improvement roadmap through the definition of future actions and KPIs. To develop the framework, an application case with a company placed in Serbia and operating in the packaging industry has been conducted. The application of the framework unveiled that there is room for improvement in developing countries to foster CE adoption, especially in the policy context. Indeed, policy incentives and instruments of public authorities would considerably support the circular transition process in companies.

Keywords: circular economy; readiness assessment; product lifecycle; manufacturing; KPI; developing country

Citation: Demko-Rihter, J.; Sassanelli, C.; Pantelic, M.; Anisic, Z. A Framework to Assess Manufacturers' Circular Economy Readiness Level in Developing Countries: An Application Case in a Serbian Packaging Company. *Sustainability* **2023**, *15*, 6982. https://doi.org/10.3390/su15086982

Academic Editor: Antonio Boggia

Received: 3 March 2023
Revised: 18 April 2023
Accepted: 19 April 2023
Published: 21 April 2023

1. Introduction

The circular economy (CE) concept was first developed in 2005 [1] with the aim of moving from a linear product lifecycle towards a closed loop capable of replacing the disposal concept with restoration [2,3]. The European Commission (EC) in 2015 [4] defined CE as a model of production and consumption, which involves sharing, leasing, reusing, repairing, refurbishing, and recycling existing materials and products as long as possible, resulting in the extension of the product's life cycle and a reduction in waste to a minimum. Unlike the traditional linear economy and its make-consume-dispose approach [5], a CE takes into account the limits of the natural resources of our planet and strives to reduce the consumption of raw materials and energy by increasing the share of renewable or recyclable resources, while reducing greenhouse gas (GHG) emissions, material losses, and waste [6]. In addition, a CE contributes to economic growth, the innovation level [7], and the creation of new jobs [8].

CE has been widely applied to a heterogeneous set of industries [9–13], involving different companies traditionally belonging to linear supply chains and pushing them to act in a circular supply chain context [14–17]. However, the adoption of CE asks for different interventions in the company (from a structural, business model [18,19], and organizational perspective [14,18], through product development [20], process optimization, and data and information management [21,22], up to the technological [23,24] and skill dimension [25]). On the other side, CE would trigger multiple benefits both internally and externally to the company's boundaries [26] and under all of the three sustainability triple bottom line perspectives (environmental, economic, and social) [27,28]. One of the most relevant industries to effectively apply CE is manufacturing [29], leading to the birth of the concept of circular manufacturing (CM) [30], defined as the concurrent adoption of different CM strategies (e.g., reuse, cleaner production, servitization), and allowing one to improve resource management, cut their consumption, and prolong and close their lifecycle loops through manufacturers' internal and external operations tailored to meet the stakeholders' needs [31]. Notwithstanding the multiple benefits occurring with the adoption of the CE paradigm, companies often do not own the needed assets, knowledge, skills, and capabilities to effectively exploit it [32]. Several methods and indexes have been detected in the literature to support companies in measuring and assessing their circularity degree [33–35]. However, a shortage of best practices, guidelines, learning experiences, frameworks, and models capable of guiding manufacturers in measuring their circular level and tracking a roadmap towards an improvement of their circular readiness is raised in the literature [21,34].

The main objective of this research is to develop a framework to assess in developing country, from the micro and macro point of view, companies' readiness to shift towards the CE paradigm and to go along a defined roadmap through the improvement of its business under different perspectives. Through an interactive research method, inspired by DRM [36], the framework, grounded on the EEA's template [8,37], was conceived and refined. Then, it was validated through an application case with a company operating in the packaging industry in Serbia (a non-EU developing country).

The packaging industry was considered since it is relevant in terms of circularity and sustainability. Plastic packaging can be especially reusable and recyclable, but packaging designers need to carefully consider the trade-offs between return rates, transport distances, difficulties, and costs in packaging sorting and cleaning. The reusability of packaging is not always the best solution. The CO_2 emissions incurred during a long-distance transport for returning and redistributing reusable packaging may have a higher negative impact on the environment [38]. In addition, the main functionalities of packaging and the protection of the enclosed products should not be jeopardized by the requirements of circularity and sustainability.

Furthermore, the research focuses on companies placed in developing countries because there is a lack of research in the literature dealing with readiness assessments of this kind of organizations to effectively embrace the CE paradigm in their business, having in mind the unregulated policy framework [39] and the lack of public instruments and measures that should fertilize a transition from linear to a circular model of economy. In particular, ref. [40] worked on developing a conceptual model to measure the change readiness for SMEs' adopting a CE, but not focusing on developed countries. Refs. [41,42] proposed a model to assess the readiness of manufacturing companies for the CE paradigm at the micro-level, in this case also not focusing on developed countries and not emphasizing the lack of a regulatory framework and support of public authorities. Ref. [43] developed a CE readiness model, composed of eight dimensions and aspects to be investigated both at micro- and macro-level (e.g., Product and Service Innovation, Manufacturing and Value Chain, Policy and Market). The model is exhaustive but the dimensions, in particular Policy and Market, could be tailored in a more specific way for companies placed in a developing country (because in such a context there are not developed regulatory frameworks for supporting circularity in the manufacturing sector, waste management at the very low level, financial support, or different financial schemes for supporting circularity, etc.). Indeed, according to [44], although the EU member countries are getting closer to the sustainable development goals (SDGs) created by the United Nations (UN) and about half of the member countries seem ready to pass to circularity, the EU regulations and policies must support stronger transition towards CE. The regulations should be harmonized for all of the EU member countries. This finding indicates that, in developing non-EU countries, there exists even more need and space for public authorities' support of transition towards CE and harmonization with EU regulations and policies, justifying the need of a tailored model to assess readiness in these kind of companies.

The paper is structured as follows. The research context is described in Section 2 (arguing about the application of the CE domain in the packaging supply chain). In Section 3, the research methodology adopted is explained. The results are shown in Section 4 and discussed in Section 5. Conclusions are presented in Section 6.

2. Research Context

This section first introduces the concept of CE and its main characteristics in relation to readiness. Then, its relevance for the packaging industry is presented.

2.1. Circular Economy: The Transition and the Readiness Level

The concept of CE is complementary to the SDGs by the UN, a collection of 17 interlinked objectives designed to serve as a shared blueprint for peace and prosperity for people and the planet now and into the future [45]. They considerably contribute to the adoption of the CE concept by companies, consultancy firms, governments, non-governmental organizations or associations, and academics [46].

Some authors indicate that companies might not be interested in CE because of the increasing number of years that a product can be used and a longer lifecycle, which can cause a significant drop in their sales and revenues in the short term (and even in the medium term depending on the type of product). To cope with this situation, policies should be defined to encourage companies to extend products' life cycles to optimize the planet's resources. In addition, education, advertising, and other qualitative incentives, such as quality labels or badges that help people differentiate ecofriendly products with a longer lifespan, could be helpful for companies and should support consumers to recognize CE-driven products on the market [47].

Furthermore, support in transitioning towards a CE is necessary at each level (macro, meso, micro) [48] because national governments and agencies, industries, or companies might not have the financial resources and knowledge to implement a CE [49]. In particular, different means (subsidies, capital support, soft loans, incentives for research on the topic, or supporting innovative business models) have been proposed to promote and ease circularity adoption [50]. Among them, ref. [51], exploring the main enhancing and inhibiting factors for a progress towards circular business models, found that relevant regulations at the European level, appropriate technologies [23,52], and increasing social and environmental awareness of consumers [53–55] and managerial capabilities [11,56] are main drivers for changes. However, companies' settings determine their predisposition towards the CE paradigm.

According to [57], organizations that focus on radical innovations and balance the efforts between technical and soft aspects are more oriented towards a CE culture. They also found that analyzing the culture orientation for CE of the organization might create a sense of urgency in leaders and employees to move towards CE as a way to obtain environmental, social, and economic benefits. On the other side, ref. [58] found that corporate environmental responsibility (CER) is positively related to the readiness for moving towards a circular business model. Perceived CE drivers act as mediators, while perceived CE barriers moderate the relationship between CER and readiness for change-acting, reducing the positive effect. CER positively influences the readiness for changes in organizations.

Researchers made some first attempts to develop a CE readiness assessment tool. Ref. [43] developed a prototype composed of different levels in the organization: organizational readiness, business model readiness, market readiness, offering readiness, and operation readiness. Ref. [59] explained the readiness for CE as the organization's capability to adapt to the related emerging business strategies. The more a company is ready to adopt a CE strategy, the more sustainable it operates and has to realize and understand its actual status in terms of CE readiness at a specific time [60]. In addition, organizations can also choose whether to be circular from an economic perspective. If they do not, they are also not sustainable. Finally, to effectively adopt CE practices in their organizations, companies need to concurrently implement CE interventions (i) bringing immediate results, dealing with what the organization already has, (ii) what needs to happen prior to the asset implementation, (iii) enabling changes (that could be existing things that need to be adapted, new things that need to be developed whether they are temporary as part of a transition as the organization designs its circular economic future, or could be new permanent things that are needed as part of that future to exist and that future that the organization wants to be a part of).

2.2. Circular Economy in the Packaging Industry

The quantity of materials used for packaging is growing continuously and in 2017 packaging waste in Europe reached a record of 173 kg per inhabitant. The aim of the EC [61] was to accelerate the reusability and recyclability of packaging on the EU market by 2030. Some of the mandatory requirements for packaging to be implemented on the EU market are that (over)packaging and packaging waste have to be reduced, and that packaging should be designed to be reused and recycled, considering restrictions on the use of some packaging materials for certain applications, reducing the complexity of packaging materials, and if it is possible to safely use some consumer goods without packaging [62].

The EC Packaging and Packaging Waste Directive (PPWD) emphasizes that packaging should protect products' integrity and consumers' health and safety; increase products' shelf-life; contribute to waste reduction; facilitate transport, efficient handling, and distribution; promote the packaged product; and provide information and convenience to consumers. Underpackaging and overpackaging should also be clarified. Underpackaging could lead to product and food waste, causing additional negative environmental, climate, and economic impacts. Measures to increase recyclability cannot jeopardize product safety and must avoid product waste. According to the PPWD, climate and environmental performance should be assessed throughout the entire lifecycle of the packaging and packaged products.

Under a CE lens, the limited ability to track all chemicals makes it difficult to control and limit combined exposure. Furthermore, for recycled single-use materials, such as paper and board which is chemically cleaned, modelling shows that even after a total stop of using a chemical (e.g., bisphenol A in receipts), it will remain in the recycled paper for an estimation of 31 years [63]. Studies show that recycled paper even accumulates persistent and hazardous chemicals [64].

3. Research Methodology

The main objective of this research is to develop a framework to assess in a developing country, from the micro and macro point of view, companies' readiness to shift towards the CE paradigm and to go along a defined roadmap through the improvement of its business under different perspectives (material input, eco-design, production, consumption, and waste recycling). To perform this, preconditions and factors that facilitate the implementation of the CE paradigm in a company operating in a developing country (outside the European Union's regulations regarding environment protection, waste management, GHG emission, etc.) have been detected. To develop the framework for assessing CE readiness in developing countries, a research process structured in three main phases (conceptualization, development, and validation) has been defined and is shown in Figure 1.

In this research we used mixed methods, including questionnaires fulfilled by a company's management, and the data presented on the websites of the surveyed company and in the company's Sustainability report, prepared according to the standards of the Global Reporting Initiative.

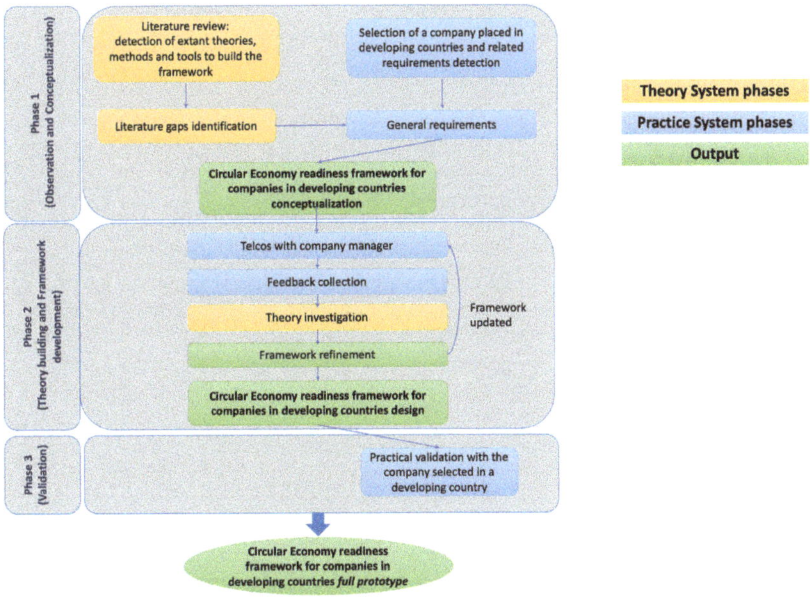

Figure 1. Research methodology adapted by [65].

3.1. Framework Conceptualization

The conceptualization phase mainly consisted of a literature review analysis. It was conducted with the Scopus database, including the following keywords: circular economy, circular economy in manufacturing, readiness, developing country. The results provided 94 documents and 4 of them were related to developing countries. The contributions were analyzed first in the title, abstract, and keywords (bringing the total to 54 contributions) and in a second round in their entire manuscript (leading to a final number of 17 papers). The main criterion used for their selection was their pertinence to the assessment of companies in terms of CE readiness. In addition, looking at the results of the literature review, it was decided to use as main reference the conceptual framework developed by the EEA for evaluating companies and their related products circularity, adjusted to the socio-economic environment of a non-EU developing country.

In addition, from the practice perspective, having in mind the increasing importance of environmental, social, and governance (ESG) issues and company's compliance to ESG standards in their operations, a company that has partly implemented CE practices in its business model and complied its activities to ESG principles was researched and detected. The main requirements of the company in terms of CE readiness were checked by triangulating the main characteristics of the EEA's framework and detecting the topics to be enhanced.

The company was chosen because of its leading role in the Serbian manufacturing industry, in terms of the responsible usage of raw materials, energy, water, reduction in GHG emission, and constant developing of employees' awareness to behave responsibly and manage resources at the workplace.

3.2. Framework and Related Protocol Development

The framework for assessing companies' readiness level in developing countries was developed based on the contributions selected with the literature review, on the EEA conceptual framework suggested by experts in the domain, and also on the requirements of the company involved in the conceptualization phase. The reason for using the conceptual framework by the EEA for this research is its comprehensiveness and cross section of the most important aspects of circularity, as well as an overview of elements that can be assessed for measuring progress towards product circularity. A questionnaire to be fulfilled by the company's management was also integrated.

Enriching the EEA conceptual framework, some new features were added to the framework based on the feedback received during this phase after conducting some telcos and showing the framework prototype to the company sustainability manager involved in this research. Indeed, to measure company circularity readiness, a Likert scale and two radar charts were used and integrated with the EEA framework structure.

3.3. Framework Validation: The Application Case

To validate the framework, it was applied to the company already involved during the conceptualization phase. The questionnaire was fulfilled by the company's manager for sustainability and her assistant. First, the researchers read the documentation produced by the company about sustainability topics (the Yearly Sustainable Report of the last two years and the SDG Progress Report). Then, a meeting with the sustainability manager and her assistants was organized to discuss the topic introduced in the reports and to make the company's employees familiar with the research framework proposed in this research. Two weeks later, a workshop was conducted for interviewing them through the framework and gathering their answers in a form. Finally, the researchers analyzed in the back office the answers gathered and assigned them a score to obtain a final result about the company readiness in terms of both the product and business model. Finally, a last workshop was organized to discuss the results with the company employees.

In this research, a five-point Likert scale was used to measure and to express the attitudes of respondents about a certain phenomenon. Indeed, the attitudes were measured by adding numbers to certain characteristics of the observed phenomenon and then scaling was carried out. The phenomena were positioned on a certain scale depending on how many characteristics the variables had, determined via a five-point Likert scale. The respondent (in the case conducted, the Sustainability manager) had to express the degree of agreement or disagreement for each individual statement on a five-point scale (1. "Strongly disagree", 2. "Disagree", 3. "Undecided", 4. "Agree", 5. "Strongly agree"). Each respondent's answer was scored, and then by calculating the average value of the points for each statement, a total score expressing the respondent's attitude was obtained.

The Company: Bosis

The family company *Bosis* (Valjevo, Serbia) was founded in 1982 as a small craft shop for screen printing, and today is a leading manufacturer of printed and laminated cardboard packaging and blister cardboard with 143 employees. Their portfolio assortment consists of both printed and laminated cardboard packaging and blister cardboard for nearly 200 satisfied clients. Caring for their employees, the environment, and the local community is exactly what makes a company recognizable not only on the domestic and regional market but also on the EU one. The owners and management of the company insist on the constant training of employees and the raising of their awareness on the importance of preserving the environment. In 2021, several actions supporting this area were organized, in accordance with Corporate Social Responsibility (CSR) company strategy that supports the SDGs. The compliance with the principles of CSR is a necessity if the company tends to participate in global supply chains. EU and multinational companies usually choose suppliers from the global supplier basis, such as EcoVadis (a *ratings platform to assess corporate social responsibility and sustainable procurement).* Bosis company has a label of Platinum supplier, which means that is among the top 1% of companies (of a total of over 100,000 companies that have passed the EcoVadis check) in all categories that meet high CSR requirements.

In addition to this, Bosis has been calculating its GHG emissions for several years, and from 2020 they also prepared an annual GHG emissions report according to the international GHG Protocol standard. At the moment, the Bosis GHG emissions report contains only SCOPE 1 (direct GHG emissions from company-owned and controlled resources, released into the atmosphere as a direct result of a company's activities) and SCOPE 2 (indirect GHG emissions released in the atmosphere from the consumption of purchased electricity, steam, heat, and cooling), but Bosis actively works to include SCOPE 3 (all indirect emissions not included in SCOPE 2, that occur in the value chain of the reporting company, including both upstream and downstream emissions).

All of this is of extreme importance both for local regulations (the law on climate change of the Republic of Serbia was adopted in 2021, but Bosis is awaiting the adoption of by-laws that will more closely define the way of reporting on GHG emissions) and for European regulations. Namely, in the EU, laws on the carbon tax have already been passed, which implies the payment of additional taxes for the import of products in proportion to their carbon footprint (i.e., the amount of GHGs released into the atmosphere in the process of their production and transport). This regulation currently covers the import of heavy industry products, but it is expected that in the near future this tax will be extended to all products imported into the EU. By working on timely data collection and reporting and reducing GHG emissions, Bosis's products will have a smaller carbon footprint and would be competitive on the EU market.

It has to be emphasized that major clients of the Bosis Company belong to the food industry. According to the EC Packaging and Packaging Waste Directive, packaging functionality is particularly important for sensitive applications being in contact with food and beverages. The safety of food products and consumers is the first priority for the packaging supply chain, and it is also what drives the search for the best sustainable solutions (e.g., for the further uptake of recycled content). Finally, producers need to remain free to choose the most appropriate packaging formats and materials for their products and their distribution systems. Because of those recommendations, Bosis deploys packaging solutions together with their clients.

Considering that Bosis has been reducing negative impacts on the environment for years, the introduction of the CE principle is a logical step to achieve the goal of zero levels of GHG emissions. The company strives to operate transparently, publishing the results of their commitment through a sustainability report aligned with the Global Reporting Initiative (GRI) methodology. The implementation of the CE concept is expected to lead in the long-term to positive financial effects, optimization of the production process, increased effectiveness and efficiency, and a positive impact on the local community, employees, and their families, which is always a company's main motive and driving force.

4. Results

In this section, the results of this research are reported. First, in Section 4.1. the framework built to explore the company selected is presented and then, in Section 4.2. the application case results are reported.

4.1. Conceptualization Phase: Literature Review

In the conceptualization phase the contributions selected were categorized into three parts.

The first category refers to the literature that emphasized the research of CE readiness of companies operating in developed countries. In this context, ref. [43] developed the CE readiness self-assessment tool as an important guidance to support the successful transition towards CE of manufacturing companies by identifying the eight key dimensions that that are necessary for manufacturing companies to make the transition to CE: (1) Organization; (2) Strategy and Business Model Innovation; (3) Product and Service Innovation; (4) Manufacturing and Value Chain; (5) Technology and Data; (6) Use, Support, and Maintenance; (7) Takeback and End-of-Life Strategies; and (8) Policy and Market. In addition, ref. [44] created a unified assessment framework to evaluate the circularity readiness of EU economies and emphasized that policies and regulations must support companies to understand that circular products and processes generate added value. At the same time, the latest research of [56] indicated that, even in a developed country (Finland), CE is in a very early phase among companies, which are mainly interested in maintenance, repair, modernization, reuse (resales), and remanufacture, while material recycling is not a core of their business model.

The second category emphasized different conceptual models and frameworks for the assessment of CE readiness, looking also at the main drivers and barriers of adopting CE business models and with a focus on characteristics that they could have. In that sense, ref. [41] proposed a readiness assessment model for manufacturing companies that frequently fail in understanding how to start a systemic transition, in which fundamental changes are needed in the design of products, production processes, business models, and supply chains. Ref. [51] explored the textile recycling sector and found that the main enhancing factors are relevant regulations at the European level, appropriate technologies and digitization, and increasing social and environmental awareness of consumers and managerial capabilities. Supply chain complexity was emphasized as one of the main inhibiting factors. Ref. [59] found that the majority of the frameworks are not readiness frameworks within a strategic business context but more about a list of CE characteristics as a tool to promote CE activities. Ref. [40] emphasized that there is no comprehensive readiness model for SMEs adopting CE, so their intention was to develop a conceptual model to measure change readiness for SMEs' adopting CEs by incorporating several factors as precursors to readiness, i.e., individual/collective difference, structural, contextual factors, and related barriers. Ref. [58] found that corporate environmental responsibility (CER) positively influences the readiness for change in organizations. Parallelly with the activities on the supply side and aimed at increasing the level of companies' readiness for implementing new business model based on CE, different activities should be realized on the demand side in terms of deploying the environmental consciousness of citizens and consumers through the programs of environmental education and similar activities of public authorities and decision-makers, as was emphasized in the research of [66].

The third category deals with the literature related to CE readiness in developing countries. Ref. [67] identified several opportunities for CE transitions in Columbia and other low- and middle-income economies: greater political coherence; a suitable fiscal framework for sustainable practices; a robust IT infrastructure; and the use of ICT by enterprises to develop CE business models. They also emphasized the necessity of promoting financing schemes and incentives to implement design-led approaches to production in the industrial sector. Innovations, education, and raising awareness would additionally support a mind-set shift. Ref. [68] researched possibilities for a transition to a circular plastic economy (CPE) in Africa and emphasized the importance of a more collaborative, multistakeholder, and multi-sectoral synergy needed to break the linear economy, supported by government's investment in capacity and skills building, education, financial incentives, and taxation to further facilitate CPE.

The main contributions from the analysis of the literature are that the transition to CE requires radical changes in products, processes, and business models and that it has to be regulatory supported by public authorities, as well as through financial schemes and tax incentives. It could be especially important for developing countries and low/middle income economies where awareness of sustainable business and corporate social responsibility are not developed enough.

4.2. The Framework to Measure Companies' CE Readiness in Developing Countries

In this research, the framework developed by the European Environment Agency (2016) was refined as a comprehensive overview of elements and preconditions for measuring the progress of manufacturing companies towards circularity. The monitoring framework covers all of the relevant dimensions of the transition: material inputs, eco-design, production, consumption, and waste recycling. Different policy questions were set, as well as indicators for measuring progress towards CE, and on the basis of the analyzed company's answers and data availability, progress was measured. An additional model for assessing progress towards product circularity was also developed by the EEA [37] and integrated in this research.

The model is composed of two main parts: the first is aimed at measuring and assessing progress towards the circularity of a product and the second part presents different perspectives from the company point of view, evaluating how to shift its business model to a circular one.

In this framework, the main perspectives to assess products' circularity (detailed in Table 1) are:

1. Product properties: technical lifetime of a product; reparability, and recycled content of a product;
2. Business/consumption model: material circularity indicator (MCI), functional lifetime of a product, and proportion of product-service system (PSS);
3. Society: policy framework;
4. Macro-scale product impacts: macro-scale impacts of circular business model and the proportion of key material losses in product cycles;
5. Environmental and economic impacts: life cycle impact analysis, exergy losses, and life cycle costing (LCC).

To realize in an easier way the maturity level based on the answers of the interviewees, researchers also defined normative answers for each question. An example of normative answers for product circularity is reported for question 1.1 ("Which is the duration of a technical lifetime of a product?"):

1. Completely unready: duration of the technical lifetime of a product is strongly shortened;
2. Partly unready: duration of the technical lifetime of a product is shortened;
3. Neither unready, nor ready: duration of the technical lifetime of a product is unchanged;
4. Partly ready: duration of the technical lifetime of a product is slightly extended;
5. Completely ready: duration of the technical lifetime of a product is strongly extended.

Instead, the second part of the framework (detailed in Table 2) deals with assessing business model circularity. Its main categories are material input, eco-design, production, consumption, and waste recycling.

Table 1. Assessing product circularity.

Dimensions	Circularity Assessment	Related Questions	
Product properties	Technical lifetime of a product	1.1	Which is the duration of a *technical* lifetime of a product?
	Reparability	1.2	Is there an ability for reusability, remanufacturing, or recyclability?
	Recycled content	1.3	What is the proportion of recycled material in new products?
Business consumption model	Material circularity indicator (MCI)	2.1	Which methodology integrates product characteristics and circular strategies available in an easy-to-use format?
	Functional lifetime of a product	2.2	What is the functional lifetime of your products?
	Proportion of product-service system in specific market	2.3	What is adoption rate of a product-service system?

Table 1. *Cont.*

Dimensions	Circularity Assessment		Related Questions
Society	Policy framework	3.1	What aspects of product circularity are stimulated or hampered by policy instruments?
		3.2	What is the size of the market that is affected by these policy instruments?
		3.3	What groups are targeted by the policy instrument?
		3.4	Are there instruments that influence the design of products such as taxes on specific products or differentiated VAT rates?
		3.5	Are there any policy measures in place favoring local production and local reuse or recycling services to shorten the transport distance between production, consumption, and reuse/recycling?
		3.6	Are there policy measures in place engaging the distribution sector in stimulating local reuse and repair?
		3.7	Are any policy measures in place favoring the separate collection of waste for reuse and/or recycling?
		3.8	Are there any instruments that support remanufacturing?
		3.9	Are there any instruments in place for stimulating the market for recyclates?
		3.10	Are there any standards on reuse/recycling or reusables/recyclates?
		3.11	Are there public procurement schemes designed to incentivize the innovators and early adopters to come up with new products/new business models that are more circular?
Macro-scale product impacts	Macro-scale impact of circular business models	4.1	Is it possible to assess the macro-scale economic and environmental impacts of circular business models?
	Proportion of key material losses in product cycles	4.2	Is it possible to obtain insights on key product flows in terms of opportunities for increasing circularity and decreasing material losses? Is it possible to assess leakage of key materials from a material cycle?
Environmental-economic aspects	Life cycle impacts	5.1	Is the LCA (life cycle assessment) methodology in usage?
	Exergy losses	5.2	Is there monitoring of exergy losses?
	LCC	5.3	Is the life cycle cost (LCC) analysis used?

Furthermore, in the case of the part of the maturity model related to the business model, to quantify and translate the qualitative answer provided by the interviewees in a quantitative rate, researchers defined normative answers. An example of normative answers for product circularity is reported for question 1.1 ("Are primary material inputs decreasing in your company?"):

1. Completely unready: primary material inputs flows are strongly increasing;
2. Partly unready: primary material inputs flows are increasing;
3. Neither unready, nor ready: primary material inputs flows are steady;
4. Partly ready: primary material inputs flows are slightly decreasing (e.g., through the recycling of wastes);
5. Completely ready: primary material inputs flows are strongly decreasing (e.g., through design modifications on the product).

Table 2. Assessing business model circularity readiness.

Categories		Policy Question
Material Input	1.1	Are primary material inputs decreasing in your company?
	1.2	Are material losses decreasing in your company?
	1.3	Is the share of recycled materials in material input increasing in your company ?
	1.4	Are the materials used in your company sustainably sourced?
Eco-Design	2.1	Are your products designed to last longer?
	2.2	Are your products designed for disassembly?
	2.3	Are recycled materials included in product design?
	2.4	Are the materials designed to be recycled, avoiding pollution from recycling loops?
Production	3.1	Does your company use fewer materials in production?
	3.2	Does your company use a lower volume and number of environmentally hazardous substances in production?
	3.3	Does your company generate less waste in production?
	3.4	Are business strategies shifting towards circular concepts such as remanufacture and service-based offers?
Consumption	4.1	Does consumption in Serbia switch patterns to less environmentally intensive types of goods and services?
	4.2	Do consumers in Serbia use products for longer?
	4.3	Does consumption in Serbia generate less waste?
Waste recycling	5.1	Is waste increasingly recycled in your company?
	5.2	How far do materials keep their value in recycling processes, avoiding down-cycling in your company?
	5.3	How far is the Serbian recycling system optimized for environmental and economic sustainability?

4.3. Application Case Results

4.3.1. Product Circularity Readiness Assessment

The assessment of Bosis' readiness in terms of product circularity went through the five dimensions introduced in Table 1 and started with the dimension "product properties", divided into:

1. Technical lifetime of a product: technical lifetime of a cardboard packaging is directly determined by the materials used and can be extended by changing the ways the products packed in cardboard packaging are distributed to customers and used.
2. Ability for reusability, remanufacturing or recyclability of a product: the cardboard packaging is fully recyclable and made from recycled materials. In its production,

high quality materials are used which comply with the standards, and, at the same time, originate from responsible and sustainable sources. All used materials are approved for packaging for the industries that have most demands regarding health and safety, such as food and confectionary industry. Almost 14% of the total portfolio are Forest Stewardship Council (FSC) certified products, having the FSC logo as a guarantee that the packaging was produced using exclusively materials originating from responsible sources. There are 142 new FSC products in their portfolio. The next steps will be aimed at the improvement of collecting, selecting, and recycling of paper and cardboard packaging.

3. Proportion of recycled materials in new products: the proportion of recycled materials in products is about 94 %. The next steps will be aimed at increasing the proportion of recycled materials and materials with a lower share of virgin fibers in paper and cardboard packaging.

Three sub-dimensions have been considered for the dimension of "consumption":

1. Material circularity indicator (MCI): the company uses MCI to detect key problems related to product circularity and the way the business model contributes to actual reuse, recycling, a longer lifespan, and more intensive uses of the packaging. In total, 100% of the paper waste generated in the production process is recycled. The design function is developing packaging, in cooperation with dedicated customers, with more sustainable materials and packaging solutions, more functions, and extended lifespans. In addition, the eco-design of packaging in the product portfolio enables more functions, communicates the values of sustainable consumption, and contains the smallest necessary amount of material. The types of material and cardboard grammage are selected carefully, and the quantity of material input is optimized, while the functionality of products is retained. The weight of packaging was reduced by reducing the thickness or quantity of the materials. The next steps will be aimed at improving the awareness of packaging buyers and designing products with extended lifecycles.
2. Functional lifetime of a product: the functional lifetime of paper and cardboard packaging depends on the lifetime of the materials used in production, which is determined by the suppliers of the materials. The next steps will be aimed at cooperating with the suppliers of the material on improving the functional lifetime of materials and packaging.
3. Proportion of PSS in a specific market: the proportion of PSS in a specific market is still not applied in the analyzed company, but there is intention to develop PSS in the future.

The "Policy framework" was assessed in terms of the impact of society through the following sub-dimensions:

1. Aspects of product circularity stimulated or hampered by policy instruments: there are only requirements for using Standards 13430 and 13428;
2. Size of the market that is affected by policy instruments: there is no official data about the size of the market affected by policy instruments;
3. Groups targeted by policy instruments: there is no official data about the number and size of the groups targeted by policy instruments;
4. Instruments that influence the design of products (e.g., taxes on specific products or differentiated VAT rates): the company is not aware of the existence of instruments that influence the design of a product;
5. Policy measures in place favoring local production and local reuse or recycling services to shorten the transport distance between production, consumption, and reuse/recycling: no existing measures. Waste cardboard and paper can be recycled only in a few plants in the country, so transportation depends on the distance of the plants;

6. Policy measures in place engaging the distribution sector in stimulating local reuse and repair: such policy measures do not exist;
7. Policy measures in place favoring the separate collection of waste for reuse and/or recycling: there are a number of policies and procedures in place (e.g., the National Plan for the reduction of packaging waste, the law on packaging and packaging waste);
8. Instruments that support remanufacturing: they are defined by procedures (e.g., regulation on the list of waste generation prevention measures);
9. Instruments in place for stimulating the market for recyclates: they are defined by procedures;
10. Standards on reuse/recycling or reusables/recyclates: there are only internal procedures for this area;
11. Public procurement schemes designed to incentivize the innovators and early adopters to come up with new products/new business models that are more circular: there are no such public procurement schemes.

The "Macro-scale product impact" has been assessed through the following sub-dimensions:

1. Macro-scale economic and environmental impact of the circular business model: it is possible to assess the macro-scale economic and environmental impacts of circular business models by taking into account all of the connections between the material inputs, outputs, and processes of the business model, the environment, and society. Products are marked in accordance with international standards so that all users, from the manufacturer to the end customer, can easily manage packaging and waste. All products are labelled with PAP 20 (paper recycling code: Cardboard) and PAP 21 (paper recycling code: Mixed Paper) depending on the type of packaging to manage the waste in the chain. The company monitors the suppliers in the chain, as well as the operators and buyers of waste. The analyzed company consumes energy and water responsibly. Energy consumption was decreased by applying more efficient equipment, such as automated press control, intelligent system of heating and air conditioning, sensors for turning on/off the lighting, and a control system with frequency regulators for the main engines great forces. The system for the expulsion waste of paper and cardboard was replaced with a new system that consumes four times less energy. The electricity consumption per processed ton of raw material also decreased. The means of internal transport which used gas and diesel engines was replaced with vehicles powered by electricity. New technologies that reduce GHG emissions were implemented. LPG has been replaced with CNG in the production of steam. Coal heating was replaced with pellet heating which reduced the GHG emissions. Water consumption per ton of raw material decreased (in the last 3 years) to about 0.586 m^3/t. Water dispersion materials which do not pollute the water were used. Chemistry for developing an offset plate with a minimum using of water was used. Equipment for the chemical preparation of water which reduces the water consumption were installed, as well as efficient equipment with a CNG gas burner for the production of water steam. The next steps will be aimed at creating a framework which would help partner companies easily assess the macro-scale economic and environmental impacts of their business models;
2. Key product flows and proportion of key material losses in product cycles: methodologies are available, but data availability is limited considering the industry and material inputs, and the results often do not communicate product-level information. Material losses have been significantly decreased during the last few years. The next steps will be aimed at improving the methodology for obtaining insights in the key product flows to decrease the key material losses and leakage.

"Environmental economic aspects" have been assessed through the three following sub-dimensions:

1. Application of the Life Cycle Assessment (LCA) methodology: the company applies the LCA methodology within the internal assessment system to better track the impact

of production to GHG emissions, water, air, and soil. The next steps will be aimed at implementing LCA analysis into the design phase of all new processes and products;
2. Exergy losses: exergy losses are not monitored;
3. Life cycle cost (LCC) analysis: it is applied by evaluating investments, especially production machine purchases.

In addition, for each answer, a value from the five-point Likert scale was assigned for the sustainability manager. Finally, by computing the average value of each of the five dimensions analyzed, a total score expressing the company's readiness was obtained (Table 3). The results are graphically presented in the radar chart shown in Figure 2.

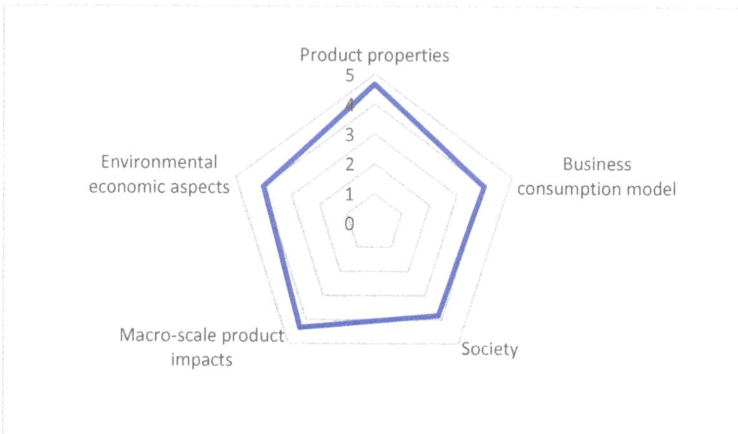

Figure 2. Radar chart for assessment of product's circularity.

Figure 2 indicates that the highest average value was obtained for the circularity assessment regarding the *Product properties* (4.67) dimension as a result of the continuous extending of a technical lifetime of a product, ability for reusability, remanufacturing or recyclability of a product, as well as increasing the share of recycled materials in new products. Regarding the *Macro-scale product impacts dimension*, the obtained average value was 4.33 as result of the continuous effort of the analyzed company to assess the economic and environmental impact of its business model, to obtain insights on key product flows as opportunities for increasing circularity and decreasing material losses, and to assess the leakage of key materials from a material cycle. The company applies LCC analysis when evaluating investments, LCA in internal assessment systems to better track the impact linked to emissions to water, air, and soil during the production, but does not monitor exergy losses, which have reflected the average value of four for the *Environmental economic aspects* dimension. The same average value was obtained for the *Business consumption model* dimension regarding the actual business model, which contributes to the reuse, recycling, a longer lifespan, and more intensive use of the products. There are also internal procedures on reuse/recycling or reusables/recyclates, but public procurement schemes are not designed to incentivize the innovators and early adopters to come up with new products/new business models that are more circular. The lowest average value was obtained for the *Society* dimension because of the lack of financial incentives and stimulative instruments and policy measures that would favor local production and local reuse or recycling services, would promote the eco-design of products, and would promote the separate collection of waste for reuse and/or recycling, stimulating the market for recyclates and promoting remanufacturing.

Table 3. Average values for each dimension of product circularity (for the respondent: sustainability manager).

Dimension	Sub-Dimension	Score
1. Product properties	1.1. Increase in technical lifetime of a product	4
	1.2. Ability for reusability, remanufacturing, or recyclability	5
	1.3. Increase in proportion of recycled material in new products	5
	Average value	**4.67**
2. Business consumption model	2.1. Contribution of actual business model to reuse, recycling, a longer lifespan	5
	2.2. Increase in functional lifetime of products	4
	2.3. Application of PSS in specific market	3
	Average value	**4.00**
3. Society	3.1. Stimulation of product circularity by policy instruments	4
	3.2. Increase in size of market affected by policy instruments	5
	3.3. Existence of groups targeted by the policy instrument	5
	3.4. Existence of instruments that influence the design of products	3
	3.5. Existence of policy measures in place favoring local production and local reuse or recycling services	2
	3.6. Existence of policy measures in place engaging the distribution sector in stimulating local reuse and repair	2
	3.7. Existence of policy measures in place favoring the separate collection of waste for reuse and/or recycling	5
	3.8. Existence of instruments that support remanufacturing	5
	3.9. Existence of instruments in place for stimulating the market for recyclates	4
	3.10. Existence of standards on reuse/recycling or reusables/recyclates	5
	3.11. Existence of public procurement schemes designed to incentivize the innovators and early adopters	2
	Average value	**3.82**
4. Macro-scale product impacts	4.1. Possibility to assess the macro-scale economic and environmental impacts of circular business models	4
	4.2. Insights on key product flows	5
	4.3. Possibility of assessment of leakage of key materials from a material cycle	4
	Average value	**4.33**
5. Environmental economic aspects	5.1. Implemention of LCA methodology	5
	5.2. Monitoring of exergy losses	2
	5.3. Application of LCC analysis by investments' evaluation	5
	Average value	**4.00**

4.3.2. Business Model Circularity

The assessment of a company's readiness to transition its business model towards a circular one has been tested through the five dimensions introduced in Table 2: material input; eco-design; production; consumption; waste recycling. For each of them, a set of sub-dimensions was investigated, detecting possible indicators and verifying the related data availability for their calculation.

In the "material input" dimension, four main sub-dimensions were explored:

1.1. Primary material inputs: the production volume and consumption of material inputs (cardboard, packaging paper) increased but at the same time the increase in the production efficiency and reduction in waste contributed to reducing the amount of material wasted. In cooperation with customers, packaging solutions were developed with a reduced quantity of material inputs, without reducing the functionality of the product (types of material and cardboard grammage were carefully selected). A possible indicator is the domestic material consumption (DMC) for which calculation data are already available.

1.2. Material losses: material losses (of paper, cardboard, energy, and water) significantly decreased during the previous years. A possible indicator is the proportion of material losses in key material cycles. For its calculation, data are not fully available to create the indicator.

1.3. Share of recycled materials in material input: the use of recycled materials increased. A possible indicator is the share of secondary raw materials in material consumption (that amounts to 94%), for which calculation data are already available.

1.4. Materials used sustainably sourced: high-quality materials, which comply with the standards and at the same time originate from responsible and sustainable sources, were chosen by the company. All used materials were approved for packaging for the industries that have most demands regarding health and safety (such as food and confectionary industry). This is confirmed by certificates on the health control of packaging, received from the suppliers of the material. The 13.91% FSC of the products' portfolio was certified with the FSC logo as a guarantee that the packaging was produced using exclusively materials originating from responsible sources (responsibly managed forests), from a transition from cellulosic to recycled materials, and from the use of recycled materials containing a lower % of virgin fibers directly results with less deforestation. A possible indicator is the share of sustainably sourced certified materials in material use (by key materials), for which calculation data are not fully available.

For the "eco-design" dimension, four topics were analyzed:

2.1. Products designed to last longer: the design function was developing products, in cooperation with its buyers, considering the practical and specific needs of the given product that was packed in packaging. The lifespan of certain types of packaging was significantly extended, while in some cases it must be shorter (as much as necessary until it fulfills its primary purpose). The eco-design of packaging enabled an extended lifespan, adds value to the product, communicates the values of sustainable consumption, and contained the smallest necessary amount of material. In this way, packaging is suitable for recycling and has more functions and an extended lifespan. One example is collective display packaging, which, in addition to transport (protects products during storage and transport), also has a marketing function (they have an attractive design and print and are suitable for displaying products on market shelves). A possible indicator is durability or lifetime compared with an industry average for a similar product; data are available to create this indicator.

2.2. Products designed for disassembly: the purpose of the packaging produced was one-time use, after which the packaging is recycled and can be used again in the production process. The packaging design enabled easy folding/unfolding and made packaging functional for easy disposal after usage. A possible indicator is time and the number of necessary tools for disassembly; no data are currently available to create this indicator.

2.3. Recycled materials included in product design: recycled materials were used in production. The proportion of recycled materials in products (January–June 2022) was 94%; 2.55% of the materials originated from natural fibers; and 70.36% were FSC-certified materials. A possible indicator is the proportion of recycled material in new products; data are available to create this indicator.

2.4. Materials designed to be recycled, avoiding pollution from recycling loops: the products were fully recyclable. The eco-design concept was developed keeping in mind the B2B2C concept, based on collecting information from the end consumers (i.e., the consumers of the product) and taking into account their requests about packaging. In the production of packaging, the minimum necessary quantity of material was used, with the highest proportion of recycled material and ecological materials, paints, and varnishes. A possible indicator is the share of materials where safe recycling options exist; data are available to create this indicator.

Regarding the "Production" dimension, four main sub-dimensions were investigated:

3.1. Use of fewer materials in production: the philosophy of smart design implies the use of a minimum quantity of material for the packaging of a specific product. By the production of sample products for customers, an optimal quantity of material inputs is applied. Sustainable consumption of packaging was promoted and a reduction in the consumption of materials was applied through customer advising. Using box compression, optimal carrying capacity of boxes and optimal use of materials was determined. A possible indicator is the material used for production compared to GDP (potentially by sector); data are available to create this indicator.

3.2. Use of a lower volume and number of environmentally hazardous substances in production: the company uses a lower volume and number of environmentally hazardous substances in production. A possible indicator is the input of substances that are classified as hazardous; data are available to create the indicator.

3.3. Generation of less waste in production: a downward trend in waste generation was marked. Automatic selection and waste collection in production were implemented and all generated waste paper was recycled. Possible indicator are waste generation (production activities) and the generation of hazardous waste in production processes; data are available to create this indicator.

3.4. Business strategies shifting towards circular concepts (such as remanufacture and service-based offers): the company strives to achieve circularity in all production and business processes and to optimize the entire production process. Currently, service-based offers do not exist. A possible indicator is the involvement of companies in circular company networks; limited data are available to create this indicator.

About the "Consumption" dimension, the following sub-dimensions were considered:

4.1. Consumption in the analyzed developing country switch patterns to less environmentally intensive types of goods and services: the trend of environmentally positive business processes, materials, and practices has taken off in Serbia as well and the number of companies that are guided by these principles in business is increasing. The possible indicators are the environmental footprint of consumption (including materials) in Serbia and the material footprint per euro spent; limited data are available to create this indicator.

4.2. Consumers in the analyzed developing country use products for longer: the purchasing power of inhabitants of Serbia is lower compared to the west, so people tend to use the products with longer lifespans. As the economic situation improves, the consumerism trend will be adopted by more people, with negative influences on the environment. The possible indicators are the actual average lifetime of the selected products and the market share of preparing for reuse and repair services related to the sales of new products; limited data are available to create this indicator.

4.3. Consumption in the analyzed developing country generates less waste: municipal waste in Serbia is not managed. Poor waste management has been identified as one of the most important barriers for successful waste management in Serbia. A possible indicator is waste generation; data are available to create this indicator.

Concerning "waste recycling", three sub-dimensions were investigated:

5.1. Increasingly recycled waste: a lot of attention has been paid to waste management. In total, 100% of the wasted cardboard and paper generated in production have been recycled. Products are marked in accordance with international standards so that all users, from the manufacturer to the end customer, can easily manage packaging and waste. All products are labelled with PAP 20 (paper recycling code: Cardboard) and PAP 21 (paper recycling code: Mixed Paper) depending on the type of packaging to manage the waste in the chain. The company monitors suppliers in the chain, as well as operators and buyers of waste. Recently, new investments in equipment have been made (system for the automatic removal of paper and cardboard waste from production and a baling press with a larger capacity which will be able to meet the increase in production capacity in the future). In terms of waste management, all employees and other stakeholders are engaged and trained on how to manage waste. Trainings and "awareness" programs are organized on a regular basis. A possible indicator is the recycling rate for different types of wastes/materials; data are available to create this indicator.

5.2. Materials keeping their value in recycling processes, avoiding down-cycling: There is some difference in using recycled and virgin materials, but it is not big; using virgin materials is better in some processes while others handle the recycled materials in a better way. Packaging is produced according to the customer specification. The possible indicators are the recycled material quality compared with the virgin material quality and the turnover of key recyclables; limited data are available to create this indicator.

5.3. Developing country recycling system optimized for environmental and economic sustainability: this is continuously being improved to include more diverse and complex material recycling processes, for which there is great interest from both private and government stakeholders. The possible indicators are the environmental effects and cost/revenues of municipal waste management in Serbia; limited data are available to create this indicator.

In addition, for each answer provided by the sustainability manager related to the second part of the framework, a score was assigned on the five-point Likert scale. Finally, computing the average value of each of the five dimensions analyzed, a total score expressing the company readiness was obtained (Table 4). The results are graphically presented in the radar chart shown in Figure 3.

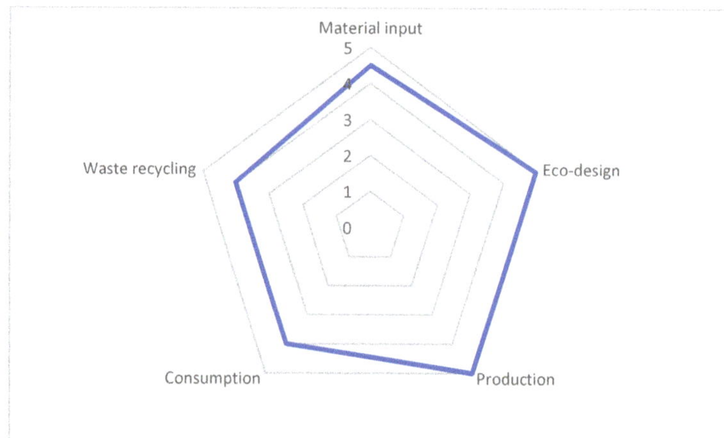

Figure 3. Radar chart for assessment of Business model circularity.

The graph presents that the *Eco-design* and *Product* dimensions have the highest average values, which indicates that the analyzed company designs its product to last longer, to be made from recycled materials, to be eligible for disassembly, to use a lower volume and number of environmentally hazardous substances, and to generate less waste in production. The average value of 4.5 for the *Material input* dimension indicates that there is a space for improvement in terms of further decrease in the material input, although the company strives to increase the efficiency of the used materials. The *Consumption* dimension with an average value of four indicates that there is a possibility for improvement from the macro point of view regarding using products with longer lifespans and generating less waste. The *Waste recycling* dimension with average value of four indicates that, although waste has increasingly been recycled in the analyzed company, there are still a lot of possibilities for the optimization of the Serbian recycling system to reach environmental and economic sustainability.

Table 4. Average values for each dimension of business model circularity.

Dimension	Sub-Dimensions	Score
1. Material input	1.1. Primary material inputs	3
	1.2. Material losses	5
	1.3. Share of recycled materials in material input	5
	1.4. Materials used sustainably sourced	5
	Average value	**4.33**
2. Eco-Design	2.1 Duration of products	5
	2.2 Possibility for dissasembly	5
	2.3. Inclusion of recycled materials	5
	2.4. Possibility for recycling and avoiding pollution from recycling loops	5
	Average value	**5.00**
3. Production	3.1. Decreasing quantity of materials used in production	5
	3.2. Decreasing volume and number of environmentally hazardous substances in production	5
	3.3. Decreasing volume of waste in production	5
	3.4. Business strategies towards circular concepts	5
	Average value	**5.00**
4. Consumption	4.1. Switch of consumption trends in analyzed country to less environmentally intensive types of goods and services	4
	4.2. Extended usage of products in analyzed country	4
	4.3. Trend of generating less waste in analyzed country	4
	Average value	**4.00**
5. Waste Recycling	5.1. Increase in waste recycling in analyzed company	5
	5.2. Retaining value of materials in recycling processes, avoiding down-cycling in analyzed company	4
	5.3. Optimization of recycling system of analyzed (developing) country for environmental and economic sustainability	3
	Average value	**4.00**

5. Discussion

In this section, the results obtained with this research are discussed, giving evidence to the next steps set by the company during the product circularity readiness assessment and also providing a set of indexes that could help the transition of the company's business model towards a full embracement of the circularity paradigm.

This research is triggered by the need to contribute to the CE paradigm that is in a very early phase among manufacturing companies [56]. Indeed, researchers highlighted the gap between the CE theory and real manufacturing practices. The results of this research are also in line with the research of Ref. [69] who found that in developing countries the background for CE development is quite different from developed countries, where there is an established waste management structure and a robust environmental policy.

5.1. Product Circularity: Next Steps for the Transition

The next steps related to the product circularity assessment can be detected for each sub-dimension analyzed. Regarding the *Product properties* dimension, possible next steps will be aimed at improving the collection, selection, and recycling of paper and cardboard packaging, as well as increasing the proportion of recycled materials and materials with a lower share of virgin fibers in paper and cardboard packaging. There is also an intention to work with customers to change the ways that the products packed in paper and cardboard packaging are sold and used. Therefore, user involvement will be a key factor in the innovation of products from a circular perspective [70,71].

Regarding *Business model consumption*, next steps will be aimed at improving the awareness of packaging buyers and designing products with extended lifecycles, followed by the extension of the functional lifetime of materials and packaging. Having in mind that PSS in specific market is still not applied, there is an intention to develop PSS [72] in the future.

Next steps regarding the dimension *Society* will be aimed at triggering bottom-up initiated discussions with public authorities by providing them with the feedback needed to develop and adapt regulatory frameworks and incentive schemes [73] to generate and boost CE in a developing country [74].

In terms of the *Macro-scale product impact*, next steps will be aimed at creating a framework which would help partner companies to easily assess the macro-scale economic and environmental impacts of their business models [75]. Additional important next steps will be aimed at improving the methodology to obtain insights in the key product flows and to decrease key material losses and leakage.

In terms of *Environmental economic aspects*, next steps will be aimed at implementing LCA analysis in the design phase of all new processes and products [33]. Exergetic efficiency gives a good indication of how efficiently materials or energy sources are used [76]. Having in mind that exergy losses are still not monitored, one of next steps will be aimed at their monitoring.

5.2. Business Model Circularity Readiness: Indexes to Lead the Transition

An assessment of the company's readiness for transitioning towards circular business models was tested through different perspectives.

Concerning *Material input*, the key performance indicator (KPI) of *Domestic material consumption (DMC)* was applied to test whether the primary material inputs decrease. Instead, to assess whether the material losses decrease in the company, *Proportion of material losses* can be used as a KPI. In addition, the *Share of secondary raw materials in material consumption* indicates proportion of recycled materials in the total material input. An important KPI to explain the overall behavior for the *Material input* sub-dimension is the *Share of sustainable-certified materials in total material use* (for key materials).

Dealing with the Eco-design dimension, one KPI is the durability or lifetime compared with an industry average for a similar product. To assess the possibility of disassembly, time and the number of necessary tools for disassembly were detected. Instead, the proportion of recycled material in new products was used to assess the inclusion of recycled materials in the product design. In a prolonged product vision, the lifecycle share of materials where safe recycling options exists is used as the KPI and indicates whether the materials are designed to be recycled, avoiding pollution from recycling loops.

Regarding *Production, the material used for production compared to GDP* (potentially by sector) could be used as a KPI to test trends for using fewer materials in production. The companies should also assess the *Input of substances that are classified as hazardous*. Instead, to assess the trend of generating less waste in production, the KPI of the *Generation of hazardous waste in production processes* can be used. Finally, with a meso perspective on production processes, the *Involvement of companies in circular company networks* should indicate whether the business strategy of a company supports shifting towards circular concepts.

Concerning *Consumption*, possible KPIs assessing different aspects of user behaviors and their impacts linked to the properties of the product provided by the company are the environmental footprint of consumption (including materials) in a developing country, the material footprint per euro spent, the actual average lifetime of selected products and the market share of preparing for reuse and repair services related to the sales of new products, and waste generation by consumers in a developing country.

From the perspective of *Waste recycling*, different KPIs have been detected to measure the amount of waste recycled in a company and the retained value of recycled materials or the optimization of recycling systems including: *Recycling rates for different types of wastes/materials*, *Recycled material quality compared with virgin material quality*, and *Turnover of key recyclables*. To assess whether the recycling systems of a (developing) country are optimized for environmental and economic sustainability, possible KPIs are *Environmental effects and cost/revenues of municipal waste management*.

5.3. Contributions to Knowledge and to Practice, and Managerial and Policy Implications

This paper impacts and contributes to both theory and practice, while also providing useful hints from a managerial and policy perspective.

From the perspective of contributions to knowledge, a framework for assessing company's readiness tailored to companies operating in developing countries was developed. This framework investigates both the product and the business model aspects, while also defining, respectively, future actions and KPIs to lead the company towards a full embracement of the CE paradigm.

This research also contributes to practice. Indeed, the framework constitutes a means to assess the level of readiness of companies willing to embrace the CE paradigm in developing countries, while also giving them the opportunity to split the assessment of the circularity level of their products from the business model perspective. The framework also supports companies to widen their view related to the CE paradigm, looking at different related dimensions (both internal and external to the company boundaries). In addition, the application of the framework proposed could be a useful tool to lead the company towards a circular-driven roadmap and to assess their progress along the time.

Finally, both company managers and decision-makers can exploit the results obtained from this research. Managers can raise their awareness about the main topics related to the CE paradigm and can use it as a support for decision-making in that domain. On the other side, public authorities and policy-makers can obtain fact-based information that could help them to facilitate CE adoption in developing countries.

6. Conclusions

This research developed and proposed a framework tailored for companies operating in developing countries to determine their CE readiness level. In detail, the model investigates the two main perspectives (product and business model) that these companies should consider assessing the readiness of their operations under a CE lens. Through an interactive research method, inspired by DRM [36], the framework, grounded on the EEA's template [8,37], has been conceived and refined. Then, the framework was validated through an application case with a company operating in the packaging industry in a non-EU developing country. The framework turned out to be an easy to use artifact constituted of two main parts, each one split in a set of categories to be investigated and then also detailed in a series of sub-dimensions. For each sub-dimension, related questions to be provided to the company employees were defined, flanked by normative answers useful to help the translation of the open answer received by the interviewees in a quantified value assignable to one of the five specific levels of maturity defined. Finally, to help the companies adopting the framework in their organizations to move towards a higher level of circularity readiness, the framework also assisted interviewees to both explore the future steps to be planned related to the product dimension and define a set of KPIs to monitor and lead the improvement under a circular perspective for their business model.

The main findings of the application case unveiled that the analyzed company had already started a path towards circularity in its business, using recycled materials; designing sustainable products with the buyers of packaging; reducing waste of energy, water, and GHG emissions; and increasing awareness of its employees and partners regarding sustainability. The framework also helped to understand that the circularity concept applied to the business of the analyzed company is part of its strategic business orientation, and the framework proposed represents a tool to support the improvement of long-term performances.

The application of the framework to the company analyzed also revealed that there is room for improvement in developing countries to foster CE adoption, especially in relation to the policy context. Indeed, policy incentives and instruments of public authorities could considerably support the circular transition process in companies [6], as well as stimulate them, their customers, and the entire community towards a more responsible business [77]. In particular, a consideration specific for Serbia can be performed due to the fact that EC is promoting CE also in the Western Balkans countries [78].

The framework proposed and the related application case conducted in this research (with a Serbian manufacturing company) belong to the research stream of implementing CE in companies' operations. The results indicated that the companies were not in a position to choose whether to behave responsibly to the environment and society. It becomes their legal duty, but it also constitutes an instrument to attract investors, to build good market reputation, to gain new customers and provide them added-value, and to improve their long-term performances.

Notwithstanding the results obtained, this research was not free from limitations. Indeed, the research results were applied only to one company, although this company is leader in the paper packaging industry of Serbia. The case selection could also be seen as one of the limitation factors of this research, even if the sample chosen was purposive, allowing an idiographic (intensive) study of an individual case [79]. Finally, during the application case, only the sustainability manager and his assistant were involved, who were called to provide their subjective perspectives.

All of these limitations open the way to further research. The framework proposed in this article may be seen as the basis for further research in the manufacturing industry of developing countries. Considering the packaging industry, this framework can also be extended to plastic packaging producers, having in mind the necessity of supporting CE adoption in the packaging industry. The application of the framework should also be extended to different companies belonging to developing countries and operating in other industries. Finally, a generic model able to systematically assess organizations' circularity readiness and maturity level placed in both developed and developing countries is still missing in the literature.

Author Contributions: Conceptualization, J.D.-R., C.S., M.P. and Z.A.; methodology, C.S.; validation, J.D.-R., C.S., M.P. and Z.A.; formal analysis, J.D.-R., M.P.; investigation, J.D.-R., C.S., M.P. and Z.A.; data curation, J.D.-R., M.P.; writing—original draft preparation, J.D.-R., C.S. and Z.A.; writing—review and editing, J.D.-R., C.S. and Z.A.; visualization, J.D.-R., C.S., M.P. and Z.A.; supervision, C.S. and Z.A. All authors have read and agreed to the published version of the manuscript.

Funding: This research received no external funding.

Institutional Review Board Statement: Not applicable.

Informed Consent Statement: Not applicable.

Data Availability Statement: All data used have been added to the manuscript.

Conflicts of Interest: The authors declare no conflict of interest.

References

1. The Ellen MacArthur Foundation History of the Ellen MacArthur Foundation. Available online: https://www.ellenmacarthurfoundation.org/our-story/milestones (accessed on 3 April 2019).
2. The Ellen MacArthur Foundation Towards the Circular Economy. Economic and Business Rationale for an Accelerated Transition. 2013. Available online: https://books.google.it/books?id=QM_4DAEACAAJ (accessed on 10 September 2022).
3. Kirchherr, J.; Reike, D.; Hekkert, M. Conceptualizing the Circular Economy: An Analysis of 114 Definitions. *Resour. Conserv. Recycl.* **2017**, *127*, 221–232. [CrossRef]

4. European Commission Closing the Loop—An EU Action Plan for the Circular Economy. *COM (2015) 614 Communication from the Commission to the European Parliament, the Council, The European Economic and Social Committee, and the Committee of the Regions*; European Commission: Brussels, Belgium, 2015; Available online: https://www.eea.europa.eu/policy-documents/com-2015-0614-final (accessed on 15 September 2022).
5. Nascimento, D.L.M.; Alencastro, V.; Quelhas, O.L.G.; Caiado, R.G.G.; Garza-Reyes, J.A.; Lona, L.R.; Tortorella, G. Exploring Industry 4.0 Technologies to Enable Circular Economy Practices in a Manufacturing Context: A Business Model Proposal. *J. Manuf. Technol. Manag.* **2019**, *30*, 607–627. [CrossRef]
6. Dwivedi, A.; Sassanelli, C.; Agrawal, D.; Moktadir, A.; Adamo, I.D. Drivers to Mitigate Climate Change in Context of Manufacturing Industry: An Emerging Economy Study. *Bus. Strateg. Environ.* **2023**, 1–18. [CrossRef]
7. Taddeo, R. Industrial Ecology and Innovation: At What Point Are We? *Editorial for the Special Issue "Industrial Ecology and Innovation." Adm. Sci.* **2021**, *11*, 93. [CrossRef]
8. European Environment Agency. Circular Economy in Europe—Developing the Knowledge Base—EEA Report No 2/2016. 2016. Available online: https://www.eea.europa.eu/publications/circular-economy-in-europe (accessed on 18 September 2022).
9. Fregonara, E.; Giordano, R.; Ferrando, D.G.; Pattono, S. Economic-Environmental Indicators to Support Investment Decisions: A Focus on the Buildings' End-of-Life Stage. *Buildings* **2017**, *7*, 65. [CrossRef]
10. Rogkas, N.; Tsolakis, E.; Kalligeros, C.; Vasileiou, G.; Vakouftsis, C.; Kaisarlis, G.; Markopoulos, A.P.; Spitas, V. Upcycling Obsolete Mechanical Equipment into Innovative Laboratory Test Rigs: A Low-Cost Solution or a Sustainable Design Approach? *Proc. Des. Soc.* **2021**, *1*, 3309–3318. [CrossRef]
11. Bocken, N.M.P.; Miller, K.; Weissbrod, I.; Holgado, M.; Evans, S. Business Model Experimentation for Circularity: Driving Sustainability in a Large International Clothing Retailer. *Econ. Policy Energy Environ.* **2017**, 85–122. [CrossRef]
12. Leal Filho, W.; Ellams, D.; Han, S.; Tyler, D.; Boiten, V.J.; Paco, A.; Moora, H.; Balogun, A.L. A Review of the Socio-Economic Advantages of Textile Recycling. *J. Clean. Prod.* **2019**, *218*, 10–20. [CrossRef]
13. Acerbi, F.; Rocca, R.; Fumagalli, L.; Taisch, M.; Acerbi, F.; Rocca, R.; Fumagalli, L.; Taisch, M. Enhancing the Cosmetics Industry Sustainability through a Renewed Sustainable Supplier Selection Model. *Prod. Manuf. Res.* **2023**, *11*, 2161021. [CrossRef]
14. Chiappetta Jabbour, C.J.; Fiorini, P.D.C.; Ndubisi, N.O.; Queiroz, M.M.; Piato, É.L. Digitally-Enabled Sustainable Supply Chains in the 21st Century: A Review and a Research Agenda. *Sci. Total Environ.* **2020**, *725*, 138177. [CrossRef]
15. Rathore, P.; Sarmah, S.P. Economic, Environmental and Social Optimization of Solid Waste Management in the Context of Circular Economy. *Comput. Ind. Eng.* **2020**, *145*, 106510. [CrossRef]
16. Taddeo, R.; Simboli, A.; Morgante, A. Implementing Eco-Industrial Parks in Existing Clusters. Findings from a Historical Italian Chemical Site. *J. Clean. Prod.* **2012**, *33*, 22–29. [CrossRef]
17. Simboli, A.; Taddeo, R.; Raggi, A. The Multiple Dimensions of Urban Contexts in an Industrial Ecology Perspective: An Integrative Framework. *Int. J. Life Cycle Assess* **2019**, *24*, 1285–1296. [CrossRef]
18. Bocken, N.M.P.; Short, S.W.; Rana, P.; Evans, S. A Literature and Practice Review to Develop Sustainable Business Model Archetypes. *J. Clean. Prod.* **2014**, *65*, 42–56. [CrossRef]
19. González Chávez, C.A.; Romero, D.; Rossi, M.; Luglietti, R.; Johansson, B. Circular Lean Product-Service Systems Design: A Literature Review, Framework Proposal and Case Studies. *Procedia CIRP* **2019**, *83*, 419–424. [CrossRef]
20. Pigosso, D.C.A.; McAloone, T.C. How Can Design Science Contribute to a Circular Economy? *Int. Conf. Eng. Des. ICED* **2017**, *5*, 299–307.
21. Acerbi, F.; Sassanelli, C.; Taisch, M. A Conceptual Data Model Promoting Data-Driven Circular Manufacturing. *Oper. Manag. Res.* **2022**, *15*, 838–857. [CrossRef]
22. Vitti, M.; Facchini, F.; Mazzilli, M.; Mossa, G.; Ranieri, L. A Circular Approach for the Management of Biogas Flows from Sewage Sludge. 2022. Available online: https://www.researchgate.net/profile/Francesco-Facchini/publication/366066797_A_circular_approach_for_the_management_of_biogas_flows_from_sewage_sludge/links/639078c5e42faa7e75a2de10/A-circular-approach-for-the-management-of-biogas-flows-from-sewage-sludge.pdf (accessed on 20 November 2022).
23. Sassanelli, C.; Rosa, P.; Terzi, S. Supporting Disassembly Processes through Simulation Tools: A Systematic Literature Review with a Focus on Printed Circuit Boards. *J. Manuf. Syst.* **2021**, *60*, 429–448. [CrossRef]
24. Trevisan, A.H.; Lobo, A.; Guzzo, D.; Gomes, L.A. de V.; Mascarenhas, J. Barriers to Employing Digital Technologies for a Circular Economy: A Multi-Level Perspective. *J. Environ. Manag.* **2023**, *332*, 117437. [CrossRef]
25. Acerbi, F.; Assiani, S.; Taisch, M. A Research on Hard and Soft Skills Required to Operate in a Manufacturing Company Embracing the Industry 4.0 Paradigm. *Proc. Summer Sch. Fr. Turco* **2019**, *1*, 1–12.
26. Birkel, H.; Müller, J.M. Potentials of Industry 4.0 for Supply Chain Management within the Triple Bottom Line of Sustainability—A Systematic Literature Review. *J. Clean. Prod.* **2021**, *289*, 125612. [CrossRef]
27. Guzzo, D.; Trevisan, A.H.; Echeveste, M.; Costa, J.M.H. Circular Innovation Framework: Verifying Conceptual to Practical Decisions in Sustainability-Oriented Product-Service System Cases. *Sustainability* **2019**, *11*, 3248. [CrossRef]
28. Di Vaio, A.; Hassan, R.; Chhabra, M.; Arrigo, E.; Palladino, R. Sustainable Entrepreneurship Impact and Entrepreneurial Venture Life Cycle: A Systematic Literature Review. *J. Clean. Prod.* **2022**, *378*, 134469. [CrossRef]
29. Bjørnbet, M.M.; Skaar, C.; Fet, A.M.; Schulte, K.Ø. Circular Economy in Manufacturing Companies: A Review of Case Study Literature. *J. Clean. Prod.* **2021**, *294*, 126268. [CrossRef]

30. Acerbi, F.; Taisch, M. A Literature Review on Circular Economy Adoption in the Manufacturing Sector. *J. Clean. Prod.* **2020**, *273*, 123086. [CrossRef]
31. Despeisse, M.; Acerbi, F. Toward Eco-Efficient and Circular Industrial Systems: Ten Years of Advances in Production Management Systems and a Thematic Framework. *Prod. Manuf. Res.* **2022**, *10*, 354–382. [CrossRef]
32. Trevisan, A.H.; Zacharias, I.S.; Castro, C.G.; Mascarenhas, J. Circular Economy Actions in Business Ecosystems Driven by Digital Technologies. *Procedia CIRP* **2021**, *100*, 325–330. [CrossRef]
33. Sassanelli, C.; Rosa, P.; Rocca, R.; Terzi, S. Circular Economy Performance Assessment Methods: A Systematic Literature Review. *J. Clean. Prod.* **2019**, *229*, 440–453. [CrossRef]
34. Vinante, C.; Sacco, P.; Orzes, G.; Borgianni, Y. Circular Economy Metrics: Literature Review and Company-Level Classification Framework. *J. Clean. Prod.* **2021**, *288*, 125090. [CrossRef]
35. Rimano, M.; Simboli, A.; Taddeo, R.; Raggi, A. Life Cycle Approaches for the Environmental Impact Assessment of Organizations: Defining the State of the Art. *Adm. Sci.* **2019**, *9*, 94. [CrossRef]
36. Blessing, L.; Chakrabarti, A. *DRM, a Design Research Methodology*; Springer: Dordrecht, The Netherlands, 2009; Available online: https://books.google.it/books?id=KdR4OmWtQdIC&printsec=frontcover&hl=it#v=onepage&q&f=false (accessed on 15 December 2022)ISBN 9781848825864.
37. European Environment Agency. *Circular by Design—Products in the Circular Ecoomy*; EEA Report, No. 6/2017; Publications Office of the European Union: Luxembourg, 2017; Available online: https://www.eea.europa.eu/publications/circular-by-design (accessed on 15 October 2022)ISBN 9789292138578.
38. Bernstad Saraiva, A.; Pacheco, E.B.A.V.; Gomes, G.M.; Visconte, L.L.Y.; Bernardo, C.A.; Simões, C.L.; Soares, A.G. Comparative Lifecycle Assessment of Mango Packaging Made from a Polyethylene/Natural Fiber-Composite and from Cardboard Material. *J. Clean. Prod.* **2016**, *139*, 1168–1180. [CrossRef]
39. D'Adamo, I.; Gastaldi, M. Perspectives and Challenges on Sustainability: Drivers, Opportunities and Policy Implications in Universities. *Sustainability* **2023**, *15*, 3564. [CrossRef]
40. Thorley, J.; Garza-Reyes, J.A.; Anosike, A. Circular Economy: A Conceptual Model to Measure Readiness for Manufacturing SMEs. *Benchmarking* **2022**, *29*, 1362–1390. [CrossRef]
41. Bressanelli, G.; Perona, M.; Saccani, N. Assessing the Readiness of Manufacturing Companies for the Circular Economy: An Analysis and an Initial Proposal. In Proceedings of the Summer School Francesco Turco, AIDI—Italian Association of Industrial Operations Professors, Virtual, 8–10 September 2021; Volume 7, pp. 343–354.
42. Bressanelli, G.; Saccani, N.; Perona, M. Evaluating the Circular Economy Readiness of Manufacturing Companies at the Micro Level. In Proceedings of the 2nd Online Symposium on Circular Economy and Sustainability, Alexandroupolis, Greece, 14–16 July 2021.
43. Pigosso, D.C.A.; McAloone, T.C. Making the Transition to a Circular Economy within Manufacturing Companies: The Development and Implementation of a Self-Assessment Readiness Tool. *Sustain. Prod. Consum.* **2021**, *28*, 346–358. [CrossRef]
44. Momete, D.C. A Unified Framework for Assessing the Readiness of European Union Economies to Migrate to a Circular Modelling. *Sci. Total Environ.* **2020**, *718*, 137375. [CrossRef]
45. United Nations. THE 17 GOALS | Sustainable Development. Available online: https://sdgs.un.org/goals (accessed on 31 January 2022).
46. Sauvé, S.; Bernard, S.; Sloan, P. Environmental Sciences, Sustainable Development and Circular Economy: Alternative Concepts for Trans-Disciplinary Research. *Environ. Dev.* **2016**, *17*, 48–56. [CrossRef]
47. Del Mar Alonso-Almeida, M.; Rodríguez-Antón, J.M.; Bagur-Femenías, L.; Perramon, J. Sustainable Development and Circular Economy: The Role of Institutional Promotion on Circular Consumption and Market Competitiveness from a Multistakeholder Engagement Approach. *Bus. Strateg. Environ.* **2020**, *29*, 2083–2814. [CrossRef]
48. Merli, R.; Preziosi, M.; Acampora, A. How Do Scholars Approach the Circular Economy? A Systematic Literature Review. *J. Clean. Prod.* **2018**, *178*, 703–722. [CrossRef]
49. Lieder, M.; Rashid, A. Towards Circular Economy Implementation: A Comprehensive Review in Context of Manufacturing Industry. *J. Clean. Prod.* **2016**, *115*, 36–51. [CrossRef]
50. Brown, P.; Bocken, N.; Balkenende, R. Why Do Companies Pursue Collaborative Circular Oriented Innovation? *Sustainability* **2019**, *11*, 635. [CrossRef]
51. Wójcik-Karpacz, A.; Karpacz, J.; Brzeziński, P.; Pietruszka-Ortyl, A.; Ziębicki, B. Barriers and Drivers for Changes in Circular Business Models in a Textile Recycling Sector: Results of Qualitative Empirical Research. *Energies* **2023**, *16*, 490. [CrossRef]
52. Basile, D.; Adamo, I.D.; Goretti, V.; Rosa, P.; Basile, D.; Adamo, I.D.; Goretti, V.; Rosa, P. Digitalizing Circular Economy through Blockchains: The Blockchain Circular Economy Index Digitalizing Circular Economy through Blockchains: The Blockchain Circular Economy Index. *J. Ind. Prod. Eng.* **2023**. [CrossRef]
53. Buil, P.; Roger-Loppacher, O.; Selvam, R.M.; Prieto-Sandoval, V. The Involvement of Future Generations in the Circular Economy Paradigm: An Empirical Analysis on Aluminium Packaging Recycling in Spain. *Sustainability* **2017**, *9*, 2345. [CrossRef]
54. Appolloni, A.; Chiappetta Jabbour, C.J.; D'Adamo, I.; Gastaldi, M.; Settembre-Blundo, D. Green Recovery in the Mature Manufacturing Industry: The Role of the Green-Circular Premium and Sustainability Certification in Innovative Efforts. *Ecol. Econ.* **2022**, *193*, 107311. [CrossRef]

55. Colasante, A.; D'Adamo, I. The Circular Economy and Bioeconomy in the Fashion Sector: Emergence of a "Sustainability Bias. " *J. Clean. Prod.* **2021**, *329*, 129774. [CrossRef]
56. Adlin, N.; Lanz, M.; Lohtander, M. The Circular Economy Competence of the Manufacturing Sector—A Case Study. *Lect. Notes Mech. Eng.* **2023**, 351–360. [CrossRef]
57. Bertassini, A.C.; Calache, L.D.D.R.; Carpinetti, L.C.R.; Ometto, A.R.; Gerolamo, M.C. CE-Oriented Culture Readiness: An Assessment Approach Based on Maturity Models and Fuzzy Set Theories. *Sustain. Prod. Consum.* **2022**, *31*, 615–629. [CrossRef]
58. Bîrgovan, A.L.; Vatca, S.D.; Bacali, L.; Szilagyi, A.; Lakatos, E.S.; Cioca, L.I.; Ciobanu, G. Enabling the Circular Economy Transition in Organizations: A Moderated Mediation Model. *Int. J. Environ. Res. Public Health* **2022**, *19*, 677. [CrossRef]
59. Waring, C.; Liyanage, K. A Systematic Literature Review of the Concept of Circular Economic Readiness and a Proposed Circular Economy Readiness Scale. *Adv. Transdiscipl. Eng.* **2022**, *25*, 245–254. [CrossRef]
60. Daú, G.; Scavarda, A.; Scavarda, L.F.; Taveira, V.J. The Healthcare Sustainable Supply Chain 4.0: The Circular Economy Transition Conceptual Framework with the Corporate Social Responsibility Mirror. *Sustainability* **2019**, *11*, 3259. [CrossRef]
61. European Parliament European Parliament and Council. *Directive 94/62/EC of 20 December 1994 on Packaging and Packaging Waste*; OPOCE: Prague, Czech Republic, 1994.
62. EUROPEN. Review of the Packaging and Packaging Waste Directive. *EUROPEN's Feedback on the Inception Impact Assessment*. 2020. Available online: https://ec.europa.eu/info/law/better-regulation/have-your-say/initiatives/12263-Reducing-packaging-waste-review-of-rules/F539807_en (accessed on 20 November 2022).
63. Pivnenko, K. The Challenge of Chemicals in Material Lifecycles. *Waste Manag.* **2016**, *56*, 1–2. [CrossRef] [PubMed]
64. Pivnenko, K.; Pedersen, G.A.; Eriksson, E.; Astrup, T.F. Bisphenol A and Its Structural Analogues in Household Waste Paper. *Waste Manag.* **2015**, *44*, 39–47. [CrossRef] [PubMed]
65. Sassanelli, C.; Terzi, S. The D-BEST Reference Model: A Flexible and Sustainable Support for the Digital Transformation of Small and Medium Enterprises. *Glob. J. Flex. Syst. Manag.* **2022**, *40171*, 345–370. [CrossRef]
66. Larchenko, L.; Kuramshina, L. Transition to a Circular Economy Model as an Alternative Option of Solving the Problem of Solid Waste Utilization. *E3S Web Conf.* **2020**, *161*, 01108. [CrossRef]
67. Garcia, C.L.; Cayzer, S. Assessment of the Circular Economy Transition Readiness at a National Level: The Colombian Case. In *The Circular Economy and the Global South*; Routledge: Oxon, UK, 2019; pp. 113–133. ISBN 9780429434006.
68. Kolade, O.; Odumuyiwa, V.; Abolfathi, S.; Schröder, P.; Wakunuma, K.; Akanmu, I.; Whitehead, T.; Tijani, B.; Oyinlola, M. Technology Acceptance and Readiness of Stakeholders for Transitioning to a Circular Plastic Economy in Africa. *Technol. Forecast Soc. Chang.* **2022**, *183*, 121954. [CrossRef]
69. Guimaraes Araújo, M.; Lovón-Canchumani, G.; Bechara Elabras Veiga, L. Circular Economy for Lubricating Oils in Brazil. In Proceedings of the Springer Proceedings in Mathematics and Statistics, Rio de Janeiro, Brazil, 22–24 February 2021; Springer: Berlin/Heidelberg, Germany, 2021; Volume 367, pp. 103–113.
70. Heyes, G.; Sharmina, M.; Mendoza, J.M.F.; Gallego-Schmid, A.; Azapagic, A. Developing and Implementing Circular Economy Business Models in Service-Oriented Technology Companies. *J. Clean. Prod.* **2018**, *177*, 621–632. [CrossRef]
71. Daae, J.; Chamberlin, L.; Boks, C. Dimensions of Behaviour Change in the Context of Designing for a Circular Economy. *Des. J.* **2019**, *21*, 521–541. [CrossRef]
72. Sassanelli, C.; Fernandes, S.D.C.; Rozenfeld, H.; Da Costa, J.M.H.; Terzi, S. Enhancing Knowledge Management in the PSS Detailed Design: A Case Study in a Food and Bakery Machinery Company. *Concurr. Eng. Res. Appl.* **2021**, *29*, 295–308. [CrossRef]
73. Tecchio, P.; McAlister, C.; Mathieux, F.; Ardente, F. In Search of Standards to Support Circularity in Product Policies: A Systematic Approach. *J. Clean. Prod.* **2017**, *168*, 1533–1546. [CrossRef]
74. Piyathanavong, V.; Garza-Reyes, J.A.; Kumar, V.; Maldonado-Guzmán, G.; Mangla, S.K. The Adoption of Operational Environmental Sustainability Approaches in the Thai Manufacturing Sector. *J. Clean. Prod.* **2019**, *220*, 507–528. [CrossRef]
75. Ethirajan, M.; Arasu, M.T.; Kandasamy, J.; Vimal, K.E.K.; Nadeem, S.P.; Kumar, A. Analysing the Risks of Adopting Circular Economy Initiatives in Manufacturing Supply Chains. *Bus. Strateg. Environ.* **2021**, *30*, 204–236. [CrossRef]
76. Huysman, S.; De Schaepmeester, J.; Ragaert, K.; Dewulf, J.; De Meester, S. Performance Indicators for a Circular Economy: A Case Study on Post-Industrial Plastic Waste. *Resour. Conserv. Recycl.* **2017**, *120*, 46–54. [CrossRef]
77. D'Adamo, I.; Mazzanti, M.; Morone, P.; Rosa, P. Assessing the Relation between Waste Management Policies and Circular Economy Goals. *Waste Manag.* **2022**, *154*, 27–35. [CrossRef]
78. European Environmental Bureau. *Guidelines on Circular Economy for the Countrie of the Western Balkans and Turkey*; European Environmental Bureau (EEB) and Institute for the Circular Economy (INCIEN): Brussels, Belgium, 2020.
79. Williamson, K. *Research Methods for Students, Academics and Professionals*; Quick Print: Wagga Wagga, Australia, 2002.

sustainability

Review

Do We Consider Sustainability When We Measure Small and Medium Enterprises' (SMEs') Performance Passing through Digital Transformation?

Isotilia Costa Melo [1,*], Paulo Nocera Alves Junior [1], Geandra Alves Queiroz [2], Wilfredo Yushimito [3] and Jordi Pereira [3,4]

1 Escuela de Ingeniería de Coquimbo (EIC), Universidad Católica del Norte (UCN), Coquimbo 1781421, Chile
2 Engineering Department, Production Engineering, State University of Minas Gerais (UEMG), Passos 31360, Brazil
3 Facultad de Ingenmiería y Ciencias (FIC), Universidad Adolfo Ibáñez (UAI), Viña del Mar 7941169, Chile
4 UPF Barcelona School of Management, Pompeu Fabra University, 08008 Barcelona, Spain
* Correspondence: isotilia.costa@ucn.cl

Abstract: Small-medium enterprises (SMEs) represent 90% of business globally. Digital Transformation (DT) affects SMEs differently from larger companies because although SMEs have more flexibility and agility for adapting to new circumstances, they also have more limited resources and specialization capabilities. Thus, it is fundamental to measure SMEs' performance considering different perspectives. Here, we describe and analyze the state-of-the-art of DT in SMEs, focusing on performance measurement. We center on whether the tools used by SMEs encompass the triple bottom line of sustainability (i.e., environmental, social, and economic aspects). To do so, in December 2021, we performed a comprehensive systematic literature review (SLR) on the Web of Science and Scopus. In addition, we also explored a novel approach for SLR: topic modeling with a machine learning technique (Latent Dirichlet Allocation). The differences and interchangeability of both methods are discussed. The findings show that sustainability is treated as a separate topic in the literature. The social and environmental aspects are the most neglected. This paper contributes to sustainable development goals (SDGs) 1, 5, 8, 9, 10, and 12. A conceptual framework and future research directions are proposed. Thus, this paper is also valuable for policymakers and SMEs switching their production paradigm toward sustainability and DT.

Keywords: digitalization; small- and medium-sized enterprises (SMEs); Industry 4.0; topic modeling; latent dirichlet allocation (LDA); triple bottom line of sustainability

Citation: Costa Melo, I.; Alves Junior, P.N.; Queiroz, G.A.; Yushimito, W.; Pereira, J. Do We Consider Sustainability When We Measure Small and Medium Enterprises' (SMEs') Performance Passing through Digital Transformation? *Sustainability* **2023**, *15*, 4917. https://doi.org/10.3390/su15064917

Academic Editors: Sergio Terzi and Claudio Sassanelli

Received: 31 January 2023
Revised: 21 February 2023
Accepted: 6 March 2023
Published: 9 March 2023

1. Introduction

Organizations confront a considerable number of challenges to their business operations. One of their initiatives to be competitive is adopting new technologies, which implies the emergence of the digital economy. Digital transformation (DT) has globally changed business practices and organizational culture [1]; it breaks boundaries, challenging the enterprises' competitiveness [2,3].

In this context, small- and medium-sized enterprises (SMEs) deserve specific attention as they represent a significant share of global business. SMEs account for 90% of all firms and 50% of employment globally [4]. Additionally, SMEs have inherent characteristics that differentiate them from larger companies [5,6]. For example, they tend to be less productive and pollute more [7]. Moreover, SMEs tend to have more flexibility and agility for adapting to new circumstances, more limited resources, and specialization capabilities [8]. These characteristics mirror SMEs' performance while facing the DT process [5,6]. Furthermore, SMEs require specific dimensions, variables, and mathematical tools for measuring their performance.

Besides measuring digital performance, measuring the economic, social, and environmental aspects of SMEs' performance is of utmost importance—i.e., the triple bottom line (TBL) of sustainability [9]. Initially, DT seems to negatively interact with the TBL's social and environmental aspects of SMEs, e.g., job losses and increased material consumption [10]. Once SMEs pollute more than larger companies [7], whether the DT improves their operational performance without any environmental consideration, it can be expected that DT encourages SMEs to pollute more. On the other hand, evidence suggests that the level of digital competitiveness of a country systematically fosters the sustainability transition of European SMEs [11]. However, a few researchers have reviewed DT performance in SMEs [6,12]. In these cases, they were focused only on one performance aspect (such as the economic or the financial) and not on the TBL (i.e., economic, social, and environmental aspects).

Therefore, a clear gap exists for SMEs' performance metrics investigations, jointly considering digital, economic, social, and environmental aspects. Consequently, we aim to answer the following main research question (RQ):

RQ: Are researchers considering the TBL while measuring the performance of SMEs passing through the DT process?

Our secondary research questions (SRQs) are the following:

SRQ 1: Which dimensions, variables, and mathematical tools are used for measuring each aspect (digital, economic, social, and environmental)?

SRQ 2: Which dimensions, variables, and mathematical tools should be explored in further investigations?

In this regard, we aimed to identify whether researchers are considering the TBL while proposing and applying approaches for measuring the performance of SMEs passing through the DT process. Additionally, we aimed to identify which dimensions, variables, and mathematical tools are used for measuring each aspect, i.e., digital, economic, social, and environmental. Finally, we aimed to identify methodological blind spots, i.e., dimensions, variables, and mathematical tools that are neglected or not fully explored in this application, and we discuss further research directions. Furthermore, our paper may derive practical implications for policymakers and SMEs interested in fostering and adopting a novel, productive paradigm toward DT and TBL.

To achieve these goals, we conducted a systematic literature review (SLR), extracting articles published before December 2021 in peer-reviewed journals from the scientific databases Scopus and Web of Science, selecting and analyzing them. Due to the novelty of the theme, we started with 331 searched papers. After applying three systematic steps, we selected 113, 74, and 35 papers in each step, respectively. Finally, 35 papers were fully and systematically analyzed. This was a limited sample size. Thus, we searched for a complementary methodological approach, enabling result comparison and new insights.

Asmussen and Møller (2019) [13] proposed a framework for explorative SLR through topic modeling. It is executed based on Latent Dirichlet Allocation (LDA), a machine-learning technique. Besides being a novelty for SLRs [13–15], this approach has several advantages. Among them, the costs and time of pre-analysis, analysis, and post-analysis are lower than manual SLRs; the categories and mapping do not need to be known in advance; and coding can be automated, reducing subjectivity [16].

Overall, our study offers several critical contributions to the extant literature. First, it is a new research theme with increasing recent interest (especially since the COVID-19 pandemic). However, most papers still do not jointly consider the environmental and social aspects of sustainability for measuring performance. These aspects are treated as a separate theme in the field. When all four aspects are considered, it is only on a strategic level. There is a gap in sustainable measurement approaches, considering the operational perspective. The relationship between digital and economic aspects is still unclear. The heterogeneity among SMEs may imply that digital aspects differ depending on sector and maturity level. The relationship between social and environmental aspects with other aspects is not explored, not even as a rough draft.

Regarding mathematical tools, we identified that structural equation modeling (SEM) and econometrics are the most used methods. Decision-Making Trial and Evaluation Laboratory (DEMATEL) and integrations with fuzzy techniques represent a methodological frontier. No paper used an approach directly associated with performance measurements, such as Stochastic Frontier Analysis (SFA) and Data Envelopment Analysis (DEA). Additionally, no paper applied Artificial Intelligence (AI) for measuring performance, probably due to the barrier of the lack of standardization among SMEs and data collection procedures. Finally, after presenting, comparing, and discussing the results from manual SLR and topic modeling, we propose further research directions and a framework with theoretical and practical implications.

This paper is structured as follows. Section 2 presents a theoretical background about SMEs, DT, and sustainability. Section 3 offers the research design used to obtain and analyze the results, which are presented and discussed in Section 4 (manual SLR) and Section 5 (LDA-based SLR). Section 6 provides the conceptual framework, summarizing the findings and implications. Section 7 concludes, states the limitations, and outlines the future research agenda.

2. Theoretical Background

2.1. Small and Medium Enterprises (SMEs)

There is no globally standardized definition of small and medium enterprises (SMEs). The most common classifications are based on a financial measure and/or the number of employees. Even the same country may have different definitions, depending on the industry. For example, in the USA, a "small enterprise" in the "Agriculture, forestry, fishing, and hunting" industry is based on annual income for all subindustries, except for the logging subindustry. In the logging subindustry, a "small enterprise" is an enterprise with less than 500 employees [17].

Another example is Brazil and Chile, two countries on the same continent, adopting standardized definitions independently of the SME industry. In both cases, they use financial and employee criteria. In Brazil, SMEs have from 20 to 249 employees. However, the more usual definition in Brazil is based on the annual income criterion according to the Statute of Micro and Small Enterprises. In this case, SMEs have a yearly income from BRL 360,000 (Brazilian currency) to BRL 3,600,000, except for SMEs in the banking sector that follow a different definition [18]. In Chile, an SME is defined as "an enterprise with 10 to 199 workers" or "an enterprise whose annual income from sales and services and other business activities is greater than 2400 UF (Chilean currency, automatically inflation corrected), but less than 100,000 UF in the last calendar year" [19]. Similar conflicts among definitions also occur among smallholder farmers and agricultural SMEs [20].

In summary, the definitions based on the number of employees are usually different in terms of number. Furthermore, some definitions may consider temporary employees, such as Japanese SMEs [21]. What is more, the definition based on financial terms is usually determined by local law, established in terms of a local currency value at the date of the law approval, without any inflationary consideration. Hence, to be comparable definitions from different countries, it may be necessary to correct inflation, convert currency, and make the definitions represent similar economic importance to each analyzed economy. To the best of our knowledge, this kind of procedure is not yet established in the literature.

The World Trade Organization [22] highlights that, in general, the lower productivity is often attributed to small businesses' inability to achieve economies of scale, difficulties they face in accessing credit or investment, lack of appropriate skills, and their informality [22]. Additionally, there are some characteristics that may affect SMEs' performance, such as industry, management, technology, technical competence in marketing and innovation [23], the level of internationalization [24], and ownership [1]. In other words, even assuming the same definition, SMEs are usually heterogeneous, and heterogeneity affects performance [25]. Specifically, in the case of SMEs in Europe, findings suggest that heterogeneity hinders the transition to the TBL of sustainability once this transition demands capabilities and capacities that are asymmetric among SMEs [11]. Heterogeneity also may influence the adoption of innovation by SMEs in Brazil [26].

Here, we assumed that all papers that used the acronym "SME" with the explanation "small and medium enterprises", "small- and medium-sized enterprises", or "small-medium enterprises" were referring to a comparable term. Furthermore, we considered that the term "SME" could encompass micro and self-employed enterprises.

2.2. Digital Transformation (DT)

After reviewing and analyzing 134 well-received definitions of DT, Gong and Ribiere (2021) [27] posed the following definition: "Digital transformation is a fundamental change process, enabled by the innovative use of digital technologies accompanied by the strategic leverage of key resources and capabilities, aiming to radically improve an entity and redefine its value proposition for its stakeholders.". In this context, an entity may be an organization, a business network, an industry, or a society. To the best of our knowledge, there is no definition of DT to SMEs as a separate entity. Thus, we adopted the definition of Gong and Ribiere (2021) [27].

Therefore, we considered DT as a synonym of "digitalization", "digital transition", and "digital innovation". However, we did not consider "digitization" and "Industry 4.0 (4.0)" as synonyms of DT. Digitization is the conversion of analog information to digital. Activities are not made more valuable by digitization. Usually, digitization is used to describe the process of digitizing internal and external procedures [28]. Although sometimes mentioned as a synonym of I4.0, the concept of DT stresses the implications for strategy and business model innovation and underlines the emerging technologies in the business model, and, in turn, the rise of cross-industry ecosystems [29]. Furthermore, the term I4.0 is mainly related to the DT process in the manufacturing sector [29]. For example, Sassanelli et al. (2020) [30] proposed a holistic methodology to evaluate a manufacturing company passing through the digitization process in terms of the level of digital and lean maturity. Readers interested in a more conceptual understanding may refer to [31].

However, when we consider a multi-industry perspective, the average size of enterprises in the service industry is typically smaller than in manufacturing [32]. Additionally, the birth rates of employer enterprises are higher in the services industry than in manufacturing [32]. Between 2008 and 2014, the employment rate in manufacturing decreased in most countries of the Organization for Economic Co-operation and Development (OECD), except in Germany and Luxemburg [33].

Therefore, we adopted the term "digital transformation" due to its industrial range and the emphasis on business model innovation and business strategy. However, we accepted papers that used the terms "Industry 4.0" or "digitization", only if the authors also jointly used the terms "digital transformation", "digital transition", "digital innovation", or "digitalization".

2.3. Triple Bottom Line (TBL)

The concept of sustainability is commonly based on three aspects: economic, social, and environmental—also named the "triple bottom line" (TBL), as proposed by Elkington (1994, 1998) [34,35]. The author encouraged organizations to measure their performance using a multidimensional perspective that integrates not only the traditional indicators but also includes environmental and social aspects. However, sustainability performance measurement is a challenge because there is no universal standard for the calculation of sustainable TBL performance [36,37]. In this perspective, as stated by Santos et al. (2019) [38], the way to measure, obtain, and analyze the appropriate environmental information can be a huge challenge for organizations.

The Circular Economy (CE) is a field related to the TBL. Although also very relevant, the concept of the CE differs from the TBL because the CE is a system-level solution framework focused on addressing resource issues (e.g., pollution, waste, biodiversity loss, and climate change) [39]. Many advancements have been achieved through the CE. For example, a conceptual data model to standardize and structure data in circular manufacturers has been proposed [40]; an SLR was made to identify the relevant information and data required to support manufacturers transitioning to the CE [41]; and an SLR was made that integrated the circular supply chain, the CE, and I4.0 [42]. The latter [42] considered TBL among the dimensions of analysis. The authors identified 19 articles encompassing TBL (from a sample of 198 papers). They concluded that 60% of the papers that considered TBL neglected the social aspect. The authors also emphasized that this result agrees with a previous paper [43]. They emphasized that the social aspect is the least investigated in supply chains and that, even when it is, the analysis is typically skimpy. In summary, the concept of the CE is focused on resources (more emphasis on economic and environmental aspects), while TBL encompasses and equally emphasizes the social aspect.

After the COVID-19 pandemic, the TBL performance became even more relevant [44]. Preliminary investigations showed that SMEs were the most affected by the pandemic and faced more difficulties from interrupting their operations. This may have caused long-term liquidity problems and affected the maintenance of jobs [45]. SMEs account for 50% of employment globally [4]. Specifically, SMEs are responsible for most female jobs [46]. During the SLR filtering process, we classified the papers based on how they approached sustainability's three aspects (TBL).

Thus, our research contributes to the United Nations (UN)'s sustainable development goals (SDGs) 1 (no poverty), 5 (gender equality), 8 (decent work and economic growth), 9 (industry, innovation, and infrastructure), 10 (reduced inequality), and 12 (responsible consumption and production).

3. Research Design

3.1. Manual SLR

This study adopts the systematic literature review (SLR) method to describe and analyze SMEs' current tools for performance measurements, specifically the mathematical tools (as well as their respective dimensions and variables) for measuring the process of DT simultaneously with the three aspects of sustainability (environmental, social, and economic). The SLR method has been considered a replicable, scientific, and transparent literature review approach that minimizes bias. It is an iterative process for identifying the extant literature about some research topics [47].

The five main steps were based on the recommendations by Tranfield et al. (2003) [47] and Moher et al. (2009) [48]. These steps were (i) Research question formulation; (ii) Search strategy; (iii) Selection and Evaluation of relevant studies; (iv) Analysis and synthesis of results; (v) Reporting the review. The full SLR process is summarized in Figure 1, which illustrates the review protocol to provide transparency to the process. In the first step, we defined the main research question (RQ) and the two secondary research questions (SRQs). They were presented in the Introduction section. Further details about the review protocol can be found in [31].

Scopus and Web of Science were chosen because they have international and wide coverage, and they are regularly updated [49]. Further justification for the dataset choice can be found in [31]. We developed strings to cover a few keywords related to the constructs from each RQ. Some keywords were identified among the three themes after a preliminary review of DT, SMEs, performance measurement, and sustainability. The final search strings were defined after running tests to ensure reliable searches. The strings can be found in [31].

Once the search strings were defined, we established the criteria for the inclusion and exclusion of papers. The search was conducted in September 2021 and repeated in December 2021. After the search, the results were exported and converted as files from the StArt software. The StArt was used during Selection (Filter 1) and Extraction (Filter 2) (Figure 1).

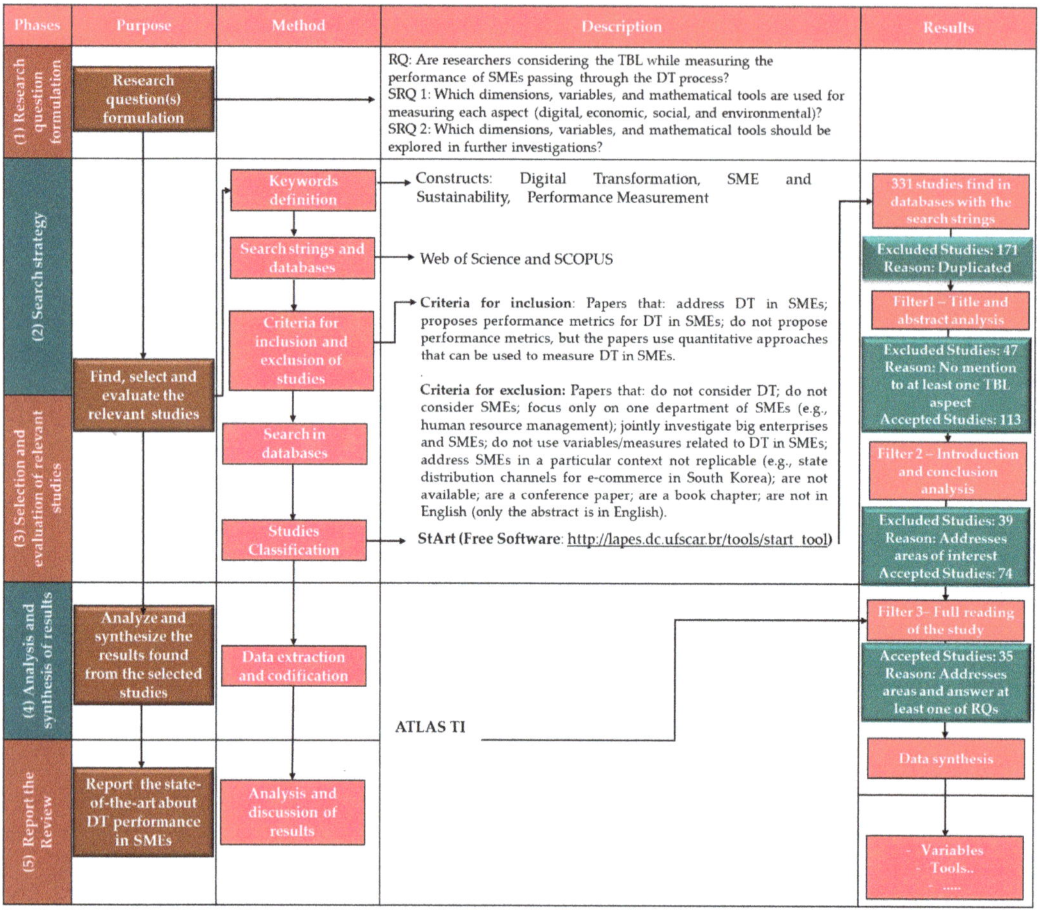

Figure 1. SLR Process.

The selection and evaluation steps consisted of three filters. The first filter consisted of reading the title and abstract of each paper found in the search, eliminating duplicated papers and papers that did not meet the inclusion criteria (Figure 1). Two independent reviewers executed all filters, double-checking the filter results until they agreed.

In the extraction step (Filter 2), the reviewers read the introduction and conclusion of the selected papers and started qualifying them. This filter was performed because, in Filter 1, the retained papers could have doubtful relevance to the interest area. If, after reading the introduction and conclusion, a paper proved not to be pertinent, it was excluded with the justification registered in the StArt software. In this step, the papers were also classified by answering 11 questions, as stated in the SLR Protocol [31]. In this filter, the questions were used to classify the selected studies and evaluate their importance for the research. Further information can be found in [31].

Data were coded in the content analysis step following the basic requirements proposed by Barnes et al. (2022) [50,51]. Specifically, a codebook was written, establishing a code for each aspect of TBL and a code for DT. Both reviewers independently read the papers and coded the text content. To determine whether a paper considers a certain aspect of TBL, it was necessary that both reviewers read it and had a consensus about coding it with the same aspects of TBL, and there were subcodes for variables and dimensions. The software ATLAS. TI was used to execute the code procedures. This software supports the organization of ideas and concepts, and it helps to cluster the tools, dimensions, and variables for measuring the impacts of the TBL and DT on the performance of SMEs.

After the papers were coded, we answered the research questions and identified the tools, dimensions, and variables for measuring the impacts of DT on the performance of SMEs.

3.2. LDA-Based SLR

Conducting a manual SLR is a time-consuming, laborious, and costly effort. Consequently, automated techniques in SLRs have increased [52]. Dinter et al. (2021) [52] analyzed 41 papers that propose an automated or semi-automated approach for SLRs. The authors concluded that selecting primary papers is the most automated stage in SLRs. However, topic modeling is still rarely applied to select primary papers in an exploratory literature review [13,15].

The LDA is a state-of-the-art [13] and the most used [53–55] topic modeling technique [53–55]. The LDA is a probabilistic method that extracts topics from a collection of papers. A topic is a distribution of terms (words) over a fixed vocabulary. The semantics and meaning of the sentences are not evaluated. However, LDA analyzes the terms in each paper and calculates the joint probability distribution between the observed (terms in the paper) and the unobserved (the hidden structure of topics) [13].

It is essential to highlight that, in general, topic modeling works best with large volumes of text data. However, the minimum number of papers required for applying topic modeling in an SLR can vary depending on the research question, the literature's nature, and the review's goals [13]. Some authors have successfully used topic modeling with relatively small datasets, depending on the research question and the scope of the review. For example, Saha (2021) [15] applied Latent Dirichlet Allocation (LDA) to 948 papers on game theory in the management literature, Asmussen and Møller (2019) [13] applied LDA to 650 papers on lean manufacturing in the management literature, Nguyen et al. (2023) [56] applied LDA to 108 papers on blockchains applications in supply chain management, and Queiroz et al. (2022) [14] applied LDA to 92 papers on DT and lean philosophy applied to SMEs.

Determining the number of topics (k) is a crucial parameter to guarantee the quality of the LDA results. Once this LDA approach is unsupervised, we do not know the relationship between the papers before the model is executed. Calculating the perplexity is normally used as cross-validation to estimate an adequate number of topics [13]. Additionally, it can be used as an indicator that the number of papers in the dataset reached the minimum threshold. Perplexity is a metric used to evaluate language models, where a low score indicates a better generalization. Lowering the perplexity is equivalent to maximizing the overall probability of papers being on a topic. Choosing the right number of topics is the art of balancing the right number while keeping the perplexity at the lowest possible level [13].

For example, Figure 2 shows the perplexity of LDA models applied to all searched papers. The line graph shows how average perplexity decreases with the increasing number of topics. In other words, the model fits better as the number of topics increases. As an illustration, a fit with $k = 15$ topics may be interesting because it is in a region where perplexity is decreasing, and it is the configuration with maximum discriminatory power, concentrating 48 of the 91 (52%) manually accepted papers on 7 topics (1,3, 7, 8, 10, 12, and 14).

Like Queiroz et al. (2022) [14], we adopted the framework proposed by Asmussen and Møller (2019) [13], followed by an onion approach. Due to the limited number of papers, it was observed that in some cases, one topic had many more papers than others. Furthermore, a topic with a high concentration of papers presents a list of words that encompass many different themes, i.e., an obstacle to adequate paper segregation. In such cases, the onion approach refers to the solution of running the LDA only with the papers on the concentrated topic and repeating this loop until the result does not present a concentration of papers on any topic. The first onion layer had 91 papers. The second onion layer had 73 papers. In both cases, the best fit was $k = 7$. The third onion layer had 64 papers. The fourth onion layer had 37 papers. In both cases, the best fit was $k = 5$.

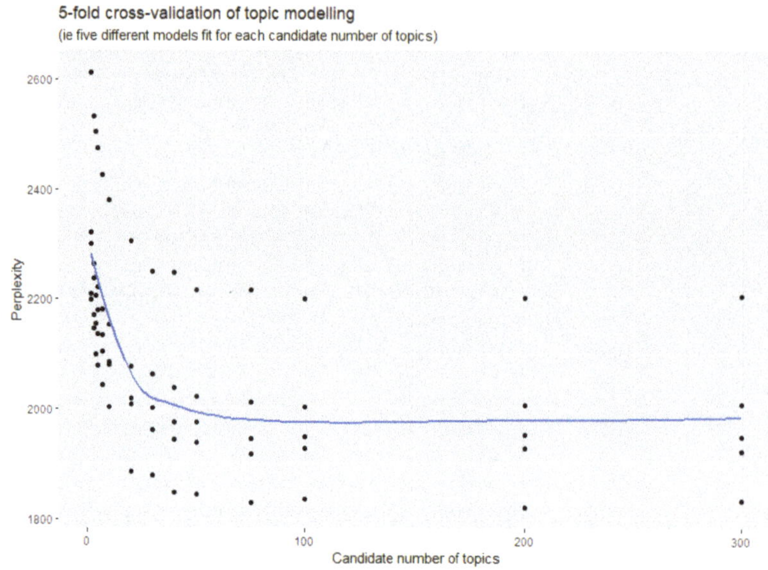

Figure 2. Perplexity of the LDA model (considering all papers).

We investigated the possibility of applying the framework of Asmussen and Møller (2019) [13] as a tool for validating the manual SLR and providing new insights. The papers that LDA used to build the topics in each onion layer were compared to the papers manually selected. LDA-based selection rejected 30 papers in Filter 1 (instead of 47 from the manual selection, i.e., 64% of the manual SLR rejected sample). Additionally, for technical issues already pointed out in the literature [13,57] and to be overpassed in future methodological developments, the proposed LDA algorithm did not read all the papers accepted by Filter 2 and 3. The algorithm read 91 of the 113 papers (81%). The algorithm used 58 of the 74 accepted by Filter 2 (78%) and 33 of the 35 accepted by Filter 3 (94%). We assumed that this level was already enough for a valid comparison and interpretation of results from both methods (manual SLR and LDA-based SLR).

4. Manual SLR Results

4.1. Research Profiling

After the selection and extraction steps, we manually profiled 35 papers. As shown in Figure 3, although there were no time restrictions, the oldest paper was published in 2016 [5]. Almost 80% (27 papers) were published in the last two years (2020–2021), and one paper was accepted to be published in 2022. This highlights the constantly growing scholarly interest in this field of study. The 35 papers were published in 27 journals, indicating that there is still no major consolidated source about the theme. Most of these are leading journals once they are indexed in Journal Citation Report (JCR)-listed journals and the Chartered Association of Business Schools (CABS) journals' ranking list. The following journals published more than one paper: *Journal of Business Research* (3), *Applied Sciences* (2), *Technological Forecasting and Social Change* (2), *Competitiveness Review: An International Business Journal* (2), and *Sustainability* (2).

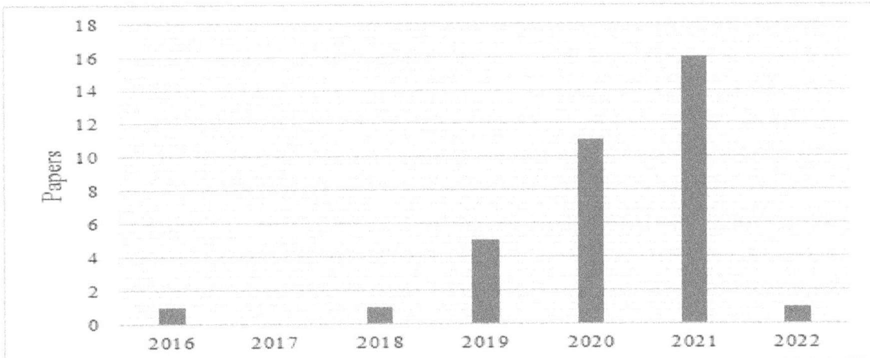

Figure 3. Year-wise publication frequency.

4.2. Data Collection Approaches

Table 1 summarizes the 35 accepted papers. As can be seen, 23 articles developed and applied surveys for collecting (primary) data. However, these surveys differ in sample size, the number of respondents per SME, and the definition of SME adopted by the authors. This indicates that the lack of consensus and standardization hardens the comparison among results.

As shown in Table 2, the following variables were the most used for characterizing SMEs in surveys: the size (number of employees), the business industry, the age, and the ownership. As discussed in Section 2.1, sizes defined by the number of employees are usually different in terms of numbers. Furthermore, in the DT process, remote and hybrid work replaces activity-based positions [10]. This implies that enterprises with fewer employees can be large enterprises (and not necessarily SMEs). In this regard, a standardized and measurable definition of a "digital SME" represents a contribution to further investigations.

In the case of ownership, the authors normally discriminate between family-owned SMEs and SMEs controlled by company groups. However, in China, it is also possible to have estate SMEs, private-public SMEs, and/or SMEs listed in stock markets [71]. These observations and the lack of a commonly accepted definition of SMEs make us question, for example, if a non-governmental organization could be considered an SME. If so, under which conditions? Additionally, in the case of family SMEs' investigation, the gender of the owner proved to be a significant variable for determining the DT performance of the SMEs. Men tend to be more engaged in DT than women [65]. There is a lack of investigation into whether the gender of the managers of non-family-owned SMEs also affects DT performance. Besides gender, the less frequented variables of characterization are sales (or income from sales), geographical location, and the marital status of the owners [81].

Furthermore, six articles worked with qualitative approaches (interviews and case studies). Seven articles worked with secondary data; these were mostly literature reviews (6), except for [1], which quantitatively investigated a dataset from the Centre of European Economic Research (ZEW)'s 2015 ICT survey. Regular surveys with standardized questionnaires enable cross-temporal studies, which are pointed to as a gap in the theme [72,82]. Additionally, one article collected data on scraping SMEs' websites [80]. Scraping may become a promising data collection approach for future investigations a having a website becomes a mandatory requirement for SMEs in most industries.

Table 1. Summary of the accepted articles, the considered aspects (Digital Transformation—DT, Economic—Econ., Social—Soc., Environmental—Env.), used methodological approaches, and procedures for data collection.

Author/Year	DT	Econ.	Soc.	Env.	Methodological Approach	Data Collection
AlMujaini et al. (2021) [58]	X	X	X		Two-step approach: Structural Equation Modelling (SEM) and Interviews	Survey, 354 respondents (owners or managers of SMEs)
AlMulhim (2021) [59]	X	X			Partial Least Square Structural Equation Modelling (PLS-SEM)	Survey, 460 respondents (of 150 SMEs)
Apostolov and Coco (2021) [60]	X	X	X	X	Case Studies	Visits and interviews (8 SMEs)
Ardito et al. (2021) [61]	X	X	X	X	Tobit regression	Survey, 369 respondents
Bouwman et al. (2019) [62]	X	X			Partial Least Square Structural Equation Modelling (PLS-SEM) and fuzzy-set Qualitative Comparative Analysis (fsQCA)	Survey, 321 respondents
Cenamor et al. (2019) [63]	X	X	X		Partial Least Square Structural Equation Modelling (PLS-SEM)	Survey, 129 respondents
Chen et al. (2016) [5]	X	X			Hierarchical Regression	In-depth interviews (8 SMEs) and survey, 46 respondents (owners or high executives of SMEs)
Chonsawat and Sopadang (2020) [64]	X				Literature Review and Case Studies	Secondary data and qualitative primary data
Denicolai et al. (2021) [9]	X	X		X	Tobit Regression	Survey, balanced sample of 438 respondents (one respondent per SME)
Dutot et al. (2021) [65]	X	X	X		Partial Least Square Structural Equation Modelling (PLS-SEM)	Survey, 197 respondents
Dutta et al. (2020) [66]	X				Literature Review and Maturity Model Assessment (descriptive statistics)	Secondary data and survey of 250 SMEs
Eller et al. (2020) [67]	X	X	X		Hierarchical Regression	Survey, 193 respondents
Fang et al. (2020) [68]	X	X	X		Acquisition–Importance Analysis (AIA) and Network Relation Map (NRM) based on Decision-Making Trial and Evaluation Laboratory (DEMATEL) Technique	Interviews with 18 experts
Gamache et al. (2020) [69]	X	X	X		Literature Review and Case Studies	Secondary data and two loops of interviews (15 SMEs and 21 SMEs)
Gruenbichler et al. (2021) [70]	X	X	X		Descriptive Statistics and Case Studies	Survey (123 respondents) and interviews
Guo et al. (2020) [71]	X	X	X		Multilinear Regression	Survey, 518 respondents (one respondent per SME)
Hassan et al. (2021) [1]	X	X	X		Bivariate Probit Regression	Secondary data, 2404 SMEs
Holopainen et al. (2020) [72]	X	X	X		Multilinear Regression	Survey, 668 respondents (one respondent per SME)

Table 1. *Cont.*

Author/Year	DT	Econ.	Soc.	Env.	Methodological Approach	Data Collection
Jelovac et al. (2021) [73]	X	X	X	X	Critical Review	Secondary data
Joensuu-Salo et al. (2018) [74]	X	X			Linear Regression	Survey, 101 respondents (one respondent per SME)
Jun et al. (2021) [75]	X	X			Correlation, Multilinear regressions, and Structural Equation Modeling (SEM)	Survey, 647 respondents (managers of SMEs)
Kamišalić et al. (2020) [76]	X	X	X		Literature Review and Case Studies	Secondary data
Kmecová et al. (2021) [77]	X	X	X		Linear Regression	Survey, 610 respondents
Kulathunga et al. (2020) [78]	X	X			Structural Equation Modelling (SEM)	Survey, 319 respondents (chief financial officers of SMEs)
Kumar et al. (2021) [79]	X		X		Literature Review, Decision-Making Trial and Evaluation Laboratory (DEMATEL)	Secondary data, experts' opinions (50 experts at the first round and 35 experts at the second round)
Lányi et al. (2021) [80]	X				Analysis of Variance (ANOVA)	Content analysis of websites of 958 SMEs
Mubarak et al. (2019) [81]	X		X		Linear Regression	Survey, 237 respondents
Nasiri et al. (2020) [82]	X		X		Structural Equation Modelling (SEM)	Survey, 280 respondents (one respondent per SME)
North et al. (2019) [83]	X	X	X		Literature Review and Case Studies	Visits and interviews (52 SMEs)
Okfalisa et al. (2021) [84]	X	X	X		Fuzzy Analytical Hierarchy Process (F-AHP) method	Experts' opinions
Ponis and Lada (2021) [85]	X	X			Descriptive Statistics	Survey, 60 respondents
Rozak et al. (2021) [86]	X	X	X		Partial Least Square Structural Equation Modelling (PLS-SEM)	Survey, 239 respondents
Troise et al. (2022) [8]	X	X	X		Partial Least Square Structural Equation Modelling (PLS-SEM)	Survey, 204 respondents
Ukko et al. (2019) [87]	X	X	X	X	Analysis of Variance (ANOVA) and Multilinear Regression	Survey, 280 respondents (one respondent per SME)
Viale Pereira et al. (2020) [10]	X	X	X		Cluster Analysis	Experts' opinion

Regarding survey sampling, five approaches were identified in the literature: random sample choice [71,82], balanced industrial representation of the SMEs in the sample [9], focus on the SMEs from a unique industry [72], focus on SMEs considered as innovative [8,58,60–62,72,75], and focus on the SMEs considered as entrepreneurial [63]. Regarding the survey respondents, as shown in Table 1, most of the research surveyed only one person per SME (usually the owner or a senior manager) [32,59,71,74]. However, one paper surveyed multiple employees from the same SME [59], and other papers did not mention how many respondents were surveyed [61,62].

Table 2. SMEs' Characterization variables.

Source	Variables
[61]	SME Size (Number of Employees), SME Age, Industry Sector, Private Ownership, Family Ownership
[63]	SME Age, SME Size (Number of Employees)
[9]	SME Size (Number of Employees), SME Age, Family Business, Company Group
[65]	SME Size (Number of Employees), Annual Revenues
[67]	SME Size (Number of Employees), Family Ownership, SME Age
[1]	SME Size (Number of Employees), Industry
[72]	EBIT to Sales, Sales, Operating Margin, Total Assets, Liabilities, SME Age, Manager's Education, SME Size (Number of Employees), ROA, Manager's Ownership, Gender
[71]	Business Model, SME Age, Ownership, Region, Listed/unlisted in Stock Market, Industry Sector
[77]	SME Size (Number of Employees), Ownership
[78]	SME Size (Number of Employees), SME Age
[86]	Business Field, SME Size (Number of Employees), Annual Sales
[8]	Industry Sector, Geographical Location, SME Size (Number of Employees)
[81]	Position of Respondent, Experience of Respondent, Marital Status of Respondent, Gender of Respondent, Age of Respondent, Income of Respondent
[85]	Determine the Primary Fashion Segment (Apparel/Footwear/Beauty/Accessories) of the Company, Age, Turnover and Headcount; Assess the Position of the Respondent Within the Fashion Organization (Department—Management Level)

4.3. Methodological Approaches

As seen in Table 1, nine papers used mixed methodological approaches. These approaches were mostly (6) literature reviews, followed by one application of the reviewed concepts/performance indicators. They strengthened and legitimated the contribution of qualitative research for management knowledge [88]. However, the additional barriers to integrating qualitative and quantitative research have already been pointed out [89].

Two papers applied exclusively qualitative approaches: a paper with case studies analyzing evidence of the challenges SMEs face while redesigning their business model due to DT [60] and a critical review aiming to design a framework for SMEs building digital trust [73]. The remaining papers (24, 69%) applied exclusively quantitative approaches. Hence, quantitative tools and approaches are predominant in the field.

Among the quantitative tools, structural modeling (SEM) (10) and econometrics (9) were the most used. Without considering DT as one of the SMEs' environmental performance factors, the literature had already discussed the application of SEM as a performance tool for SMEs [90]. In this regard, the results indicated that SEM and regressions are consolidated tools in the field. On the other hand, cluster analysis, analysis of variance (ANOVA), Decision-Making Trial and Evaluation Laboratory (DEMATEL), and integrations with fuzzy techniques were used more than once. This indicates these tools are emerging in the field. No paper used an approach directly associated with performance measurements, such as

Stochastic Frontier Analysis (SFA) and Data Envelopment Analysis (DEA). This suggests that these tools are still not explored in the field and remain an open field for possibilities.

4.4. Variables and Dimension

Regarding the aspects, as seen in Table 1, 33 papers directly considered the DT aspect, 28 considered the economic aspect, 22 considered the social aspect, and 5 papers considered the environmental aspect. Only four papers [60,61,73,87] considered the four aspects simultaneously. Hence, there is a gap in the studies considering the TBL in this context.

4.4.1. DT

Table 3 shows the identified variables and dimensions regarding DT. The dimensions were attributed by the authors of each paper. Only nine authors classified the used variables into dimensions. We identified 192 variables and 34 dimensions.

Table 3. DT variables and dimensions.

Source	Variables	Dimensions
[58]	Digital Transformation	-
[59]	Digital Transformation, Smart Technologies	-
[61]	Digital Orientation	-
[62]	Resources for Business Model Experimentation, Business Model Strategy Implementation Practices, Business Model Experimentation Practices	-
[63]	Platform Integration, Platform Reconfiguration	Platform Orientation
[5]	Portal Training, Portal Usage	Portal Usefulness
	Operational Friendliness, Industry Benchmark Information Bilingual Information	Portal Interface
	Portal Maintenance Service, B2B Function, Cloud Computing	Service-Orientation Portal Function
[64]	Business Model, Business Strategy, Digital Transformation, Leadership, Organizational Structure, Supply Chain Management	Organizational Resilience
	Infrastructure, Financial Resource and Investment, Standardization	Infrastructure System
	Logistics System, Collaborative Robot, Customized Product, Industrial Automation, Industrial Internet	Manufacturing System
	Cloud Manufacturing, Data Acquisition, Data Connected, Real Time Data	Data Transformation
	Big Data Analytics, Information System, Tracking System, Predictive Maintenance, Cybersecurity	Digital Technology
[65]	IT Governance, IT Strategy	-
[66]	Horizontal Integration, Vertical Integration, End-to-end Integration	System Integration
	Internet of Things, Big Data, Cloud Computing, Autonomous Robots, Simulation, Cyber Physical Systems, Augmented Reality, Additive Manufacturing	-
[67]	Information Technology, Digitalization, Digital Strategy	-
[69]	Operational Use of Data, Strategic Use of Data	Measurement System
	Ecosystem and Architecture, Mastery of Technologies, Cybersecurity, Intelligence, Autonomy, and Automation	Technology Management
	Collection of Data, Integrity and Quality of Data, Data Delivery	Data Management
	Customization, Engagement and Loyalty, Cocreation and Open Innovation, E-Commerce and SMAC	Customer Experience

Table 3. *Cont.*

Source	Variables	Dimensions
[70]	Suitable Performance Measurement (PM) System	-
[71]	Digital Artifact, Digital Platform, Digital Infrastructure, Digital Business Model, Digital Management Model	Overall Digitalization Degree
	Internal R&D, External Acquisition	Digitalization Method
	Big Data, AI, Mobile, Cloud Computing, IoT, Social, Platform Development	Digital Technology Adoption
	Rate of Online Business	Business Mode
	Digital Transformation	Long-term Crisis Responses
[1]	Cloud Computing, Social Media Network, Perceived Usefulness, Security and Privacy Concerns	-
[72]	Internet is the Source of Innovation	
[73]	High Performance Computing, Technological (Corporate Digital Responsibility)	-
[74]	Digitalization	-
[75]	Digital Platform Capability	-
[76]	Phoneline Connection, Fixed BB/NGA Connection, Fast BB Connection, Ultra-fast BB Connection	Fixed-line Broadband—Connectivity
	4G Connection, 5G Connection	Mobile Broadband—Connectivity
	Proprietary Website, E-marketing Activity, Social Media Presence	Online Presence
	Online Sales, E-commerce Turnover, Cross-border E-commerce, Digital/electronic Catalog, Online Communication with Customers, Customer Engagement in Product Customization	E-commerce—Online Presence
	B2B E-business Activity (Online Activity), B2G E-business Activity (Online Activity), E-banking (Online Activity), Online Purchases	Online Activity
	Intranet, Electronic records, Automatically Generated Invoices, Electronic Information Sharing	ICT Infrastructure
	Big Data, Cloud Services, Integrated or Specialized Systems or Tools, Business Intelligence or Knowledge Base, Decision Support Tool	Advanced Technologies—ICT Infrastructure
	Robots and 3D Printing, Automation, Product Identification Throughout the Supply Chain (Unique, Automated), Digital Supply Chain Management and Supplier Relationships	Production Technologies—ICT Infrastructure
	Security Policy, Data Protection Policy, Regulatory Quality, Assessment Effectiveness, Software or Hardware Upgrades	ICT Policy
	Computer or Mobile Device Use, Internet Use, E-mail or IM Use, Standard Application or Office Software Usage, Video Calls or Conferences	ICT Usage
[77]	Digitization of Analogue Data, Digitization of Biometric Data, Digital Interaction Platforms, Networking, Big Data Analytics, Rapid Analytics, Predictive Analytics, Use of Social Networks for the Recruitment of Employees	-
[78]	Technological Literacy	-
[79]	Security of Data, Efficient Data management system, Reliable and Affordable Big Data Analytic technologies, Egression of IoT-specific Operating Systems, Trust on IoT Systems, Customer and Supplier Relationship Management, Collaborations Between Heterogeneous IoT Systems	-
[80]	WebIXbin, Online Presence Category, Competitiveness Index, Competitiveness Index, Category	-
[81]	Big Data, Interoperability, Internet of Things, Cyber-Physical Systems	-

Table 3. *Cont.*

Source	Variables	Dimensions
[82]	Digital Transformation, Smart Technologies	-
[83]	Searching for Digitally Enabled Growth Opportunities, Understanding and Developing Digital Customer Needs, Sensing Technology Driven Opportunities, Use of External Sources Digital Innovation	Sensing Digitally Enabled Growth Potentials
	Digitally Enabled Growth Strategy, Digital Leadership, Digital Mindset	Developing a Digitally Enabled Growth Strategy and Mindset
	Digitally Enabled Business Models, Digital Market Presence, Digital Customer Experience, Agile Deployment of Digitalization Initiatives	Seizing Digitally Enabled Growth Potentials
	Digital Processes, Digital Technology and Security	Managing Resources for Digital Transformation
[84]	Technical Infrastructures	IT Perspectives
[10]	Smart Cities'/Regions' Rebounds, SMEs vs. Global ICT-based Company, Digital Divide, Digital Literacy	-
[85]	General Standpoint of the Respondent on DT in the Sector, Respondents' Opinion on DT in their Organization and Whether They Have Been Involved in a DT Process, Respondent's Opinion on Which Digital Technologies he/she Expects to Impact his/her Company's Digital Strategy in the Coming Years and to What Extent, Respondent's Opinion on Which are the Major Barriers to Implementing their Company's DT Strategy, Respondent's Opinion on the Importance of Specific Digital Practices/Initiatives, Respondents' Opinion on the Benefits their Company has Ripped from DT Initiatives, Respondent's Opinion on the Effectiveness of his/her Company's DT Strategy.	-
[86]	ICT Utilization, Social Media Engagement	
[8]	IT Infrastructure Flexibility, Application Digital Technology	
[87]	Managerial Capabilities, Operational Capabilities	

The dimensions were:

- Platform Orientation [63], for understanding the competitiveness of SMEs in platforms. This dimension is associated with external operations.
- Portal Usefulness, Portal Interface, and Service-Orientation Portal Function [5], for understanding the effects of DT on SMEs' performance. This is the oldest identified paper. The word "portal" can be understood as what was later named "platform". These dimensions are associated with external activities.
- Organizational Resilience, Infrastructure System, Manufacturing System, Data Transformation, and Digital Technology [64], for defining readiness indicators for SMEs' DT. The authors focused on the manufacturing industry; the adopted dimensions are associated with internal activities. Instead of "platform", the authors used the words "system", "technology", and "data" for the use of internal platforms.
- System Integration [66], for understanding the DT priorities of Indian SMEs in the manufacturing sector. Again, the dimension is associated with internal activities, and the word used is "system".
- Measurement System, Technology Management, Data Management, and Customer Experience [69], for understanding the DT priorities of Canadian SMEs in the manufacturing industry. Three dimensions are associated with internal activities and use the words "system", "technology", and "data". The "Customer Experience" dimension is associated with external activities, and it is linked to the measure "customization". There are no variables related to platforms for communicating with customers. The use of platforms is a general practice for SMEs in retailing but is not taken into consideration in the literature in the manufacturing industry.

- Overall Digitalization Degree, Digitalization Method, Digital Technology Adoption, Business Mode, and Long-term Crisis Responses [71], for investigating the response of the SMEs to the pandemic. These dimensions are associated with internal and external operations in a multi-industrial context.
- Fixed-line Broadband—Connectivity, Mobile Broadband—Connectivity, Online Presence, E-commerce—Online Presence, Online Activity, ICT Infrastructure, Advanced Technologies—ICT Infrastructure, Production Technologies—ICT Infrastructure, ICT Policy, and ICT Usage [76], for understanding the DT of fiber-based SME manufacturers in Europe. The dimensions are associated with internal and external activities as well as a strategy (ICT Policy). However, the dimensions associated with external activities are not associated with the use of a shared selling platform (common for retailing SMEs) but with the "Proprietary Website" and "B2B E-business Activity (Online Activity)". Although common for SMEs in the manufacturing industry, these investments are unusual and unaffordable for the DT of SMEs in other industries (such as service and retailing).
- Sensing Digitally Enabled Growth Potentials, Seizing Digitally Enabled Growth Potentials, and Managing Resources for Digital Transformation [83], for understanding the maturity level of SMEs regarding DT.
- IT Perspectives [84], for measuring the effects of factors influencing the readiness of SMEs towards DT. This dimension is linked to the variable "Technical Infrastructure", which may be associated with internal activities. As it is a multi-industrial perspective, the term "infrastructure" can represent investments in generally used technologies or advanced ones (usually prohibitive for SMEs in some industries).

In summary, we conclude that the literature explores dimensions and variables for the DT in this context. However, there are some considerations that should be highlighted. First, most papers focused on measuring performance at a micro-level (the SMEs' perspective only). There is a gap in measurements considering a macro perspective (e.g., market, legislation, etc.). Second, the variables and dimensions can be divided into those focused on the SME's internal activities and those focused on the SME's external activities (e.g., communication with customers and suppliers). In general terms, papers that investigate manufacturing tend to consider the DT internally in the SMEs, while papers that investigate other industries tend to consider the DT as an enabler for the relationships and communications external to the SMEs. It is critical to highlight here that the DT measures (e.g., digital platform metrics) have an integrative potential, connecting internal and external activities.

We observed that, although manufacturing is a decreasing sector with a smaller representation in the employment rate worldwide [40], it tends to be more investigated in the literature. Additionally, SMEs from the manufacturing industry may be more interested in and have more resources for investing in the DT process than SMEs from other industries. The manufacturing industry tends to associate DT with the words "system", "data", "technology", and "ICT", while other industries tend to use the words "platform", "portal", "website", and "marketplace". This may be a consequence of the fact that (i) there is heterogeneity among SMEs, and (ii) authors work with different definitions of SME.

As well as a standardized definition of SME, there is a lack of a clear definition of "platform", and its differentiation from "portal", "website", "marketplace", and "social media". Finally, there is a lack of variables and dimensions that consider cybersecurity and data protection.

4.4.2. Economic

Table 4 shows the identified economic variables and dimensions. We identified 122 variables and 16 dimensions. In other words, there are 36% fewer economic variables than DT ones. This indicates that economic aspects are somewhat less explored than DT in the context of SMEs' performance.

Table 4. Economic variables and dimensions.

Source	Variables	Dimensions
[58]	Organizational Innovativeness, Organizational Performance	-
[59]	Firm Performance	-
[61]	Product Innovation, Process Innovation, Market Orientation Trade, Market Orientation Business, Market Internationalization, Cost Strategy, Collaboration, Obstacles	-
[62]	Overall Performance	-
[63]	Entrepreneurial SMEs' Performance, International, Orientation, Exploration Orientation, Exploitation Orientation	-
[5]	Finance, Customer, Process, Learning	Organizational Performance—Balanced Score Card
[9]	Export Intensity, Investment in Innovation	-
[65]	Organizational Performance	-
[67]	Financial Performance	-
[68]	Product Innovation Capability, R&D and Manufacturing Capability, Marketing Capability, Branding Capability	Professional Competence
	Stable Cash Flow, Merge and Acquisition, International Operation Experience	Operation Management
	Fundraising, Obtain Key Resource, Obtain Market Information	Critical Resources
	Approve by International Certification, Approve by Local Sales Certification, Approve by Health Insurance, Apply Intellectual Property	Regulatory System
	Build up Reputation, Connect Channels, Influence by National Image, Understand Different Culture	Market Expansion
[69]	Vision and Strategy, Technological Watch, New Business Models, Commitment and Exemplarity	Leadership
	Change Management, Agile Manufacturing and Innovation, Investment and Available Resources, Lean and Continuous Improvement	Culture and Organization
[70]	Define and/or Visualize Business Objectives and Strategies, Allocate Resources for Implementation (Within Budget), Define Key Objectives in Consideration of Vision/Objectives, Define KPI for All Areas, Modify Existing Management System (Processes and Org. Structure) and Incentive System, Implement Measures and Use KPI, Initiate Change Process (Change Management), Set Up a Secondary Org. for Project Implementation, Identify Risks for Each Activity During the Implementation, Review Objectives, KPIs, and Measures, Evaluate Project Implementation	-
[71]	Production Recovery	Short-term Crisis Responses
	Strategic Change	Long-term Crisis Responses
	Cost Control Status, Cash Flow Status, Revenue Status in the First Quarter, Predicted Performance	Performance
[1]	Implementation Cost, Innovativeness, Productivity, Exporter, End Customer	-

Table 4. *Cont.*

Source	Variables	Dimensions
[72]	EBIT to sales, ROA, Total Assets, Liabilities to Assets	-
	Desire to Growth, Sales Change in Five Years, Sum Variable Measuring Organization Structure and Cooperation, Sum Variable Measuring Organization Structure, Family's Share (%) and Cooperation as Projects, Sum Variable Measuring Size Measured as Growth in Sales, Market Share and Profitability, Sum Variable Measuring Environment, Cooperation with Resources and Subcontracting, Sum Variable Measuring Environment, Customers with Voucher or with Agreement, Sum Variable Measuring Strategy to Grow with Boost sales, New Products and Expansion Internationally, Sum Variable Measuring Strategy to Meet the Competition by New Investments and Better Quality, Sum Variable Measuring Strategy to Grow with Investments in New Customers and New Products, Business Plan Exists, Bank Financing	
[73]	Economic	Corporate Digital Responsibility
[74]	Firm Performance, Market Orientation, Marketing Capability	-
[75]	Improvisational Capability, Organizational Readiness, Innovation Performance	-
[77]	R&D Department, ICT Investment in R&D, Patents or Trademarks, In-house Innovation Capacity, Innovative Collaboration	R&D Infrastructure
[78]	Financial Literacy, Enterprise Risk Management, SME Performance, SME Sector	-
[83]	Digital Investments	Managing Resources for Digital Transformation
[84]	Financial Resources	IT Perspectives
	Business Activities, Transaction, Marketing, Management, Micro-Environment, Macro-Environment	Economical Perspective
[10]	Corruption, Economic Value of Data	-
[86]	Organizational Agility, SMEs Performance	-
[8]	Firm Innovativeness, Coupling, Organizational Agility, Product Innovation, Process Innovation, Financial Performance	-

The identified dimensions were:

- Organizational Performance—Balanced Score Card (BSC) [5], for understanding the effects of DT on SMEs' performance based on BSC.
- Professional Competence, Operation Management, Critical Resources, Regulatory System, and Market Expansion [68], for establishing DT strategies for Med-Tech SMEs.
- Leadership and Culture and Organization [69], for understanding the DT priorities of Canadian SMEs in the manufacturing industry.
- Short-term Crisis Responses, Long-term Crisis Responses, and Performance [71], for investigating the response of the SMEs to the pandemic.
- Corporate Digital Responsibility [73], for building digital trust while implementing high-performance computing (HPC) in SMEs. The variable linked to this dimension is "Economic" and related to operations.
- R&D Infrastructure [77], for identifying and evaluating indicators of DT in SMEs and determining critical factors of DT.
- Managing Resources for DT [83], for understanding the maturity level of SMEs regarding DT.
- IT Perspectives and Economical Perspectives [84], for measuring the effects of factors influencing the readiness of SMEs towards DT.

Economic dimensions are linked with variables of more than one aspect of interest in our paper. For example, the social aspect of the TBL is also professional competence, leadership, culture, and digital responsibility. The dimensions of R&D Infrastructure, Managing Resources for DT, and IT Perspectives and Economic Perspectives are linked to economic and DT variables. However, any economic variable that clearly connects micro- and macro levels (or internal and external activities in the SME) was not observed.

In summary, the economic dimensions tended to be more strategic than operational. The economic dimensions brought to light some environmental factors (macro-level) that serve as enablers and driving forces, such as regulatory systems, market orientation, and internationalization. It also showed that the pandemic's short- and long-term impacts are a focus of interest. In this regard, future studies should explore more operational variables of the economic aspect. Furthermore, they should understand the dependency of the economic aspect related to the others and how it is affected by the environmental aspect.

4.4.3. Social

Table 5 shows the identified social variables and dimensions. We identified 74 variables and 11 dimensions. In other words, there are 62% fewer social variables than DT ones. This indicates that social aspects are more neglected. The same phenomenon was registered in the supply chain literature, considering the circular economy [43].

Table 5. Social variables and dimensions.

Source	Variables	Dimensions
[58]	Adaptative Learning, Culture, Experimental Learning, People and Networking	-
[61]	Sustainability Workforce, Sustainability Community, Sustainability Human Rights	-
[63]	Internal Communication, Coordination, Relationship Skill, Partner Knowledge	Network Orientation
[65]	Family Harmony, Gender of Respondent, Age of Respondent, Family Generation of Respondent	-
[67]	Employee Skills	-
[68]	Motivate All Teams to Transform	Operation Management
	Train Multifunction Team Member	Critical Resources
[69]	Acquisition and Development of Skills, External Openness and Collaboration, Internal Communication	Culture and Organization
[70]	Sensitize Shareholders and/or Top Management to the Need for Performance Measurement (PM), Identify Required Knowledge for Performance Measurement (PM) (Knowledge Management), Provide Employees with Information (Communication Management), Raise Acceptance Among Employees and Review the Change Management Process, Evaluate Communication Within the Company	-
[71]	Employee Protection (Short-term Crisis Responses), Donation (Short-term Crisis Responses)	-
[1]	Skilled Labor	-

Table 5. *Cont.*

Source	Variables	Dimensions
[72]	Manager's Education, Sum Variable Measuring Manager's Age and Experience in Industry (Social), Sum Variable Measuring Manager in Practical Work (%), Sum Variable Measuring Environment as Customer and Educational Institutions, Sum Variable Measuring Environment, Cooperation with Other Companies, Educational Institutions, and R&D Institution, Sum Variable Measuring Culture, Competition Require Lower Quality in Service or Problems to Find Qualified Employees, Sum Variable Measuring Culture, Service Innovations Internal or External From the Company, Sum Variable Measuring Culture, Firm's Age and International Employees.	-
[73]	Moral, Socio-Cultural	Corporate Digital Responsibility
	Integrity, Credibility, Security, Reliability, Transparency	Digital Trust
	Responsible Corporate Digital Governance	-
[76]	ICT Department, Employment of STEM Graduates, Employment of Business Specialists, Telework	Human Resources
	ICT Training, Self-learning, Expertise Reuse	Employee Skills—Human Resources
[77]	GDPR (Employee Protection)	-
[78]	Financial Literacy	-
[79]	Availability of In-house Trained Manpower, Fear of Unemployment, Top Management Allegiance	-
[82]	Relationship Performance (Assesses Internal Collaboration Over the Last Three years. Assesses External Collaboration Over the Last Three Years.)	-
[83]	Digitally Empowered Employees	Developing a Digitally Enabled Growth Strategy and Mindset
	Digital Skills and Learning	Managing Resources for Digital Transformation
[84]	Education, Culture	IT Perspectives
[10]	Legal Systems, Biases Due to Digital Data, Autonomous Decision-making Acceptance, Trust in Unknown Digital Information, Ethical Dilemmas, Social Media's Democracy Threat, Vulnerable Group, Governmental Capacity	-
[86]	Digital Skill	-
[8]	Relational Capability	-

The identified dimensions were:

- Network Orientation [63], for understanding the competitiveness of SMEs in platforms.
- Operation Management and Critical Resources [68], for establishing DT strategies for Med-Tech SMEs.
- Culture and Organization [69], for understanding the DT priorities of Canadian SMEs in the manufacturing sector.
- Corporate Digital Responsibility and Digital Trust [73], for building digital trust while implementing high-performance computing (HPC) in SMEs.
- Human Resources and Employee Skills—Human Resources [76], for understanding DT of fiber-based SME manufacturers in Europe.
- Developing a Digitally Enabled Growth Strategy and Mindset and Managing Resources for Digital Transformation [83], to understand SMEs' maturity level regarding DT.
- IT Perspectives [84], for measuring the effects of factors influencing the readiness of SMEs towards DT.

In summary, the observed social variables and dimensions were mostly strategic and not operational. They relate to (and sometimes depend on) DT and economic variables and dimensions. It is worth noting that no paper considered work satisfaction. Moreover, no paper considered the salary of employees. Finally, no paper considered that SMEs might be hiring employees from other countries to work remotely. The only paper that considered job security during the DT was focused on the response to the pandemic crisis and not on a long-term response to job impacts caused by the DT as a historical process. Hence, these topics stand out as future directions for research.

4.4.4. Environmental

Table 6 shows the identified environmental variables and dimensions. We identified six variables and two dimensions. There are 97% fewer environmental variables than DT ones. Hence, the environmental aspect is the most neglected.

Table 6. Environmental variables and dimensions.

Source	Variables	Dimensions
[61]	Environmental Orientation Sustainability Market	-
[60]	Environmental Sustainability Readiness	-
[73]	Environmental	Corporate Digital Responsibility
	Sustainability	Digital Trust
[87]	Sustainability Strategy	-

The identified environmental variables were:

- Environmental Orientation and Sustainability Market [61], for understanding the relationship between digital and environmental orientations to enhance innovation outcomes. Innovations are assumed to be mandatorily related to Digital Transformation.
- Environmental Sustainability Readiness [60], for investigating the impact of AI on the international performance of SMEs and investigating how the relationship between internationalization and DT affects sustainability.
- Environmental linked to the dimension of Corporate Digital Responsibility and Sustainability linked to the dimension of Digital Trust [73], for building digital trust while implementing high-performance computing (HPC) in SMEs.
- Sustainability Strategy [87], for understanding the role of sustainability in the relation between digital business strategy and financial performance.

In summary, all papers adopted strategic environmental variables and dimensions. There is a lack of operational and tactical perspectives. Papers that considered the environmental aspect were focused on sustainability. There is a gap in considering the environment as one of the aspects of any SME's performance without the need for a specific focus on sustainability.

5. LDA-Based SLR

5.1. *Topic Modeling for Papers' Initial Selection*

Our first purpose was to investigate if the LDA could discriminate against the 30 papers initially rejected by the SLR (Filter 1), serving as support for the SLR validation. We applied the LDA for all papers of the sample (121). We investigated different numbers of topics (k = 5, 7, 10, 15, 20). Then, we compared the number of papers approved by the SLR that was allocated to each topic for each k. As can be seen in Table 7, k = 15 represents the configuration with maximum discriminatory power and with a decreasing perplexity level (Figure 2), concentrating 48 of the 91 (52%) initially accepted papers on 7 topics. We consider this evidence that the human subjectivity of selecting papers was constrained enough by the SLR methodology. Thus, the results are comparable with the non-human method. Both methods are not interchangeable but serve as mutual support.

Table 7. Distribution of the topics with all papers.

K	5	7	10	15	20
Topics with 100% of accepted papers	T4	-	T5	T1, T3, T7, T8, T10, T12, T14	T1, T5, T7, T8, T11, T12, T16, T17, T18, T19
Number of papers	24	-	5	48	34

Our second purpose was to investigate the possibilities of LDA for providing new insights into the SLR interpretation. Table 8 provides the topics that are 100% composed of accepted papers. The terms that are exclusive to these topics are in green. Thirty terms only appear in these topics. Among them, we highlight the terms "covid", "pandem", "crisis", "respons", "measur", "impact", all in T10. This suggests that interest in the theme increased due to the pandemic, and the papers that treat the pandemic may correspond to a separate focus of interest inside the field.

Table 8. Topics composed by papers accepted in Filter 1 ($k = 15$).

T1	T3	T7	T8	T10 **	T12	T14 *
digit	digit	adopt	capabl	covid	perform	work
busi	knowledg	organis	market	smes	innov	sustain
smes	firm	technolog	perform	pandem	busi	safeti
transform	technolog	smes	firm	crisi	model	health
technolog	smes	inform	orient	busi	manag	studi
valu	capabl	factor	smes	respons	studi	environment
organiz	innov	market	technolog	firm	suppli	cultur
capabl	extern	competit	busi	measur	chain	osh
matur	inform	portal	agil	impact	organiz	employe
level	platform	cloud	manag	economi	firm	relat

The topic with * is focused on the environmental and social aspects and the topic with ** is focused on the pandemics. The terms that are exclusive to these topics are in green.

Additionally, the terms "platform" (T3), "portal", and "cloud" (T7) indicate that these topics are relevant for the DT variables. The terms "sustain" and "environment" (T14) are not exclusive of the topics produced with accepted papers, but they are in the same topic (T14) with the exclusive terms "work" "safeti", "health", "cultur", and "employe", which are all related to social sustainability. There is no other topic with terms related to any social or environmental aspects of sustainability. This corroborates the conclusion that social and environmental aspects are under-investigated in the literature and treated as a separate theme in the field. It is also worth noting that the pandemic and social sustainability are treated in different topics (T10 and T14). This suggests that research about the pandemic may be neglecting the social impacts (such as unemployment).

5.2. Topic Modeling for Papers' Filtering: The Onion Approach

Filters 2 and 3 required more human interpretation. We compared LDA results with the human filters. To do this, we adopted what we call the onion approach, as explained in Section 3.2 "LDA-based SLR". Tables 9–12 show the results of each onion layer.

Table 9. First onion layer (91 papers).

T1	T2	T3	T4	T5 *	T6 **	T7 *
industri	digit	innov	capabl	orient	covid	port
product	manag	knowledg	perform	firm	pandem	work
manufactur	busi	famili	innov	market	crisi	data
system	technolog	enterpris	busi	perform	smes	sustain
chain	smes	perform	model	environment	busi	servic
data	research	industri	agil	sustain	respons	health
evalu	studi	model	organiz	innov	distribut	safeti
suppli	develop	extern	strategi	green	enterpris	forest
oper	compani	search	firm	capabl	impact	environment
indic	model	dih	relationship	suppli	market	iot

The topic with * is focused on the environmental and social aspects and the topic with ** is focused on the pandemics. The orange column was used to show which topic was used to perform the LDA of the next onion layer.

The orange column was used to show which topic was used to perform the LDA of the next onion layer. The terms "digit", "technolog", "smes", busi", and "firm" appear in all layers, indicating that the search strings and filters achieved the research focus well. Additionally, the terms "market", "orient", "manag", "capabl", "knowledg", and "innov" appear in all layers. This corroborates that market orientation (external activities) may be a driving force of the DT. Moreover, managerial aspects are the most frequently investigated. Market orientation is usually investigated considering the capabilities' perspective, and this theme is deeply correlated with innovation and knowledge.

Table 10. Second onion layer (73 papers).

T1	T2	T3 *	T4	T5	T6	T7
capabl	digit	sustain	perform	market	capabl	industri
innov	busi	environment	variabl	onlin	firm	product
matur	manag	port	studi	media	knowledg	manufactur
market	smes	green	chain	distribut	agil	chain
orient	technolog	capit	suppli	communic	entrepreneuri	compani
knowledg	research	cultur	effect	custom	famili	design
social	innov	variabl	factor	social	orient	technolog
crisi	model	smsps	portal	compani	effect	suppli
organiz	develop	intellectu	adopt	competit	intern	process
servic	studi	tool	signific	enterpris	innov	system

The topic with * is focused on the environmental and social aspects. The orange column was used to show which topic was used to perform the LDA of the next onion layer.

Table 11. Third onion layer (64 papers).

T1	T2	T3 *	T4	T5
perform	capabl	sustain	firm	digit
innov	digit	environment	market	technolog
busi	innov	port	social	mes
manag	firm	manag	innov	busi
studi	orient	research	distribut	manag
effect	busi	green	knowledg	industri
model	research	adopt	custom	compani
strategi	knowledg	cultur	strategi	product
firm	manag	iot	action	process
research	smes	social	servic	develop

The topic with * is focused on the environmental and social aspects. The orange column was used to show which topic was used to perform the LDA of the next onion layer.

Table 12. Fourth onion layer (37 papers).

T1	T2	T3	T4	T5
digit	valu	innov	industri	firm
busi	market	capabl	knowledg	work
manag	chain	market	innov	crisi
technolog	onlin	competit	technolog	respons
smes	aspect	matur	smes	chang
compani	resourc	servic	product	covid
research	smes	technolog	manufactur	dynam
process	enterpris	orient	extern	entrepreneur
transform	develop	iot	firm	small
develop	indic	knowledg	adopt	smes

Table 9 has the terms related to sustainability concentrated in T5 and T7. Similarly, Table 10 has them in T1, T3, and T4. Table 11 has them concentrated in T3, and Table 12 has no more terms related to environmental sustainability but has "work" in the same topic with "covid", correlating the pandemic with social sustainability. Only Tables 7 and 9 have terms related to the pandemic. In both cases, they are isolated in a unique topic. This indicates that these themes, "pandemic" and "sustainability", are treated as separate themes in the field and separate from each other.

The terms "environment", "sustain", "model", "perform", and "studi" appear in Tables 9–11, evidencing the adequacy of the SLR. The terms "transform" and "chang" only appear in Table 12. The term "entrepreneur" appears in Tables 10 and 12, suggesting that, as well as innovation, entrepreneurship is frequently correlated to DT in SMEs. Hence, the variables for DT should also consider measuring innovation, entrepreneurship, and market orientation capability.

Additionally, "famili" appears in Tables 9 and 10, indicating that SMEs are usually correlated with the family business, and the term "manufactur" appears in Tables 9–11, corroborating the result that the manufacturing industry is more studied than other industries. The terms "suppl" and "chain" appear in Tables 9 and 10, suggesting that "supply chain" is also more studied than other themes. The term "agile" appears in Tables 9 and 10, suggesting that agile management is also associated with SMEs. Finally, the term "matur" appears in Tables 10 and 12, indicating that DT is frequently understood based on maturity levels. In this way, it is demonstrated that the LDA served as a useful support tool while executing an SLR.

6. Framework for Measuring SMEs' Performance

Figure 4 provides the framework proposed here for those (researchers, policymakers, and SMEs) interested in measuring SMEs' performance, considering the TBL and DT. SMEs' performance is subject to internal (micro-level) and external factors (macro-level). As discussed in Section 4.3, SMEs are subject to different definitions and heterogeneity. Thus, this may be considered in any performance investigation due to its possible moderator effect. Table 2 provides the variables identified in the literature for quantifying heterogeneity.

As discussed in Section 4.4.1. DT, papers divide performance metrics into those related to internal and external activities. This is also supported by the LDA evidence discussed in Section 5.2. However, the DT aspect has an integrative potential, using a shared digital platform among different stakeholders. Depending on the level of the SMEs' digital maturity, it is possible to measure DT performance by considering internal and external activities jointly. This was also evidenced by the LDA results in Section 5.2. Similarly, depending on the business environment characteristics, it is possible to measure the DT performance considering micro- and macro-levels jointly. Given this, although not found in the literature (Table 3), cybersecurity and data protection procedures are important variables for enabling this integration.

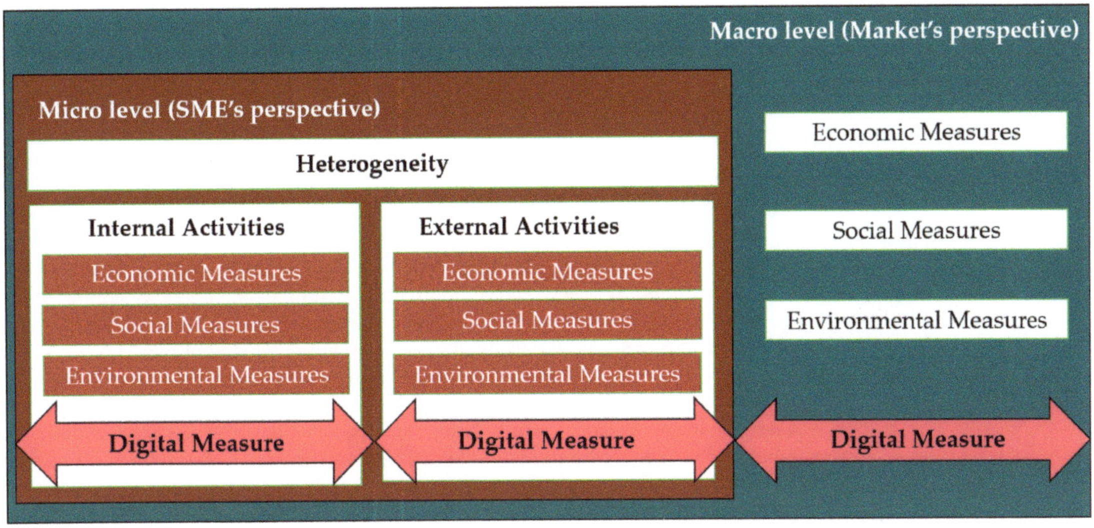

Figure 4. Framework for measuring the sustainable and digital performance of SMEs.

All aspects of TBL and DT affect each other and affect final performance. Researchers are recommended to deeply investigate these relationships. However, our results indicate that, differently from the DT aspect, economic, social, and environmental aspects do not have a strong potential to enable the integration between internal and external activities and between micro- and macro-levels. Regarding the TBL aspects, performance can be measured considering different variables for internal/external activities and micro-/macro-levels.

The variables found for measuring the economic, social, and environmental aspects are registered in Tables 5–7, respectively. For example, policymakers interested in fostering innovation among SMEs may be interested in tracking digital and economic measures. In this case, at the macro-level, the digital variable can be the use (or not) of a certain platform connecting regulatory agencies, SMEs, suppliers, and other stakeholders. Furthermore, policymakers can track a variable representing the financing approved for each SME to innovate (economic measure). At the micro-level, for attending to this policy, SMEs can, for example, track the economic dimension "R&D Infrastructure", through the variables related to internal activities ("R&D Department", "ICT Investment in R&D", "Patents or Trademarks", "In-house Innovation Capacity") and external activities ("Innovative Collaboration") as registered in [77]—Table 5.

However, to keep the same illustration while promoting an innovation policy, policymakers may also be interested that this policy positively impacts jobs (in quantity or quality). From the macro-level perspective, this can be measured based on labor protection requests (as [71] in Table 6) or regional unemployment rates. At the micro-level, the SMEs can measure this through variables for internal activities (such as "Acquisition and Development of Skills", as in [69], Table 6) and variables for external activities (such as "External Openness and Collaboration", as in [68], Table 6).

Finally, policymakers may also be interested in fostering an innovation policy while guaranteeing negative impacts on jobs and the environment will be restricted. From the macro-level perspective, environment restriction can be measured through a certification system for SMEs, while SMEs can measure whether their suppliers (external activities) and themselves (internal activities) are attending to the agreed environmental targets. This framework can be used for guiding the creation of performance metrics at operational, tactic, and strategic levels.

Finally, variables' relationships can be investigated through SEM and econometrics. Furthermore, performance indicators can be proposed based on DEA, SFA, or among other methods (Table 1). In this way, it will be possible to highlight benchmarks and best practices, as well as determine if targets are achieved. Although here it is recommended to measure performance considering at least ten variables (i.e., six at the micro-level, three at the macro-level, and a DT variable integrating levels, as in Figure 4), it is essential to highlight that the variable choice depends on the goal and a context view.

7. Conclusions

The tools, dimensions, and variables for measuring and investigating the impacts of DT on the performance of SMEs are an increasing topic of interest, but the body of knowledge is still developing. The number of papers on the theme is still small. Moreover, the lack of a commonly accepted definition of SME and the heterogeneity among SMEs are obstacles to the comparison of results among different papers. Consequently, this also represents an obstacle to the building of the body of knowledge. What is more, there is a lack of a definition of what DT is for an SME and how it could differ from DT in larger companies.

Among the analyzed papers, we identified that quantitative tools are predominant, mainly structural equation modeling (SEM) and econometrics. Decision-Making Trial and Evaluation Laboratory (DEMATEL) and integrations with fuzzy techniques may represent the current methodological frontier on the theme. However, it is worth noting that Stochastic Frontier Analysis (SFA) and Data Envelopment Analysis (DEA), methods focused on performance measurement, still need to be explored, representing complete fields of new future research possibilities. Furthermore, Artificial Intelligence (AI) is not explored, probably because of data unavailability. We strongly recommend the standardization of definitions and data collection procedures for enabling the application of AI methods.

We collected and classified the used variables and dimensions. Then, we could conclude that the TBL is still neglected. The joint analysis of the manual and the LDA-based SLRs indicated that environmental and social sustainability are treated as separate themes in the field, and they are not integrated with investigations about DT and economic performance. The variables and dimensions of DT are the most explored in the literature. They vary depending on digital maturity level. Potentially, the heterogeneity among SMEs (such as their sizes and industries) is affecting the maturity of the DT process. The manufacturing industry is more investigated and may have specific characteristics. The economic variables and dimensions are the second-most investigated. In both cases, we identified operational and strategic variables and dimensions, but the relationship between economic performance and DT remains unexplained. Third, social and environmental variables and dimensions are significantly less investigated. When they are treated in the literature, they tend to represent only a strategic level.

Therefore, many future research directions were pointed out in this text. Among them, we pinpoint the standardization of the definition of "SME"; the standardization of SMEs' data collection procedures; investigations into cross-national and cross-temporal scenario;, the development of a systematic taxonomy of the DT and TBL variables and dimensions considering operational, tactical, and strategical levels; investigations about how TBL and DT aspects interact and influence each other; and the development of a quantitative approach for measuring SMEs' sustainable and digital performance based on tools such as Data Envelopment Analysis (DEA) and Stochastic Frontier Analysis (SFA).

Regarding the SLR methodology, manual and LDA-based SLRs demonstrated to be useful and practical approaches for the comparison of results and novel insights. However, in our investigation, the LDA-based SLR could not substitute for the manual SLR completely. As with all research, our study is not without limitations. One of them is that the used LDA algorithm was not able to read all papers. Future applications should improve the algorithm proposed by Asmussen and Møller (2019) [13] until it is able to read all papers. The improved algorithm should also be able to deal with pre-defined expressions composed of two or more terms, such as "supply chain", instead of "suppl" and "chain" as two different terms in the topic construction. Beyond improving the algorithm for paper selection, it is important to emphasize that further research is needed to automate other steps of the SLR, such as the planning and reporting steps [66].

Author Contributions: Conceptualization, I.C.M., G.A.Q., W.Y., and P.N.A.J.; methodology, I.C.M., G.A.Q. and P.N.A.J.; software, I.C.M., G.A.Q. and P.N.A.J.; validation, I.C.M., G.A.Q. and P.N.A.J.; formal analysis, I.C.M., G.A.Q. and P.N.A.J.; investigation, I.C.M., G.A.Q. and P.N.A.J.; resources, W.Y. and J.P.; data curation, I.C.M., G.A.Q. and P.N.A.J.; writing—original draft preparation, I.C.M., G.A.Q. and P.N.A.J.; writing—review and editing, I.C.M., G.A.Q., P.N.A.J. and W.Y.; visualization, I.C.M., G.A.Q., P.N.A.J.; supervision, W.Y. and J.P.; project administration, I.C.M.; funding acquisition, W.Y. and J.P. All authors have read and agreed to the published version of the manuscript.

Funding: This research received no external funding.

Conflicts of Interest: The authors declare no conflict of interest.

References

1. Hassan, S.S.; Reuter, C.; Bzhalava, L. Perception or Capabilities? An Empirical Investigation of the Factors Influencing the Adoption of Social Media and Public Cloud in German SMEs. *Int. J. Innov. Manag.* **2021**, *25*, 2150002. [CrossRef]
2. Grover, V.; Kohli, R. Revealing Your Hand: Caveats in Implementing Digital Business Strategy. *MIS Q.* **2013**, *37*, 655–662.
3. Lyytinen, K.; Rose, G.M. The Disruptive Nature of Information Technology Innovations: The Case of Internet Computing in Systems Development Organizations. *MIS Q.* **2003**, *27*, 557. [CrossRef]
4. World Bank. World Bank SME Finance: Development News, Research, Data | World Bank. Available online: https://www.worldbank.org/en/topic/smefinance (accessed on 8 January 2022).
5. Chen, Y.-Y.K.; Jaw, Y.-L.; Wu, B.-L. Effect of Digital Transformation on Organisational Performance of SMEs. *Internet Res.* **2016**, *26*, 186–212. [CrossRef]
6. González-Varona, J.; López-Paredes, A.; Poza, D.; Acebes, F. Building and Development of an Organizational Competence for Digital Transformation in SMEs. *J. Ind. Eng. Manag.* **2021**, *14*, 15. [CrossRef]
7. Siegel, R.; Antony, J.; Garza-Reyes, J.A.; Cherrafi, A.; Lameijer, B. Integrated Green Lean Approach and Sustainability for SMEs: From Literature Review to a Conceptual Framework. *J. Clean. Prod.* **2019**, *240*, 118205. [CrossRef]
8. Troise, C.; Corvello, V.; Ghobadian, A.; O'Regan, N. How Can SMEs Successfully Navigate VUCA Environment: The Role of Agility in the Digital Transformation Era. *Technol. Forecast. Soc. Chang.* **2022**, *174*, 121227. [CrossRef]
9. Denicolai, S.; Zucchella, A.; Magnani, G. Internationalization, Digitalization, and Sustainability: Are SMEs Ready? A Survey on Synergies and Substituting Effects among Growth Paths. *Technol. Forecast. Soc. Chang.* **2021**, *166*, 120650. [CrossRef]
10. Viale Pereira, G.; Estevez, E.; Cardona, D.; Chesñevar, C.; Collazzo-Yelpo, P.; Cunha, M.A.; Diniz, E.H.; Ferraresi, A.A.; Fischer, F.M.; Cardinelle Oliveira Garcia, F.; et al. South American Expert Roundtable: Increasing Adaptive Governance Capacity for Coping with Unintended Side Effects of Digital Transformation. *Sustainability* **2020**, *12*, 718. [CrossRef]
11. Chatzistamoulou, N. Is Digital Transformation the Deus Ex Machina towards Sustainability Transition of the European SMEs? *Ecol. Econ.* **2023**, *206*, 107739. [CrossRef]
12. Pfister, P.; Lehmann, C. Returns on Digitisation in SMEs—A Systematic Literature Review. *J. Small Bus. Entrep.* **2021**, 1–25. [CrossRef]
13. Asmussen, C.B.; Møller, C. Smart Literature Review: A Practical Topic Modelling Approach to Exploratory Literature Review. *J. Big Data* **2019**, *6*, 93. [CrossRef]
14. Queiroz, G.A.; Alves Junior, P.N.; Costa Melo, I. Digitalization as an Enabler to SMEs Implementing Lean-Green? A Systematic Review through the Topic Modelling Approach. *Sustainability* **2022**, *14*, 14089. [CrossRef]
15. Saha, B. Application of topic modelling for literature review in management research. In *Interdisciplinary Research in Technology and Management*; CRC Press: Boca Raton, FL, USA, 2021; pp. 1–8. ISBN 9781003202240.
16. Quinn, K.M.; Monroe, B.L.; Colaresi, M.; Crespin, M.H.; Radev, D.R. How to Analyze Political Attention with Minimal Assumptions and Costs. *Am. J. Pol. Sci.* **2010**, *54*, 209–228. [CrossRef]
17. U.S. Small Business Administration (SBA) SBA Table of Size Standards. 2017. Available online: https://www.sba.gov/sites/default/files/2019-08/SBA%20Table%20of%20Size%20Standards_Effective%20Aug%2019%2C%202019_Rev.pdf (accessed on 8 January 2022).
18. OECD Brazil | Financing SMEs and Entrepreneurs 2020: An OECD Scoreboard | OECD iLibrary. Available online: https://www.oecd-ilibrary.org/sites/8153da8d-en/index.html?itemId=/content/component/8153da8d-en (accessed on 4 April 2022).
19. Library of Chilean Congress. Biblioteca del Congreso Chileno] Estatuto de Las PYMES—Ley Fácil—Biblioteca Del Congreso Nacional de Chile. Available online: https://www.bcn.cl/leyfacil/recurso/estatuto-de-las-pymes (accessed on 6 January 2022).
20. Alves Junior, P.N.; Costa Melo, I.; Yamanaka, L.; Severino, M.R.; Rentizelas, A. Supporting the Bidding Decisions of Smallholder Farmers in Public Calls in Brazil. *Agriculture* **2021**, *12*, 48. [CrossRef]
21. Government of Japan Small and Medium Enterprise Agency. Available online: https://www.chusho.meti.go.jp/sme_english/outline/07/01.html (accessed on 8 January 2022).

22. World Trade Organization Levelling the Trading Field for SMEs. 2017. Available online: https://www.wto.org/english/news_e/spra_e/spra135_e.htm (accessed on 8 January 2022).
23. Kim, S.-S. Sustainable Growth Variables by Industry Sectors and Their Influence on Changes in Business Models of SMEs in the Era of Digital Transformation. *Sustainability* **2021**, *13*, 7114. [CrossRef]
24. Lu, J.W.; Beamish, P.W. The Internationalization and Performance of SMEs. *Strateg. Manag. J.* **2001**, *22*, 565–586. [CrossRef]
25. Dosi, G.; Lechevalier, S.; Secchi, A. Introduction: Interfirm Heterogeneity–Nature, Sources and Consequences for Industrial Dynamics. *Ind. Corp. Chang.* **2010**, *19*, 1867–1890. [CrossRef]
26. Marinho, B.F.D.; Costa Melo, I. Fostering Innovative SMEs in a Developing Country: The ALI Program Experience. *Sustainability* **2022**, *2*, 13344. [CrossRef]
27. Gong, C.; Ribiere, V. Developing a Unified Definition of Digital Transformation. *Technovation* **2021**, *102*, 102217. [CrossRef]
28. Verhoef, P.C.; Broekhuizen, T.; Bart, Y.; Bhattacharya, A.; Qi Dong, J.; Fabian, N.; Haenlein, M. Digital Transformation: A Multidisciplinary Reflection and Research Agenda. *J. Bus. Res.* **2021**, *122*, 889–901. [CrossRef]
29. Culot, G.; Nassimbeni, G.; Orzes, G.; Sartor, M. Behind the Definition of Industry 4.0: Analysis and Open Questions. *Int. J. Prod. Econ.* **2020**, *226*, 107617. [CrossRef]
30. Sassanelli, C.; Rossi, M.; Terzi, S. Evaluating the Smart Maturity of Manufacturing Companies along the Product Development Process to Set a PLM Project Roadmap. *Int. J. Prod. Lifecycle Manag.* **2020**, *12*, 185. [CrossRef]
31. Costa Melo, I.; Queiroz, G.A.; Alves Junior, P.N.; de Sousa, T.B.; Yushimito, W.; Pereira, J. Sustainable Digital Transformation in Small and Medium Enterprises (SMEs): A Review on Performance. *Heliyon* **2020**, *9*, e13908. [CrossRef]
32. OECD. *Entrepreneurship at a Glance 2011*; OECD: Paris, France, 2011; ISBN 9789264095762.
33. OECD. *Entrepreneurship at a Glance 2017*; OECD: Paris, France, 2017; ISBN 9789264279926.
34. Elkington, J. Towards the Sustainable Corporation: Win-Win-Win Business Strategies for Sustainable Development. *Calif. Manag. Rev.* **1994**, *36*, 90–100. [CrossRef]
35. Elkington, J. Accounting for the triple bottom line. *Meas. Bus. Excell.* **1998**, *2*, 18–22. [CrossRef]
36. Helleno, A.L.; de Moraes, A.J.I.; Simon, A.T. Integrating Sustainability Indicators and Lean Manufacturing to Assess Manufacturing Processes: Application Case Studies in Brazilian Industry. *J. Clean. Prod.* **2017**, *153*, 405–416. [CrossRef]
37. Slaper, T.F.; Hall, T.J. The Triple Bottom Line: What Is It and How Does It Work? *Indiana Bus. Res. Cent.* **2011**, *86*, 4–8.
38. Santos, J.; Muñoz-Villamizar, A.; Ormazábal, M.; Viles, E. Using Problem-Oriented Monitoring to Simultaneously Improve Productivity and Environmental Performance in Manufacturing Companies. *Int. J. Comput. Integr. Manuf.* **2019**, *32*, 183–193. [CrossRef]
39. Ellen MacArthour Foundation. What Is Circular Economy? Available online: https://ellenmacarthurfoundation.org/topics/circular-economy-introduction/overview (accessed on 16 February 2023).
40. Acerbi, F.; Sassanelli, C.; Taisch, M. A Conceptual Data Model Promoting Data-Driven Circular Manufacturing. *Oper. Manag. Res.* **2022**, *15*, 838–857. [CrossRef]
41. Acerbi, F.; Sassanelli, C.; Terzi, S.; Taisch, M. A Systematic Literature Review on Data and Information Required for Circular Manufacturing Strategies Adoption. *Sustainability* **2021**, *13*, 2047. [CrossRef]
42. Taddei, E.; Sassanelli, C.; Rosa, P.; Terzi, S. Circular Supply Chains in the Era of Industry 4.0: A Systematic Literature Review. *Comput. Ind. Eng.* **2022**, *170*, 108268. [CrossRef]
43. Walker, A.M.; Vermeulen, W.J.V.; Simboli, A.; Raggi, A. Sustainability Assessment in Circular Inter-Firm Networks: An Integrated Framework of Industrial Ecology and Circular Supply Chain Management Approaches. *J. Clean. Prod.* **2021**, *286*, 125457. [CrossRef]
44. Ranjbari, M.; Shams Esfandabadi, Z.; Zanetti, M.C.; Scagnelli, S.D.; Siebers, P.-O.; Aghbashlo, M.; Peng, W.; Quatraro, F.; Tabatabaei, M. Three Pillars of Sustainability in the Wake of COVID-19: A Systematic Review and Future Research Agenda for Sustainable Development. *J. Clean. Prod.* **2021**, *297*, 126660. [CrossRef]
45. Rodrigues, M.; Franco, M.; Sousa, N.; Silva, R. COVID 19 and the Business Management Crisis: An Empirical Study in SMEs. *Sustainability* **2021**, *13*, 5912. [CrossRef]
46. Lopez-Nicolas, C.; Nikou, S.; Molina-Castillo, F.-J.; Bouwman, H. Gender Differences and Business Model Experimentation in European SMEs. *J. Bus. Ind. Mark.* **2020**, *35*, 1205–1219. [CrossRef]
47. Tranfield, D.; Denyer, D.; Smart, P. Towards a Methodology for Developing Evidence-Informed Management Knowledge by Means of Systematic Review. *Br. J. Manag.* **2003**, *14*, 207–222. [CrossRef]
48. Moher, D.; Liberati, A.; Tetzlaff, J.; Altman, D.G. Preferred Reporting Items for Systematic Reviews and Meta-Analyses: The PRISMA Statement. *PLoS Med.* **2009**, *6*, e1000097. [CrossRef]
49. Chadegani, A.A.; Salehi, H.; Yunus, M.M.; Farhadi, H.; Fooladi, M.; Farhadi, M.; Ebrahim, N.A. A Comparison between Two Main Academic Literature Collections: Web of Science and Scopus Databases. *Asian Soc. Sci.* **2013**, *9*. [CrossRef]
50. Barnes, S.; Rutter, R.N.; la Paz, A.I.; Scornavacca, E. Empirical Identification of Skills Gaps between Chief Information Officer Supply and Demand: A Resource-Based View Using Machine Learning. *Ind. Manag. Data Syst.* **2021**, *121*, 1749–1766. [CrossRef]
51. Krippendorff, K. *Content Analysis: An Introduction to Its Methodology*, 2nd ed.; Sage Publications: Thousand Oaks, CA, USA, 2004; Volume 13.
52. van Dinter, R.; Tekinerdogan, B.; Catal, C. Automation of Systematic Literature Reviews: A Systematic Literature Review. *Inf. Softw. Technol.* **2021**, *136*, 106589. [CrossRef]

53. Lancichinetti, A.; Sirer, M.I.; Wang, J.X.; Acuna, D.; Körding, K.; Amaral, L.A.N. High-Reproducibility and High-Accuracy Method for Automated Topic Classification. *Phys. Rev. X* **2015**, *5*, 011007. [CrossRef]
54. Saura, J.R.; Ribeiro-Soriano, D.; Zegarra Saldaña, P. Exploring the Challenges of Remote Work on Twitter Users' Sentiments: From Digital Technology Development to a Post-Pandemic Era. *J. Bus. Res.* **2022**, *142*, 242–254. [CrossRef]
55. Gupta, A.; Li, H.; Farnoush, A.; Jiang, W. Understanding Patterns of COVID Infodemic: A Systematic and Pragmatic Approach to Curb Fake News. *J. Bus. Res.* **2022**, *140*, 670–683. [CrossRef]
56. van Nguyen, T.; Cong Pham, H.; Nhat Nguyen, M.; Zhou, L.; Akbari, M. Data-Driven Review of Blockchain Applications in Supply Chain Management: Key Research Themes and Future Directions. *Int. J. Prod. Res.* **2023**, 1–23. [CrossRef]
57. Olorisade, B.K.; de Quincey, E.; Brereton, P.; Andras, P. A Critical Analysis of Studies That Address the Use of Text Mining for Citation Screening in Systematic Reviews. In Proceedings of the 20th International Conference on Evaluation and Assessment in Software Engineering, Klagenfurt am Wörthersee, Austria, 1–3 June 2016; pp. 1–11.
58. AlMujaini, H.; Hilmi, M.F.; Abudaqa, A.; Alzahmi, R. Corporate Foresight Organizational Learning and Performance: The Moderating Role of Digital Transformation and Mediating Role of Innovativeness in SMEs. *Int. J. Data Netw. Sci.* **2021**, *5*, 703–712. [CrossRef]
59. AlMulhim, A.F. Smart Supply Chain and Firm Performance: The Role of Digital Technologies. *Bus. Process Manag. J.* **2021**, *27*, 1353–1372. [CrossRef]
60. Apostolov, M.; Coco, N. Digitalization-Based Innovation—A Case Study Framework. *Int. J. Innov. Technol. Manag.* **2021**, *18*, 2050025. [CrossRef]
61. Ardito, L.; Raby, S.; Albino, V.; Bertoldi, B. The Duality of Digital and Environmental Orientations in the Context of SMEs: Implications for Innovation Performance. *J. Bus. Res.* **2021**, *123*, 44–56. [CrossRef]
62. Bouwman, H.; Nikou, S.; de Reuver, M. Digitalization, Business Models, and SMEs: How Do Business Model Innovation Practices Improve Performance of Digitalizing SMEs? *Telecomm. Policy* **2019**, *43*, 101828. [CrossRef]
63. Cenamor, J.; Parida, V.; Wincent, J. How Entrepreneurial SMEs Compete through Digital Platforms: The Roles of Digital Platform Capability, Network Capability and Ambidexterity. *J. Bus. Res.* **2019**, *100*, 196–206. [CrossRef]
64. Chonsawat, N.; Sopadang, A. Defining SMEs' 4.0 Readiness Indicators. *Appl. Sci.* **2020**, *10*, 8998. [CrossRef]
65. Dutot, V.; Bergeron, F.; Calabrò, A. The Impact of Family Harmony on Family SMEs' Performance: The Mediating Role of Information Technologies. *J. Fam. Bus. Manag.* **2021**, *12*, 1131–1151. [CrossRef]
66. Dutta, G.; Kumar, R.; Sindhwani, R.; Singh, R.K. Digital Transformation Priorities of India's Discrete Manufacturing SMEs—A Conceptual Study in Perspective of Industry 4.0. *Compet. Rev. Int. Bus. J.* **2020**, *30*, 289–314. [CrossRef]
67. Eller, R.; Alford, P.; Kallmünzer, A.; Peters, M. Antecedents, Consequences, and Challenges of Small and Medium-Sized Enterprise Digitalization. *J. Bus. Res.* **2020**, *112*, 119–127. [CrossRef]
68. Fang, I.-C.; Chen, P.-T.; Chiu, H.-H.; Lin, C.-L.; Su, F.-C. Establishing the Digital Transformation Strategies for the Med-Tech Enterprises Based on the AIA-NRM Approach. *Appl. Sci.* **2020**, *10*, 7574. [CrossRef]
69. Gamache, S.; Abdul-Nour, G.; Baril, C. Evaluation of the Influence Parameters of Industry 4.0 and Their Impact on the Quebec Manufacturing SMEs: The First Findings. *Cogent Eng.* **2020**, *7*, 1771818. [CrossRef]
70. Gruenbichler, R.; Klucka, J.; Haviernikova, K.; Strelcova, S. Business Performance Management in Small and Medium-Sized Enterprises in the Slovak Republic: An Integrated Three-Phase-Framework for Implementation. *J. Compet.* **2021**, *13*, 42–58. [CrossRef]
71. Guo, H.; Yang, Z.; Huang, R.; Guo, A. The Digitalization and Public Crisis Responses of Small and Medium Enterprises: Implications from a COVID-19 Survey. *Front. Bus. Res. China* **2020**, *14*, 19. [CrossRef]
72. Holopainen, R.; Niskanen, M.; Rissanen, S. The Impact of Internet and Innovation on the Profitability of Private Healthcare Companies. *J. Small Bus. Entrep.* **2020**, 1–25. [CrossRef]
73. Jelovac, D.; Ljubojević, Č.; Ljubojević, L. HPC in Business: The Impact of Corporate Digital Responsibility on Building Digital Trust and Responsible Corporate Digital Governance. *Digit. Policy Regul. Gov.* **2021**, *24*, 485–497. [CrossRef]
74. Joensuu-Salo, S.; Sorama, K.; Viljamaa, A.; Varamäki, E. Firm Performance among Internationalized SMEs: The Interplay of Market Orientation, Marketing Capability and Digitalization. *Adm. Sci.* **2018**, *8*, 31. [CrossRef]
75. Jun, W.; Nasir, M.H.; Yousaf, Z.; Khattak, A.; Yasir, M.; Javed, A.; Shirazi, S.H. Innovation Performance in Digital Economy: Does Digital Platform Capability, Improvisation Capability and Organizational Readiness Really Matter? *Eur. J. Innov. Manag.* **2021**, in press. [CrossRef]
76. Kamišalić, A.; Šestak, M.; Beranič, T. Supporting the Sustainability of Natural Fiber-Based Value Chains of SMEs through Digitalization. *Sustainability* **2020**, *12*, 8121. [CrossRef]
77. Kmecová, I.; Stuchlý, J.; Sagapova, N.; Tlustý, M. SME Human Resources Management Digitization: Evaluation of the Level of Digitization and Estimation of Future Developments. *Pol. J. Manag. Stud.* **2021**, *23*, 232–248. [CrossRef]
78. Kulathunga, K.M.M.C.B.; Ye, J.; Sharma, S.; Weerathunga, P.R. How Does Technological and Financial Literacy Influence SME Performance: Mediating Role of ERM Practices. *Information* **2020**, *11*, 297. [CrossRef]
79. Kumar, R.; Sindhwani, R.; Singh, P.L. IIoT Implementation Challenges: Analysis and Mitigation by Blockchain. *J. Glob. Oper. Strateg. Sourc.* **2021**. [CrossRef]
80. Lányi, B.; Hornyák, M.; Kruzslicz, F. The Effect of Online Activity on SMEs' Competitiveness. *Compet. Rev. Int. Bus. J.* **2021**, *31*, 477–496. [CrossRef]

81. Mubarak, M.F.; Shaikh, F.A.; Mubarik, M.; Samo, K.A.; Mastoi, S. The Impact of Digital Transformation on Business Performance: A Study of Pakistani SMEs. *Eng. Technol. Appl. Sci. Res.* **2019**, *9*, 5056–5061. [CrossRef]
82. Nasiri, M.; Ukko, J.; Saunila, M.; Rantala, T. Managing the Digital Supply Chain: The Role of Smart Technologies. *Technovation* **2020**, *96*, 102121. [CrossRef]
83. North, K.; Aramburu, N.; Lorenzo, O.J. Promoting Digitally Enabled Growth in SMEs: A Framework Proposal. *J. Enterp. Inf. Manag.* **2019**, *33*, 238–262. [CrossRef]
84. Okfalisa, O.; Anggraini, W.; Nawanir, G.; Saktioto, S.; Wong, K.Y. Measuring the Effects of Different Factors Influencing on the Readiness of SMEs towards Digitalization: A Multiple Perspectives Design of Decision Support System. *Decis. Sci. Lett.* **2021**, *10*, 425–442. [CrossRef]
85. Ponis, S.T.; Lada, C. Digital Transformation in the Greek Fashion Industry: A Survey. *Int. J. Fash. Des. Technol. Educ.* **2021**, *14*, 162–172. [CrossRef]
86. Rozak, H.; Adhiatma, A.; Fachrunnisa, O.; Rahayu, T. Social Media Engagement, Organizational Agility and Digitalization Strategic Plan to Improve SMEs' Performance. *IEEE Trans. Eng. Manag.* **2021**, 1–10. [CrossRef]
87. Ukko, J.; Nasiri, M.; Saunila, M.; Rantala, T. Sustainability Strategy as a Moderator in the Relationship between Digital Business Strategy and Financial Performance. *J. Clean. Prod.* **2019**, *236*, 117626. [CrossRef]
88. Bluhm, D.J.; Harman, W.; Lee, T.W.; Mitchell, T.R. Qualitative Research in Management: A Decade of Progress. *J. Manag. Stud.* **2011**, *48*, 1866–1891. [CrossRef]
89. Bryman, A. Barriers to Integrating Quantitative and Qualitative Research. *J. Mix. Methods Res.* **2007**, *1*, 8–22. [CrossRef]
90. Hussey, D.M.; Eagan, P.D. Using Structural Equation Modeling to Test Environmental Performance in Small and Medium-Sized Manufacturers: Can SEM Help SMEs? *J. Clean. Prod.* **2007**, *15*, 303–312. [CrossRef]

sustainability

MDPI

Article

Servitization 4.0 as a Trigger for Sustainable Business: Evidence from Automotive Digital Supply Chain

Anja Jankovic-Zugic [1], Nenad Medic [1,*], Marko Pavlovic [2], Tanja Todorovic [1] and Slavko Rakic [1]

1 Faculty of Technical Sciences, University of Novi Sad, Trg Dositeja Obradovica 6, 21000 Novi Sad, Serbia
2 The Institute for Artificial Intelligence of Serbia, Fruskogorska 1, 21000 Novi Sad, Serbia
* Correspondence: medic.nenad@uns.ac.rs

Abstract: The COVID-19 pandemic strengthens the use of digital services in the supply chains of manufacturers and suppliers in the automotive industry. Furthermore, the digitalization of the production process changed how manufacturing firms manage their value chains in the era of Industry 4.0. The automotive sector represents the ecosystem with rapid digital transformation, which provides a strong relationship between manufacturing firms in supply chains. However, there are many gaps in understanding how digital technologies and services could better shape relations between manufacturers and suppliers in the automotive industry. Accordingly, this study investigates the relations in deliveries of digital services in supply chains of the automotive industry. The data set was obtained through annual reports of the automotive firms, both from suppliers and manufacturers, between 2018 and 2020. From the network perspective, throughout the years, authors have used Social Network Analysis (SNA) method. SNA evaluates the relationship between actors (i.e., manufacturers and suppliers) in the use of services in their business models. The research results demonstrate how suppliers influence car manufacturers to deliver digital services to their customers. Finally, this study provides information that the combination of digital technologies with product-related services enables a stronger relationship between manufacturers and suppliers in the manufacturing ecosystem. These relations support the manufacturing ecosystem to survive the influence of different environments.

Keywords: digital servitization; digital technologies; digital supply chain; automotive industry; Industry 4.0; social network analysis

Citation: Jankovic-Zugic, A.; Medic, N.; Pavlovic, M.; Todorovic, T.; Rakic, S. Servitization 4.0 as a Trigger for Sustainable Business: Evidence from Automotive Digital Supply Chain. *Sustainability* **2023**, *15*, 2217. https://doi.org/10.3390/su15032217

Academic Editors: Claudio Sassanelli and Sergio Terzi

Received: 23 December 2022
Revised: 18 January 2023
Accepted: 22 January 2023
Published: 25 January 2023

1. Introduction

Despite the fact that digital transformation is one of the leading terms of today, for a complete understanding of this field, it is necessary to cross theory and practice. Although there is a large amount of literature on digital transformation [1], scholars have not yet reached a consensus on the definition of "digital transformation" [2]. Existing studies have defined digital transformation mainly from the perspectives of technology and value. Some scholars have also defined digital transformation from the perspective of strategic change [3]. The term "digital transformation" refers to a company-wide shift that results in the development of new business models [4]. Through business models, digital transformation applies digital components to establish a new value chain [5]. The function of digital servitization is a relevant subject for research in digital transformation [6]. The use of digital technology can hasten the transition from product to service-based business models [6]. The bulk of the studies examined pertain to digital transformation in a larger sense, concentrating on the creation of value using various digital technologies. The main drivers for digital transformation are technologies such as IoT, Big Data, AI, and Cloud. Digital transformation, on the other hand, might be considered an essential component of many company tasks, such as sales, marketing, and supply chains [7]. From all the above, it is clear that companies implementing digital transformation must be able to undergo major changes. It is important to understand the potential of digital technologies and how digital transformation is changing competitive scenarios [8]. Digital transformation affects all sectors, in particular the automotive industry. Automotive supply chains are international and, consequently, very complicated. They have a large network of relationships between manufacturers, suppliers, and customers worldwide [9]. Thus, firms that do not employ creative consumer solutions cannot compete in the market. Strategy cycles in the automotive industry are shortening due to differentiation and rapidly changing customer demands [10]. Furthermore, the continually changing reality of the COVID-19 outbreak compels firms to quickly modify their approaches in order to maintain their businesses. Long lockdowns in China occurred during the COVID-19 pandemic, resulting in a scarcity of spare parts for its European partners [11]. According to these issues, 95% of all German automotive businesses are relocating their staff to temporary positions [12]. As a result of the delays, material concerns, and a lack of transportation choices, this issue impacted the worldwide automobile supply chains as well [11]. Based on these concerns, digital technologies provide new ways to address human resource and supply chain issues. For example, some digital services, such as online collaboration platforms, were frequently utilized to resolve concerns with face-to-face meetings [13]. In light of these considerations, three main questions arise.

RQ1: Which digital technologies encourage the relations between automotive manufacturers and suppliers?

RQ2: Which product-related services encourage the relations between automotive manufacturers and suppliers?

RQ3: How does the COVID-19 pandemic affect relations between manufacturers and suppliers in the use of digital technologies and product-related services?

To answer these research questions, the paper provides a business model in the automotive industry from suppliers through manufacturers and to customers. The structure of the paper is as follows. The theoretical foundation for the investigation is presented in the next part. The third section describes the research methodology. In the findings section, results followed by the discussion are presented. Finally, the main implications of the research are derived, including the concluding remarks and propositions for further research directions.

2. Research Context

2.1. Industry 4.0

Industry 4.0 was created in 2011 in Germany to concentrate attention on the influence of technology in future production systems [14]. Various Industry 4.0 concepts have lately surfaced. The term "fourth industrial revolution" [15,16] has been used to characterize new production systems supported by digital technologies and networking [17]. The previous research provided a theoretical framework for comprehending Industry 4.0. Accordingly, the notion of Industry 4.0 encompasses several business characteristics that are supported by evolving technologies [18].

These aspects are underpinned by basic technologies such as cloud computing, big data, and IoT. In this view, the digital transformation of firms is viewed as the transition process from traditional to smart business [17]. Although the phrase or notion of Industry 4.0 is widely used, coherence in what it represents is sometimes lacking. Furthermore, related technologies have not yet been unified; thus, a full and mutually exclusive classification is still lacking.

Russmann et al. (2015) analyze Industry 4.0 through nine technologies that will be used in this paper [14,19]. Below is a brief overview of each technology and its use and utility if applied to services [14,20]:

- Big Data Analytics—the full analysis of accessible data in order to make better real-time decisions. When applied to a service, it allows for the creation of a more in-depth understanding of client behavior and preferences.

- Collaborative Robots—robots are used by manufacturers in a variety of sectors to perform complicated tasks. They are self-sufficient, adaptable, and cooperative. They take the role of people in totally rule-based work processes.
- Artificial Intelligence—businesses work on computer simulations of human intelligence processes using technology.
- Internet of Things (IoT)—an interconnected network of machines and products. Multidimensional communication between networked things.
- Cyber Security—entails safe, dependable communications as well as advanced machine and user identification and access control.
- Cloud Computing—communication in real time for manufacturing systems. Increased distant data sharing by the corporation reduces response times from all networked data consumers to a few milliseconds.
- 3D Printing—allows firms to develop and manufacture specific components.
- Augmented and Virtual Reality—a range of services are supported by augmented-reality-based systems, including the selection of components at a warehouse and the transmission of repair instructions via mobile devices.
- Digital Twin—computer models that depict the condition of the network at any given time, in real time.

Industry 4.0 technologies enable businesses to better comprehend what value represents to consumers by collecting a large quantity of data on their behavior and product consumption [17,21]. Large quantities of data, paired with rising computing capacity, are causing profound changes in industrial firms [22]. The phenomenal progress and acceptance of digital technologies have greatly altered customers' perceptions of product innovation and delivery speed [23]. In this scenario, product manufacturers must adapt to the demand-pull measure of service innovation, while simultaneously investing heavily in advanced technologies and connectivity to be more competitive [17]. Industry 4.0 enabling technologies provide mitigation potential for the multiple hazards that the automobile industry encountered during the COVID-19 epidemic. Climate change, increasing urbanization, digitization, and electrification, on the other hand, modify social requirements and customer preferences toward vehicle mobility [24]. The combination of digitization and services allows car manufacturers to form a green value chain with suppliers in order to obtain better market outcomes. Furthermore, these technologies can help businesses reduce the risk of trouble so that they can keep operating. This is especially true in the event of a pandemic, when a lack of employees is one of the many crucial elements possibly damaging operations, particularly supply chain activities [9].

2.2. Digital Servitization

Industry 4.0 and servitization are some of the most current innovations that are changing industrial businesses [25–27]. Industry 4.0 is typically associated with bringing value to the production process, whereas servitization is primarily concerned with providing value to customers [17]. Both Industry 4.0 and servitization arose from distinct study areas, the former from engineering and the latter from management science [15]. The notion of servitization was first established in 1988 in response to the requirement to assure and provide integrated products and services in order to provide additional value [28]. Today, servitization refers to a phenomenon that involves technology [29] that assists or improves the service delivered [30]. The adoption of IoT technology enables manufacturers to provide new kinds of services, enhancing servitization through digitalization [31], introducing the idea of "digital servitization". This is described as the creation of new services or the enhancement of current ones by employing digital technologies. These may be used to allow advanced business models and develop information from data in order to obtain a competitive advantage [7]. One of the most difficult servitization challenges is selecting the new value proposition, which has a significant influence on the whole value architecture of the business model [32].

Pournader et al. (2020) defined the COVID 19 epidemic as a "crisis-as-process" rather than a "crisis-as-event" [33]. Accordingly, firms must prepare in the long term for how they will employ digital technologies in order to develop their business model and adapt to the new situation. One of the primary trends in firm strategy is the expansion of product service content [34]. Product-related services are those that are closely tied to the items in the products. In this paper, the authors investigate product-related services in the automotive industry. Automobile production is a classic example of a business that provides a product–service combination.

Car manufacturers provide a wide spectrum of services to their consumers, including finance, maintenance, and availability, among others. This trend of developing car-related services has lately been accelerated using digital technologies. They want to provide clients with new sorts of "telematics services" [35]. For the purposes of this research, the following product-related services are used:

- Spare parts—exchangeable components stored in an inventory and employed to restore damaged equipment.
- Maintenance—includes performing functional tests, and maintaining and replacing relevant machinery.
- Training—implies informing or instructing to help and improve knowledge.
- Leasing—is a sort of funding that may be obtained from an outside firm if there is insufficient cash at the time.
- Renting and Pay Per Use—is an arrangement in which a payment is paid for the momentary use of another's products or services.
- Full-service contract—a long-term arrangement between the firm and its customers.

Previous study indicates that product-related services impacted by digital technologies may have a favorable impact on manufacturing enterprises' financial performance [6]. Furthermore, the digital product–service system may be created by combining digital technology with product-related services. Based on digitalization, digital product–service systems might assist car manufacturers in reducing their environmental impact while increasing their financial performance.

2.3. Digital Supply Chain

Academic studies on the digital supply chain phenomena are still in their early phases. This field has lately evolved as a result of technological advancements and the complexity of the international market. Supply networks must deal with ever-changing client demands as well as a wide range of external disruptions. Smart goods combined with advanced supply chain services pave the path for a paradigm change in supply chain management [36]. The Digital Supply Chain is a comprehensive examination of the platforms and models that enable the design and administration of digitally related supply chains. Digital technologies have a huge influence on the value chains in the automobile sector [37]. Ivanov et al. (2020) consider that the supply chain could only be as useful as the digital technologies that power it [38]. To fulfill the dynamic expectations of customers in a highly competitive market, the supply chain must be efficient and cost-effective [39]. This necessitates a high degree of digitalization and automation in the company's supply chain.

IT technologies with transparency and visibility of the information shape supply chain resilience [40]. Supply networks were already under strain before the COVID-19 pandemic. Increasingly complicated supply networks, globalization, and outward factors have all contributed to supply chain disruptions in recent years [41]. Nonetheless, no recent occurrence has shown the fragility of supply chains in the same way as the COVID-19 outbreak [33]. During the COVID-19 pandemic, suppliers were unable to satisfy their supply requirements [38]. With the growth of social media, unfavorable experiences may be quickly shared with a huge audience [9]. This information may quickly taint a company's reputation. The development of the pandemic has heightened the urgency of creating supply networks that can be better sustained. Supply chain redesign to achieve Circular Economy goals is still in its early stages [42]. The Circular Economy in the supply chain has gained traction as a result of the COVID-19 pandemic [38]. Circular Economy systems may be supported by Industry 4.0 principles, which can influence the digital supply chain [43]. According to Hussain and Malik (2020), the Circular Economy is related to supply chain resilience and capacities [44]. Companies are beginning to adopt e-commerce platforms that preserve links between manufacturers and users to enhance supply chains throughout the epidemic [45]. These new relationships between manufacturers and customers help to promote the re-manufacturing process, which affects suppliers in the automobile sector [46].

Several firms from several manufacturing sectors are involved in the automobile supply chain. Worldwide automakers have a policy in place for recycling and reusing their products. As a result of these factors, the automotive supply chain is being driven to alter its resources and operations in order to meet environmental standards [47]. Connecting Industry 4.0 with the Circular Economy can enhance supply chain partners' operational and logistical issues for achieving long-term sustainability [43].

3. Methodology

To answer the research questions presented in the introduction, according to the previous research, authors propose to use Social Network Analysis (SNA) as a method to address findings. SNA is an often used technique in social science; however, in the last decade, this method has increased its application in manufacturing research. Furthermore, the goal of this research is to identify connections in the automobile industry's supply chain. According to this aim, with this method, this study provides a network perspective of the automotive supply chain.

3.1. Data Collection

The data for this empirical study originate from the annual reports from 2018 and 2020 in the automotive industry. Previous studies show that annual reports provide relevant information for research in the automotive industry [48]. For this research, authors use data from automotive manufacturers such as Volkswagen Group, Ford Motor, General Motors, BMW, and Toyota Motor. From the automotive suppliers, the authors use data from firms such as Magna, Continental, ZF, Lear, and Bosch. The authors choose these 10 companies because they are in the top 5 by revenue in both groups (i.e., manufacturers and suppliers) in the automobile industry worldwide. Every annual report is directly provided by the company website and provides financial, innovation, technology, and other useful information. All this information gives a good overview of the automotive manufacturers' and suppliers' overall performance before and during the COVID-19 pandemic.

To obtain data from these annual reports, authors use a snowball sampling method to find all digital technologies and product-related services which are connected to the firms. Moreover, the snowball method helps authors to find which automobile manufacturers are closely related to automobile suppliers. In the application of snowing ball methods, authors search for the relations between manufacturers, suppliers, and their use of digital technologies and product-related services. This study has two sets of actors: firms (manufacturers and suppliers) and resources (digital technologies and product-related services). Authors labeled firms with a combination of letters and numbers ranging from AM1 to AM5 for automotive manufacturers and from AS1 to AS5 for automotive suppliers. The digital technologies are labeled with the following marks: DT1—"*3D-Printing*", DT2—"*Collaborative robots*", DT3—"*Artificial Intelligence*", DT4—"*Big Data Analytics*", DT5—"*Cloud Computing*", DT6—"*Cyber Security*", DT7—"*Internet of Things*", DT8—"*Augmented and Virtual Reality*", and DT9—"*Digital twin*". Product-related services are marked as follows: TS1—"*Spare parts*", TS2—"*Maintenance*", TS3—"*Training*", TS4—"*Leasing*", TS5 —"*Renting*", TS6—"*Pay-per-use*", and TS7—"*Full-service contract*".

3.2. Data Analysis

To visualize relations between automotive manufacturers and suppliers, the authors used SNA graphs with the metrics, which describe the cohesion of the network. For the data analysis, the authors use centrality measures [49]. Degree centrality shows how many direct connections each manufacturer has to suppliers via digital technologies and product-related services in the network. Eigenvector centrality quantifies a firm's impact by counting the number of ties it has to other companies in the network via digital technology and product-related services. Eigenvector Centrality then considers how well linked a business is, how many interconnections its connections have, and so on via the network. Closeness centrality scores each firm based on its 'closeness' to all other firms in the network. The number of times a business stands on the shortest path between other firms in the network in the usage of digital technology and product-related services is measured by betweenness centrality. To measure the centrality of the automotive network, authors use a sociogram table as a base. Table 1 shows a sociogram of the automotive network.

Table 1. Sociogram of the automotive industry.

	DT1	DT2	DT3	...	DT8	DT9	TS1	...	TS7
AM1	1	1	1	1	1	1	1	1	1
AM2	1	1	1	1	1	1	1	1	1
...	0	1	1	1	1	0	1	1	1
AM5	1	1	1	1	1	0	1	1	1
AS1	1	1	1	1	1	0	1	1	0
...	0	1	1	1	1	0	1	1	1
AS4	0	1	1	1	1	0	1	1	1
AS5	1	1	1	1	1	1	1	1	0

In the sociogram, the binary data describe whether a firm (manufacturer or supplier) did (+1) or did not (0) use the resources (digital technology or product-related services) in the automotive network.

4. Results

Figures 1 and 2 show the structure of the automotive network before and during the period of the COVID-19 pandemic. Figure 1 shows the automotive network in 2018.

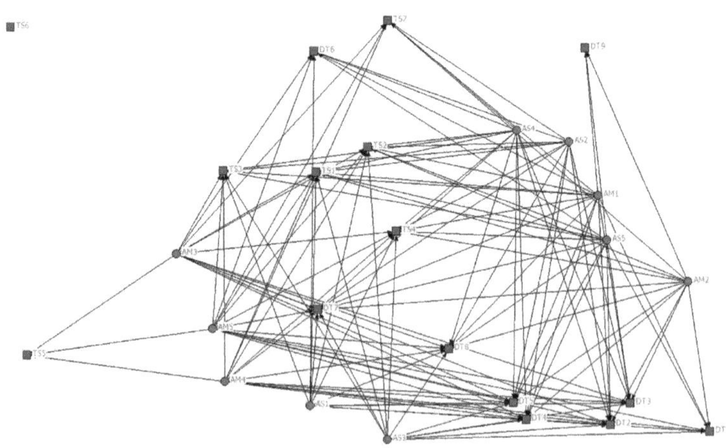

Figure 1. The automotive network in 2018.

In the graphs, the blue squares represent digital technologies and product-related services. The blue square from AM1 to AM5 represents automotive manufacture, and the blue square from AS1 to AS5 is for automotive suppliers. The red circles represent automotive manufacturers and suppliers. The red circles from DT1 to DT9 represent digital technologies, and the red circles from TS1 to TS7 represent product-related services. The density of the network has a value of 0.788, the network average geodesic distance has a value of 1.667, and the network diameter has a value of 4. Figure 2 shows the automotive network in 2020.

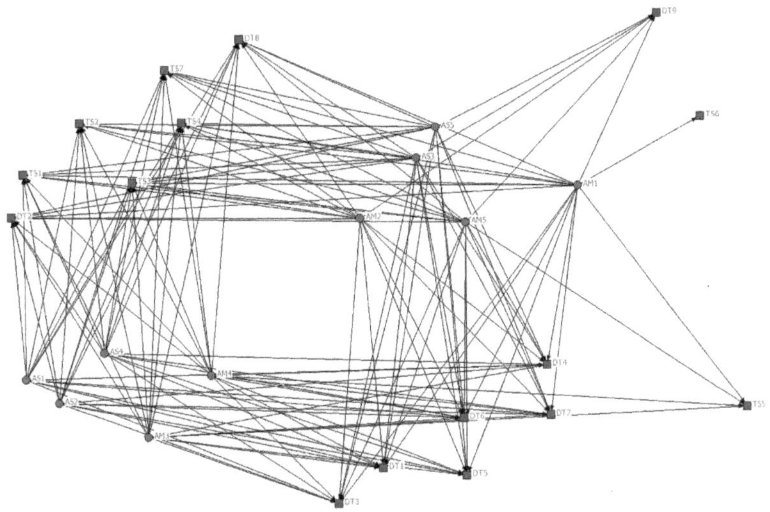

Figure 2. The automotive network in 2020.

The degree of the network in 2020 has a value of 0.875, the network average geodesic distance has a value of 1.631, and the network diameter has a value of 3. According to the description of the network in 2018 and 2020, authors find the difference in all three categories, which shows the power of the networks. The degree measure of the network grows from 0.788 to 0.875 and shows that in the COVID-19 pandemic period, manufacturers have more opportunities to be connected with suppliers via digital technologies and product-related services. The network average geodesic distance went from 1.667 in 2018 to 1.631 in 2020. This value shows that firms are closely connected in the automotive ecosystem. Furthermore, the value of the network diameter goes down from 4 to 3. This value shows that manufacturers and suppliers in the automotive networks have more opportunities to implement new digital technologies and product-related services during the COVID-19 pandemic than before the COVID-19 pandemic. Tables 2 and 3 show the centrality measures of the networks in the automotive industry in 2018 and 2020.

Table 2. Centrality measures of the networks in the automotive industry in 2018.

Firm	Degree Centrality	Eigenvector Centrality	Closeness Centrality	Betweenness Centrality
AM1	**0.875**	**0.339**	**0.971**	**0.065**
AM2	0.75	0.295	0.872	0.046
AM3	0.813	0.32	0.919	0.053
AM4	0.813	0.32	0.919	0.053
AM5	**0.875**	**0.339**	**0.971**	0.064
AS1	0.75	0.311	0.872	0.03
AS2	0.75	0.312	0.872	0.03
AS3	0.688	0.292	0.829	0.022
AS4	0.75	0.312	0.872	0.03
AS5	**0.813**	**0.32**	**0.919**	0.054

The results from 2018 show that auto manufacturers AM1 and AM5 had the highest value in all the centrality measures, except betweenness centrality, where AM1 had better results than AM5. All other manufacturers also had very strong relations in the use of digital technologies and product-related services. Furthermore, results show that automotive suppliers had fewer average relations with automotive manufacturers in the use of digital technologies and product-related services. Only AS5 had similar values as automotive manufacturers. The results from 2020 show different values in the centrality measures. All automotive manufacturers and suppliers had the strongest relations between themselves in the use of digital technologies and product-related services. Moreover, AM1 had relations with all suppliers in the use of these resources. In addition, in the COVID-19 era, all suppliers implement more digital technologies and product-related services to be more competitive in the automotive value chains. Table 4 shows the eigenvector of digital technologies.

Table 3. Centrality measures of the networks in the automotive industry in 2020.

Firm	Degree Centrality	Eigenvector Centrality	Closeness Centrality	Betweenness Centrality
AM1	**1**	**0.33**	**1**	**0.14**
AM2	0.875	0.318	0.895	0.039
AM3	0.875	0.316	0.895	0.043
AM4	0.875	0.316	0.895	0.043
AM5	**0.938**	**0.328**	**0.944**	**0.057**
AS1	0.813	0.306	0.85	0.027
AS2	0.813	0.306	0.85	0.027
AS3	**0.875**	**0.318**	**0.895**	**0.039**
AS4	0.813	0.306	0.85	0.027
AS5	**0.875**	**0.318**	**0.895**	**0.039**

Table 4. Eigenvector of digital technologies in automotive industry in 2018 and 2020.

Digital Technology	2018	2020
DT1	0.207	**0.273**
DT2	**0.295**	**0.273**
DT3	**0.295**	**0.273**
DT4	**0.295**	**0.273**
DT5	**0.295**	**0.273**
DT6	0.21	**0.273**
DT7	**0.295**	**0.273**
DT8	**0.295**	**0.273**
DT9	0.089	0.139

The results of the eigenvector centrality from 2018 show that Collaborative Robots, Artificial Intelligence, Big Data Analytics, Cloud Computing, the Internet of Things, and Augmented and Virtual Reality had a higher number of links that connected automotive manufacturers and automotive suppliers. On the other hand, results from 2020 show that all digital technologies, except Digital Twin, had strong relations with automotive manufacturers and suppliers. These results show that from 2018 to 2020, automotive manufacturers and suppliers created a digital ecosystem to survive the influence of the environment. Table 5 shows the eigenvector of the product-related services industry in 2018 and 2020.

The results of the eigenvector centrality of the product-related services show a similar situation as the eigenvector centrality of digital technologies. The 2018 results show that Spare parts, Maintenance, and Leasing had strong relations with automotive manufacturers and suppliers. On the other hand, results from 2020 show similar eigenvector values for all product-related services, except Renting and Pay-per-use services.

Table 5. Eigenvector of product-related services in automotive industry in 2018 and 2020.

Traditional Services	2018	2020
TS1	**0.295**	**0.273**
TS2	**0.295**	**0.273**
TS3	0.267	**0.273**
TS4	**0.295**	**0.273**
TS5	0.091	0.111
TS6	0	0.028
TS7	0.209	**0.273**

5. Discussion

5.1. Theoretical Implications

The manuscript presents connections in the supply chain of the automotive industry. The data for this study are collected from the annual reports from 2018 and 2020 in the mentioned industry, to find the relations between manufacturers, suppliers, and their use of digital and service resources.

One of the goals of this research is to show the structure of the automotive network before (in 2018) and during the COVID-19 pandemic (in 2020). The degree of the network during the pandemic has a greater value compared to the situation before, which means that in that time interval, the relations between producers and suppliers have become stronger. Automotive manufacturers increasingly realize that improved supplier integration leads to improved performance for the supply chain as a whole [50]. The results from 2018 show that two auto manufacturers had the highest value in all the centrality measures, but all other manufacturers had very strong relations in the use of digital technologies and product-related services as well. However, the results from 2020 show that all automotive manufacturers and suppliers had the strongest relations between themselves in the use of digital technologies and product-related services. In addition, the results show which digital technologies have a higher number of links that connect automotive manufacturers and automotive suppliers. In 2018 these technologies were *Collaborative Robots, Artificial Intelligence, Big Data Analytics, Cloud Computing, the Internet of Things, and Augmented and Virtual Reality*. On the other hand, results from 2020 show that all digital technologies, except Digital Twin, had strong relations with automotive manufacturers and suppliers. This research backs up earlier studies that illustrate the benefits of digital solutions for the automobile sector [51].

The results for product-related services show a similar situation. The 2018 results show that the product-related services *Spare parts, Maintenance, and Leasing* have strong relations with automotive manufacturers and suppliers. On the other hand, results from 2020 show high values for all product-related services and have strong relations with automotive manufacturers and suppliers, except for Renting and Pay-per-use services. Hereafter, the outcomes of this study suggest that digital technology and product-related services have a significant impact on the resilience of manufacturing firms. These findings fill the gaps in the literature about business, sustainability, and digital supply chains in the COVID-19 pandemic era [49]. The ideas of Industry 4.0 assist manufacturing organizations in overcoming the hazards posed by the COVID-19 pandemic. Digital technology, in particular, assists manufacturing organizations in resolving human resource difficulties, which mostly influence supply chain operations [9]. Given the COVID-19 pandemic, supply chain resilience has taken center stage. The Circular Economy and digital supply chains help to ensure social and environmental sustainability [43]. The Circular Economy has served as a catalyst for transformation in the automobile industry, increasing the robustness of business models [40]. From the methodology perspective, this research supports previous related works which employ SNA as a method in manufacturing research [49,52]. These findings open new questions about the implementation of social methods in manufacturing research to obtain relations between different segments.

5.2. Practical Implications

For RQ1: *"Which digital technologies encourage the relations between automotive manufacturers and suppliers?"*, the results show that the digital technologies that boost the relationship between car manufacturers and suppliers are as follows: 3D-Printing, Collaborative robots, Artificial Intelligence, Big Data Analytics, Cloud Computing, Cyber Security, Internet of Things, and Augmented and Virtual Reality.

These findings indicate that the aforementioned digital technologies have the best connections in the industrial ecosystem and strengthen the interaction between vehicle manufacturers and suppliers. For RQ2: *"Which product-related services encourage the relations between automotive manufacturers and suppliers?"*, Spare parts, Maintenance, Training, Leasing, and Full-service contract are product-related services that have the best connections in the COVID-19 pandemic era and boost the relationship between car manufacturers and suppliers. For RQ3: *"How does COVID-19 affect relations between manufacturers and suppliers in the use of digital technologies and product-related services?"*, the results show that manufacturers and suppliers in the automotive networks are more involved in the implementation of new digital technologies and product-related services during the COVID-19 pandemic than before the COVID-19 pandemic. Firms that use more digital solutions have a higher chance of surviving environmental issues. When comparing the situation in 2018 and 2020, it was observed that during the pandemic, all the digital technologies listed in this study had strong relationships with car manufacturers and suppliers. A similar situation exists with product-related services, where also during the pandemic period (2020), all product-related services (except for two) listed in this study had strong relationships with car manufacturers and suppliers.

With this information, managers in the automotive industry could shape their circular business models based on digital services to achieve better market success. During the World Economic Crisis in 2008, the automotive industry transformed its business from traditional producers to service providers (especially Maintenance and Spare parts). The reason for this transformation was the lower level of the economic power of the customers. On the other hand, during the COVID-19 crisis, the automotive industry employs different digital technologies in combination with traditional services to achieve resilience in the environment. The future trends of managers in the automotive industry could be to provide solutions for the customers that support relations with their suppliers, such as electric cars, which enable the Circular Economy. Digital solutions could be a trigger for new business models in the automotive industry. For example, a combination of Maintenance with the IoT and Cloud Computing could result in a predictive maintenance service for customers. Predictive maintenance could enable new information for the suppliers about the need for spare parts to better organize their business. Furthermore, customers could better organize old parts, which could be disposed of by the manufacturers or suppliers based on big data analyses.

6. Conclusions

This study investigates the relations in supply chains of the automotive industry and involves a mixed-method approach, using SNA methods and annual reports of the automotive firms. The purpose of this article is to assess the interaction between manufacturers and suppliers in terms of resource use in respective business models. To achieve the greatest position in the manufacturing ecosystem, firms couple digital technology with product-related services. Worldwide automakers have a policy in place for recycling, reusing, and recovering their products. For these reasons, the automotive supply chain is being compelled to modify its activities to accomplish the governmental guidelines and achieve sustainability goals. The present stage of development of the world automotive industry in the conditions of digitalization indissolubly and everywhere relates to the implementation of new technologies. Today, the automotive industry is moving towards complete digitization, which leads to the development of electric vehicles. Furthermore, the outcomes demonstrate possibilities for firms to improve their processes, making them more sustainable and resilient to challenges such as the COVID-19 pandemic. The emergence of the pandemic has further accelerated the need to make supply chains more sustainable. Due to the COVID-19 pandemic, this Circular Economy supply chain resilience linkage has gained momentum. The mentioned digital technologies, which also affect the digital supply chain, can support the Circular Economy system, which is reflected in the resilience and capabilities of the supply chain. Finally, this study provides information that the combination of digital technologies with product-related services enables a stronger relationship between manufacturers and suppliers in the manufacturing ecosystem. These relations support the manufacturing ecosystem to survive the influence of different environments.

The main limitation of this study is the data set. This study only used annual reports of the automotive firms and analyzed the five most dominant manufacturers. For future research, authors could use the results from a whole consortium to show a wider picture, as well as to expand the research with other actors. It is also desirable to include the annual report from 2022, to investigate possible differences and to try to establish a two-year trend of progress. Moreover, future research could include more information about Industry 5.0 concepts in the supply chain of the automotive industry. With this information, production managers could receive a clearer picture of the sustainability and resilience aspects, rather than only of the value of digitalization in the automotive industry.

Author Contributions: Conceptualization, A.J.-Z., N.M., M.P. and S.R.; methodology, T.T., M.P., N.M. and A.J.-Z.; analyzed the data: A.J.-Z. and S.R.; investigation, N.M., M.P. and T.T.; writing—original draft preparation, A.J.-Z., N.M., M.P., T.T. and S.R.; writing—review and editing, A.J.-Z., N.M. and S.R. All authors have read and agreed to the published version of the manuscript.

Funding: This research received no external funding.

Data Availability Statement: Not applicable.

Acknowledgments: The results presented in this paper are part of the research within the project "Implementation of the results of scientific and research activities in the field of Industrial Engineering and Management in the DIEM teaching processes with the aim of their continuous improvement", Department of Industrial Engineering and Management, Faculty of Technical Sciences in Novi Sad, University of Novi Sad, Republic of Serbia.

Conflicts of Interest: The authors declare no conflict of interest.

References

1. Llopis-Albert, C.; Rubio, F.; Valero, F. Impact of Digital Transformation on the Automotive Industry. *Technol. Forecast. Soc. Change* **2021**, *162*, 120343. [CrossRef]
2. Chatzopoulos, C.G.; Weber, M. Digitization and Lean Customer Experience Management: Success Factors and Conditions, Pitfalls and Failures. *Int. J. Ind. Eng. Manag.* **2021**, *12*, 73–84. [CrossRef]
3. Weiqing, L.; Xiaoli, J. Digital Transformation: A Review and Research Framework. *Eur. Manag. J.* **2022**, *5*, 21–27. [CrossRef]
4. Lakhani, K.R.; Iansiti, M. How Connections, Sensors, and Data Are Revolutionizing Business. *Harv. Bus. Rev.* **2014**, *92*, 19.
5. Zott, C.; Amit, R. Business Model Innovation: How to Create Value in a Digital World. *NIM Mark. Intell. Rev.* **2017**, *9*, 18–23. [CrossRef]
6. Rakic, S.; Pero, M.; Sianesi, A.; Marjanovic, U. Digital Servitization and Firm Performance: Technology Intensity Approach. *Eng. Econ.* **2022**, *33*, 398–413. [CrossRef]
7. Paschou, T.; Rapaccini, M.; Adrodegari, F.; Saccani, N. Digital Servitization in Manufacturing: A Systematic Literature Review and Research Agenda. *Ind. Mark. Manag.* **2020**, *89*, 278–292. [CrossRef]
8. Medić, N.; Anišić, Z.; Lalić, B.; Marjanović, U.; Brezocnik, M. Hybrid Fuzzy Multi-Attribute Decision Making Model for Evaluation of Advanced Digital Technologies in Manufacturing: Industry 4.0 Perspective. *Adv. Prod. Eng. Manag.* **2019**, *14*, 483–493. [CrossRef]
9. Spieske, A.; Birkel, H. Improving Supply Chain Resilience through Industry 4.0: A Systematic Literature Review under the Impressions of the COVID-19 Pandemic. *Comput. Ind. Eng.* **2021**, *158*, 107452. [CrossRef]
10. Piccinini, E.; Hanelt, A.; Gregory, R.W.; Kolbe, L.M. Transforming Industrial Business: The Impact of Digital Transformation on Automotive Organizations. *Int. Conf. Inf. Syst. Explor. Inf. Front.* **2015**, *5*, 1–20.
11. Pató, B.S.G.; Herczeg, M.; Csiszárik-Kocsir, Á. The COVID-19 Impact on Supply Chains, Focusing on the Automotive Segment during the Second and Third Wave of the Pandemic. *Risks* **2022**, *10*, 189. [CrossRef]

12. Available online: https://www.mckinsey.com/industries/automotive-and-assembly/our-insights/reimagining-the-auto-industrys-future-its-now-or-never (accessed on 5 January 2023).
13. Yadav, G.; Luthra, S.; Jakhar, S.K.; Mangla, S.K.; Rai, D.P. A Framework to Overcome Sustainable Supply Chain Challenges through Solution Measures of Industry 4.0 and Circular Economy: An Automotive Case. *J. Clean Prod.* **2020**, *254*, 120112. [CrossRef]
14. Rüßmann, M.; Lorenz, M.; Gerbert, P.; Waldner, M.; Justus, J.; Harnisch, M. Industry 4.0: The Future of Productivity and Growth in Manufacturing Industries. *Boston Consult. Group* **2015**, *9*, 54–89.
15. Liao, Y.; Deschamps, F.; Loures, E.D.F.R.; Ramos, L.F.P. Past, Present and Future of Industry 4.0—A Systematic Literature Review and Research Agenda Proposal. *Int. J. Prod. Res.* **2017**, *55*, 3609–3629. [CrossRef]
16. Marjanovic, U.; Rakic, S.; Lalic, B. *Digital Servitization: The Next "Big Thing" in Manufacturing Industries*; Springer International Publishing: Cham, Switzerland, 2019; Volume 566, ISBN 9783030299996.
17. Frank, A.G.; Mendes, G.H.S.; Ayala, N.F.; Ghezzi, A. Servitization and Industry 4.0 Convergence in the Digital Transformation of Product Firms: A Business Model Innovation Perspective. *Technol. Forecast. Soc. Change* **2019**, *141*, 341–351. [CrossRef]
18. Frank, A.G.; Dalenogare, L.S.; Ayala, N.F. Industry 4.0 Technologies: Implementation Patterns in Manufacturing Companies. *Int. J. Prod. Econ.* **2019**, *210*, 15–26. [CrossRef]
19. Kamp, B.; Gamboa, J.P. Industry 4.0 Technologies, Skills and Training and Their Influence on Servitization in Industrial Firms Industry 4.0 Technologies, Skills and Training and Their Influence on the Servitization of Industrial Firms. *Servitization A Pathw. Towards A Resilient Prod. Sustain. Future* **2021**, *10*, 174.
20. Ennis, C.; Barnett, N.; De Cesare, S.; Lander, R.; Pilkington, A. A Conceptual Framework for Servitization in Industry 4.0: Distilling Directions for Future Research. *SSRN Electron. J.* **2020**, *4*. [CrossRef]
21. Fosso Wamba, S.; Akter, S.; Edwards, A.; Chopin, G.; Gnanzou, D. How "big Data" Can Make Big Impact: Findings from a Systematic Review and a Longitudinal Case Study. *Int. J. Prod. Econ.* **2015**, *165*, 234–246. [CrossRef]
22. Janković, A.; Adrodegari, F.; Saccani, N.; Simeunović, N. Improving Service Business of Industrial Companies through Data: Conceptualization and Application. *Int. J. Ind. Eng. Manag.* **2022**, *13*, 78–87. [CrossRef]
23. Lee, J.; Kao, H.A.; Yang, S. Service Innovation and Smart Analytics for Industry 4.0 and Big Data Environment. *Procedia CIRP* **2014**, *16*, 3–8. [CrossRef]
24. Wittmann, J. Electrification and Digitalization as Disruptive Trends: New Perspectives for the Automotive Industry? In *Phantom Ex Machina*; Springer International Publishing: Cham, Switzerland, 2017; pp. 137–159.
25. Rakic, S.; Visnjic, I.; Gaiardelli, P.; Romero, D.; Marjanovic, U. *Transformation of Manufacturing Firms: Towards Digital Servitization*; Springer International Publishing: Cham, Switzerland, 2021; Volume 631, IFIP; ISBN 9783030859015.
26. Acerbi, F.; Sassanelli, C.; Terzi, S.; Taisch, M. A Systematic Literature Review on Data and Information Required for Circular Manufacturing Strategies Adoption. *Sustainability* **2021**, *13*, 2047. [CrossRef]
27. Acerbi, F.; Sassanelli, C.; Taisch, M. A Conceptual Data Model Promoting Data-Driven Circular Manufacturing. *Oper. Manag. Res.* **2022**, *15*, 838–857. [CrossRef]
28. Vandermerwe, S.; Rada, J. Servitization of Business: Adding Value by Adding Services Sandra. *Eur. Manag. J.* **1988**, *6*, 314–324. [CrossRef]
29. Rabetino, R.; Harmsen, W.; Kohtamäki, M.; Sihvonen, J. Structuring Servitization-Related Research. *Int. J. Oper. Prod. Manag.* **2018**, *38*, 350–371. [CrossRef]
30. Opresnik, D.; Taisch, M. The Value of Big Data in Servitization. *Int. J. Prod. Econ.* **2015**, *165*, 174–184. [CrossRef]
31. Coreynen, W.; Matthyssens, P.; Van Bockhaven, W. Boosting Servitization through Digitization: Pathways and Dynamic Resource Configurations for Manufacturers. *Ind. Mark. Manag.* **2017**, *60*, 42–53. [CrossRef]
32. Ayala, F.; Frank, G. The Moderating Role of Service Suppliers Managing Servitization in Product Companies: The Moderating Role of Service Suppliers. *Int. J. Oper. Prod. Manag.* **2019**, *39*, 43–74. [CrossRef]
33. Pournader, M.; Shi, Y.; Seuring, S.; Koh, S.C.L. Blockchain Applications in Supply Chains, Transport and Logistics: A Systematic Review of the Literature. *Int. J. Prod. Res.* **2020**, *58*, 2063–2081. [CrossRef]
34. Magnusson, P.R. *Customer-Oriented Product Development: Experiments Involving Users in Service Innovation*; Stockholm School of Economics: Stockholm, Sweden, 2003; ISBN 9172586184.
35. Lenfle, S.; Midler, C. The Launch of Innovative Product-Related Services: Lessons from Automotive Telematics. *Res. Policy* **2009**, *38*, 156–169. [CrossRef]
36. Pflaum, A. Introduction to The Digital Supply Chain of the Future: Technologies, Applications and Business Models Minitrack The Digital Supply Chain of the Future: Technologies, Applications and Business Models Minitrack. 2017. Available online: https://aisel.aisnet.org/cgi/viewcontent.cgi?article=1531&context=hicss-50 (accessed on 3 January 2023).
37. Psychiatry, C.; Jalali-farahani, S.; Chin, Y.; Taib, N.; Amiri, P. Disordered Eating and Its Association with Overweight and Health-Related Quality of Life among Adolescents in Selected High Schools of Tehran Disordered Eating and Its Association with Overweight and Health-Related Quality of Life among Adolescents in Sel. *Child Psychiatry Hum. Dev.* **2014**, *46*, 485–492. [CrossRef]
38. Ivanov, D. Viable Supply Chain Model: Integrating Agility, Resilience and Sustainability Perspectives—Lessons from and Thinking beyond the COVID-19 Pandemic. *Ann. Oper Res.* **2020**, *319*, 1411–1431. [CrossRef] [PubMed]

39. Fatorachian, H.; Kazemi, H. The Management of Operations A Critical Investigation of Industry 4.0 in Manufacturing: Theoretical Operationalisation Framework. *Prod. Plan. Control.* **2018**, *7287*, 1–12. [CrossRef]
40. Nandi, S.; Sarkis, J.; Hervani, A.; Helms, M. Do Blockchain and Circular Economy Practices Improve Post COVID-19 Supply Chains? A Resource-Based and Resource Dependence Perspective. *Ind. Manag. Data Syst.* **2021**, *121*, 333–363. [CrossRef]
41. Lechler, S.; Canzaniello, A.; Roßmann, B.; von der Gracht, H.A.; Hartmann, E. Real-Time Data Processing in Supply Chain Management: Revealing the Uncertainty Dilemma. *Int. J. Phys. Distrib. Logist. Manag.* **2019**, *49*, 1003–1019. [CrossRef]
42. Taddei, E.; Sassanelli, C.; Rosa, P.; Terzi, S. Circular Supply Chains in the Era of Industry 4.0: A Systematic Literature Review. *Comput. Ind. Eng.* **2022**, *170*, 108268. [CrossRef]
43. Bag, S.; Yadav, G.; Wood, L.C.; Dhamija, P.; Joshi, S. Industry 4.0 and the Circular Economy: Resource Melioration in Logistics. *Resour. Policy* **2020**, *68*, 101776. [CrossRef]
44. Hussain, M.; Malik, M. Organizational Enablers for Circular Economy in the Context of Sustainable Supply Chain Management. *J. Clean Prod.* **2020**, *256*, 120375. [CrossRef]
45. Zhang, X.; Chen, H.; Liu, Z. Operation Strategy in an E-commerce Platform Supply Chain: Whether and How to Introduce Live Streaming Services? *Int. Trans. Oper. Res.* **2022**, 1–29. [CrossRef]
46. Zhang, X.; Li, Q.; Liu, Z.; Chang, C.-T. Optimal Pricing and Remanufacturing Mode in a Closed-Loop Supply Chain of WEEE under Government Fund Policy. *Comput. Ind. Eng.* **2021**, *151*, 106951. [CrossRef]
47. Pinho Santos, L.; Proença, J.F. Developing Return Supply Chain: A Research on the Automotive Supply Chain. *Sustainability* **2022**, *14*, 6587. [CrossRef]
48. MacGregor Pelikánová, R. Corporate Social Responsibility Information in Annual Reports in the EU—A Czech Case Study. *Sustainability* **2019**, *11*, 237. [CrossRef]
49. Sofic, A.; Rakic, S.; Pezzotta, G.; Markoski, B.; Arioli, V.; Marjanovic, U. Smart and Resilient Transformation of Manufacturing Firms. *Processes* **2022**, *10*, 2674. [CrossRef]
50. Harrison, A.; Van Hoek, R. *Logistics Management and Strategy*; McGraw-Hill/Irwin: Boston, MA, USA; ISBN 9780273712763.
51. Dalenogare, L.S.; Benitez, G.B.; Ayala, N.F.; Frank, A.G. The Expected Contribution of Industry 4.0 Technologies for Industrial Performance. *Int. J. Prod. Econ.* **2018**, *204*, 383–394. [CrossRef]
52. Zivlak, N.; Rakic, S.; Marjanovic, U.; Ciric, D.; Bogojevic, B. The Role of Digital Servitization in Transition Economy: An SNA Approach. *Teh. Vjesn.—Tech. Gaz.* **2021**, *28*, 1912–1919. [CrossRef]

sustainability

MDPI

Review

Synergies and Trade-Offs between Lean-Green Practices from the Perspective of Operations Strategy: A Systematic Literature Review

Geandra Alves Queiroz [1,*], Ivete Delai [2], Alceu Gomes Alves Filho [2], Luis Antonio de Santa-Eulalia [3] and Ana Lúcia Vitale Torkomian [2]

1 Department of Engineering, Production Engineering, University of Minas Gerais State (UEMG), Passos 37900-106, MG, Brazil
2 Department of Industrial Engineering, Federal University of São Carlos, São Carlos 13565-905, SP, Brazil
3 Department of Information Systems and Quantitative Methods in Management, École de Gestion, Université de Sherbrooke, Sherbrooke, QC J1K 2X9, Canada
* Correspondence: geandraqueiroz@gmail.com

Abstract: In the operations management and sustainability literature, the integration of Lean and Green manufacturing is considered one of the great solutions to balancing operational gains and environmental sustainability. This literature focuses mainly on the integration between them. However, there are no studies investigating how this integration is related to the Operations Strategy content: competitive priorities and decision areas. Thus, this study aims to contribute to reducing this research gap by providing a more in-depth understanding of the relationships between Lean-Green practices from the point of view of the Operations Strategy. We identify synergies and potential trade-offs between competitive priorities and changes in decision areas when Lean-Green practices are implemented. We performed a systematic literature review to answer two questions: Does the implementation of Lean and Green practices affect operations' competitive priorities, causing synergies or trade-offs? What decision area(s) are modified with the implementation of each practice? This systematic review analyzed 338 selected articles. Competitive priorities, decision areas, Lean practices, Green practices and Lean-Green practices were identified and discussed, highlighting trade-offs, synergies and changes in decision areas. The results suggest that Lean and Green are synergistic in most practices, but they must be managed according to the Operations Strategy, especially as their focuses are essentially different and trade-offs may occur.

Keywords: lean manufacturing; green manufacturing; competitive priorities; decision areas; sustainability

Citation: Queiroz, G.A.; Delai, I.; Alves Filho, A.G.; Santa-Eulalia, L.A.d.; Torkomian, A.L.V. Synergies and Trade-Offs between Lean-Green Practices from the Perspective of Operations Strategy: A Systematic Literature Review. *Sustainability* **2023**, *15*, 5296. https://doi.org/10.3390/su15065296

Academic Editors: Sergio Terzi and Claudio Sassanelli

Received: 10 January 2023
Revised: 6 March 2023
Accepted: 13 March 2023
Published: 16 March 2023

1. Introduction

Environmental problems, such as climate change, pollution, the reduction of natural resources and loss of biodiversity, tend to collapse the planet [1]. Facing this context, society, governments, investors, and companies themselves have increasingly demanded the elimination or reduction of the environmental impacts of products and production processes [2,3]. To meet these demands, companies seek to adopt programs and practices that reduce such impacts but, at the same time, provide for the achievement of their competitive production priorities [4,5]. An Operations Strategy (OS) aligned with market requirements is essential, as it can determine the company's competitive advantage [6].

Lean and Green Manufacturing practices have been seen as a solution to improve and balance all the competitive priorities of OS; the relationship between these two strands has been explored by the Lean-Green (LG) integration literature. The study of Leong et al. [6] pointed out that the LG approach can obtain the maximum operational performance without compromising the environment. Recent research [7,8] presents cases from results in both operational gains and environmental aspects, such as lead time reduction, and water and wastes reduction. Complementary, Caldera et al. [9] consider that LG enable the transition to sustainable business. Many studies [10–12] demonstrate that Lean aims to reduce wastes in the value chain, which can contribute to reducing costs and defects and to the increasing of natural resources (e.g., water, energy and materials) efficiency. Lean and Green are complementary, and Lean enables the development of environmental management capability helping to "green" the organization [13–15]. However, Lean does not take into account environmental impacts directly. Thus, organizations need to implement Green tools into the Lean management to fill this gap [14]. To refer to Lean and Green integration, the term "Lean-Green" has been used [16–19] and will be adopted in this work.

Research about the Lean-Green has addressed three main topics. Some studies explore the relationship between Lean and Green; they present the synergies and differences among them. The authors argue that the approaches are very compatible but some trade-offs may appear and should be considered [13,20–29]. Another group of authors proposes tools and frameworks to integrate Lean and Green, describing requirements and steps to implement them, as well as barriers and enablers to doing so [2,14,18,22,30–40]. There is also a group focused on the implementation of LG. In general, these studies show improvements in cost reduction, quality and environmental performance; mainly through reduction in energy consumption and waste generation [7,10,41–43]. However, there are studies that show some negative results in the environment, such as the increase of emissions when Just-in-Time (JIT) is implemented [13,20,44–50].

Despite these efforts, an integrated and holistic understanding of how LG are linked with all OS content (namely, competitive priorities and decision areas) is still missing. Chatha and Butt [51] presented an extensive literature review about OS, providing a historical overview and the current status of this topic; but the results did not show any study that discussed LG and the entire OS content. There are only a few studies in the literature that somehow correlated OS with LG: Longoni and Cagliano [52] provided evidence about how the cross-functional executive involvement and worker involvement, in the formulation and implementation of the OS supporting the strategic alignment of Lean and sustainability; Suifan et al. [53] analyze the trade-offs between LG and through a multi-criteria decision-making shows that competitive priorities can differ in each approach; and, Queiroz et al. [26] present the competitive priorities and the Lean-Green practices adopted for automotive suppliers However, both studies still do not provide a wide understanding about the relationships between LG and OS.

In addition, it is important to mention there are some systematic reviews that focus on Lean-Green [4,36,54–59]. However, these studies do not consider the relationship of Lean-Green and Operations Strategy content (competitive priorities and decision areas) to understand the trade-offs, synergies and changes in the decision areas. Considering the relevance of the operations function–for the organization's competitiveness, the current stakeholders' requirements, and the lack of studies on LG practices from the perspective of the consolidated background of OS–this study aims to provide a broad perspective on the possible impacts of adopting LG practices on the content of OS, highlighting the occurrence of synergies and tradeoffs between competitive priorities and the changes promoted in decision areas. Thus, this work seeks to take a step towards systematizing contributions from studies that have addressed issues related to constructs in the field of OS—competitive priorities and decision areas—when LG practices are implemented, and to update and expand this part of part of a doctoral thesis developed in 2021 by Queiroz et al. [60]. With this objective, a Systematic Literature Review (SLR) and a content analysis of 338 articles were carried out from 1996 to the present moment. The results can contribute to a better understanding of the trade-offs and implementation practices aligned with corporate sustainability objectives, allowing the development of well-positioned production systems that can meet new market demands and be consistent with the global strategy of the company.

This paper is organized as follows. We present a brief theoretical background about OS and LG in Section 2. Next, Section 3 explains the research design, the process of data collection and the method used to select studies and perform content analysis. Subsequently, Section 4 discusses our findings, which include descriptive evidence regarding the sample of articles and the results from the content analysis to answer the research questions. Lastly, Section 5 highlights the main implications, and proposes some avenues for future research.

2. Theoretical Background

Operation Strategy is the set of decisions that seeks to balance production resources with market needs to contribute to the overall strategy of organizations [61]. Skinner [62] published the first study discussing OS; it was emphasized that the production function should be considered strategic and as a source of competitive advantage. The implementation of an adequate OS, including the development of production function capabilities, plays a crucial role for companies in the business environment, and must be in line with the way the company seeks to create competitive advantages [63].

The success of an OS is related to the definition of its content, which is composed of competitive priorities (CP) and actions to be implemented in decision areas (DA) [64]. CP are related to the performance objectives that the production function adopts to align itself to the company's competitive strategy [62], which are: cost, quality, delivery, flexibility and service [62,65]. Moreover, as indicated by Longoni and Cagliano [5], "environment" can be considered another competitive priority of the operations. These priorities are achieved through a pattern of decisions and actions implemented in the set of DA of the company, such as facilities, capacity, technology, supply chain, human resources, quality, production planning and control, product development, performance measurement systems and organization [62,66].

The content of OS can be seen through the lens of the "strategic choices" paradigm (one of the three proposed by [64,67] in which strategic decisions in the processes and infrastructures of organizations guide the implementation of practices (or actions) and changes in decision areas aimed at improving the performance of operations and gaining competitive advantages. This is the perspective chosen in this work to examine the impacts promoted by the adoption of LG practices in decision areas and competitive production priorities.

Lean Manufacturing (LM) is considered one of the most used approaches to improve operations performance and increase competitiveness [68]. LM is a set of principles and practices that aims to eliminate all kinds of waste in an organization [69]. It is an approach that goes beyond a production management strategy, and it can be considered a management philosophy [70] and an integrated socio-technical system [71]. The main LM practices are 5s, Kaizen, Value Stream Map (VSM), Just in Time (JIT), SMED, Total Productive Maintenance (TPM), Kanban, Standardized Work, Visual Management and 5 Why's (root cause analysis) [72].

Regarding the concept of Green Manufacturing (GM) or Sustainable Manufacturing (SM), it emerged in the 1990s as a philosophy and operational approach to reduce the negative environmental impacts of products [4]. It concerns the search to reduce pollution, energy consumption and the generation of toxic substances through the development of new processes in the manufacturing phase [73–75]. According to research [75,76], GM, or SM, encompasses different tools to reduce the environmental impacts generated by production processes, such as: Cleaner Production, (Life Cycle Assessment (LCA), Environmental Management System (EMS), Circular Economy (CE), Eco-design/Design for Environment, Green/Sustainable Supply Chain, and 3R (Recycling, Remanufacturing and Reuse) [75].

LG integration has been considered the approach that supports achieving the sustainability performance (economic, environmental and social dimensions) of a production system [77]. There are many proposals in the literature, like frameworks for LG integration, cases that show positive and negative environmental results from Lean implementation, and some integrated tools such as Environmental Value Stream Mapping—E-VSM [78], 7s—that is, 5s plus S (safety) and S (sustainability) [79], and Green Lean Six Sigma [33]. Figure 1 presents the constructs that will be discussed in this SLR.

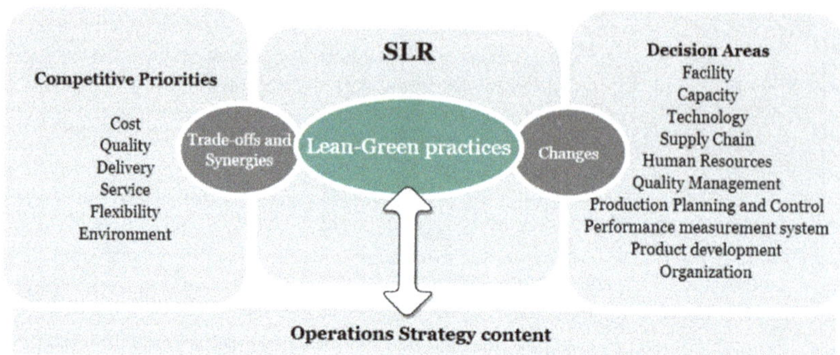

Figure 1. Relationships between the concepts, research method and results. Source: created by authors.

A SLR was carried out to identify contributions in the literature that highlighted the impacts of the implementation of LG practices on operations' competitive priorities and on decision areas.

3. Research Design

This SLR followed the three macro stages proposed by Denyer and Tranfield [80], as well as the Prisma Statement Flow Diagram proposed by Moher et al. [81]. The SLR process is detailed in Figure 2, which illustrates the summary of the SLR protocol to ensure the transparency and reliability of the process. Initially, the SLR protocol was elaborated and validated jointly by all authors. Throughout the SLR development, meetings between the authors were held to evaluate the results and resolve any disagreement.

3.1. Research Question Formulation

We established the SLR research question needed to achieve the aim of the project, which was to understand how Lean-Green are related to the OS content. Considering this, the research questions addressed in this review are:

RQ1. Does the implementation of Lean and Green practices affect operations' competitive priorities, causing synergies or trade-offs?

RQ2. What decision area(s) are modified with the implementation of each practice?

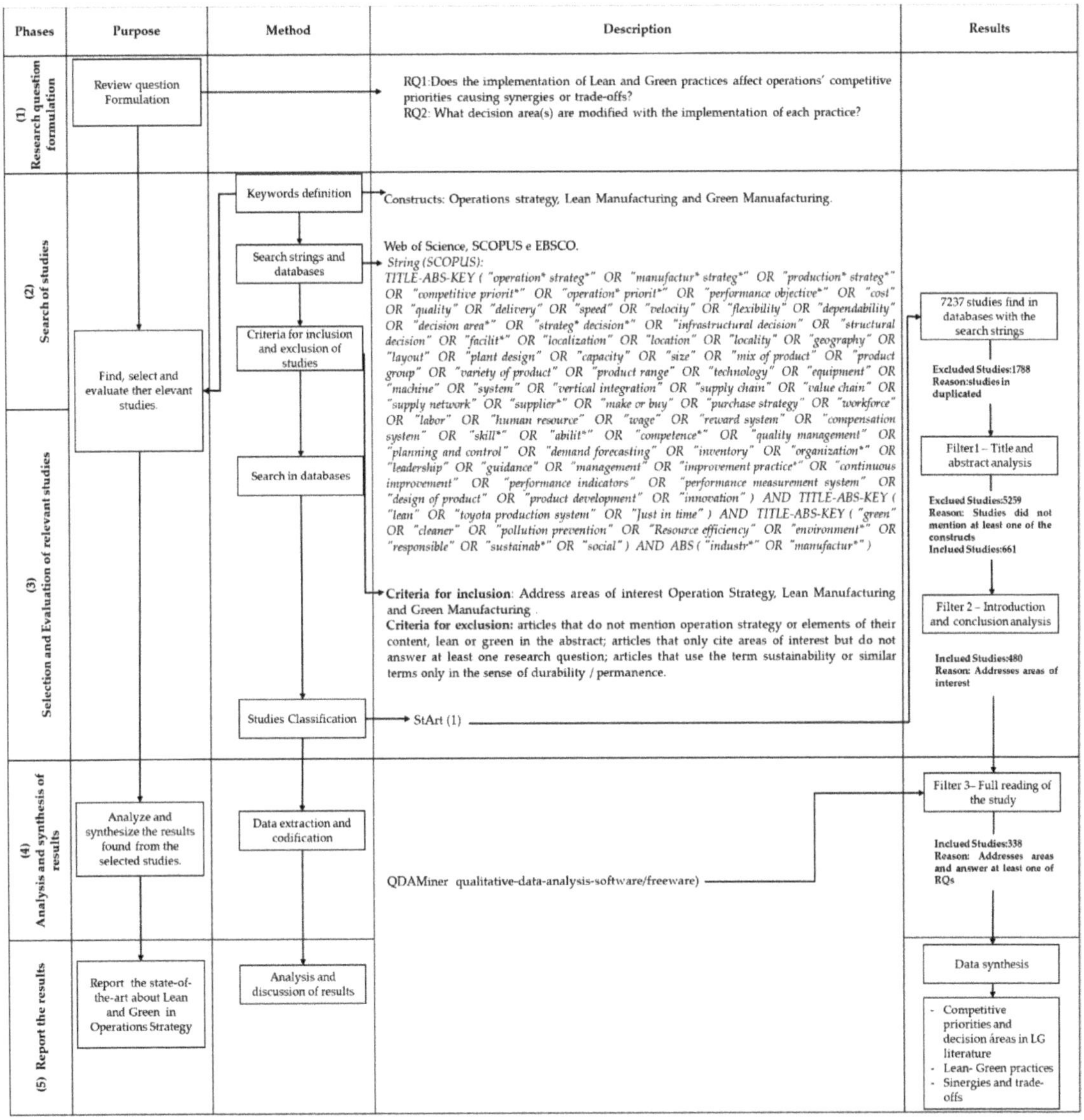

Figure 2. Summary of SLR Protocol. Source: created by authors. http://lapes.dc.ufscar.br/tools/start_tool (accessed on 21 November 2018). https://provalisresearch.com/products/qualitative-data-analysis-software/ (accessed on 10 July 2019).

3.2. Search Strategy

Studies were searched in three databases chosen according to their scientific scope to provide better results. The Scopus from Elsevier and The Web of Science from Thomson Reuters Institute of Scientific Information were chosen as they are regularly updated and have a wide breadth of coverage in most scientific subjects [82]. Seeking to improve the scope of searches, we also included the EBSCO because it is an extensive database in management.

After a preliminary review of the literature in OS [5,51,52,62,83], LM [3,72] and GM [73,74,84], the strings were developed to conduct the search in the databases. We covered several keywords related to the constructs from the RQs: OS, CP, DA, LM and GM. The search method consisted of the use of strings, defined in such a way as to return results that simultaneously contained at least one keyword referring to each construct. In the search string we considered all the synonyms of the constructs, and we detailed all the CPs and the DAs. A more detailed string was chosen for a wider range of articles since there are articles that focus only on one CP or one DA. Furthermore, we did not limit the field "year" to obtain the largest number of articles on the theme. The search in the databases was first carried out in January 2019 and then updated in November 2020 and again in February 2023. In all databases, we did the search in title, abstract and keywords by focusing on journal articles (excluding books and conference papers). Figure 2 presents the strings used in the searches and the summary of the SLR protocol.

3.3. Selection and Evaluation of Relevant Studies

Studies were selected through three filters. The initial search resulted in 7237 studies; it is worth mentioning that 1788 duplicates were excluded. In sequence was applied to the first Filter with the reading of titles and, when necessary, the reading of abstracts. This filter was applied to 5449 papers with the support of StArt software version 3.0.3 BETA. The process was manual, and it took two months to be concluded. There were studies that were very simple to exclude because some titles were completely out of the domain area researched; and for titles that were related with RQs, it was necessary to read the abstract and analyse them. To select the articles for this review, we applied two sets of inclusion and exclusion criteria, presented in Figure 2. The first one was applied during the screening phase; it searched title, abstract and keywords, and included articles that presented at least one term of each construct, related with content of the OS, LM and GM. For example, a document that presented one CP, like "quality", one Lean practice (LP), like "Kanban", and one Green practice (GP), like "Life Cycle Analysis", was selected for the next step. The articles that did not mention at least one decision area or competitive priority or any GP, were excluded.

Then, in Filter 2 (eligibility) we read the (each) paper's introduction and conclusion, and included those that fulfilled the search inclusion criteria: full content access, written in English, published in scientific peer-reviewed journals and discuss at least one element of each construct. For example, the study of [85] cites "Lean" but does not discuss any practice, referring only to some aspect of OS and GM. We also found cases of using the term "sustainability" just to refer to the stability of the practices implemented, as the study of ref. [86]. Studies like these were also excluded following the exclusion criteria "Articles that use the term sustainability or similar terms only in the sense of durability/permanence".

The third filter was then applied to the full paper using the same inclusion and exclusion criteria as Filter 2. In this filter, 142 papers were excluded. As a result, a total of 338 were selected for content analysis to answer the proposed RQs. The main reason for the excluded papers was because the papers mentioned the three constructs but did not answer the research questions. One example is the study of ref. [87] that only cites the LP "JIT" as an example of an initiative in operations. Therefore, considering this sample of 338 papers, the next topic will present how the analysis of this work was done.

3.4. Analysis and Synthesis of the Results

After Filter 3, an analysis of 338 papers was done in full, aimed at extracting specific information from studies related to the research topic. In this filter, the articles were analyzed in a descriptive manner, seeking to generate a classification of the articles by year, journal, country of empirical studies, research method, industrial sector, and main research focus. Moreover, a content analysis was made by following the recommendations of [88] seeking to answer the research questions (RQs). The QDA Miner Software (Version 5) was used as a tool to facilitate the analysis process (individual papers and cross-papers).

According to ref. [89] et al., this software collaborates in the organization of ideas and comparison between the cited. The analysis steps are presented in Figure 3.

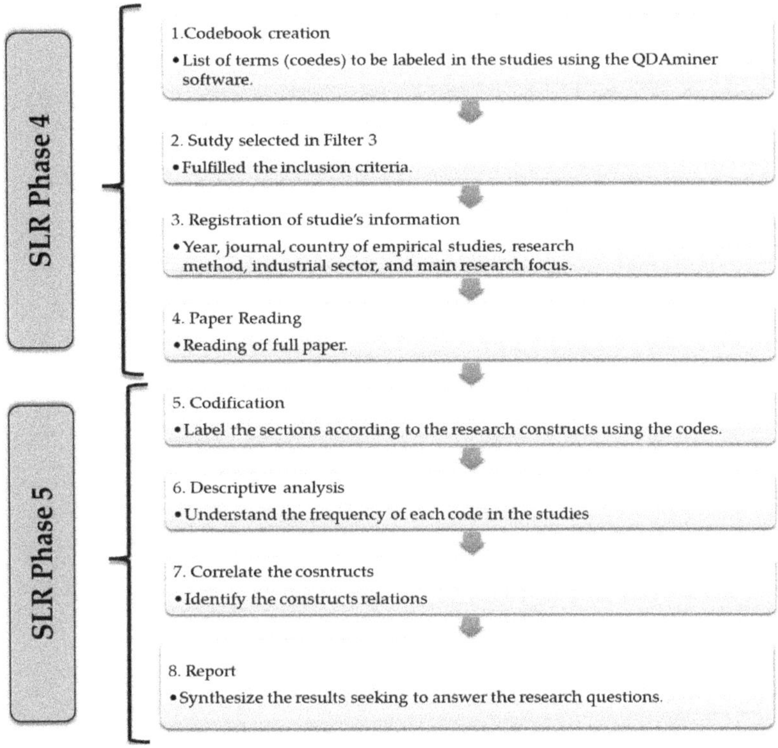

Figure 3. SLR Analysis steps.

The data were coded following the basic requirements proposed by Krippendorff [88]. The codes are very important to help identify relationships and establish connections among the many studies that write about the same topics [72,90]. We used the concept-based coding, extracting data from the texts and others that emerged from the reading related with database coding, as suggested by Gibbs [90]. The codebook, attached in the supplementary material, was defined based on the constructs found in the literature of OS content (competitive priorities and decision areas), LM and GM. In this codebook we specified what the initial codes are from the literature and those from the final process.

The content analysis process started looking for the LP and GP, and the CP and DA in the studies. Later, the frequency analysis of the constructs was carried out seeking which LG practices and elements from OS are discussed in the LG literature. This whole process was supported and supervised by three senior researchers. Once the papers were codified, and the content of OS and the practices from LM and GM were identified, it was possible to find the relationship between them all. In the supplementary material are specified all the competitive priorities, decision areas, lean practices, green practices and lean-green practices found and studied where they were addressed. The next topic presents the analysis results (Step 5 of SLR) and discussion.

4. Results and Discussion

The results concerning LG from the perspective of OS are discussed in two parts. First, we describe the sample (Section 4.1), and then how Lean-Green is related to the OS content (Section 4.2).

4.1. Descriptive Analysis

It is observed that half of the studies were published in the last four years (Figure 4). The first two publications were Florida [12] and Ferrone [44], and the year with the largest number of publications was 2019. This growth may be attributed to two main reasons. First, the need to integrate sustainability issues in productive systems has awakened the interest of academia in studying practices that focus on that. Secondly, based on the initial proposals of LG integration, studies have focused on their validation and implementation.

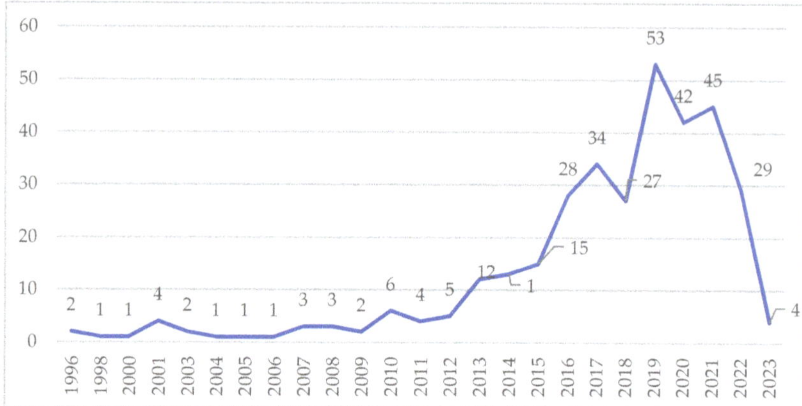

Figure 4. Historical evolution of articles analyzed. Source: created by the authors.

In terms of the number of publications per journal, studies were found in 153 different journals, and the journals with more than 1% of publications presented in Figure 5 50.15% of the studies analyzed. *The Journal of Cleaner Production* is the outlet with the highest number of publications on this topic, with 61 publications, which corresponds to 18% of the total articles. The second journal is *Sustainability* with 5.93%. Figure 5 presents this ranking and the journals that have up to three articles.

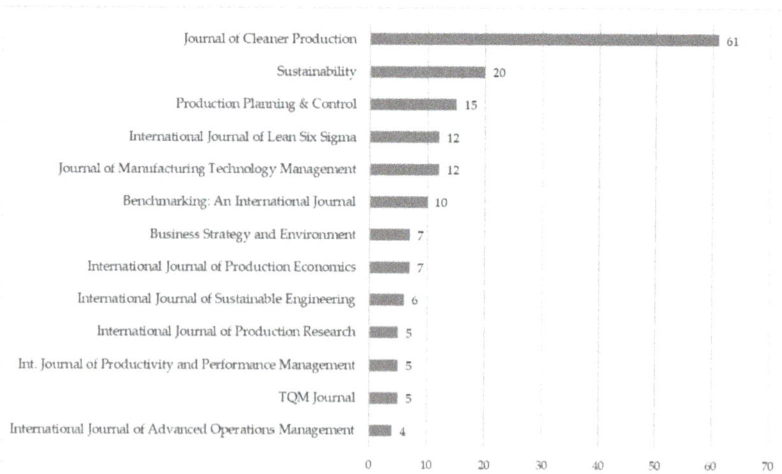

Figure 5. Number of articles published by journal. Source: created by the authors.

Figure 6 presents the research methods used in the studies and some characteristics of the samples for the empirical papers. There is a rising trend to adopt empirical studies (79.88%) that applied mainly case studies (33.72% of the sample) and survey (26.03%).

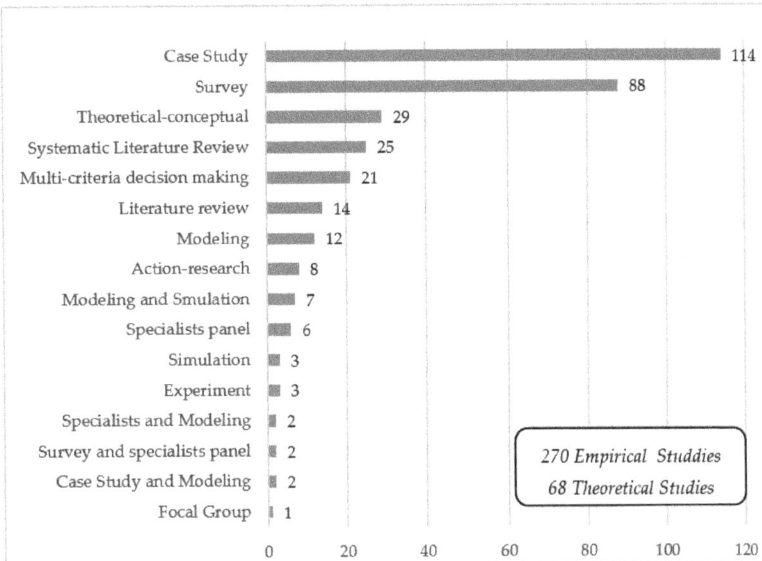

Figure 6. Research Method.

Furthermore, Figure 7 presents the classification of the empirical studies regarding the industrial sector and the country, where the study was done and the size of the organizations. Regarding the industrial sector where empirical research was done, the three most frequent sectors were automotive, metal-mechanical, civil construction and electro-electronics, corresponding to 36.66% of the empirical studies.

As for the location, the majority (around 45%) of the research was done in India, Brazil, Malaysia, the United States and China. UNCTAD (United Nations Conference on Trade and Development) [91] developed the status classification. Only the US is classified as a developed economy, while the other ones are classified as developing economies. Therefore, these data can mean that countries that focus on developing their economies are seeking practices to improve competitiveness sustainably. Figure 8 presents a global map with the papers' distributions, and Table 1 presents a distribution of the papers per country of empirical studies.

Regarding the size of the companies in the empirical studies, Figure 9 representd that only 100 studies identified it, with 12% studying large companies, 9.62% studying SMEs companies, and 4% the small companies; 7.4% were done with a mixed sample including small, medium and large, and another 2.2% represented medium companies.

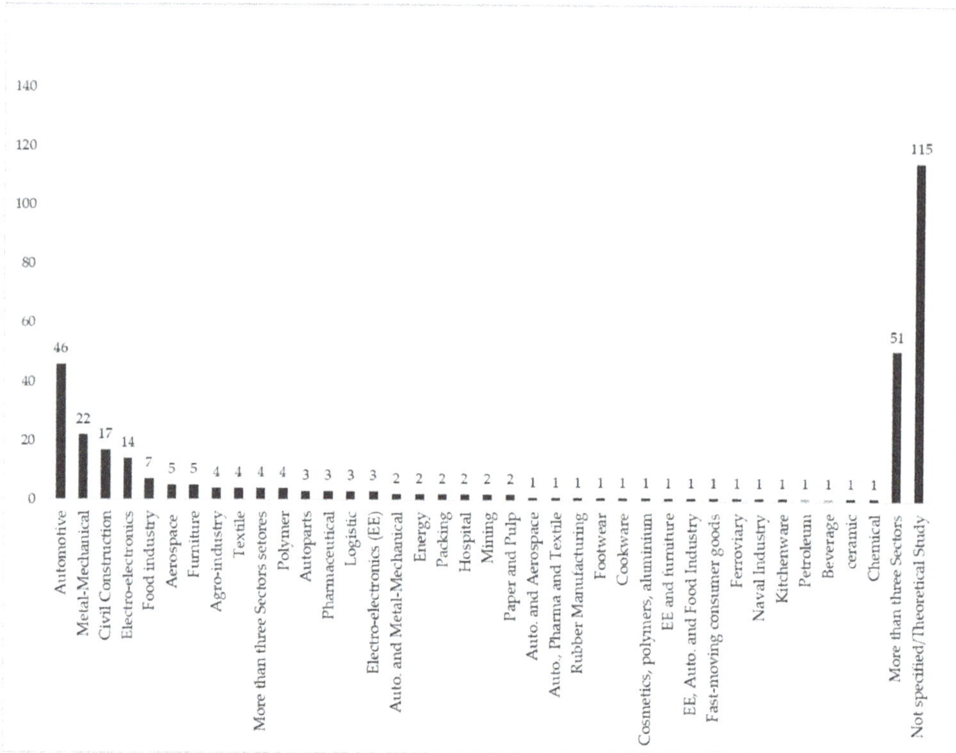

Figure 7. Industrial Sector of empirical studies.

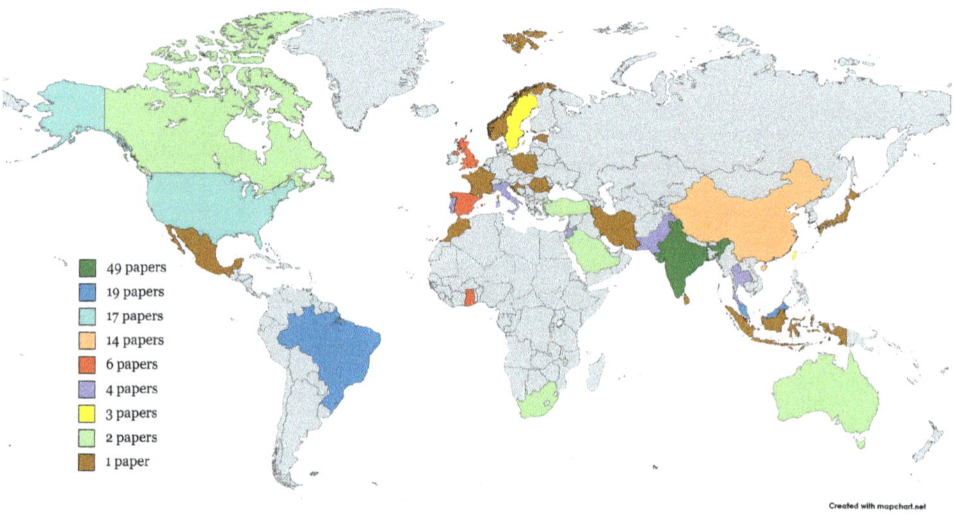

Figure 8. Papers per country of empirical studies.

Table 1. Distribution of the papers per country of empirical studies.

Country of Study	Total of Studies	Total of Studies (%)
Not specified or theoretical study	136	40.24%
India	49	14.50%
Brazil	19	5.62%
Malaysia	19	5.62%
United States	17	5.03%
China	14	4.14%
Several countries	9	2.66%
United Kingdom	6	1.78%
Spain	6	1.78%
Ghana	6	1.78%
Portugal	4	1.18%
Singapore	4	1.18%
Jordan	4	1.18%
Pakistan	4	1.18%
Thailand	4	1.18%
Italy	4	1.18%
Sweden	3	0.89%
Taiwan	3	0.89%
Australia	2	0.59%
Canada	2	0.59%
Various	2	0.59%
South Africa	2	0.59%
Saudi Arabia	2	0.59%
Turkey	2	0.59%
North Africa	1	0.30%
France	1	0.30%
Estonia	1	0.30%
Poland	1	0.30%
Romania	1	0.30%
Faroe Islands	1	0.30%
Norway	1	0.30%
Japan	1	0.30%
Croatia	1	0.30%
Wales	1	0.30%
Sri Lanka	1	0.30%
Indonesia	1	0.30%
Mexico	1	0.30%
Iran	1	0.30%
Morocco	1	0.30%

Most studies aimed to present steps, frameworks and guidelines to integrate the two approaches. Another presented one or more specific hybrid tools that are integrated tools, like E-VSM, that take the VSM from Lean and the environmental aspects from Green. The other major focus of the LG literature is to try to discuss the main links between these approaches, the synergies and trade-offs between them, highlighting the negative and positive impacts that each one has on the other. However, there is no study that discusses LG from the OS perspective.

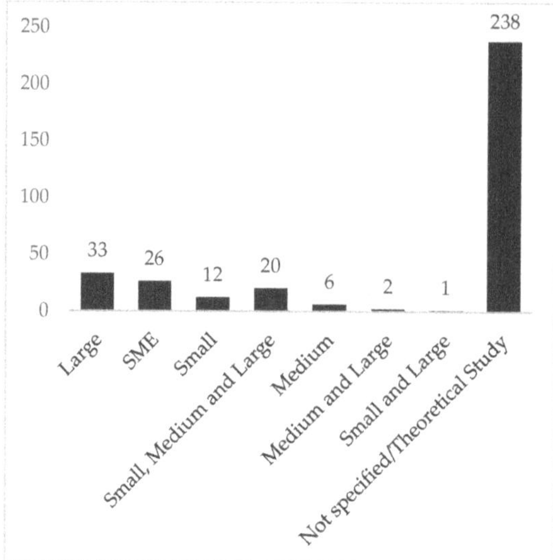

Figure 9. Size of the companies studied.

4.2. Discussion

First, it is relevant to present the CP and DA that are discussed in the LG literature. The priority "cost" was the most frequent in these studies, and it is related to one of the main goals of Lean Manufacturing, i.e., cost reduction [15,92]. Cost reduction used to be the motivation of companies that aim to become green: they seek to reduce the costs of materials and energy inputs as well as the costs of waste disposal [93].

Regarding the priority "environment", most of the studies discuss the reduction of the negative environmental impacts from the production process that implements Lean. These studies consider this competitive priority as a new one aiming to reduce mainly energy consumption, emissions, water consumption, waste generation and toxic substances. Also, some researchers call the negative environmental impacts in Lean operations "Green waste" or "Environmental waste" [10,11,76,94].

Next, "quality" and "delivery" are both focuses of Lean implementation [46,53]. Quality is discussed, as in [12,95], as a reduction or elimination of defects which can decrease the consumption of raw materials and the costs of production. The priority delivery is addressed when the studies discuss order lead-time, which is one of the metrics considered in Lean implementations. According to Dües et al. [96], customer satisfaction is driven by the reduction of lead time and, as indicated by [30], one of the results from Lean is to allow faster deliveries.

The other priorities found in the studies, but less frequently in Lean systems, were social, service and flexibility. The priority social is related to improving workers' health, safety and morale, improving the local supply, and reducing corruption risk [33]. Service is mentioned as the improvements in customer satisfaction through Lean and Green implementation [96]. Regarding flexibility, the capacity to increase the mix of products that can be targeted by LP is mentioned [31,97].

In addition to the CP, we observed the DA of OS cited in these studies. It is possible to note that Performance Measurement System is the most frequent decision area in these studies. This can be explained because the inclusion of an environmental performance indicator is one of the highest requirements to consolidate GP in industrial operations [18,37,76], and the performance measurement systems in LM are an important aspect [98]. Aligned with this, as highlighted by Leong et al. [6], it is necessary to use support technologies that allow the operational data to be registered for an effective improvement process.

Regarding Supply Chain, the literature discusses mainly supplier relationship (collaboration, selection, purchasing) and logistics [99]; trade-offs or synergies from the JIT deliveries and the integration of the supply chain to implement an LCA; or it analyzes environmental impact from the Life Cycle perspective. Human Resources is another much-discussed decision area, which includes essential changes and the importance of employee involvement in training to implement and to sustain both Lean and Green [30,47,98].

Quality Management is a decision area that is also studied. It includes the implementation of quality programs, like six sigma or Total Quality Control or ISO standard [100]. Technology is normally related to the improvement of equipment to reduce resources consumption, which includes information systems and Industry 4.0 initiatives [6,47]. Product development, on the other hand, addresses improving the design of the product, aiming to improve environmental performance since the beginning of the project [56,96]. Organization area is related to issues of the leadership structure required for Lean and Green integration to attain success in the implementation of the practices [30,33]. Production, Planning and Control discussions include the impacts of a schedule to achieve environmental optimization. Facilities are mentioned to address layout change or facilities projects, like the positive environmental impacts of i cellular manufacturing implementation because of less motion and transportation [101]. Finally, capacity, the least addressed decision area, was discussed regarding the size of batch and capacity planning [9].

Regarding how LG affects the operations' competitive priorities, causing trade-offs synergies and decision area(s) modified with the implementation of LG, we found the relationships between LG and OS are studied in three different ways by the studied authors. The first group focuses on discussing the positive and negative environmental effects of LP on GM; thus Lean and Green are analyzed separately. The other one discusses the GP adopted as a complement to Lean practices by Lean-oriented companies to address the negative environmental impacts of their operations. In this group, Lean and Green are analyzed separately. Finally, in contrast to the previous groups, the third one discusses the effects of Lean and Green together through the analysis of the so-called hybrid practices, or Lean-Green practices, on some Operations Strategy aspect.

4.2.1. Lean Practices

It was possible to find a wide range of cited Lean practices, covering 26 different practices in total, as presented in Figure 10. They are used in these studies to show how they can help to achieve better environmental performance, such as by reducing energy consumption, the consumption of materials, wastes generation and emissions; and to show the benefits already known from Lean, such as the reduction of lead time and cost, as well as the improvement of quality and productivity. Moreover, there are some practices, like VSM and TQM, that are mentioned as a foundation for integrating Lean with Green and make a hybrid practice showing synergies between them. These practices are presented in Section 4.2.3. However, some studies mention that there are practices that can increase negative environmental impacts.

As regards the synergy with the priority "environment", all the LPs are considered synergic because of their main focus on waste reduction. Several studies, e.g., [11,102], have demonstrated that Lean Practices can bring environmental benefits and that this can be attributed to the more efficient use of resources (e.g., water and other inputs). Similarly [10,12,41,42,96,103–106] argue that LP implementation can offer significant advantages and synergies with the green performance of companies, which are mainly related to the reduction in consumption of materials, energy and water.

However, some studies show that LP can present certain trade-offs with the priority "environment" [15,52,73,76]. One of the most frequent trade-offs cited refers to delivery and environment. LP may negatively impact the environment, since the JIT (JIT) delivery process results in more deliveries, and then more emissions from the vehicles [5,15,17,47].

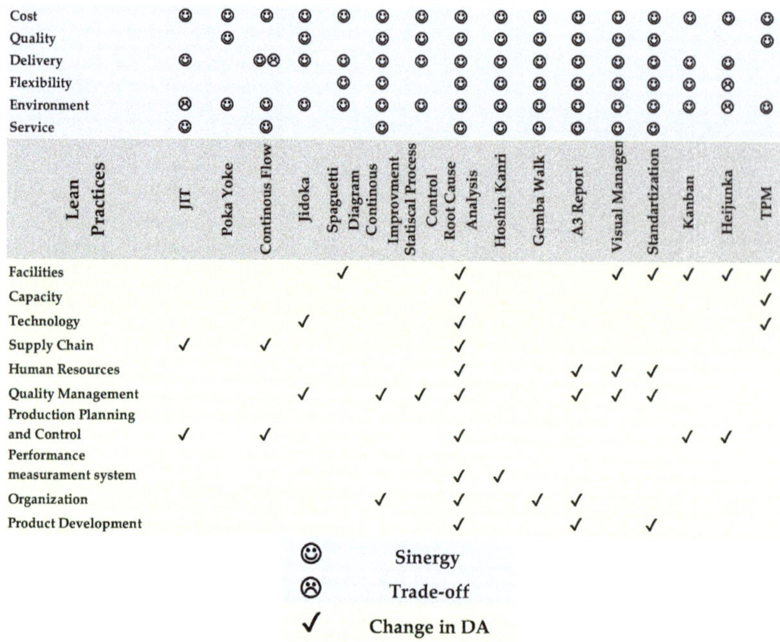

Figure 10. Lean Practices in Operations Strategy. Source: created by the authors.

Another trade-off identified is related to environment and quality. Pil and Rothenberg [107] exemplify that sometimes more resources consumption is necessary to maintain a good quality of product. They show that in the paint process, a large amount of water is necessary to have good quality. Therefore, quality-oriented practices such as TQM and Six Sigma can generate this trade-off, which means the reduction of resource consumption can be limited due to technical requirements of the process and the product.

Moreover, flexibility and environment can present trade-offs. Small batches allow more product variety, but they may increase the number of setups [96]. In this sense, to maintain the programming level in variety and volume of items produced, Heijunka is used. Consequently, more setups are necessary. Therefore, as pointed out by [96,103], more cleaning products are required and an increased disposal of unused process material. That way, as exposed by [108], the increase in the disposal of chemical products results in unrecyclable sewage and waste chemical reagents, eventually increasing the environmental burden. In the same way, the increase in consumption of these products can be caused by the practice 5s, which focuses on improving the cleaning and organization of the work environment and on the elimination of unnecessary items which can contribute to quality and cost [23,109,110].

4.2.2. Green Practices

Figure 11 presents six practices focusing on reducing the environmental impact in production process, product design and supply chain. They are presented as a complement to LP in companies, or they are part of a framework seeking to integrate Lean and GP. There are also some studies that highlight that it is essential to implement GP in Lean systems to resolve trade-offs with other competitive priorities. Also, there are cases that show their application is done separately from Lean, aiming only to reduce environmental impacts and comply with the legislation.

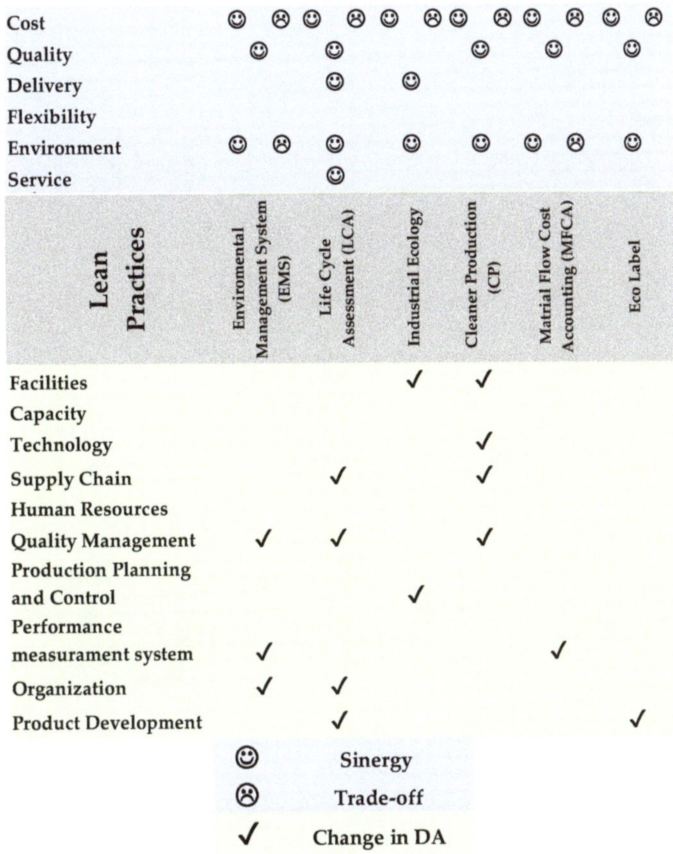

Figure 11. Green Practices and Operations Strategy. Source: created by the authors.

In terms of synergies between the competitive priorities from GP, it was possible to note that all of them contribute to and focus on the efficient use of resources; as a result, the priority cost can be improved. It is increasingly related to the cost reduction of material; water and energy consumption as well as waste disposal costs have dramatically increased over the past decade [93]. Furthermore, there are empirical studies that present the cost reduction through improvements in environmental aspects [33,34,111–113].

A study [114] argues that GP can reduce material and production costs, reduce transportation and logistics costs, increase product quality and reduce warehouse costs. Moreover, improving the environment can lead to cost reductions from any punitive or restrictive measures that may be introduced through legislation [46,113]. Mor et al. [48] points out that these practices are not only good for sustainability but also good for business value. Jabbour [23] concludes that the implementation of the EMS, the more frequent practice in the studies, can have positive impacts in various areas of organization performance.

Lastly, it is important to present the practice Material Flow Accounting (MFCA), a method to measure environmental performance by an accounting approach for estimating the output-input ratio and material flow using physical and monetary units [108]. Additionally, Dues et al. [96] argue that GP has a positive influence on LP.

Although several studies argue that there are cost savings with the implementation of GP, it is possible to have certain trade-offs. These conflicts happen when the green improvement means an increase in costs, or requires a huge investment. For example, when the results from an LCA, or an improvement from an EMS, identify that it is necessary to buy less harmful or more efficient equipment, the financial return on investment can be negative. Another example is related to the supply chain; when to reduce the CO_2 emission, the load is consolidated, to seek to transport with the least number of deliveries possible, resulting in fewer emissions and fuel consumption. Thus, it can generate an increase in delivery lead time because some loads would have to wait to fill up the load.

4.2.3. Lean-Green Practices

This topic explains the LG Practices and it shows the links with OS in Figure 8. They are described in more detail because they represent a consolidation of LG integration. These practices take the LP as a foundation and include the environment as a target of the improvements. This integration is one of the main steps for a LG practice and it helps to solve the trade-offs. In Figure 12 we can represent the Lean-Green practices relations with trade-offs, synergies and changes in DA.

Lean Practices	**E-VSM**	**7S/5S**	**Root cause analysis for**	**Overall Environment**	**Lean and Green Supply Chain**	**Kaizen with green**	**Sustainable BSC**	**Green Lean Six Sigma**	**Total Quality Environmental**	**Green MRP**	**Lean 3R**
Cost	☺	☺	☺	☺	☺	☺	☺	☺	☺	☺	☺
Quality	☺	☺☹	☺	☺		☺	☺	☺	☺		☺
Delivery	☺		☺		☺☹	☺				☺	
Flexibility	☺		☺			☺			☺	☺	
Environment	☺	☺☹	☺	☺	☺☹	☺	☺	☺☹	☺☹	☺☹	☺
Service	☺		☺	☺	☺	☺			☺		
Facilities			✓						✓		
Capacity			✓								
Technology			✓	✓						✓	
Supply Chain	✓		✓		✓						✓
Human Resources		✓	✓			✓					
Quality Management		✓	✓	✓		✓		✓	✓		
Production Planning and Control			✓							✓	
Performance measurament system	✓		✓			✓	✓				
Organization		✓	✓			✓	✓	✓			
Product Development			✓						✓		✓

☺	**Sinergy**
☹	**Trade-off**
✓	**Change in DA**

Figure 12. Lean-Green Practices and Operations Strategy. Source: created by the authors.

E-VSM (Environmental Value Stream Map)

E-VSM practice, also called Green Value Stream Map, Extended Value Stream Map or Sustainable Value Stream Map, was the most frequent LG practice in the studies analyzed. It aims to map the processes, present the main process indicators, and highlight the problems in the current state. This tool is based on Lean, specifically in VSM, and it was created to guide improvement processes and facilitate communication between those involved. The environmental aspects, such as energy, water and materials consumption and emissions, are included in E-VSM, which helps to collect and analyze these environmental data [76,78,115,116].

Regarding CP, this tool seeks to highlight performance indicators related to delivery, measuring lead-time of operations, quality, showing the rate of rework and defects in operations; and environmental performance indicators such as consumption of materials, energy, water, emissions, among others. Cost is widely cited, since the general objective of an E-VSM is to reduce all types of waste, including environmental waste, seeking cost reduction or, in some cases, to improve operational efficiency [25,37,116,117]. There is no trade-off observed with this practice; E-VSM has demonstrated a very synergic relationship by focusing on both Lean and Green.

In relation to the decision areas, this practice had a great impact on the decision area "performance measurement system". It is fundamental to measure and map the environmental impacts of the processes. The proposal is that sustainability indicators–raw material consumption, water consumption, energy consumption, waste generation, gas emissions, worker safety, ergonomic aspects and noise levels–should be integrated in the traditional VSM, as exposed by Helleno et al. [118] based on the proposal of [56,117,119,120]. In summary, as the study [119] discusses, the E-VSM seeks to provide a wealth of information that can facilitate communication and management through Lean and Green indicators. The study [121] highlights that it is necessary to determine a sustainability indicator set for the E-VSM tool so it would support thee decision making process.

The Supply Chain is another area of decision which is related to the E-VSM practice [122] since it can be used to obtain a global view of the entire chain [56] (Caldera et al., 2017)–from the extraction of the raw material to the final disposal–helping in the measurement, monitoring and visualization of possible improvements throughout the value chain. A point highlighted by [25] is the importance of integrating process information and understanding the client's value, i.e., capturing stakeholders' expectations following a life cycle perspective [3]. Therefore, when considering and integrating the entire value chain, it is essential to use an LCA that contemplates the product/service life cycle 'from cradle to grave', which is the concept that is included in the E-VSM to measure the environmental performance of the entire supply chain [123]. It is also important that circular economy strategies (recycling, remanufacturing, etc.) are considered a driver for the sustainable development. To adopt CE, it would be necessary to understand the productive process and to get information from it [1,124]. Thus, -E-VSM can support CE by drawing the process and presenting the data from it. The data collection and information sharing can be a great barrier to implement this strategy, but the literature has shown that I4.0 strategies can overcome this [125,126].

7s

This practice is an extension of the 5s tool that includes safety in some cases, is called 6s, and can be further expanded with one more s of sustainability. As far as priorities are concerned, this practice is used for cost reduction, quality and environmental improvement through waste reduction and environmental organization [79,110]. Despite the presence of synergies to improve the cost, quality and environment, 7s may create trade-offs when increasing the consumption of cleaning products. Regarding the decision areas, it is noted that Quality Management in the improvement programs and Human Resources in the training are changed with the inclusion of safety and environmental aspects.

In summary, the study presented by [79] shows that this program is the basis for improvement programs in the working environment. The study by [110] states that the objective of this practice is to improve quality, increase sales, reduce cost and provide a quality environment. Duarte and Cruz Machado [19] explain that this is a standardized work methodology used by organizations to collaborate to achieve Green objectives.

Root Cause Analysis for Environmental Problems

According to [109], the practice of root cause analysis can be extended to identify potential causes of environmental problems in processes, as already done for problems related to other competitive priorities.

Pinto and Mendes [127] point out that this practice should be structured to allow the visualization of environmental problems and should be structured in cycles and improvement programs, such as PDCA and Kaizen. Galeazzo et al [45] comment that by adopting this approach together with quality managers it was possible to understand and question environmental problems and thus identify possible solutions. In summary, these studies infer that just like recurrent problems of quality or process inefficiency to reduce cost, environmental problems should be solved using the same reasoning as the others, and they can be included as variables for the cause of the other problems. Thus, it was found that these practices can change the way to solve the problems in Quality Management and support the achievement of cost, quality and environment.

OEEE

OEE is an abbreviation for the term Overall Equipment Effectiveness, which means the overall efficiency of the equipment in dealing with an indicator that monitors the efficiency of the manufacturing process. The OEEE, according to Domingo and Aguado, 2015, means Overall Environmental Equipment Effectiveness, and it incorporates the concept of sustainability based on a calculation of the environmental impact in the life cycle. According to the authors, using this indicator can allow decision making by integrating sustainability and making a comparative analysis of environmental impact when analyzing improvements implemented in the processes of the organization. Thus, it was observed that OEEE has a relationship with the area of Performance Measurement System, and aims at efficiency improvements in which one monitors the performance in productivity and quality that imply cost reduction.

Further, this hybrid indicator is linked and can change the LP TPM and SMED. The first practice, TPM (Total Productive Maintenance), is a set of techniques to ensure the reliability and productivity of all machines in the production process [128]. The SMED (Single-Minute Exchange of Dies) or Quick Tool Change has as a principle to perform setup in less time, allowing the increased flexibility and/or availability of the machine in the production flow.

Lean and Green Supply Chain

This practice, presented in the study by Sant'Anna et al. [129]), was defined as the combination of the three approaches Lean, Green and Supply Chain, seeking cooperation for cost reduction, consumer focus, quality and environmental management through the ISO9000 and ISO14001 standards and risk management. In addition, the authors emphasize that the integration of the three approaches must be done to meet legal requirements, since Lean may not be sufficient to achieve them, and environmental management can collaborate in this direction. Thus, this practice seeks to include environmental performance in the decision area of supply chain and quality, as well as the objective of cost, quality and environmental performance [97]. This practice also involves supplier selection, procurement, third-party logistics and transportation, that aim to minimize the environmental impact of the product [17].

One practice that is widely used to improve the delivery and reduce cost in the supply chain is the JIT. As we cited before, this practice can present trade-offs between the competitive priorities "delivery" and "environment" because the studies show that to be better in delivery it is necessary to increase the consumption of fuels and then increase emissions for the environment [15,17,49,52]. Furthermore, it is important to highlight that LCA is a widely-used practice to measure environmental impacts in the supply chain. Thus, the decision area "Supply Chain" is modified by the practices Green Supply Chain, JIT and

LCA, and has to start to consider the trade-offs between delivery and the environment in its decisions.

Kaizen with Green Goals

Lean uses kaizen for improvement processes. In this search, studies have been identified that deal with kaizen integrated with GP. They show that kaizen can be integrated into ISO14001 continuous improvement cycles, which can help involve employees and find innovative solutions to the problems identified [56,130].

Kaizen with Green goals can affect mainly, in a synergic way, cost, quality and environment, because it includes environmental objectives that can reduce energy and material consumption. Thus, environmental metrics, such as energy consumption (analyzing monthly energy bills), and water consumption within a period, are used to determine costs [76]. In summary, as US EPA [131] shows, Kaizen activities should be aimed at reducing environmental costs in addition to tracking them, and that the tools used in the improvement cycle are mainly traditional Lean tools. Thus, this practice implies changes in improvement programs (Quality Management) and training for human resources (Human Resources), and more involvement of employees and the inclusion of the environment as a priority in the implemented improvements, measured through environmental performance indicators.

The study about a Lean and Green model [76] highlights that Kaizen is a great way to promote green improvements in organizations. As a study of Oliveira et al. [132] shows, Kaizen is a practice characterized by the involvement of staff for a specific goal, and can be a good strategy to identify and improve environmental aspects.

Lean with Environmental Indicators

First, it is important to show the study by [16] that shows an adaptation in the BSC strategic planning tool which includes sustainability in financial, client, internal processes and learning and growth perspectives: SBSC—Sustainable Balanced Score Card. In summary, the BSC is about several interrelated indicators, and indicators from these perspectives. According to the authors, including all these aspects in strategic planning can enable better performance management of the organization by senior management, and can be a basic model to structure the integration of Lean and Green. The study of these authors highlights the inclusion of the social perspective, which is not the focus of this research, and the environmental perspective.

The authors stress the importance of performance measures and incentives for investments in information systems, coordination and autonomy for decision making. Thus, the use of this model for LG integration changes the system of performance measurement, information systems (technology) for measuring, recording and sharing information, and it seeks to monitor the results obtained in all competitive priorities and in the value chain from "Cradle to Grave".

In addition, the most frequent indicators related to environmental priority are energy consumption (kw/h), water consumption (m^3 consumed per period), kg of materials consumed, kg of waste, kg of CO_2 emitted, among others. Several studies highlight the importance of including environmental performance indicators, since this is what will allow the integration with lean systems and the monitoring of environmental gains [17,33,37,76,118,133].

As we mentioned in the E-VSM topic, the I4.0 tools have supported the performance measurement. An example is the technology of wireless measurement systems used to capture data in industries, which collect accurate information such as energy consumption and costs, water consumption, gas emission, etc. [7,29]. From the analysis of how much is being consumed it is possible to obtain a more precise knowledge of the processes in relation to critical resources and thus make improvements attacking the root cause of the problem [7].

Lean 3R

Lean 3R is related to the remanufacturing, recycling and reuse of products and/or materials used in the processes [134]. It is defined as a product recovery process that uses energy and available resources, reduces the waste associated with the processes, and therefore can increase the overall efficiency of the process. The advantages associated with it can include the reduction of lean waste, such as overproduction, inventory, lead-time, unnecessary movement, waiting time and transportation [135]. Thus, there is a synergy between the environment and cost. To be able to apply lean remanufacturing, this must be considered from product and process Design [136]. The GP Design for Environment and the LCA are very useful for this practice [15]. Moreover, this is very much aligned with the strategies of the Circular Economy or Circular Manufacturing, which can help to reduce resources consumption and to extend lifecycles by remanufacture, recycle, resource efficiency, waste management, etc. [1].

Green Lean Six Sigma

This practice is a methodology that allows the search for environmental performance, applying robust tools for analysis and problem solving. Green Lean Six Sigma utilizes traditional aspects of Lean and Six Sigma while providing the tools needed to identify, implement and structure improvements that have a positive impact on the environment [115].

It is based on six pillars: leadership and people, Green and Lean six sigma tools, continuous improvement, strategic planning, interaction with stakeholders and results, and knowledge management; refs. [33,137] highlight that this proposal has a great impact on product development, contributing to cost reduction, process optimization, and enabling sustainability. Ruben et al. [84] demonstrate that the Lean Six Sigma basis allows the reduction of process variations and thus helps in the reduction of defects and waste generation during the production process. In addition, the study by Sreedharan V et al. [133] indicates that the use of these concepts contributes to the competitive priorities cost, quality and environment, through the reduction of environmental impacts and increases in the level of service provided. It is also noted that this practice mainly modifies the following decision areas: Quality Management and Human Resources.

Total Quality Environmental Management—TQEM

This practice is a sub-development of the practice Total Quality Management- TQM, which extends the principles of quality management to include manufacturing practices and processes that affect environmental quality [12,15,34,47,107]. TQM can help increase quality [35] and reduce production defects, which consequently reduce the consumption of raw materials and energy use [95,138]. However, trade-offs may arise when the production systems need more resources to improve the quality of the product [20,107].

Green MRP

The Green MRP tool is essentially a conventional Material Requirements Planning system that has been modified to include environmental considerations with the objective of minimizing the environmental impact of the generated waste, and seeking to increase the planning potential for the components and process waste, i.e., optimizing production planning in order to reduce possible problems related to the environment [139].

This practice is directly related to the decision area "Production Planning and Control". The production schedule and delivery schedule can help environmental performance by minimizing net energy consumption and defining shorter delivery routes to reduce the emissions of CO_2 [46]. Thus, this tool aims to balance better production planning with cost and environmental performance. Therefore, this tool directly influences the decision areas "Production Planning and Control", "Capacity" and "Technology".

Closing Remarks from Lean-Green Practices

The main aspects from the hybrid practices that integrate Lean and Green are that they focus on simultaneously considering the environmental performance and the other performance goals with the same degree of weighting in most cases, which helps to solve the trade-offs that can emerge in production systems. Because these tools make environmental indicators visible, and train employees in environmental priorities, they make the environment an important variable in the process of continuous improvement and decision making.

Another important aspect from hybrid tools and LG integration is that simultaneous implementation of LG methods multiply performance parameters, and this can result in significant cost reduction [111]. Furthermore, when Lean and Green are implemented together, they can create a more significant positive impact on organization than when implemented separately [96,140].

Finally, it is important to observe that the hybrid LG tools emerge from Lean. This feature can infer that Lean is used as a foundation for the management of production systems to implement the OS. Furthermore, GP works as a complement of Lean to help to support the OS that the environment is a competitive priority. Lean can also be insufficient to be fully Green.

4.3. Summary and Research Agenda

Most of the studies found are empirical and focus on understanding the relationship between Lean and Green and how they can be integrated. Their results suggest that Lean and Green objectives are different in essence, i.e., while LP aims to reduce cost and lead time (delivery) and to improve quality and flexibility, GP seeks to reduce waste related to environmental impacts [20,33,48,49,53,111,117]. Fahat et al. [141] cite that Lean waste is different from Green waste. Pinto and Mendes [127], in agreement with [73], point to the fact that Lean and Green objectives are different and generate different impacts on business performance. GP has the direct and clear objective of reducing the environmental impact of processes, while LP will directly impact on cost reduction, lead time and quality improvement; and the improvement of environmental aspects can be achieved indirectly.

From these results it is possible to understand that Lean and Green are synergic in most of their practices, but some trade-offs exist. In this sense, it is important to have the OS well defined to support the strategy and the targets of the organization. When the environment is a CP, Lean can be insufficient to solve the trade-offs; thus it becomes necessary include GP directly in the decision areas. Further, when LP and GP are implemented together, it becomes possible to leverage the performance of an organization more than when they are implemented separately.

Moreover, these results reveal to us that there are many trade-offs that have not yet been explored, as well as synergy relationships that are often limited to waste reduction and efficient use of resources. Furthermore, the strategic perspective of LG is still missing. Given these gaps, future research is suggested:

- Verifying the existence of unidentified trade-offs between the Lean and Green, as well as understanding how, and how much, the competitive priorities are affected; and evaluating quantitatively the relationship between individual practices and the contribution for the competitive priorities.
- Exploring in-depth the changes and the contribution from LG implementation in the decision areas; and exploring the particularities of each practice to be implemented.
- Empirical studies examining how companies frame their competitive priorities and decision areas in different levels of Lean and Green implementation and in different industrial sectors. Further, the researchers in operations management may quantify the performance of the companies related with the operation strategy adopted.
- Quantifying the synergy between the LG empirically, exploring different industrial sectors as well as different operations strategies.

- Identifying and comparing the reasons for OS formulation and Lean-GP adoption, and discussing their strategic role.
- Understanding how the Circular Economy strategies can be supported by Lean and Green Manufacturing practices.
- In terms of research methods, based on theoretical propositions identified qualitatively by most of the researched literature, quantitative approaches can be developed, such as surveys and modeling, to confirm them.

5. Conclusions

The main intention of this work was to provide, from studies reported in the literature about LP and GP, an initial and holistic analysis of the adoption of such practices from the perspective of the OS, presenting synergies and trade-offs between competitive priorities; as well to discuss the main changes in decision areas from LG implementation.

The study of these links between the entire content of the OS in the LG literature is still recent and complex. The systematic review of the literature performed indicates that the analyzed articles do not cover the entire content of the OS and that the studies are still exploratory. As discussed, there is a tendency to favor a few practices, or a few competitive priorities; or only economic results and environmental impacts, namely in terms of cost reduction and energy consumption, waste generation and emissions resulting from the implementation of LG. Through this systematic literature review, it was possible to answer the proposed RQs based on a content analysis performed with 338 papers. The results of this article have the potential to help managers and policymakers gain a holistic understanding of how they can implement integrated practices and learn about existing practices to improve the environmental impact of lean systems.

This work has some limitations related to its search strategy, including the three selected databases and only journal publications in English. In addition, it focuses only on the immediate links between the content of OS and LG, excluding the enabling and hindering variables of practices and OS adoption.

Author Contributions: Conceptualization, G.A.Q., I.D. and A.G.A.F.; methodology, L.A.d.S.-E. and A.L.V.T.; validation, G.A.Q. and I.D. analysis, G.A.Q. and A.G.A.F.; investigation, G.A.Q. and A.G.A.F.; resources, G.A.Q., A.G.A.F., I.D. and L.A.d.S.-E.; writing—original draft preparation, G.A.Q., A.G.A.F., I.D. and L.A.d.S.-E. and A.L.V.T.; writing—review and editing. All authors have read and agreed to the published version of the manuscript.

Funding: This research was supported by Coordination for the Improvement of Higher Education Personnel (Coordenação de Aperfeiçoamento de Pessoal de Nível Superior CAPES), scholarship number 88887.194801/2018-00, 88887.514124/2020-00 and 88882.426297/2019-01.

Institutional Review Board Statement: Not applicable.

Informed Consent Statement: Not applicable.

Data Availability Statement: Not applicable.

Conflicts of Interest: The authors declare no conflict of interest.

References

1. Acerbi, F.; Sassanelli, C.; Terzi, S.; Taisch, M. A Systematic Literature Review on Data and Information Required for Circular Manufacturing Strategies Adoption. *Sustainability* **2021**, *13*, 2047. [CrossRef]
2. Erdil, N.O.; Aktas, C.B.; Arani, O.M. Embedding Sustainability in Lean Six Sigma Efforts. *J. Clean. Prod.* **2018**, *198*, 520–529. [CrossRef]
3. Vinodh, S.; Ramesh, K.; Arun, C.S. Application of Interpretive Structural Modelling for Analysing the Factors Influencing Integrated Lean Sustainable System. *Clean Technol Environ. Policy* **2016**, *18*, 413–428. [CrossRef]
4. Garza-Reyes, J.A. Lean and Green-a Systematic Review of the State of the Art Literature. *J. Clean. Prod.* **2015**, *102*, 18–29. [CrossRef]
5. Longoni, A.; Cagliano, R. Environmental and Social Sustainability Prioritiesn: Their Integration in Operations Strategies. *Int. J. Oper. Prod. Manag.* **2015**, *35*, 216–345. [CrossRef]
6. Leong, W.D.; Teng, S.Y.; How, B.S.; Ngan, S.L.; Rahman, A.A.; Tan, C.P.; Ponnambalam, S.G.; Lam, H.L. Enhancing the Adaptability: Lean and Green Strategy towards the Industry Revolution 4.0. *J. Clean. Prod.* **2020**, *273*, 122870. [CrossRef]

7. Viles, E.; Santos, J.; Muñoz-Villamizar, A.; Grau, P.; Fernández-Arévalo, T. Lean–Green Improvement Opportunities for Sustainable Manufacturing Using Water Telemetry in Agri-Food Industry. *Sustainability* **2021**, *13*, 2240. [CrossRef]
8. Goyal, A.; Vaish, D.C.; Agrawal, R.; Choudhary, S.; Nayak, R. Sustainable Manufacturing through Systematic Reduction in Cycle Time. *Sustainability* **2022**, *14*, 6473. [CrossRef]
9. Caldera, H.T.S.; Desha, C.; Dawes, L. Evaluating the Enablers and Barriers for Successful Implementation of Sustainable Business Practice in 'Lean' SMEs. *J. Clean. Prod.* **2019**, *218*, 575–590. [CrossRef]
10. Garza-Reyes, J.A.; Kumar, V.; Chaikittisilp, S.; Tan, K.H. The Effect of Lean Methods and Tools on the Environmental Performance of Manufacturing Organisations. *Int. J. Prod. Econ.* **2018**, *200*, 170–180. [CrossRef]
11. Vinodh, S.; Arvind, K.R.; Somanaathan, M. Tools and Techniques for Enabling Sustainability through Lean Initiatives. *Clean Technol. Environ. Policy* **2011**, *13*, 469–479. [CrossRef]
12. Florida, R. Lean and Green: The Move to Environmentally Conscious Manufacturing. *Calif. Manag. Rev.* **1996**, *39*, 80–105. [CrossRef]
13. Ghobakhloo, M.; Azar, A.; Fathi, M. Lean-Green Manufacturing: The Enabling Role of Information Technology Resource. *Kybernetes* **2018**, *47*, 1752–1777. [CrossRef]
14. Ng, R.; Low, J.S.C.; Song, B. Integrating and Implementing Lean and Green Practices Based on Proposition of Carbon-Value Efficiency Metric. *J. Clean. Prod.* **2015**, *95*, 242–255. [CrossRef]
15. Salvador, R.; Piekarski, C.M.; Francisco, A.C. de Approach of the Two-Way Influence Between Lean and Green Manufacturing and Its Connection to Related Organisational Areas. *Int. J. Prod. Manag. Eng.* **2017**, *5*, 73. [CrossRef]
16. Zhu, X.; Zhang, H.; Wu, C.; Huang, Z. An Economic Model of Integration Framework of Lean Production and Green Manufacturing Based on Sustainability Balanced Scorecard. *Bol. Tec. Tech. Bull.* **2017**, *55*, 263–269.
17. Mollenkopf, D.; Stolze, H.; Tate, W.L.; Ueltschy, M. Green, Lean, and Global Supply Chains. *Int. J. Phys. Distrib. Logist. Manag.* **2010**, *40*, 14–41. [CrossRef]
18. Thanki, S.; Govindan, K.; Thakkar, J. An Investigation on Lean-Green Implementation Practices in Indian SMEs Using Analytical Hierarchy Process (AHP) Approach. *J. Clean. Prod.* **2016**, *135*, 284–298. [CrossRef]
19. Duarte, S.; Cruz Machado, V. Green and Lean Implementation: An Assessment in the Automotive Industry. *Int. J. Lean Six Sigma* **2017**, *8*, 65–88. [CrossRef]
20. Rothenberg, S.; Pil, F.K.; Maxwell, J. Lean, Green, and the Quest for Superior Environmental Performance. *Prod. Oper. Manag.* **2001**, *10*, 228–243. [CrossRef]
21. Marudhamuthu, R. The Development of Green Environment Through. *J. Eng. Appl. Sci.* **2011**, *6*, 104–111.
22. Hajmohammad, S.; Vachon, S.; Klassen, R.D.; Gavronski, I. Lean Management and Supply Management: Their Role in Green Practices and Performance. *J. Clean. Prod.* **2013**, *39*, 312–320. [CrossRef]
23. Jabbour, C.J.C.; De Sousa Jabbour, A.B.L.; Govindan, K.; Teixeira, A.A.; De Souza Freitas, W.R. Environmental Management and Operational Performance in Automotive Companies in Brazil: The Role of Human Resource Management and Lean Manufacturing. *J. Clean. Prod.* **2013**, *47*, 129–140. [CrossRef]
24. Resta, B.; Dotti, S.; Gaiardelli, P.; Boffelli, A. How Lean Manufacturing Affects the Creation of Sustainable Value: An Integrated Model. *Int. J. Autom. Technol.* **2017**, *11*, 542–551. [CrossRef]
25. Souza, J.P.E.; Alves, J.M. Lean-Integrated Management System: A Model for Sustainability Improvement. *J. Clean. Prod.* **2018**, *172*, 2667–2682. [CrossRef]
26. Queiroz, G.A.; Filho, A.G.A.; Costa Melo, I. Competitive Priorities and Lean–Green Practices—A Comparative Study in the Automotive Chain' Suppliers. *Machines* **2023**, *11*, 50. [CrossRef]
27. Díaz-Reza, J.R.; García-Alcaraz, J.L.; Figueroa, L.J.M.; Vidal, R.P.I.; Muro, J.C.S.D. Relationship between Lean Manufacturing Tools and Their Sustainable Economic Benefits. *Int. J. Adv. Manuf. Technol.* **2022**, *123*, 1269–1284. [CrossRef]
28. Afum, E.; Sun, Z.; Agyabeng-Mensah, Y.; Baah, C. Lean Production Systems, Social Sustainability Performance and Green Competitiveness: The Mediating Roles of Green Technology Adoption and Green Product Innovation. *J. Eng. Des. Technol.* **2023**, *21*, 206–227. [CrossRef]
29. Muñoz-Villamizar, A.; Santos, J.; Grau, P.; Viles, E. Toolkit for Simultaneously Improving Production and Environmental Efficiencies. *Central Eur. J. Oper. Res.* **2021**, *29*, 1219–1230. [CrossRef]
30. Alves, J.R.X.; Alves, J.M. Production Management Model Integrating the Principles of Lean Manufacturing and Sustainability Supported by the Cultural Transformation of a Company. *Int. J. Prod. Res.* **2015**, *53*, 5320–5333. [CrossRef]
31. Domingo, R.; Aguado, S. Overall Environmental Equipment Effectiveness as a Metric of a Lean and Green Manufacturing System. *Sustainability* **2015**, *7*, 9031–9047. [CrossRef]
32. Vinodh, S.; Ben Ruben, R.; Asokan, P. Life Cycle Assessment Integrated Value Stream Mapping Framework to Ensure Sustainable Manufacturing: A Case Study. *Clean Technol. Environ. Policy* **2016**, *18*, 279–295. [CrossRef]
33. Cherrafi, A.; Elfezazi, S.; Chiarini, A.; Mokhlis, A.; Benhida, K. The Integration of Lean Manufacturing, Six Sigma and Sustainability: A Literature Review and Future Research Directions for Developing a Specific Model. *J. Clean. Prod.* **2016**, *139*, 828–846. [CrossRef]
34. Prasad, S.; Khanduja, D.; Sharma, S.K. An Empirical Study on Applicability of Lean and Green Practices in the Foundry Industry. *J. Manuf. Technol. Manag.* **2016**, *27*, 408–426. [CrossRef]
35. Raghu Kumar, B.R.; Agarwal, A.; Sharma, M.K. Lean Management—A Step towards Sustainable Green Supply Chain. *Compet. Rev.* **2016**, *26*, 311–331. [CrossRef]

36. Farias, L.M.S.; Santos, L.C.; Gohr, C.F.; de Oliveira, L.C.; Amorim, M.H. da S. Criteria and Practices for Lean and Green Performance Assessment: Systematic Review and Conceptual Framework. *J. Clean. Prod.* **2019**, *218*, 746–762. [CrossRef]
37. Ben Ruben, R.; Vinodh, S.; Asokan, P. Implementation of Lean Six Sigma Framework with Environmental Considerations in an Indian Automotive Component Manufacturing Firm: A Case Study. *Prod. Plan. Control* **2017**, *28*, 1193–1211. [CrossRef]
38. Kaswan, M.S.; Rathi, R. Green Lean Six Sigma for Sustainable Development: Integration and Framework. *Env. Impact Assess. Rev.* **2020**, *83*, 106396. [CrossRef]
39. Singh, R.K.; Kumar Mangla, S.; Bhatia, M.S.; Luthra, S. Integration of Green and Lean Practices for Sustainable Business Management. *Bus Strat. Environ.* **2022**, *31*, 353–370. [CrossRef]
40. Gandhi, J.; Thanki, S.; Thakkar, J.J. An Investigation and Implementation Framework of Lean Green and Six Sigma (LG&SS) Strategies for the Manufacturing Industry in India. *TQM J.* **2021**, *33*, 1705–1734. [CrossRef]
41. Chiarini, A. Sustainable Manufacturing-Greening Processes Using Specific Lean Production Tools: An Empirical Observation from European Motorcycle Component Manufacturers. *J. Clean. Prod.* **2014**, *85*, 226–233. [CrossRef]
42. Salah, S.A.; Mustafa, A. Integration of Energy Saving with Lean Production in a Food Processing Company. *J. Mach. Eng.* **2021**, *21*, 118–133. [CrossRef]
43. Kazancoglu, Y.; Ekinci, E.; Ozen, Y.D.O.; Pala, M.O. Forum Reducing Food Waste Through Lean and Sustainable Operations: A Case Study From The Poultry Industry. *RAE* **2020**, *61*, 1–18.
44. Ferrone, B. Environmental Business Management Practices for a New Age. *Change Manag.* **1996**, *5*, 1–7. [CrossRef]
45. Galeazzo, A.; Furlan, A.; Vinelli, A. Lean and Green in Action: Interdependencies and Performance of Pollution Prevention Projects. *J. Clean. Prod.* **2014**, *85*, 191–200. [CrossRef]
46. Ball, P. Low Energy Production Impact on Lean Flow. *J. Manuf. Technol. Manag.* **2015**, *26*, 412–428. [CrossRef]
47. Campos, L.M.S.; Vazquez-Brust, D.A. Lean and Green Synergies in Supply Chain Management. *Supply Chain Manag.* **2016**, *21*, 627–641. [CrossRef]
48. Mor, R.S.; Singh, S.; Bhardwaj, A. Learning on Lean Production: A Review of Opinion and Research within Environmental Constraints. *Oper. Supply Chain Manag.* **2016**, *9*, 61–72. [CrossRef]
49. Sartal, A.; Martinez-Senra, A.I.; Cruz-Machado, V. Are All Lean Principles Equally Eco-Friendly? A Panel Data Study. *J. Clean. Prod.* **2018**, *177*, 362–370. [CrossRef]
50. Azevedo, S.G.; Carvalho, H.; Cruz-Machado, V. LARG Index: A Benchmarking Tool for Improving the Leanness, Agility, Resilience and Greenness of the Automotive Supply Chain. *Benchmarking* **2016**, *23*, 1472–1499. [CrossRef]
51. Chatha, K.A.; Butt, I. Themes of Study in Manufacturing Strategy Literature. *Int. J. Oper. Prod. Manag.* **2015**, *35*, 604–698. [CrossRef]
52. Longoni, A.; Cagliano, R. Cross-Functional Executive Involvement and Worker Involvement in Lean Manufacturing and Sustainability Alignment. *Int. J. Oper. Prod. Manag.* **2015**, *35*, 1332–1358. [CrossRef]
53. Suifan, T.; Alazab, M.; Alhyari, S. Trade-off among Lean, Agile, Resilient and Green Paradigms: An Empirical Study on Pharmaceutical Industry in Jordan Using a TOPSIS-Entropy Method. *Int. J. Adv. Oper. Manag.* **2019**, *11*, 69–101. [CrossRef]
54. Sassanelli, C.; Rossi, M.; Pezzotta, G.; Augusto, D.; Pacheco, J.; Terzi, S. *Defining Lean Product Service Systems Features and Research Trends through a Systematic Literature Review*; Inderscience Publishers: Geneva, Switzerland, 2019; Volume 12.
55. de Carvalho, A.C.V.; Granja, A.D.; da Silva, V.G. A Systematic Literature Review on Integrative Lean and Sustainability Synergies over a Building's Lifecycle. *Sustainability* **2017**, *9*, 1156. [CrossRef]
56. Caldera, H.T.S.; Desha, C.; Dawes, L. Exploring the Role of Lean Thinking in Sustainable Business Practice: A Systematic Literature Review. *J. Clean. Prod.* **2017**, *167*, 1546–1565. [CrossRef]
57. Gaikwad, L.; Sunnapwar, V. An Integrated Lean, Green and Six Sigma Strategies: A Systematic Literature Review and Directions for Future Research. *TQM J.* **2020**, *32*, 201–225. [CrossRef]
58. Saarani, P.S.N.; Abdul Tharim, A.H.; Che Ahmad, A.; Mohamed Salleh, R. A Systematic Literature Review (SLR) on the Strategies of Managing Waste in Relative to Green Building (GB) Practice. *Pertanika J. Sci. Technol.* **2022**, *30*, 1363–1380. [CrossRef]
59. Sharma, V.; Raut, R.D.; Mangla, S.K.; Narkhede, B.E.; Luthra, S.; Gokhale, R. A Systematic Literature Review to Integrate Lean, Agile, Resilient, Green and Sustainable Paradigms in the Supply Chain Management. *Bus Strategy Environ.* **2021**, *30*, 1191–1212. [CrossRef]
60. Queiroz, G.A. Estratégia de Operações e Práticas Lean-Green: Um Estudo de Casos Múltiplos Em Empresas Do Setor Automotivo. Ph.D Thesis, Federal University of São Carlos, São Carlos, Brazil, 2021.
61. Slack, N.; Lewis, M. *Operations Strategy*, 3rd ed.; Pearson Education Limited: London, UK, 2011; ISBN 9780273740445.
62. Skinner, W. Manufacturing—Missing Link in the Corporate Strategy. *Harv. Bus Rev.* **1969**, *47*, 136–145.
63. Hilmola, O.P.; Lorentz, H.; Hilletofth, P.; Malmsten, J. Manufacturing Strategy in SMEs and Its Performance Implications. *Ind. Manag. Data Syst.* **2015**, *115*, 1004–1021. [CrossRef]
64. Voss, C.A. Paradigms of Manufacturing Strategy Re-Visited. *Int. J. Oper. Prod. Manag.* **2005**, *25*, 1223–1227. [CrossRef]
65. Garvin, D.A. Manufacturing Strategic Planning. *Calif.Manag. Rev.* **1993**, *35*, 85–106. [CrossRef]
66. Wheelwright, S.C. Manufacturing Strategy: Defining the Missing Link. *Strateg. Manag. J.* **1984**, *5*, 77–91. [CrossRef]
67. Voss, C.A. Alternative Paradigms for Manufacturing Strategy. *Int. J. Oper. Prod. Manag.* **1995**, *15*, 5–16. [CrossRef]
68. Losonci, D.; Demeter, K. Lean Production and Business Performance: International Empirical Results. *Compet. Rev.* **2013**, *23*, 218–233. [CrossRef]

69. Mostafa, S.; Dumrak, J.; Soltan, H. A Framework for Lean Manufacturing Implementation. *Prod. Manuf. Res.* **2013**, *1*, 44–64. [CrossRef]
70. Bhasin, S.; Burcher, P. Lean Viewed as a Philosophy. *J. Manuf. Technol. Manag.* **2006**, *17*, 56–72. [CrossRef]
71. Shah, R.; Ward, P.T. Defining and Developing Measures of Lean Production. *J. Oper. Manag.* **2007**, *25*, 785–805. [CrossRef]
72. Shah, R.; Ward, P.T. Lean Manufacturing: Context, Practice Bundles, and Performance. *J. Oper. Manag.* **2003**, *21*, 129–149. [CrossRef]
73. Kleindorfer, P.R.; Singhal, K.; Van Wassenhove, L.N. Sustainable Operations Management. *Prod. Oper. Manag.* **2005**, *14*, 482–492. [CrossRef]
74. Pathak, P.; Singh, M.P. Sustainable Manufacturing Concepts: A Literature Review. *Int. J. Eng. Technol. Manag. Res.* **2017**, *4*, 1–13. [CrossRef]
75. Silva, D.A.L.; da Silva, E.J.; Ometto, A.R. Green Manufacturing: Uma Análise Da Produção Científica e de Tendências Para o Futuro. *Production* **2015**, *26*, 642–655. [CrossRef]
76. Pampanelli, A.B.; Found, P.; Bernardes, A.M. A Lean & Green Model for a Production Cell. *J. Clean. Prod.* **2014**, *85*, 19–30. [CrossRef]
77. Bhattacharya, A.; Nand, A.; Castka, P. Lean-Green Integration and Its Impact on Sustainability Performance: A Critical Review. *J. Clean. Prod.* **2019**, *236*, 117697. [CrossRef]
78. Aguado, S.; Alvarez, R.; Domingo, R. Model of Efficient and Sustainable Improvements in a Lean Production System through Processes of Environmental Innovation. *J. Clean. Prod.* **2013**, *47*, 141–148. [CrossRef]
79. Anvari, A.; Zulkifli, N.; Yusuff, R.M. Evaluation of Approaches to Safety in Lean Manufacturing and Safety Management Systems and Clarification of the Relationship between Them. *World Appl. Sci. J.* **2011**, *15*, 19–26.
80. Denyer, D.; Tranfield, D. Producing a Systematic Review. In *The SAGE Handbook of Organizational Research Methods*; Buchanan, D.A., Bryman, A., Eds.; Sage Publications Ltd.: Thousand Oaks, CA, USA, 2009; pp. 671–689.
81. Moher, D.; Liberati, A.; Tetzlaff, J.; Altman, D.G. Preferred Reporting Items for Systematic Reviews and Meta-Analyses: The PRISMA Statement. *PLoS Med.* **2009**, *6*, e1000097. [CrossRef]
82. Aghaei Chadegani, A.; Salehi, H.; Md Yunus, M.M.; Farhadi, H.; Fooladi, M.; Farhadi, M.; Ale Ebrahim, N. A Comparison between Two Main Academic Literature Collections: Web of Science and Scopus Databases. *Asian Soc. Sci.* **2013**, *9*, 18–26. [CrossRef]
83. Slack, N. *Operations Strategy*; Bookboon: London, UK, 2015; ISBN 9788776818289.
84. ben Ruben, R.; Vinodh, S.; Asokan, P. Lean Six Sigma with Environmental Focus: Review and Framework. *Int. J. Adv. Manuf. Technol.* **2018**, *94*, 4023–4037. [CrossRef]
85. Šišková, V.; Dlabač, J. Environmental Management System as the Part of the Methodology for Optimization of Assembly Workplaces. *Int. J. Math. Model. Methods Appl. Sci.* **2013**, *7*, 428–435.
86. Pham, D.T.; Thomas, A.J. Fit Manufacturing: A Framework for Sustainability. *J. Manuf. Technol. Manag.* **2011**, *23*, 103–123. [CrossRef]
87. Hanson, J.; Melnyk, S.A.; Calantone, R. Core Values and Nation-States. *Oppor. Chall. Biling.* **2011**, *87*, 69–85. [CrossRef]
88. Krippendorff, K. *Content Analysis: An Introduction to Its Methodology*, 2nd ed.; Sage Publications: Thousand Oaks, CA, USA, 2004; Volume 13.
89. Hutchisona, A.J.; Johnstonb, L.H.; Breckona, J.D. Using QSR-NVivo to Facilitate the Development of a Grounded Theory Project: An Account of a Worked Example. *Int. J. Soc. Res. Methodol.* **2010**, *13*, 283–302. [CrossRef]
90. Gibbs, G. *Análise de Dados Qualitativos*; Artmed, Ed.; Porto Alegre: Artmed, Brazil, 2009; 98p.
91. UNCTAD—United Nations Conference on Trade and Development. Country Classification. Available online: https://unctadstat.unctad.org/en/classifications.html (accessed on 10 January 2023).
92. Thanki, S.J.; Thakkar, J.J. Value–Value Load Diagram: A Graphical Tool for Lean–Green Performance Assessment. *Prod. Plan. Control.* **2016**, *27*, 1280–1297. [CrossRef]
93. Gupta, S.; Dangayach, G.S.; Singh, A.K.; Meena, M.L.; Rao, P.N. Implementation of Sustainable Manufacturing Practices in Indian Manufacturing Companies. *Benchmarking* **2018**, *25*, 2441–2459. [CrossRef]
94. Choudhary, S.; Nayak, R.; Dora, M.; Mishra, N.; Ghadge, A. An Integrated Lean and Green Approach for Improving Sustainability Performance: A Case Study of a Packaging Manufacturing SME in the UK. *Prod. Plan. Control.* **2019**, *30*, 353–368. [CrossRef]
95. Belhadi, A.; Touriki, F.E.; El Fezazi, S. Benefits of Adopting Lean Production on Green Performance of SMEs: A Case Study. *Prod. Plan. Control.* **2018**, *29*, 873–894. [CrossRef]
96. Dües, C.M.; Tan, K.H.; Lim, M. Green as the New Lean: How to Use Lean Practices as a Catalyst to Greening Your Supply Chain. *J. Clean. Prod.* **2013**, *40*, 93–100. [CrossRef]
97. Ruiz-Benitez, R.; López, C.; Real, J.C. Environmental Benefits of Lean, Green and Resilient Supply Chain Management: The Case of the Aerospace Sector. *J. Clean. Prod.* **2017**, *167*, 850–862. [CrossRef]
98. Siegel, R.; Antony, J.; Garza-Reyes, J.A.; Cherrafi, A.; Lameijer, B. Integrated Green Lean Approach and Sustainability for SMEs: From Literature Review to a Conceptual Framework. *J. Clean. Prod.* **2019**, *240*, 118205. [CrossRef]
99. Azfar, K.R.W. Application of Lean Agile Resilient Green Paradigm Framework on China Pakistan Economic Corridor: A Case Study. *Mehran Univ. Res. J. Eng. Technol.* **2017**, *36*, 621–634. [CrossRef]
100. Puvanasvaran, A.L.; Perumal, A.; Tian, R.K.S.; Vasu, S.A.L.; Muhamad, M.R. Integration Model of ISO 14001 with Lean Principles. *Am. J. Appl. Sci.* **2012**, *9*, 1974–1978. [CrossRef]
101. Diaz-Elsayed, N.; Jondral, A.; Greinacher, S.; Dornfeld, D.; Lanza, G. Assessment of Lean and Green Strategies by Simulation of Manufacturing Systems in Discrete Production Environments. *CIRP Ann. Manuf. Technol.* **2013**, *62*, 475–478. [CrossRef]

102. Vais, A.; Miron, V.; Pedersen, M.; Folke, J. "Lean and Green" at a Romanian Secondary Tissue Paper and Board Mill—Putting Theory into Practice. *Resour. Conserv. Recycl.* **2006**, *46*, 44–74. [CrossRef]
103. King, A.A.; Lenox, M.J. Lean and Green? An Empirical Examination of the Relationship between Lean Production and Environmental Performance. *Prod. Oper. Manag.* **2001**, *10*, 244–256. [CrossRef]
104. Crosby, B.; Badurdeen, F. Integrating Lean and Sustainable Manufacturing Principles for Sustainable Total Productive Maintenance (Sus-TPM). *Smart Sustain. Manuf. Syst.* **2022**, *6*, 20210025. [CrossRef]
105. Teixeira, P.; Sá, J.C.; Silva, F.J.G.; Ferreira, L.P.; Santos, G.; Fontoura, P. Connecting Lean and Green with Sustainability towards a Conceptual Model. *J. Clean. Prod.* **2021**, *322*, 129047. [CrossRef]
106. Abualfaraa, W.; AlManei, M.; Kaur, R.; Al-Ashaab, A.; McLaughlin, P.; Salonitis, K. A Synergetic Framework for Green and Lean Manufacturing Practices in SMEs: Saudi Arabia Perspective. *Sustainability* **2023**, *15*, 596. [CrossRef]
107. Pil, F.K.; Rothenberg, S. Environmental Performance as a Driver of Superior Quality. *Prod. Oper. Manag.* **2003**, *12*, 404–415. [CrossRef]
108. Fu, X.; Guo, M.; Zhanwen, N. Applying the Green Embedded Lean Production Model in Developing Countries: A Case Study of China. *Environ. Dev.* **2017**, *24*, 22–35. [CrossRef]
109. Ben Ruben, R.; Asokan, P.; Vinodh, S. Performance Evaluation of Lean Sustainable Systems Using Adaptive Neuro Fuzzy Inference System: A Case Study. *Int. J. Sustain. Eng.* **2017**, *10*, 158–175. [CrossRef]
110. Ho, S.; Hashim, A.G.B.M.; Idris, M.A.M. Applicability of SIRIM Green 5-S Model for Productivity & Business Growth in Malaysia. *TQM J.* **2015**, *27*, 185–196. [CrossRef]
111. Miller, G.; Pawloski, J.; Standridge, C. A Case Study of Lean, Sustainable Manufacturing. *J. Ind. Eng. Manag.* **2010**, *3*, 11–32. [CrossRef]
112. Torielli, R.M.; Abrahams, R.A.; Smillie, R.W.; Voigt, R.C. Using Lean Methodologies for Economically and Environmentally Sustainable Foundries. *China Foundry* **2010**, *2*, 710–726.
113. Parmar, P.S.; Desai, T.N. Ranking the Solutions of Sustainable Lean Six Sigma Implementation in Indian Manufacturing Organization to Overcome Its Barriers. *Int. J. Sustain. Eng.* **2020**, *14*, 304–317. [CrossRef]
114. Wiengarten, F.; Fynes, B.; Onofrei, G. Exploring Synergetic Effects between Investments in Environmental and Quality/Lean Practices in Supply Chains. *Supply Chain Manag.* **2013**, *18*, 148–160. [CrossRef]
115. Swarnakar, V.; Singh, A.R.; Antony, J.; Kr Tiwari, A.; Cudney, E.; Furterer, S. A Multiple Integrated Approach for Modelling Critical Success Factors in Sustainable LSS Implementation. *Comput. Ind. Eng.* **2020**, *150*, 106865. [CrossRef]
116. Gholami, H.; Jamil, N.; Mat Saman, M.Z.; Streimikiene, D.; Sharif, S.; Zakuan, N. The Application of Green Lean Six Sigma. *Bus Strategy Environ.* **2021**, *30*, 1913–1931. [CrossRef]
117. Cherrafi, A.; Elfezazi, S.; Govindan, K.; Garza-Reyes, J.A.; Benhida, K.; Mokhlis, A. A Framework for the Integration of Green and Lean Six Sigma for Superior Sustainability Performance. *Int. J. Prod. Res.* **2017**, *55*, 4481–4515. [CrossRef]
118. Helleno, A.L.; de Moraes, A.J.I.; Simon, A.T.; Helleno, A.L. Integrating Sustainability Indicators and Lean Manufacturing to Assess Manufacturing Processes: Application Case Studies in Brazilian Industry. *J. Clean. Prod.* **2017**, *153*, 405–416. [CrossRef]
119. Faulkner, W.; Badurdeen, F. Sustainable Value Stream Mapping (Sus-VSM): Methodology to Visualize and Assess Manufacturing Sustainability Performance. *J. Clean. Prod.* **2014**, *85*, 8–18. [CrossRef]
120. Sulaiman, M.A.; Ito, T.; Rahman, M.A.A.; Ishak, A.; Bin Salleh, M.R.; Mohamad, E. Cleaner Production Value Stream Mapping at a Chromium Plating Plant: A Case Study. *Int. J. Agil. Syst. Manag.* **2019**, *12*, 245. [CrossRef]
121. Lee, J.K.Y.; Gholami, H.; Saman, M.Z.M.; bin Ngadiman, N.H.A.; Zakuan, N.; Mahmood, S.; Omain, S.Z. Sustainability-Oriented Application of Value Stream Mapping: A Review and Classification. *IEEE Access* **2021**, *9*, 68414–68434. [CrossRef]
122. Müller, E.; Stock, T.; Schillig, R. A Method to Generate Energy Value-Streams in Production and Logistics in Respect of Time- and Energy-Consumption. *Prod. Eng.* **2014**, *8*, 243–251. [CrossRef]
123. Lucato, W.C.; Vieira, M.; Da Silva Santos, J.C. Eco-Six Sigma: Integration of Environmental Variables into the Six Sigma Technique. *Prod. Plan. Control.* **2015**, *26*, 605–616. [CrossRef]
124. Acerbi, F.; Sassanelli, C.; Taisch, M. A Conceptual Data Model Promoting Data-Driven Circular Manufacturing. *Oper. Manag. Res.* **2022**, *15*, 838–857. [CrossRef]
125. John, L.; Sampayo, M.; Peças, P. Lean & Green on Industry 4.0 Context—Contribution to Understand L&G Drivers and Design Principles. *Int. J. Math. Eng. Manag. Sci.* **2021**, *6*, 1214–1229. [CrossRef]
126. Rahardjo, B.; Wang, F.-K.; Yeh, R.-H.; Chen, Y.-P. Lean Manufacturing in Industry 4.0: A Smart and Sustainable Manufacturing System. *Machines* **2023**, *11*, 72. [CrossRef]
127. Pinto, M.J.A.; Mendes, J.V. Operational Practices of Lean Manufacturing: Potentiating Environmental Improvements. *J. Ind. Eng. Manag.* **2017**, *10*, 550–580. [CrossRef]
128. Feld, W.M. *Lean Manufacturing: Tools, Techniques, and How to Use Them*, 1st ed.; Francis, T., Ed.; CRC Press: Boca Raton, FL, USA, 2000; ISBN 9781420025538.
129. Sant'Anna, P.R.; Bouzon, M.; Tortorella, G.L.; Campos, L.M.S. Implementation of Lean and Green Practices: A Supplier-Oriented Assessment Method. *Prod. Eng.* **2017**, *11*, 531–543. [CrossRef]
130. Singh, P. Lean in Healthcare Organization: An Opportunity for Environmental Sustainability. *Benchmarking* **2019**, *26*, 205–220. [CrossRef]
131. US EPA. The Lean and Environment Toolkit. Prevention 2007. Available online: https://www.epa.gov/sites/default/files/2013-10/documents/leanenvirotoolkit.pdf (accessed on 10 January 2023).

132. Oliveira, J.A.d.; Devós Ganga, G.M.; Godinho Filho, M.; Silva, D.A.L.; dos Santos, M.P.; Aldaya Garde, I.A.; Penchel, R.A.; Esposto, K.F.; Ometto, A.R. Environmental and Operational Performance Is Not Always Achieved When Combined with Cleaner Production and Lean Production: An Overview for Emerging Economies. *J. Environ. Plan. Manag.* **2022**, *65*, 1530–1559. [CrossRef]
133. Raja Sreedharan, V.R.; Sandhya, G.; Raju, R. Development of a Green Lean Six Sigma Model for Public Sectors. *Int. J. Lean Six Sigma* **2018**, *9*, 238–255. [CrossRef]
134. Duarte, S.; Cruz-Machado, V. Green and Lean Supply-Chain Transformation: A Roadmap. *Prod. Plan. Control.* **2019**, *30*, 1170–1183. [CrossRef]
135. Maqbool, Y.; Rafique, M.Z.; Hussain, A.; Ali, H.; Javed, S.; Amjad, M.S.; Khan, M.A.; Mumtaz, S.; Haider, S.M.; Atif, M. An Implementation Framework to Attain 6R-Based Sustainable Lean Implementation—A Case Study. *IEEE Access* **2019**, *7*, 117561–117579. [CrossRef]
136. Vasanthakumar, C.; Vinodh, S.; Vishal, A.W. Application of Analytical Network Process for Analysis of Product Design Characteristics of Lean Remanufacturing System: A Case Study. *Clean Technol. Env. Policy* **2017**, *19*, 971–990. [CrossRef]
137. Kumar, S.; Luthra, S.; Govindan, K.; Kumar, N.; Haleem, A. Barriers in Green Lean Six Sigma Product Development Process: An ISM Approach. *Prod. Plan. Control.* **2016**, *27*, 604–620. [CrossRef]
138. Sunk, A.; Kuhlang, P.; Edtmayr, T.; Sihn, W. Developments of Traditional Value Stream Mapping to Enhance Personal and Organisational System and Methods Competencies. *Int. J. Prod. Res.* **2017**, *55*, 3732–3746. [CrossRef]
139. Melnyk, S.A.; Sroufe, R.P.; Montabon, F.L.; Hinds, T.J. Green MRP: Identifying the Material and Environmental Impacts of Production Schedules. *Int. J. Prod. Res.* **2001**, *39*, 1559–1573. [CrossRef]
140. Cherrafi, A.; Garza-Reyes, J.A.; Kumar, V.; Mishra, N.; Ghobadian, A.; Elfezazi, S. Lean, Green Practices and Process Innovation: A Model for Green Supply Chain Performance. *Int. J. Prod. Econ.* **2018**, *206*, 79–92. [CrossRef]
141. Fahad, M.; Naqvi, S.A.A.; Atir, M.; Zubair, M.; Shehzad, M.M. Energy Management in a Manufacturing Industry through Layout Design. *Procedia Manuf.* **2017**, *8*, 168–174. [CrossRef]

sustainability

Review

Synergy between Circular Economy and Industry 4.0: A Literature Review

Carlos Andrés Tavera Romero [1,*], Diego F. Castro [1], Jesús Hamilton Ortiz [2], Osamah Ibrahim Khalaf [3] and Miguel A. Vargas [1]

1 COMBA I+D Research Group of Universidad Santiago de Cali, 760036 Santiago de Cali, Colombia; diego.castro05@usc.edu.co (D.F.C.); miguel.vargas02@usc.edu.co (M.A.V.)
2 Closemobile R&D Telecommunications LS, 28070 Madrid, Spain; jesushamilton.ortiz@gmail.com
3 Al-Nahrain Nanorenewable Energy Research Center, Al-Nahrain University, Baghdad 10001, Iraq; usama.ibrahem@coie-nahrain.edu.iq
* Correspondence: carlos.tavera00@usc.edu.co; Tel.: +57-311-3348381

Abstract: Recent research has shown that there is a correlation between the circular economy (CE) and Industry 4.0 (I4.0). In addition, other research papers have analyzed the way that CE uses the different I4.0 technologies to transfer from the existing linear economy to CE; however, there are still gaps in the literature regarding the challenges and impacts that society and individuals must face to be ready for the transition from a linear to a circular economy. These challenges seek to guarantee the sustainability and sustainable development of the different business models that mobilize products and services through supply chains. Here, we conducted a review and compilation of the latest bibliography of circular economy and Industry 4.0 theory. The objective of this work is to present the evolutionary relationship between CE and I4.0, as well as its multi-step model of analysis. This research is relevant because its topics are timely and pertinent, especially for academics. Further, at the time that this research was performed, none of the countries were concerned about the impact that technological changes have on the human being and on society, and up to now we do not currently have studies that show how people are being trained to face the transition from the linear economy, which is common in most societies, toward a CE.

Keywords: circular economy; Industry 4.0; sustainability

Citation: Tavera Romero, C.A.; Castro, D.F.; Ortiz, J.H.; Khalaf, O.I.; Vargas, M.A. Synergy between Circular Economy and Industry 4.0: A Literature Review. *Sustainability* **2021**, *13*, 4331. https://doi.org/10.3390/su13084331

Academic Editor: Claudio Sassanelli

Received: 26 February 2021
Accepted: 6 April 2021
Published: 13 April 2021

1. Introduction

During the industrial revolution, technology was the most important subject. OCDE and circular economy (CE) plans give more relevance to the human component, the relationship between the environment and technology development, and the construction of Society 5.0. The aim of this study was to discuss the role of human society in economic and community progress.

The economic model focused on leveraging resources and reusing waste to create goods and services for mass consumption is currently known as the CE. A CE model seeks to reduce the consumption and emission of resources and waste, with the aim of increasing the useful life of products and services, while, at the same time, becoming eco-friendly and optimizing the energy taken from the different sources available.

This model is based on the same evolution of nature that is currently taking place in the modern world, with a large impact on the use of technologies. Although CE has been documented since the beginning of the 20th century, it is only recently that we find abundant literature assessing its relationship with Industry 4.0 (I4.0).

I4.0 is the fourth industrial revolution, based on the digital transformation adopted by current societies. Some of these societies show evidence that they are heading toward environmental conservation, described as an ecological transition [1].

The concepts of CE and I4.0 were born independently; however, they are moving and evolving together, becoming industrial paradigms that, in recent years, have shifted their focus to a restorative industrial production model. The research is in agreement that CE models mostly attempt to design and produce goods and services that reduce waste, supplemented by the implementation of digital technologies that may support their development and are currently framed within the concept of I4.0 [2].

In this context, different studies found in the literature have emphasized the close relationship between sustainability and sustainable development, framed as CE principles and I4.0. Considering I4.0 as an industrial engine and incentive that has increased social awareness of technological effects, decreased the depletion of ecological resources, and driven growing industries toward profitability, there is a strong practice and theoretical relevance in the study of the interconnection among people, technologies, and resources [3].

Other approaches have based their research on logistics and supply chains, assessing case studies and applying different methodologies to reach meaningful conclusions. This is how the authors in [4] developed a literature review with the purpose of identifying the use of service providers known as third party logistics (3PL) in supply chains. These 3PLs provide reverse logistics solutions with an adequate technological infrastructure that provides access to the product and service supply chains by collecting used products for their subsequent transportation, reprocessing, or even redistribution.

In this sense, Rajput and Singh [5] addressed the reduce, reuse, recycle (3R) model used to support the transformation of the supply chain from linear to circular, based on CE and using emerging technologies (I4.0) as a sustainable and inclusive technological innovation. For its part, in [6], a company was used to verify the application of the framework developed through a hybrid method that proposed using I4.0 and CE as the guiding principles to overcome the challenges faced by supply chains. Few studies have focused on CE and its evolutionary relationship with I4.0 from the human aspect, that is, the social impacts and personal development of human beings linked to the areas of application of these industrial principles of production.

The CE is seen as a model based on the evolution of nature. Therefore, humans must be considered as the main actors since they are always impacted in any change or evolutionary adoption. For this reason, A.M. Gómez, F.A. González, and A. Luque [7] proposed a holonic framework where, through CE and the use of digital and technological I4.0 enablers, the loss of social, economic, and natural resources generated by supply chain growth may be mitigated and reversed. The integrated CE and I4.0 framework allows the supply chain to co-evolve in the environment through adaptation and integration processes oriented toward the mitigation and reversal of metabolic ruptures. By contrast, Kuba and Milichovský [8] sought to establish the relationship between worker age groups, their theoretical conceptualization of I4.0, and the skills and abilities they acquire in their profession. In this study, they aimed to identify the future expectations of technological changes.

This paper is organized as follows: Section 2 conceptualizes both CE and I4.0. Section 3 presents our literature review with its corresponding analysis. In Sections 4 and 5, we discuss the findings and our analysis of the synergy between CE and I4.0. Finally, in Sections 6 and 7, we denote our proposals for future work and our conclusions from this literature review.

2. Topic Conceptualization

This review is based on two main notions, the CE and I4.0. In this section, their concepts and definitions are presented.

2.1. Circular Economy

The concept of CE became popular on the premise of the economic growth experienced by China in the 1990s and the limited natural resources available to produce goods and services. The studies that were conducted at that time on CE concluded that manufacturing systems were lacking in social aspects and sustainability. There is currently a lack of clarity in the justification of how CE may fulfill the three pillars of sustainability from a social perspective [5]. A good definition of CE has been provided by the Ellen MacArthur Foundation, which defines CE as a "restorative and regenerative system by design," whose purpose is to keep products, components, and materials at their maximum usefulness and value [9].

For their part, Rajput and Singh [5] stated that CE is configured as a closed supply chain that seeks restorative and regenerative use, which implies that the system is conceptualized within the "end-of-life" or "lifespan" of the products and services at an industrial level, thus eliminating toxic materials, and reusing and eliminating waste through the explicit implementation of design models, product systems, and materials design. In terms of the supply chain, the purpose of CE is to make resources more efficient and optimize environmental results. In addition, CE is known for using an economic model aimed at minimizing the consumption of materials to instead focus on their intelligent use. CE models transition from a classical open linear economic model to a closed circular model, centered on providing positive and balanced economic, environmental, and social impacts [2].

The organizational transition to CE is progressing, denoting that there is still a gap between the linear and circular models. Linear economic models find their accent on the remaining linear supply chains, wherein raw materials continue to be wasted [6]. Studies addressing the factors linked to linear economies have indicated that they are unsustainable and demonstrate the urgent need for adopting a new paradigm-shifting model that can foster the necessary transition towards a CE [5].

2.2. Industry 4.0

I4.0, also known as the technological revolution, encompasses big data, industrial automation (robotics), simulations, integration systems, the internet of things (IoT), cybersecurity, cloud computing, additive manufacturing, and augmented reality as the main comprehensive factors of technological work aimed at continuous improvement. As a concept, I4.0 combines information and communication technologies (ICTs) with production and manufacturing processes [10]. Figure 1 provides an overview of how I4.0 integrates major technologies, and the pillars of I4.0 are outlined following.

Figure 1. Technological integration in Industry 4.0.

2.2.1. Big Data

Big data is a set of technological tools created with the capacity to capture, store, manage, process, and analyze a large variety and high volumes of data. Big data was created because conventional tools failed to perform these tasks in viable timeframes [5], and it supports product development processes, supply and demand, production logistics, communications, radio frequency (RFID), and product identification (sensors and barcodes), and so on, through IoT implementation [10].

2.2.2. Industrial Automation (Robotics)

Industrial automation refers to incorporating artificial intelligence (AI) into software. It possesses machine learning capabilities for systematically performing repetitive tasks, previously performed by humans, on a large scale.

2.2.3. Simulations

Simulations are used to model systems extracted from the real environment to assess their behavior under different conditions and understand their responses to different direct and indirect agents that may influence the system results [11]. Due to their flexibility, simulations are a fundamental pillar in I4.0, as they may be implemented in any field, such as manufacturing, services, design, and healthcare. I4.0 finds software-based process simulation beneficial by virtue of the implementation of analytical processes.

2.2.4. Integration Systems

This concept refers to linking system components (vertical integration) or two or more systems (horizontal integration) to provide the interfaces required to associate physical and virtual system objects. It is also known as end-to-end integration.

2.2.5. Internet of Things (IoT)

First defined in 1999 by Kevin Ashton, the IoT is conceived as a system that transfers products, service, process, activity, or task data in real time, thus supporting dynamic information management [12]. On the other hand, Lopez de Sousa et al. [10] defined IoT as the interconnection between electronic systems that exchange and collect data in real time, generating added value to the final object.

2.2.6. Cybersecurity

Cybersecurity helps organizations create secure business controls, as well as the methods used to assess those controls, to improve the effectiveness of industrial systems. Cybersecurity means using intelligent interconnected cyber–physical systems to automate industrial operations through secure networks, and providing security protocols that offer protection against cyber threats.

2.2.7. Cloud Computing

Cloud computing is a digital infrastructure that facilitates access through the internet to computing and processing engines located in remote servers. Currently, the use of cloud computing has been promoted to implement electronic commerce services, facilitating the interaction between customers and vendors characterized by resource virtualization and the automation of logistics and business processes [10].

2.2.8. Additive Manufacturing (3D)

Additive manufacturing is based on the generation of product parts without having to acquire and use specialized tools while reducing waste, thus differentiating manufacturing from conventional processes. Production is based on three-dimensional (3D) digital designs, using conventional raw materials without the requirement of preconceived molds [13]. 3D printers are the main resources associated with additive manufacturing.

2.2.9. Augmented Reality (AR) and Virtual Reality (VR)

Augmented reality (AR) is a technology used by I4.0 as an approach in which people use digital tools to access virtual spaces superimposed on actual physical spaces through virtual information data. AR is positioned between the physical world and virtual reality (VR), without replacing the real world, but allowing data to be collected from simulations in order to better understand real-world systems.

3. Literature Search and Selection Procedure and Criteria

The criteria used for searching and selecting papers, as well as the implicit process developed for these purposes, are described below. As reference, we used Snyder's [14] "Literature Review as a Research Methodology: General Description and Guidelines". The research question for this study was: "How has the synergy between I4.0 and CE impacted societies in the last 3 years?" The main steps performed for finding and selecting academic papers for review are described below.

3.1. Search Criteria

To guarantee suitable content and approach transparency, only certain specific databases were considered. The search criteria defined for the research question were "circular economy and Industry 4.0", "sustainability and circular economy", and "circular economy and society".

During the review, only formal literature (scientific publications, academic papers, industrial case studies, and conference proceedings) was considered. The queries were made in the following international databases: IEEE, Scopus, and ScienceDirect. The search was conducted considering only documents written in English and published between 2018 and 2020.

The total process was carried out in four steps. First, in the identification stage, we began with 4484 papers based on the above search criteria. After removing duplicate papers, we had 2261 remaining. In the screening stage, we only selected papers from the last three years, which reduced the number to 1884. In the eligibility stage, we filtered by the paper structure, citations, methodology, and internationalism, and the result was 179 papers. After that, we made other selections by the main topic "industry and economy", and the number of papers became 124. In the end, after selection of the papers by the subjects CE and I4.0, there were 41 papers.

3.2. Search Strategy

Through the advanced search of the consulted databases, three search strings were considered to collect documents for subsequent review. We used Boolean operators, such as AND, OR, and NOT, in these search strings. Three databases were used to search and filter the relevant papers: Science Direct, IEEE, and Wiley Interscience. Figure 2 illustrates the search process.

3.3. Content Analysis

The results of the search conducted in terms of the number of papers published per semester are shown in Figure 3. A total of 41 papers were reviewed, of which 95% focused on the main topic of the study. If scientific journals with an assessment indicator of relative importance are considered (Ranking Q1 in Scimago), we found that 83% of the journals were categorized and interested in publishing these topics. Figure 4 shows the main journals dedicated to the publication of the topics discussed in this paper.

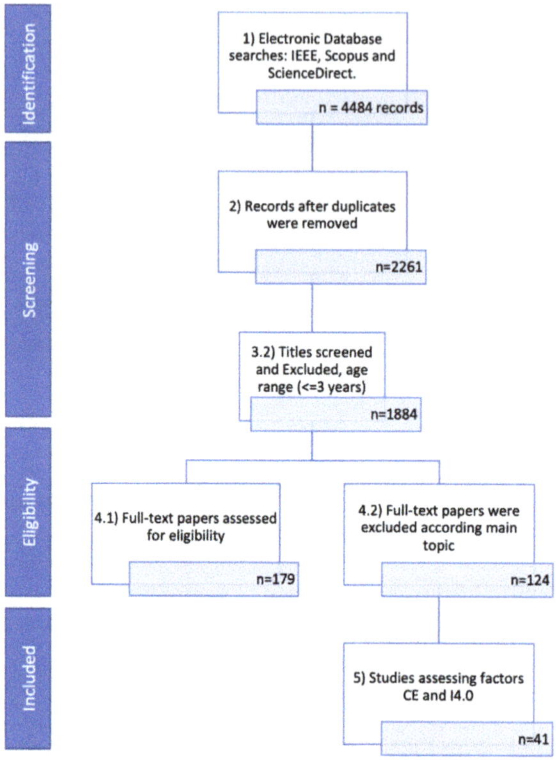

Figure 2. Search process.

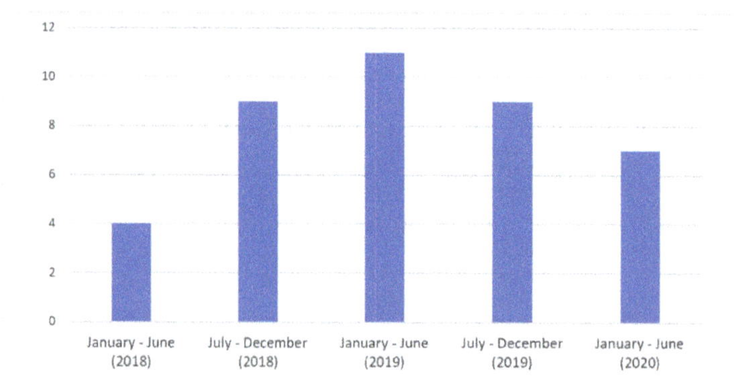

Figure 3. Historical series of published papers.

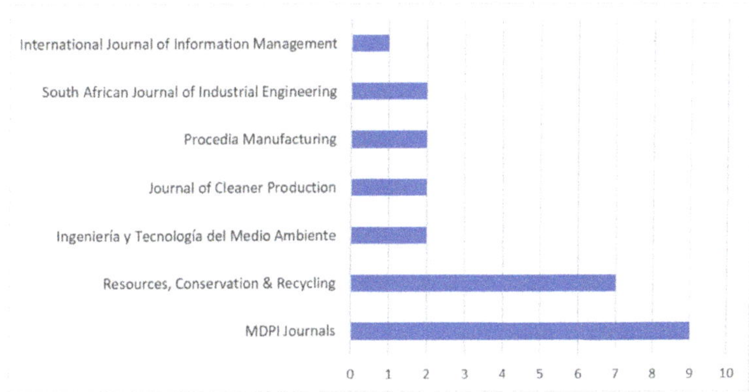

Figure 4. Main scientific journals.

Twelve percent of all papers reviewed were published in the form of conference proceedings. Regarding the countries in which the authors' affiliation institutions are located, Figure 5 highlights India, Italy, Spain, Brazil, and the United Kingdom, accounting for 54% of the total number of published works. Figure 6 lists the topics and their classification by group of published papers, wherein the influence from I4.0 influences and its different technological applications within the context of CE and sustainability were considered as the first selection criterion. The multidisciplinary factor was the main supporting aspect of the research analysis.

CE and I4.0 are topics that have been highly explored in recent years. The publication trends on research topics revolving around CE and I4.0 showed an increase toward 2020. Our analysis also considered the methodologies used by the authors, whether the case studies were taken from real activities, the transition toward the CE, and the correlation between CE and I4.0.

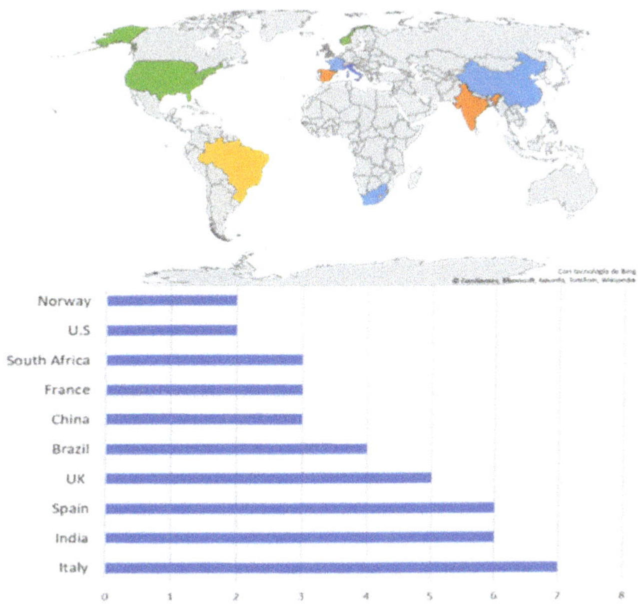

Figure 5. Main publishing countries.

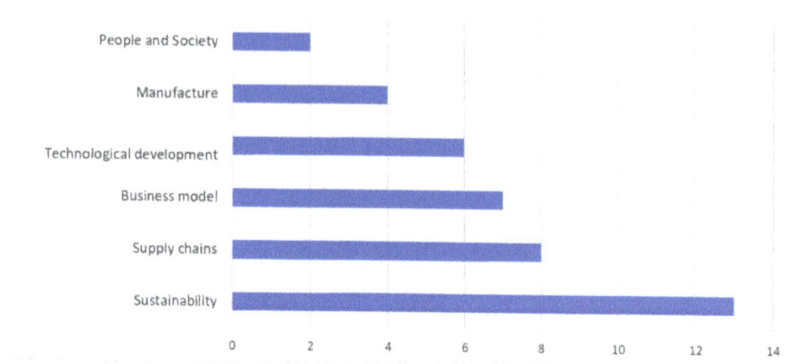

Figure 6. Topics classified by group of papers.

In terms of methodology, most authors developed their research through case studies and literature reviews. In their papers, many of the authors addressed the binding relationship between CE and I4.0, highlighting the latter as the main protagonist with the application of its technological phases in the transition that organizations and their supply chains must undertake to evolve from a linear to a circular economy.

For the review of the selected research papers, we also identified and systematized the most relevant aspects found in the results, discussions, and conclusions of the papers selected for this literature review. Each of the main topics of the reviewed publications was identified and coded (see Table 1), and the I4.0 technologies referenced by each publication were also considered.

Table 1. Topics and abbreviations.

Central Topic	Main Topic	Abbreviation
Circular Economy (CE)	Supply chain management	SCM
	Technological development	TD
	Manufacturing	MAN
	Business model	BM
	Sustainability	SUS
	People and society	PS
Industry 4.0 (I4.0)	Big data	BD
	Industrial automation (robotics)	IA
	Simulation	SIM
	Integration systems	IS
	Internet of things (IoT)	IOT
	Cybersecurity	CYB
	Cloud computing	CC
	Additive manufacturing (3D)	AM
	Augmented reality (AR)	AR

4. Results

The information of the selected papers is presented in a table, including the main topic, reference number, type, and objective. Subsequently, the relevant data were extracted after a detailed reading, grouping different publications according to their common topic, as shown in Table 2.

Table 2. Classification of publications.

Main Topic	Reference	Type	Objective
Supply Chain	[4]	Review	Literature systematic review on 3PLS operators in reverse logistics.
	[7]	Framework	Proposal for a holonic multiscale framework for integrated sustainable supply chain management that allows evolution taking environment into account.
	[15]	Framework	Proposes the thematic framework that must be considered in research studies seeking to address the opportunities and challenges that the I4.0 supply chain must face in CE environments.
	[16]	Case study	Analysis of the supply chain of an organization to meet the requirements of Industry 4.0 and enable the circular economy, based on the concept of 6Rs.
	[17]	Framework	Analysis of the sustainable supply chain in the health sector to propose the conceptual framework for the transition of the circular economy.
	[18]	Mathematical model	This paper proposes a route for reverse logistics in a sustainable reverse supply chain by implementing the principles of I4.0 and the ReSOLVE model of circular economy (CE).
	[5]	Review	Understanding the hidden connection between the circular economy (CE) and Industry 4.0 in the context of the supply chain.
	[6]	Framework	This paper proposes a solution framework supported in I4.0 and CE to overcome the challenges of supply chain 4.0 and the circular economy.
Technological Development	[19]	Review	This paper proposes an integrative framework for the formulation of policies for future research through a systematic review of literature in terms of CE and I4.0 fields.
	[20]	Case study	An assessment of a management model for the analysis capacity of Industry 4.0 technologies based on different costing models (ABC and TDABC).
	[21]	Review	This paper presents a results summary of project K, recycling and waste recovery 4.0 publications, considering the systems, methods and technologies used in the industry and waste collection.
	[11]	Case study	This paper presents a lab showing how I4.0-based technologies can support CE practices through virtual testing.
	[22]	Case study	An assessment of differences and similarities in the implementation of big data in the cryptocurrency sector and municipal waste management.
Manufacturing	[23]	Case study	This paper defines an approach to product design based on user experience considering smart production as the foundation of I4.0 in the circular economy of the glass recycling industry.
	[24]	Case study	This paper implements tools to define the balance point between sustainability and the circular economy in the environment for the manufacturing of ceramic tiles, and demonstrates how new business opportunities can be created through the evolution of a linear business model toward a circular model, thanks to I4.0.
	[25]	Framework	This paper examines the technical limitations of implementing zero-waste manufacturing technologies ZWM in dense urban areas in Singapore.
	[26]	Framework	An assessment of circular economy implementation in electrical and electronic equipment production, considering the waste amount generated, as well as raw materials and the energy balance.

Table 2. *Cont.*

Main Topic	Reference	Type	Objective
Business Model	[10]	Framework	This paper assesses how I4.0 can support CE strategies, providing bases for decision-making in sustainable operations management.
	[13]	Framework	This paper assesses how I4.0 integrates with the circular economy (CE) to establish a business model that reuses and recycles waste such as scrap metal or electronic waste.
	[9]	Framework	This paper develops a conceptual framework, based on literature and a case study of a company, identifying the functionalities of I4.0 that implement a business model focused on the use of the industrial household appliance.
	[27]	Case study	This paper explores how I4.0 integrates with the circular economy (CE) to establish a business model that reuses and recycles waste such as scrap metal or electronic waste.
	[2]	Framework	This paper develops an innovative framework that highlights the links between I4.0 and CE and the unveiling of future fields of research.
	[28]	Framework	This paper identifies how purchases 4.0 and digital transformations are related, in addition to how digital transformation impacts the intention to optimize the hiring process in the circular economy.
	[29]	Case study	This paper addresses challenges and opportunities for the collaborative economy based on original evidence compiled from cases of the Brazilian manufacturing industry that use natural metallic resources.
Sustainability	[30]	Framework	This paper presents the existing gaps between data and sustainability as the optimal solution for the existence of an industrial symbiosis.
	[12]	Framework	This paper proposes reducing or eliminating the source of impacts to create environmental value through the sustainability of a global ecosystem.
	[31]	Framework	This paper evaluates sustainability in smart products connected to each other, through methodological development oriented to I4.0.
	[32]	Case study	This paper describes a mobile equipment model with the ability to provide sustainability to conventional videographic production equipment by applying CE principles.
	[33]	Case study	This paper explores the transition from linear economy to circular economy through a procedure to introduce the sustainability principles (environmental, economic, and social).
	[34]	Framework	This paper addresses the concept of sustainable development and the need for its introduction in the circular economy activity.
	[35]	Case study	This paper exemplifies a model taken from the literature that considers automobile manufacturing and shows how we can deal with the problem using standard techniques of performance evaluation in the circular economy.
	[1]	Case study	This paper evaluates the sustainability of agri-food processes and their environmental impacts using big data technologies to improve management in the supply chain.
	[36]	Review	Literature review on emerging digital technologies from I4.0 focused on sustainability through remanufacturing.
	[37]	Mathematical model	This paper investigates the relationship of economic development measured as economic growth, energy use, trade and foreign direct investment, and the environment in eleven emerging countries in Eastern Europe and Asia.
	[38]	Case study	This paper explores the feasibility of a redistributed business model for manufacturers employing new manufacturing technologies (I4.0) as part of a sustainable and circular production and consumption system.

Table 2. *Cont.*

Main Topic	Reference	Type	Objective
Sustainability	[39]	Framework	This paper presents a scheme of risks in the Industry 4.0 situation that is associated to the triple bottom line of sustainability.
	[3]	Review	An assessment of the literature that jointly considers the concepts of sustainability, sustainable development, and sociotechnics, as well as their transitions with I4.0.
People and Society	[40]	Framework	This paper examines the role of CE in addressing COVID-19 by relating issues both in the present and in the future using strategies to close, reduce, and slow resource loops as a life sustainability strategy.
	[41]	Framework	An assessment of the relationship between happiness, co-creation, and sustainable development as a theoretical framework based on the well-established construct of value co-creation.
	[8]	Case study	This paper aims to discover the possible relationships between the age of workers and knowledge of the concept of Industry 4.0, the threats of new technologies, and the future expectation of technological changes.

The literature review of the papers selected allowed us to establish a reciprocal relationship between CE and I4.0—in other words, to describe the way I4.0 and its technologies exerted a direct effect on CE, or vice versa, and how CE and its different applications promoted the functionality of I4.0 technologies. The studies that have been published to date show a wide range of technological applications to the various contexts of organizations and their manufacturing and marketing processes through supply chains. Out of the 41 papers found in the searched databases that were directly related to the topic at hand and, therefore, selected for our literature review, 32.5% (13) were published in 2018, 50% (20) were published in 2019, and the remaining 17.5% (7) in 2020.

The papers analyzed in this review were classified by type, as shown Figure 7. These papers focused mostly on presenting frameworks, and few of them focused on mathematical models. A total of 35% of the reviewed papers were study cases applied to frameworks proposed in previous years; however, 45% of the research presented frameworks that proposed potential study cases for future works. Only 15% of the papers were reviews, which suggests developing more reviews for works published in recent years.

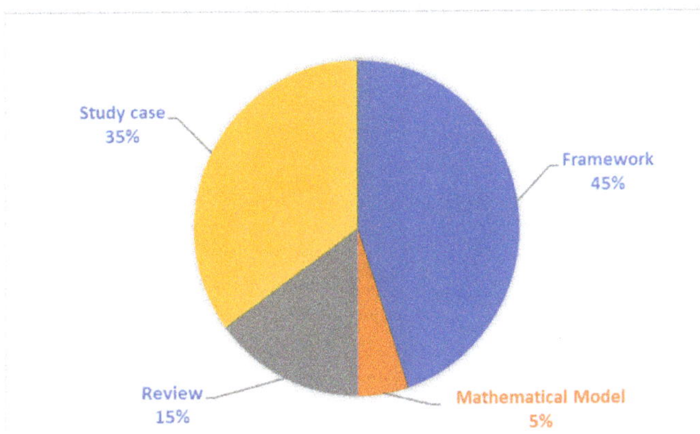

Figure 7. Classification of papers.

The 10 countries with the highest concentration of publications, in which 77% of the reviewed publications were found, are shown in Figure 8. Figure 8 shows the thematic preferences developed by the top 10 countries as identified and classified in the group of reviewed papers.

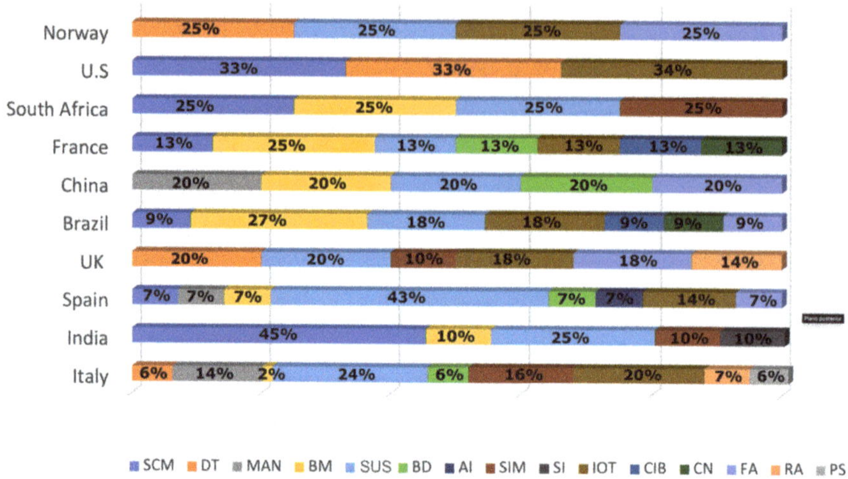

Figure 8. Development topic by country.

In this review, all papers focused their research on CE through sustainability and technological applications of I4.0. The top three topics related to CE were sustainability (SUS), with 23 papers; supply chain management (SCM), with 10 papers; and business models (BM), with 11 papers. Except the US, all countries had related papers on SUS with percentages in the range of 13% to 43%, for example Spain. India, the US, and South Africa led in the percentages of papers developing topics on SCM, at 25% to 45% of their total studies. Finally, Brazil, South Africa, and France showed much higher percentages of papers on BM, from 25% to 27%. On the other hand, the top topic related to I4.0 was IOT, with 17 papers. Other I4.0 topics had fewer related papers. This shows that the studies had specific approaches to CE, while they selected a more varied set of tools for I4.0.

Regarding the subject of people and society (PS), we identified that [8,40,41] are some of the recent research works dedicated to understanding how CE directly impacts individuals and how they should prepare to face the changes required by on-going transition from linear to circular economic models. Table 3 shows the relationship between the topics identified and categorized for each one of the different selected research studies, which highlight SCM, technological development (TD), manufacturing (MAN), SUS, and PS as the main topics of CE. In turn, these issues are contrasted against the technologies envisioned in I4.0 to identify the relationship between CE and I4.0 described in each paper.

The analysis demonstrated that CE is mainly supported by SUS, SCM, and MAN using IOT and big data (BD) as the main I4.0 technologies to propose a linear-to-circular model transition that may support sustainability or sustainable economic development. In their literature review, Rosa et al. [2] summarized that papers addressing the relationship between CE and I4.0 should be viewed in a generic way. Herein, the authors also claimed that any organization undergoing a transition from a linear to a circular economy must implement technological development within its value chain through I4.0.

In their literature review, Okorie et al. [19] mentioned a growing trend in CE and I4.0 publications when considering areas of individual research, especially during the 2012–2018 period. At the time, the US and China led the research, in terms of engineering and informatics. In addition, Kerin and Pham [36], based on their own literature review, also viewed the IOT and AR as necessary technologies for manufacturing.

Table 3. Relationship between CE and I4.0.

	Circular Economy						Industry 4.0								
Reference	**SCM**	**TD**	**MAN**	**BM**	**SUS**	**PS**	**BD**	**IA**	**SIM**	**IS**	**IOT**	**CYB**	**CC**	**AF**	**AR**
[1]	X				X		X								
[2]				X							X				
[3]					X										
[4]	X										X				
[5]	X									X					
[6]	X				X										
[7]	X				X										
[8]		X									X			X	
[9]				X			X				X				
[10]				X							X	X	X		
[11]		X							X						X
[12]					X						X				
[13]				X	X									X	
[15]	X										X				
[16]	X				X										
[17]	X				X						X				
[18]	X				X										
[19]		X									X				
[20]		X													
[21]		X						X							
[22]		X					X								
[23]			X											X	
[24]		X		X						X					
[25]			X								X				
[26]			X				X				X				
[27]				X											
[28]				X					X						
[29]				X			X								
[30]					X		X								
[31]					X		X	X							
[32]					X									X	
[33]				X	X						X				
[34]	X				X										
[35]			X		X										
[36]				X	X						X				X
[37]					X						X			X	
[38]				X	X									X	
[39]		X			X						X			X	
[40]	X				X	X									
[41]					X	X									
Total	11	7	5	10	23	2	8	4	3	2	17	2	2	8	3

4.1. Contributions per Journal

Table 4 displays information on the total number of papers published in scientific journals, detailing the percentages for each journal with respect to the total of 41 publications that were reviewed. The MDPI journals and Resources, Conservation & Recycling journal accounted for 34.88% of the papers reviewed. The other 23 journals, wherein one or two papers were published, provided 35.12% of the papers reviewed.

Table 4. Contributions per journal.

Journal Title	Publications No.	%
MDPI journals	8	18.60
Resources, Conservation & Recycling	7	16.28
Ingeniería y Tecnología del Medio Ambiente	2	4.65
International Journal of Information Management	2	4.65
Journal of Cleaner Production	2	4.65
Procedia Manufacturing	2	4.65
South African Journal of Industrial Engineering	2	4.65
19th International Multidisciplinary Scientific GeoConference	1	2.33
Computers & Industrial Engineering	1	2.33
Computers in Industry	1	2.33
CrossMark	1	2.33
Economics of Food Industry	1	2.33
International Journal of Production Research	1	2.33
Journal of Business Research	1	2.33
Journal of Engineering	1	2.33
Journal of Manufacturing Technology Management	1	2.33
Journal Risk and Financial Management	1	2.33
Littera Scripta	1	2.33
MobileHCI-2018	1	2.33
Plastics Technology	1	2.33
Resources Policy	1	2.33
Socio-Economic Planning Sciences	1	2.33
Trakia Journal of Sciences	1	2.33
ValueTools '20	1	2.33
Waste Management	1	2.33

4.2. Contributions per Country

There were 22 countries considered in the analysis, which we associated to the nationality of the authors of the research papers reviewed. According to Table 5, the top five countries were Italy, India, Spain, the United Kingdom, and Brazil, which accounted for 53% of the total number of nationalities represented in the research papers addressing CE and I4.0 topics. The second group of 17 countries represents 47%, from which China, France, and South Africa stand out.

Table 5. Contributions per country.

Country	No.	%	Country	No.	%
Italy	7	16.67	Bulgaria	1	2.38
India	6	14.29	Canada	1	2.38
Spain	6	14.29	Czech Republic	1	2.38
United Kingdom	5	11.90	Denmark	1	2.38
Brazil	4	9.52	Germany	1	2.38
China	3	7.14	Ireland	1	2.38
France	3	7.14	Lithuania	1	2.38
South Africa	3	7.14	New Zealand	1	2.38
United States	2	4.76	Poland	1	2.38
Norway	2	4.76	Singapore	1	2.38
Austria	1	2.38	Ukraine	1	2.38

5. Discussion

Based on the papers reviewed, the authors are in agreement that CE and I4.0 exhibit correlated behavior since they depend on each other to achieve the transition from a linear to circular model to guarantee sustainability in supply chains. In our review, we found very few studies that address the impact of CE in society, as well as the mechanisms used to measure the effects that CE implementation generates in people. In fact, only three studies came close. The first one used ideas derived from CE principles to promote a collaborative supply chain model for the healthcare sector, engaging people and society to change the focus from consumer to co-creator due to the crisis generated by COVID-19 [40].

In the second study, Cosimato et al. [41] claimed that individual, community, or global happiness can be approached as a key component for circular logic-based sustainable development. To support this theoretical research, these authors proposed an extreme case study, highlighting the way in which a specific Italian prison community approached the individual and collective aspects of happiness, both in the present and in the future. Finally, the third study examined the possible relationships between age and awareness of the Industry 4.0 concept, considering the fundamental skills and capacities requiring the use of new technologies and the expectation of future technological changes as threats [8].

In order to measure the maturity level of industry 3.0, and to migrate to industry 4.0 based on circular economy, it is important to consider the following elements: decentralization, interoperability, ability to adapt to changes, real-time working capabilities, virtualization, and service orientation [6,35].

6. Future Work

In this review, we established a relationship between CE and other areas of study, such as SCM, TD, MAN, BM, SUS, and PS, in the last three years. These relationships must be further studied in greater depth using different methodologies. To this end, this work updates the state-of-the-art for future research and offers concrete information on the research advances regarding CE and its impact in areas other than those that have been recently documented in published works.

For future research, we recommend emphasizing the strategies and the social and individual dynamics implemented in different countries in terms of their economic development, so that the impact from I4.0 technologies and the transition toward sustainability is easier to assimilate and incorporate into daily life, thus allowing processes that develop in harmony with the environment and the protection of the planet.

7. Conclusions

This literature review allowed us to identify the different theories and concepts revolving around CE and I4.0 through a comprehensive work of understanding, analysis, and synthesis. Several of the papers reviewed provided arguments from case studies based on industrial or organizational processes. Other papers focused on literature reviews to find methodologies and arguments that supported other hypotheses.

This research identified relationships and concordances in the studies reviewed; however, there are still areas of research in which the implementation of CE principles through I4.0 in the organizational value chains is not evidenced. It is important to mention that very few studies addressed the impacts from a transition to CE supported by I4.0 on individuals and society. Although it is true that CE is based on eco-conceptions, industrial and territorial ecology, functional economy, second use, reuse, repair, recycling, and valuation, we still lack studies that show that these pillars positively benefit individuals by guaranteeing sustainability or the development of sustainable life from a social and political perspective.

According to the works reviewed, it is still uncertain how much the transition to the CE may cost society and what strategies these actors have available to avoid social failure. In emerging economy countries, adopting the different I4.0 technologies may involve major challenges, in addition to those related to the dynamics required by the system. Here, we show that it is feasible to build an intelligent industry (I4.0) based on a CE model. The disruptive technologies that are the basis of I4.0 will be managed as active elements for a circular economy business model and, in turn, achieve the 17 targets set by the OECD for 2030. In I4.0, the CE model allows for measuring the income generated by chain production waste in a way that can increase the ROI while decreasing the impact on the environment.

Author Contributions: Project administration, Funding acquisition and Writing—review & editing, C.A.T.R.; Investigation and Writing—original draft, D.F.C.; Resources and Methodology, J.H.O. and O.I.K.; Validation and Supervision, M.A.V. All authors have read and agreed to the published version of the manuscript.

Funding: This research has been funded by Dirección General de Investigaciones of Universidad Santiago de Cali under call No. 01-2021.

Institutional Review Board Statement: Not applicable.

Informed Consent Statement: Not applicable.

Data Availability Statement: The data that support the findings of this study are available from the corresponding author upon reasonable request.

Conflicts of Interest: The authors declare that they have no conflicts of interest to report regarding the present study.

References

1. Belaud, J.-P.; Prioux, N.; Vialle, C.; Sablayrolles, C. Big data for agri-food 4.0: Application to sustainability management for by-products supply chain. *Comput. Ind.* **2019**, *111*, 41–50. [CrossRef]
2. Rosa, P.; Sassanelli, C.; Urbinati, A.; Chiaroni, D.; Terzi, S. Assessing relations between Circular Economy and Industry 4.0: A systematic literature review. *Int. J. Prod. Res.* **2019**, *58*, 1662–1687. [CrossRef]
3. Asiimwe, M.M.; Kock, I.H. An analysis of the extent to which industry 4.0 has been considered in sustainability or socio-technical transitions. *S. Afr. J. Ind. Eng.* **2019**, *30*, 41–51. [CrossRef]
4. Tombido, L.L.; Louw, L.; Van Eeden, J. A systematic review of 3pls' entry into reverse logistics. *S. Afr. J. Ind. Eng.* **2018**, *29*, 235–260. [CrossRef]
5. Rajput, S.; Singh, S.P. Connecting circular economy and industry 4.0. *Int. J. Inf. Manag.* **2019**, *49*, 98–113. [CrossRef]
6. Yadav, G.; Luthra, S.; Jakhar, S.K.; Mangla, S.K.; Rai, D.P. A framework to overcome sustainable supply chain challenges through solution measures of industry 4.0 and circular economy: An automotive case. *J. Clean. Prod.* **2020**, *254*, 120112. [CrossRef]
7. Gómez, A.M.; González, F.A.; Luque, A. A holonic framework for managing the sustainable supply chain in emerging economies with smart connected metabolism. Resource. *Conserv. Recycl.* **2018**, *141*, 219–232. [CrossRef]
8. Kuba, F.M. Industry 4.0 and its impact on employees' age. *Littera Scr.* **2019**, *12*, 1–12.
9. Bressanelli, G.; Adrodegari, F.; Perona, M.; Saccani, N. Exploring How Usage-Focused Business Models Enable Circular Economy through Digital Technologies. *Sustainability* **2018**, *10*, 639. [CrossRef]

10. de Sousa, A.B.L.; Chiappetta, C.J.; Filho, M.G.; Roubaud, D. Industry 4.0 and the circular economy: A proposed research agenda and original roadmap for sustainable operations. *Springer Sci.* **2018**, *270*, 273–286.
11. Rocca, R.; Rosa, P.; Sassanelli, C.; Fumagalli, L.; Terzi, S. Integrating Virtual Reality and Digital Twin in Circular Economy Practices: A Laboratory Application Case. *Sustainability* **2020**, *12*, 2286. [CrossRef]
12. García, A.; González, F.A.; Roldán, A.C. Propuesta de marco de trabajo para la evaluación de la sostenibilidad de productos desde el paradigma de la economía circular basada en industria 4.0 (parte 1). Cod. 8631, Ingeniería y tecnología del medio ambiente. *Rev. DYNA* **2018**. [CrossRef]
13. Nascimento, D.L.M.; Alencastro, V.; Quelhas, O.L.G.; Caiado, R.G.G.; Garza-Reyes, J.A.; Rocha-Lona, L.; Tortorella, G. Exploring Industry 4.0 technologies to enable circular economy practices in a manufacturing context: A business model proposa. *J. Manuf. Technol. Manag.* **2019**, *30*, 607–627. [CrossRef]
14. Snyder, H. Literature review as a research methodology: An overview and guidelines. *J. Bus. Res.* **2019**, *104*, 333–339. [CrossRef]
15. Singh, S.P.; Singh, R.K.; Gunasekaran, A.; Ghadimi, P. Supply Chain Management, Industry 4.0, and the Circular Economy. *Resour. Conserv. Recycl.* **2018**, *142*, 281–282.
16. Manavalan, E.; Jayakrishna, K. An Analysis on Sustainable Supply Chain for Circular Economy. *Procedia Manuf.* **2019**, *33*, 477–484. [CrossRef]
17. Daú, G.; Scavarda, A.; Scavarda, L.F.; Portugal, V.J.T. The Healthcare Sustainable Supply Chain 4.0: The Circular Economy Transition Conceptual Framework with the Corporate Social Responsibility Mirror. *Sustainablity* **2019**, *11*, 3259. [CrossRef]
18. Dev, N.K.; Shankar, R.; Qaiser, F.H. Industry 4.0 and circular economy: Operational excellence for sustainable reverse supply chain performance. *Resour. Conserv. Recycl.* **2020**, *153*, 104583. [CrossRef]
19. Okorie, O.; Salonitis, K.; Charnley, F.; Moreno, M.; Turner, C.; Tiwari, A. Digitisation and the Circular Economy: A Review of Current Research and Future Trends. *Energies* **2018**, *11*, 3009. [CrossRef]
20. Mortensen, S.T.; Nygaard, K.K.; Madsen, O. Outline of an Industry 4.0 Awareness Game. *Procedia Manuf.* **2019**, *31*, 309–315. [CrossRef]
21. Sarc, R.; Curtis, A.; Kandlbauer, L.; Khodier, K.; Lorber, K.; Pomberger, R. Digitalisation and intelligent robotics in value chain of circular economy oriented waste management—A review. *Waste Manag.* **2019**, *95*, 476–492. [CrossRef] [PubMed]
22. Limba, T.; Novikovas, A.; Stankevičius, A.; Andrulevičius, A.; Tvaronavičienė, M. Big Data Manifestation in Municipal Waste Management and Cryptocurrency Sectors: Positive and Negative Implementation Factors. *Sustainability* **2020**, *12*, 2862. [CrossRef]
23. Lin, K.-Y. User experience-based product design for smart production to empower industry 4. *0 in the glass recycling circular economy. Comput. Ind. Eng.* **2018**, *125*, 729–738. [CrossRef]
24. Garcia, F.E.; González, R.; Ferrari, A.M.; Volpi, L.; Pini, M.; Siligardi, C.; Settembre, D. Identifying the Equilibrium Point between Sustainability Goals and Circular Economy Practices in an Industry 4.0 Manufacturing Context Using Eco-Design. *Soc. Sci.* **2019**, *8*, 241. [CrossRef]
25. Kerdlap, P.; Choong, J.S.; Ramakrishna, S. Zero waste manufacturing: A framework and review of technology, research, and implementation barriers for enabling a circular economy transition in Singapore. Resources. *Conserv. Recycl.* **2019**, *151*, 104438. [CrossRef]
26. Palka, D. Closed circuit economy in the context of implementing new technologies. In Proceedings of the 19th International Multidisciplinary Scientific GeoConference SGEM, Albena, Bulgaria, 28 June–7 July 2019.
27. Sterev, N. New industrial business models: From linear to circular economy approach. *Trakia J. Sci.* **2019**, *17*, 511–523. [CrossRef]
28. Bag, S.; Wood, L.C.; Mangla, S.K.; Luthra, S. Procurement 4.0 and its implications on business process performance in a circular economy. *Resour. Conserv. Recycl.* **2020**, *152*, 104502. [CrossRef]
29. Jabbour, C.J.C.; Fiorini, P.D.C.; Wong, C.W.; Jugend, D.; Jabbour, A.B.L.D.S.; Seles, B.M.R.P.; Pinheiro, M.A.P.; da Silva, H.M.R. First-mover firms in the transition towards the sharing economy in metallic natural resource-intensive industries: Implications for the circular economy and emerging industry 4.0 technologies. *Resour. Policy* **2020**, *66*, 101596. [CrossRef]
30. Tseng, M.-L.; Tan, R.R.; Chiu, A.S.; Chien, C.-F.; Kuo, T.C. Circular economy meets industry 4.0: Can big data drive industrial symbiosis? *Resour. Conserv. Recycl.* **2018**, *131*, 146–147. [CrossRef]
31. García, A.; González, F.A.; Roldán, A.C. Propuesta de marco de trabajo para la evaluación de la sostenibilidad de productos desde el paradigma de la economía circular basada en industria 4.0 (parte 2). Cod. 8718, Ingeniería y tecnología del medio ambiente. *Rev. DYNA* **2018**. [CrossRef]
32. Forlastro, G.; Chiesa, I.; Gena, C.; Cietto, V. IoT for the Circular Economy: The case of a mobile set for video-makers. In Proceedings of the MobileHCI'18: 20th International Conference on Human-Computer Interaction with Mobile Devices and Services Adjunct, Barcelona, Spain, 3–6 September 2018.
33. Garcia-Muiña, F.E.; González-Sánchez, R.; Ferrari, A.M.; Settembre-Blundo, D. The Paradigms of Industry 4.0 and Circular Economy as Enabling Drivers for the Competitiveness of Businesses and Territories: The Case of an Italian Ceramic Tiles Manufacturing Company. *Soc. Sci.* **2018**, *7*, 255. [CrossRef]
34. Sedikova, I. Development of conceptual principles of the circular economy. *Food Ind. Econ.* **2019**, *11*. [CrossRef]
35. Gribaudo, M.; Manini, D.; Pironti, M.; Pisano, P. Circular Economy: A Performance Evaluation Perspective. In Proceedings of the VALUETOOLS'20: 13th EAI International Conference on Performance Evaluation Methodologies and Tools, Tsukuba, Japan, 18–20 May 2020.
36. Kerin, M.; Pham, D.T. A review of emerging industry 4.0 technologies in remanufacturing. *J. Clean. Prod.* **2019**, *237*, 237. [CrossRef]

37. Koilo, V. Evidence of the Environmental Kuznets Curve: Unleashing the Opportunity of Industry 4.0 in Emerging Economies. *J. Risk Financ. Manag.* **2019**, *12*, 122. [CrossRef]
38. Turner, C.J.; Moreno, M.; Mondini, L.; Salonitis, K.; Charnley, F.; Tiwari, A.; Hutabarat, W. Sustainable production in a Circular Economy: A business model for re-distributed manufacturing. *Sustainability* **2019**, *11*, 4291. [CrossRef]
39. Birkel, H.S.; Veile, J.W.; Müller, J.M.; Hartmann, E.; Voigt, K.-I. Development of a Risk Framework for Industry 4.0 in the Context of Sustainability for Established Manufacturers. *Sustainability* **2019**, *11*, 384. [CrossRef]
40. Wuyts, W.; Marin, J.; Brusselaers, J.; Vrancken, K. Circular economy as a COVID-19 cure? *Resour. Conserv. Recycl.* **2020**, *162*, 105016. [CrossRef]
41. Cosimato, S.; Faggani, M.; Prete, M. The co-creation of value for pursuing a sustainable happiness: The analysis of an Italian prison community. *Socio-Econ. Plan. Sci.* **2020**. [CrossRef]

sustainability

Article

Nurture: A Novel Approach to PSS-Rebound Effect Identification

Salman Alfarisi [1], Yuya Mitake [2], Yusuke Tsutsui [3], Hanfei Wang [1] and Yoshiki Shimomura [1,*]

1 Faculty of Systems Design, Tokyo Metropolitan University, Tokyo 191-0065, Japan; alfarisi-salman@ed.tmu.ac.jp (S.A.)
2 Research into Artifacts, Center for Engineering, School of Engineering, The University of Tokyo, Tokyo 113-8656, Japan; mitake@race.t.u-tokyo.ac.jp
3 Faculty of Computer Science and Systems Engineering, Okayama Prefectural University, Okayama 719-1197, Japan; tsutsui@cse.oka-pu.ac.jp
* Correspondence: yoshiki-shimomura@tmu.ac.jp

Abstract: The product–service system is a significant research subject related to business model innovation and sustainability. However, the product–service system feature has affected the consumption behaviour, affecting nurture. The authors identified an apparent knowledge gap in the prior literature concerning nurture in the product–service system. This study examined whether nurture should be a prominent issue in the product–service system since certain features can significantly affect the achievement of set targets by generating a rebound effect. This study demonstrated that the business model system is complex, with interconnected solutions and issues. Solutions are not implemented in isolation, therefore, each decision affects the system. This study employed feedback system thinking using system dynamics. To validate its findings against the actual situation, this study employed car-sharing as a case study. The findings of this study indicate that the variable of nurture is a significant indicator of profit growth but generates a deterioration in the environmental and social performance of product–service system implementation, which leads to a rebound effect of the product–service system.

Keywords: nurture; product–service system; rebound effect; sustainability; dematerialisation; system dynamics

Citation: Alfarisi, S.; Mitake, Y.; Tsutsui, Y.; Wang, H.; Shimomura, Y. Nurture: A Novel Approach to PSS-Rebound Effect Identification. *Sustainability* **2023**, *15*, 7359. https://doi.org/10.3390/su15097359

Academic Editor: Claudio Sassanelli

Received: 17 February 2023
Revised: 22 April 2023
Accepted: 26 April 2023
Published: 28 April 2023

1. Introduction

The challenge of reducing the environmental implications of the products from a life cycle perspective has plagued industrial enterprises over the past few decades. In the 1990s, numerous authors in an arena dominated by environmentalists contended that society would face a near-certain catastrophe unless methods were discovered to decouple economic expansion from environmental pressure [1]. During the same period, the business literature also increased interest in functional business models. The product–service system (PSS) is considered one strategy [2]. A PSS can be regarded as a market offer that includes additional services to a product's standard functionality [3]. For some authors, sustainability in terms of social, economic, and environmental factors is also included in the idea of a PSS [4]. The PSS is a research issue strongly related to business model innovation and sustainability; this subfield of research has garnered a growing amount of interest from many research streams, as Boons and Lüdeke-Freund [5] demonstrated. The basic principle is that a PSS will have less of an environmental impact than a conventional transaction in which a business manufactures products but then shifts the ownership and use responsibilities to the customer. This business model pushes companies to focus on the services and experiences that accompany the product rather than just the physical product itself. By doing this, companies may lessen their reliance on physical materials and resources, which is a crucial aspect of attaining dematerialisation.

While PSS research has been well-established for more than two decades, there is still a growing interest and a need to investigate certain facets. The idea of PSS has been extensively discussed in the literature (see, for instance, [6–8]), although the industry's adoption of such concepts seems to be relatively slow. The notion that PSS equals sustainability is a myth [9]. Nicola [10] and Vezzoli [11] concur that PSS can cause second-order effects, which they refer to as secondary effects. However, further study has revealed that the phenomenon resembles the term the "rebound effect" [12]. Tukker [13] argued that the lower cost of product access through renting and leasing (use-oriented services) may lead to undesirable effects such as shortened product lifetimes due to user carelessness [9,14]. This refers to the manner in which a series of interconnected consequences on attitudes and behaviours determine product longevity throughout the consumption phase, known as "nurture" [15]. According to Mugge et al. [1], a strong emotional connection between users and their products, or strong product attachment, can lengthen the product's lifespan. Additionally, users will take better care of their products, hence minimising the probability of failure [16].

Although the rebound effect has been recognised by several earlier researchers [17,18], there are less well-documented variables in the literature that have not been analysed for their impact on the emergence of the rebound effect. In particular, the problem identified in this study is the uncertain driver of the rebound effect during the implementation of a PSS, which has tended to be a qualitative assumption without legitimate evidence. Therefore, this study intends to assess the implementation of PSS by considering the variable of nurture and analysing its contribution to the emergence of rebound effects that can impede the attainment of targeted sustainability goals. Several prior studies support this premise, which justifies the inclusion of nurture as a variable of interest. For instance, Alfarisi et al. [12] examined car-sharing as a case study and demonstrated that some characteristics of a PSS can serve as the primary motivating factor behind the rebound effect such as the absence of ownership, affecting product attachment. Product attachment refers to a consumer's intense emotional bond or attachment to an object [19]. When the emotional connection of ownership is lost, there is a greater tendency to mistreat the object. This finding is also supported by [20] concerning emotional attachment to products and indicates that strong emotions associated with personal identity and a sense of belonging have an effect on product longevity. While there are a number of potential causes for the rebound effect, this study restricted it to technological factors (improvements in efficiency), as Herring and Sorrel [21] argued that efficiency is the most important factor in predicting the occurrence of the rebound effect.

This study employed feedback system thinking, which posits that issues and solutions are causally interconnected inside the system itself, to provide a comprehensive analysis of this issue. Sarmiento et al. [22] stated in their research that management processes and systems are crucial to consider in the context of business innovation. Thus, to optimise the potential of complex PSSs, a comprehensive understanding of the system structure is required. According to Morecroft [23], when a decision is made, there will be a consequence of the cumulative effect of earlier decisions and actions that develop in the system itself. This influence is frequently unnoticed and frequently overlooked but produces equally significant problems. In addition, due to the complexity of this assessment, the scope of this study's assessment of sustainability attainment covered indicators such as the amount of pollution, the use of natural resources, the profit ratio, and the quality of life.

The rest of this is organised as follows. Section 2 examines the theoretical foundations for developing the research questions and identifying the gaps. Section 3 describes the case study's background and the processes performed to construct the model in detail. Section 4 presents the findings. Section 5 examines the study's findings and clarifies the answers to the research questions. Section 6 summarises our findings, conclusions, and future research directions.

2. Literature Review

2.1. PSS-Rebound Effect

The product-service system (PSS) is a developing topic of study and industrial practice that focuses on the purposeful and planned coupling of products and services [24]. The ultimate purpose of PSS is to enhance a company's competitiveness and profitability [25], and one of the objectives of a PSS is to reduce product consumption through alternative scenarios of product use rather than acquisition. Customers who seldom drive, for instance, may not need to purchase automobiles but rather utilise a car-sharing system [26]. Unfortunately, as numerous scholars have observed, PSSs are not intrinsically more sustainable than products [27]. Several studies on sustainable PSSs in the previous decade have presented fairly isolated concepts, manuals, and case studies. Case study research was frequently driven by normative sustainability objectives and did not investigate the causes of poor PSS implementation [9].

Frequent emphasis is placed on the potential for PSSs to improve environmental performance by dematerialisation [9]. Even if the net resource consumption and impact are reduced, to date, social issues have mostly been overlooked in PSS sustainability research. In fact, PSS sustainability must incorporate the three fundamental pillars of sustainability (social, economic, and environmental), such that the evaluation is based on the achievement of these three aspects including the assessment of the PSS rebound effect. Initially, the rebound effect was utilised in the context of energy efficiency, but subsequent research indicated that energy efficiency must be coupled with sufficiency. Sufficiency is widely defined as minimising the consumption of products and services to better satisfy individual desires and contribute to communal objectives [21]. Instead of selling items, a PSS focuses on selling usage, incorporating the concepts of effectiveness and sufficiency. The rebound effect of the PSS happens when increases in production and consumption offset increases in the production and service efficiency and sufficiency. Researchers have recognised the complexity of RE, with direct or first-order influences resulting in indirect or second-order effects [21,28].

Inadequate identification and minimisation of the rebound effect in a system would not only impede the achievement of a PSS but also result in catastrophic failure. Kjaer et al. [29] emphasised the importance of recognising and evaluating the rebound effect during PSS implementation. Alfarisi et al. [12] built a framework to analyse the possibility of a systemic rebound impact during design by including mitigating actors. The systemic rebound effect is inevitable in implementing PSS; Vezzoli et al. [11] refer to it as an unwanted effect. Unfortunately, stakeholders have misused the rebound effect as an excuse for inaction.

The findings of the literature review on potential rebound effect drivers are summarised in Table 1. The rebound effect, according to Maxwell and Andrew [30], is induced by an unanticipated increase in consumption due to environmental efficiency interventions. According to Vivanco et al. [31], the driver is a change in efficiency that leads to a change in consumption and production factors as a result of a change in price elasticity. Using a causal loop diagram, Laurenti et al. [32] stated that the incremental innovation–obsolescence cycle is a mutually reinforcing feedback loop and identified that incremental innovation leads to a shorter product life, which then increases consumption, which is the driver of the rebound effect. In addition, Alfarisi et al. [12] demonstrated that non-ownership is the primary cause of the rebound effect. Non-ownership is believed to influence product attachment and result in changes in behaviour. Liedtke et al. [33] argued that the potential for PSS to change production and consumption systems in a manner that enables a sustainable transition must be carefully evaluated. The most likely driver of rebound effects is unanticipated user behaviour or the inappropriate implementation of potentially sustainable efficiency innovations. Other PSS researchers such as Kuo and Wang [34] and Gottberg et al. [35] concurred with Kjaer et al.'s [36] assertion that changes in consumption practices are well-known as the driver of the rebound effect while the primary driver, according to Mylan [37], is a more specific factor, namely attitude, which can influence the users' consumption behaviour during the consumption phase.

Table 1. Potential rebound effect drivers.

PSS Case	Identified Driver	Authors
Energy efficiency in cars, heating/cooling, household appliances, lighting	Behavioural responses	[30]
Car non/ownership	Changes in consumption and production	[31]
Use-oriented (car sharing) and result-oriented (photocopy machine)	Incremental innovation	[32]
Car-sharing	Non-ownership lead to behavioural change	[12]
Heating and space heating	Unanticipated user behaviour	[33]
Bike-sharing	Changes in consumption practices	[34–36]
Energy efficient lighting and low temperature laundry	Attitude	[37]

Based on these findings, the authors reached the reasonable conclusion that almost all researchers, with the exception of Laurenti et al. [32], concurred that behaviour change is the main driver of the rebound effect. Based on this finding, Alfarisi et al. [12] conducted additional research into the root causes of these behavioural modifications. The findings highlight the absence of a sense of belonging and consequently, the loss of "willingness to keep", which leads to careless use of the product during the consumption phase, as Mylan [37] further explains in detail.

2.2. *Nurture*

The product lifetime (PL) has been focused on in innovation, technology, processes, and systems approaches and has a strong bond with manufacturers [38]. The PL is also the consequence of acts and practices that improve the qualities and functions of products, which Cox et al. [15] referred to as "nurture". 'Nurture' is controlled by functional product durability and reflects a set of interrelated effects on the attitudes and behaviours that determine a product's lifetime during the consumption phase. The concept of 'nurture' appears to be primarily divided into individual and social environment-based factors [39]. At the individual level, the role that products play in satisfying personal needs is of critical importance in terms of the functional utility provided by a product, emotional attachment to belongings, and strong feelings related to personal identity and a sense of belonging in society [40]. Important external influences include pricing, information, product quality, and availability. In its simplest form, "willingness to keep" is inextricably linked to the consumers' perceptions of value, which result from the interaction of multiple individual and societal forces and the nature of the commodity itself [15].

In a consensus study conducted by Cox et al. [15] on thirty product types utilising the traditional business model, it was concluded that consumers wanted goods to last (i.e., not break) as long as they wanted them to last, but not necessarily longer. Consumers rated durability (a product designed to endure a long time) and functional reliability (a product that performs reliably without breaking down regardless of how long it is designed to last) differently. Functional reliability was essential for all items (even those expected to be kept for a short period), whereas consumers valued durability primarily for products they planned to keep for more than a few years. Thus far, the literature has presented the subject of nurture within the context of the traditional business model, where product ownership is shifted from the producers to consumers. Not all consumers exhibit a "willingness to keep" attitude, even when the goods are owned. In the context of a PSS, where there is no shifting of goods from the producer/service provider to the customer, this phenomenon must be a research priority as the emotional link of ownership is lost, and the propensity to treat items incorrectly increases.

2.3. Feedback System Thinking

Too frequently, decoupling efforts are ultimately undermined by unanticipated responses to the initial interventions. For example, an increase in consumption can be driven by a decrease in price as a result of advances in material and energy efficiency [20]. Increasing consumption generates negative environmental externalities such as waste and pollution. Eventually, the accumulation of waste and pollution has significant social consequences. This system behaviour results from ripple effects propagating across the system's structure. Ripple effects arise when one event generates consequences that propagate across the system and produce other ripple effects [32]. The primary cause and its repercussions are typically separated in time and place. A system's structure is composed of feedback loops and causal links generated by the interaction between the system's components.

A fundamental principle of systems including the PSS is that the behaviour of a system is essentially dictated by the attributes of the whole and not by the properties of its variables. The interactions of system variables within a closed boundary consequently form the examined types of behaviour. By comprehending the link between variables, it is possible to forecast the system's behaviour, making it simpler to propose modifications. Unfortunately, identifying variables that affect the rebound effect needs to be adequately studied, making it difficult. Traditional assessment efforts typically ignore that these single units are embedded in a much larger socio-technical system, which is subject to dynamic interactions with causal links and responses (feedback loops) from numerous socio-aspects, technical aspects, and economic aspects over time [41]. In practice, a system's variables are interdependent, and feedback system thinking may be the most effective method for explicating this complexity. Feedback system thinking is typically circular, beginning with a problem, moving to a solution, and then going back to the problem [23]. The crucial point is that issues do not simply appear and demand solutions. They result from the cumulative effect of earlier decisions and actions, which are sometimes intended, but frequently have unintended consequences. Typically, a difficulty manifests as a disparity between an important objective and the present circumstance. Those accountable for accomplishing the objective arrive at a solution in the form of a choice that results in actions and outcomes that alter the current situation. Numerous feedback is nearly undetectable in practice. They manifest themselves through unexpected side effects, resistance to change, and unexpected outcomes.

Several research studies used causal loop diagrams (CLD) and system dynamics (SD) to solve PSS issues, either as an evaluation or performance measurement approach. Generally, Sassanelli et al. [42] demonstrated, through a review of the relevant literature, that the process modelling approach is one of the methods commonly used to evaluate the system performance of a business model, which in this instance focused on a circular economy that seeks to close the circle of linear product life cycles. System dynamics is one of the PSS's three effective modelling methods [43]. While Grüneisen et al. [44] attempted to represent PSSs in system dynamics to enhance the knowledge of the PSS by combining multiple multidisciplinary fields, further research is required. Lee et al. [45] employed system dynamics (SD) to examine the dynamics from a triple bottom line (TBL) perspective to cover the multidimensionality of PSS sustainability, and their findings were positive. Lee et al. [46] focused on measuring the functional performance of a PSS using a dynamic approach in a separate study. As suspected by Vezzoli [11], Nicola [10], Tukker [13], and Cherry and Pigeon [14], studies on the utilisation of system dynamics in prior research do not appear to have provided evidence of the emergence of rebound effects. Therefore, the authors assumed that the variables contributing to the rebound effect's appearance had not been accounted for in the simulations performed. Before integrating variables were deemed to have a major effect on the appearance of the rebound effect in PSS implementation, this study examined the literature. According to the authors, this is the first study to evaluate the attainment of sustainability and identify the emergence of rebound effects using nurture in system dynamics.

2.4. Research Gap and Question

Since prior research has provided significant evidence of the rebound effect in PSS deployment, this study revealed a knowledge gap in establishing the sustainability of a PSS. Consequently, the main concern of this research is "how may the rebound effect arise during PSS implementation?" To address this question, an evaluation of the variables with the potential to induce the rebound effect must be conducted. It was discovered that "nurture" was hypothesised as a significant determinant due to the loss of emotional attachments between customers and goods that were not owned, leading to a decline in "willingness to keep". This research was then divided into the following sub-research questions:

1. Does "nurture" significantly influence the emergence of the rebound effect in PSS implementation?
2. What dimensions of the system are changed by "nurture"?

This study used a feedback system thinking approach to obtain comprehensive results.

3. Materials and Methods

This study illustrates the proposed methodology with a case study of a car-sharing system. The assessment procedure adheres to the standard steps of SD: conceptualisation, formulation, testing, and analysis [45,46]. Particular emphasis was placed in this study on conceptualisation. The proposed methodology divides the conceptualisation into three steps. The first phase, establishing the indicators, addresses the explanation of the situation. To do this, the sustainability of the PSS in each dimension perspective was specified, and the necessary perspectives on the measurable indicators were proposed. The second and third steps involve model construction. The difficulty of drawing the system models, even for experts, has been illustrated [47]. Moreover, the more complex a model for conceptualising, the more difficult it is to comprehend. In SD modelling, it is evident that a modest model provides advantages over a large model [48]. Consequently, a prevalent tendency in SD has been the use of modest models to improve comprehension. The model that was developed in this work is the sustainability of a PSS from each dimension perspective; as it is a huge and complex system with multidimensional characteristics, it can be overwhelming to evaluate everything at once. The sequential approach is illustrated in Figure 1. Following this section is a detailed explanation of each section, based on the general technique of SD, with an emphasis on the unique characteristics applicable to analysing the PSS sustainability attainment and rebound effect potential.

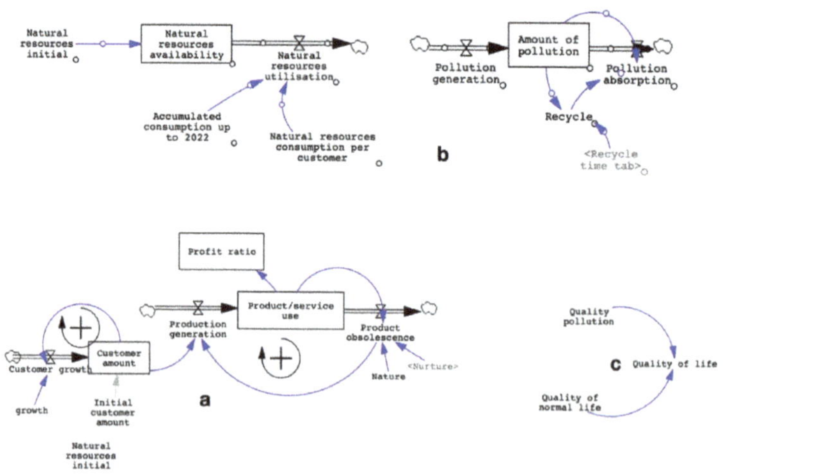

Figure 1. The partial model of PSS sustainability: (**a**) environmental dimension; (**b**) economic dimension; (**c**) social dimension.

3.1. Materials

3.1.1. Case Study Background

As sustainability has become a worldwide issue, interest in sustainable transportation systems has increased significantly. Compared to the studies performed on the perspectives and motivations of people involved in the implementation and use of car sharing in developed countries, there is limited research for those in developing countries. For instance, Java Island, the political and economic hub of Indonesia, was home to around 57% of the country's population of over 271 million in 2015 [49]. Motorcycles continue to outnumber automobiles, but the recent trend of purchasing automobiles appears to be continuing [50]. Despite the fact that bus lines including bus rapid transit (BRT) and four-wheeled minibuses or minivans (called Angkot) and taxis including unofficial two-wheeled taxis (called Ojek) are the backbone of public transportation in the Java region, the public transport system is still not fully distributed in some areas.

This background is significant for a start-up company in Indonesia wishing to launch a transportation business and adopt a more sustainable business strategy. For reasons of discretion, we refer to the company as "Me Share". This company is located in Jakarta, which remains on the Indonesian mainland of Java Island. The rapid rise of car-sharing in Indonesia is indicated by the compound annual growth rate (CAGR) of car-sharing in Indonesia, which reached 60.42 USD million by 2022, an increase of 6.24% per year [51]. In response to the increasing market competition, Me Share promises various advantages including:

- Flexible: Everyone can use the car whenever they need to.
- Simple: Reservation until the car door opens and closes; only a smartphone is required.
- Easy: Cashless and credit card payment is guaranteed to be secure.
- Well: The car is cleaned and maintained regularly so it is always ready for use.
- Affordable: Special promotions every time.

Me Share utilises a round-trip car-sharing approach, defined as a shared vehicle that begins and ends at the same location. Me Share, as a start-up company, is a service provider and collaborates with automobile manufacturers as car providers. However, as the number of customers increased, the company experienced a dilemma in which the pace of automobile obsolescence exceeded the planned life span. On numerous occasions, vehicles have been discovered to be inoperable due to engine problems. The high rate of vehicle failures has resulted in increased vehicle sales and a great demand for spare parts. Although this enhances the profit area, the company discovered a contradiction between the intended aim of car-sharing and its implementation, notably the attainment of sustainability. Considering this issue, the authors attempted to uncover probable contributory variables. Due to the business model's complexity and the system's interconnection, a system dynamics technique was applied in this study. Following the modelling process's principles, the boundary selection in this case was as follows:

- Theme: Assess the effect of nurture on the emergence of rebound effects and the attainment of sustainability in the car-sharing model business.
- Variables: Variables that contribute to the three elements of sustainability were considered. In Section 3.2.1, the three dimensions are established, and in Section 3.2.2, the variables used for each dimension are provided.
- Time horizon: This study utilised company-specific historical data from 2018 through to 2021. Then, modelling was employed to forecast the business circumstances for the years 2022–2027.

This research comprised a descriptive analysis of the company-provided dataset. The descriptive analysis provides a summary of the dataset and by using suitable statistical analysis, directed the authors to analyse potential problem drivers using a system dynamics simulation.

3.1.2. Data Collection

The purpose of this study was to investigate the existence of a rebound effect triggered by nurture with respect to the parameter of the dimension of sustainability in PSS implementation. To achieve this objective, a mixed-methods strategy using both primary and secondary data was employed. By analysing available documents, primary data were gathered directly from the company. The company's collection of primary data provided extensive insights into its operations and performance outcomes. By analysing documents such as financial statements, it is possible to gain a thorough comprehension of the factors influencing the company's performance. However, it was evident that the company lacked certain data, necessitating the use of secondary data sources. Previous research has demonstrated the significance of data integration in modelling to identify value chains, and that this process should begin internally. However, Acerbi et al. [52], after proposing a classification of data and information, suggested that awareness of the need to use both internal and external data to succeed in this path is necessary, as the absence of certain data can limit the turnover of resources, and external data can be utilised if the appropriate data are available.

Therefore, to cover this void, this study collected secondary data based on strongly related references by following the rule of the pedigree matrix. This approach involves meticulous data curation and normalisation to ensure normal distribution conformity. Using this approach, the research was able to compile a comprehensive dataset that provides a broader perspective on the company's performance.

For the reader's convenience, the data presentation was divided into three tables, each arranged according to a sustainability dimension. The data constituting the economic dimension are displayed in Table 2. The environmental dimension dataset derived from the secondary data is presented in Table 3, the social dimension dataset is displayed in Table 3, and the social dimension dataset is presented in Table 4.

Table 2. Dataset for the economic dimension.

Variable	Amount	Reference
Initial customer amount (ICA)	500,000	Company's existing data
Growth (G)	0.5 per 0.125 year	Company's existing data
Nature	10 years	Company's existing data
Nurture	[(2022,0)-(2032,1)],(2022,0),(2023,0.1),(2024,0.1),(2025,0.4),(2027,0.5),(2028,0.6),(2032,0.9),(2032,1)	Company's existing data
Available product/service in 2022	1,000,000	Company's existing data

Table 3. Dataset for the environment dimension.

Variable	Amount	Reference
Natural resources initial (NRI)	8.5×10^{13} kg	[53]
Natural resources consumption per customer (NRC)	975 kg	[54]
Accumulated consumption up to 2022 (AC 2022)	3×10^{13} kg	[55]
Production emission of car	4.56 kg CO_2 Eq./kg	[56]
Car life cycle emission	49,559 kg CO_2 Eq.	[57]
Recycling capacity	12,328,643,752 kg CO_2 Eq.	[58]

Table 4. Dataset for the social dimension.

Variable	Amount	Reference
Pollution contribution to quality of life (PCQ)	0.167	[59]

3.2. *Methods*

3.2.1. Define Indicators

According to research conducted by Alfarisi et al. [12], the premise of this study is that the PSS works well in one dimension but can generate problems in other dimensions in the context of sustainability. Therefore, it is essential to explain the definition of the sustainability dimensions. Depending on the objective and specific measurement purpose, several sustainability-based indicators have been proposed and implemented. Focusing on the firm's activities, numerous sets of indicators for analysing and reporting sustainability have been proposed and widely used by businesses worldwide [60]. Labuschagen et al. [61] proposed a framework for hierarchical business sustainability indicators at the industry level. In addition, research has been undertaken on measuring the sustainability of technology [62]; the accepted sets of indicators adhered differently to the specific goal of measurement.

According to Roy [63], the concept of sustainability cannot be easily described in operational terms; rather, it must be intuitively understood. Despite varying definitions and meanings, sustainable development refers to quality development that "promotes harmony between humans and between humans and nature" [64]. To assess the sustainability of the PSS and the systemic rebound effect, each dimension's sustainability—environmental, economic, and social sustainability—must be specified according to Table 5. Each definition reflects the systematic characteristics of a PSS and the contextual variables of PSS adoption as a replacement for an ownership-based consumption pattern.

Table 5. Definition of PSS sustainability attainment.

PSS Sustainability Dimension	Definition
Environmental	The production and consumption patterns of PSS are capable of limiting the depletion of natural resources due to dematerialisation and minimising the pollution existing than existing products.
Economic	PSS is able to preserve the company's economic motive in a sustainable manner.
Social	PSS can preserve the quality of life without sacrificing social rights.

Indicators for the Me Share system's sustainability were defined for the three dimensions. In this study, only the four indicators listed in Table 6 are included. For environmental sustainability, two indicators were considered: natural resource consumption and the amount of pollution, with an emphasis on reduction as a relative term rather than an absolute quantity. The profit ratio was also assessed for economic sustainability based on the assumption that the car-sharing system is geared toward private car owners. Among the different public welfare-related indicators, quality of life was seen as an indicator of social sustainability.

Table 6. Definition of PSS sustainability attainment.

PSS Sustainability Dimension	Indicator
Environmental	Amount of pollution Natural resources consumption
Economic	Profit ratio
Social	Quality of life

3.2.2. Build the Partial Model

Much of the art of SD modelling is in identifying and describing the feedback process that affects the system's behaviour [47]. Developing a model, however, requires extensive knowledge and a comprehensive understanding of the system. Assessing a PSS's sustainability entails analysing various aspects, which becomes more complex as the system expands; this is how sustainability is measured. The step-by-step method is useful to comprehend a huge and complicated system from several viewpoints: first, create a partial model, and then an integrated one.

The integrated models include the fundamental and common notion of indicators for measuring PSS sustainability as well as typical correlations between them, indicating the relevance of SD and sustainability for measuring PSS attainment. From the definition of PSS sustainability shown in Table 5, several critical aspects for establishing indicators were obtained; these were used to develop the integrated models for partial systems. Based on the defined indicators, straightforward causal links were identified within each dimension. The indicator value change was attributed to the variable 'usage of the car-sharing system'.

Economic Dimension

The parameters were derived from product/service consumption and profitability. The parameters were derived from the consumption and profitability of the product/service. Since the number of customers will affect the production generation rate, the first stage is to estimate the number of customers for the following five years. Over the next five years, the expected number of customers is calculated using an integral customer growth function with a growth value of 50% and a current initial customer count of 500,000. System behaviour demonstrates that customer growth and customer number are interdependent in this situation. This pattern indicates a relationship is reinforcing: the larger the number of customers, the greater the customer growth, and vice versa. In addition, normally, the number of products/services produced is proportional to the number of customers, but since the PSS has been shown to significantly increase vehicle utility, it means that the same vehicle can be used by four different customers, resulting in a much smaller amount of production generation than the conventional business model. The originality and novelty of this study lie in the fact that it models the product obsolescence rate not only through planned obsolescence/nature but also nurture, which refers to the influences of attitudes and behaviours that affect a product's lifetime during the consumption phase. Intriguingly, nurture is a new variable that can alter the overall simulation results, where the rebound effect is ultimately identified. Due to the absence of emotional attachments and a sense of belonging, the contrast in features between the PSS and conventional business models—where there is no transfer of ownership in car-sharing—significantly alters consumer behaviour towards the product/service. In this study, historical data were used to model nurture. High retentiveness (following guidance, affixing a sense of belonging and providing simple technical care) can enhance the vehicle's lifespan by ten years. Medium willingness to retain (guideline followed, no additional care) can lengthen the vehicle's lifespan by five to six years while with a low desire to retain (use carelessly without regard for the instructions), the vehicle only lasts one to two years. The rate of product obsolescence impacts the use of the product or service within the producer's expected lifespan. The profit is calculated using the profit ratio method [65], where the ratio value is derived by dividing the profit region's area by the profit region's area plus the loss region's area, yielding a ratio of 0.9238, hereafter known as the profit area constant. In addition, the profit ratio is the result of the ratio's value and the quantity of the product/service used. Table 3 presents the input data utilised for the economic dimension. In addition, Equations (1) and (2) demonstrate the equations needed to determine the customer number and profit ratio, respectively. Figure 1a depicts the economic dimension model

$$\text{Customer number} = \text{ICA} \int \text{customer growth}. \tag{1}$$

$$\text{Profit ratio} = \text{profit area constant} \times \text{product/service use} \tag{2}$$

Environmental Dimension

The parameters and their relationships were derived from the perspective of the existing PSS. The environmental dimension was considered from both the upstream and downstream perspectives. Upstream is the use of natural resources that affect the availability in nature, while downstream is the amount of pollution generated. However, as a commitment, the car-sharing system seeks to minimise pollution by increasing the utility of cars that can decrease the production number and by controlling the amount of pollution with a recycling policy. This model considers the recycling policy with the term "pollution absorption policy". In car-sharing, the material is concentrated iron, as it provides 64% of the car's weight [54]. Approximately 975 kg of iron is required for a standard-sized family vehicle. Although the Earth has a relatively significant iron content, its rapidly increasing use has to be a concern because iron is a non-renewable resource. The estimated quantity of an identified resource that meets the specified minimum physical and chemical criteria in relation to current mining and production practices including those for the grade, thickness, quality, and depth is 180 billion metric tonnes of crude ore with 85 billion metric tonnes of iron ore [53]. The data show that the average utilisation of iron ore is two billion metric tonnes per year. This study implies that around 30 billion metric tonnes of iron ore have been consumed since the 1950s, when the world industry peaked. "Natural resource utilisation" refers to the current utilisation of iron compared to its accumulated usage in the period. Additionally, this sum fluctuates as the number of customers grows. Since it is known that a PSS can increase vehicle utility by up to four times, natural resource consumption follows this premise. The gap between natural resources that are available in nature and those that have been consumed is termed "natural re-source availability". Table 4 presents the input data utilised for the environment dimension. Equations (3) and (4) present the calculation of natural resource availability and its derivatives.

$$\text{Natural resources utilization (NRU)} = (\text{Customer number} \times \text{NRC}) + \text{AC 2022} \quad (3)$$

$$\text{Natural resources availability (NRA)} = \text{NRI} - \text{NRU} \quad (4)$$

The simulation downstream focuses on the amount of pollution discharged into the environment. The value of "pollution generation" is calculated by multiplying the number of products/services used by the amount of pollution produced by a medium-sized vehicle throughout the period of its life cycle [56,57]. The units for measuring the pollutant output are kg CO_2 Eq. Since pollution is measured in kg CO_2 Eq., this simulation's recycling process utilised the same units to make detecting the amount of recovered pollution easier. The recycling strategy is evaluated based on the annual CO_2 quota that Indonesia can accommodate, which is 12,328,643,752 kg CO_2 Eq., and then the quota becomes a function for the subsequent pollution recycling procedure (Kojima, 2017). The amount of pollution is the difference between the amount of pollution generated and the amount of pollution absorbed. Equation (5) presents the calculation for the amount of pollution. The environmental dimension model is shown in Figure 1a.

$$\text{Amount of pollution} = \text{Pollution generation} - \text{pollution absorption} \quad (5)$$

Social Dimension

Typical factors such as the number of customers, availability of natural resources, and pollution that have been identified in the literature as being associated with quality of life were provided. The quality of life is obtained by dividing the "quality of normal life" by the "quality pollution". The ideal quality of normal life's is 100%. "Quality pollution" refers to the extent to which pollution affects the quality of life. "Natural resources availability" divided by "product/service use" multiplied by "amount of pollution" multiplied by 0.167 provides the value of quality pollution. As stated by Fuller et al. [59], pollution contributes to one out of every six deaths worldwide; hence, a value of 0.167 was used to assess the impact of pollution on the quality of life. Table 5 presents the input data along with Equations (6) and (7), which indicate the quality pollution and quality of life assessments. The social dimension model is illustrated in Figure 1c.

$$\text{Quality pollution} = \left(\frac{\text{NRA}}{\text{product/service use}}\right) \times (\text{Amount of pollution} \times \text{PCQ}) \quad (6)$$

$$\text{Quality of life} = \frac{\text{Quality of normal life}}{\text{Quality pollution}} \quad (7)$$

3.2.3. Model Integration

The partial models for the three sustainability dimensions were incorporated into the final model. To integrate these three partial models, additional variables and relationships were introduced based on the overlapping variables and relationships in the component models. Many works have identified the interrelationships among the three dimensions of sustainability [11,66]. Figure 1 depicts the integrated model illustrating typical relationships between the three pillars of sustainability for PSS sustainability. The employed linkage can be set with parameters and causal relationships based on the indicators defined for each component of PSS sustainability. As depicted in Figure 2, the partial models were incorporated into the unified model at this step. The causal structure was completed by identifying and utilising the linking factors between the dimensions. The integrated models identified positive (reinforcing) feedback loops of the upstream environment and economic dimensions.

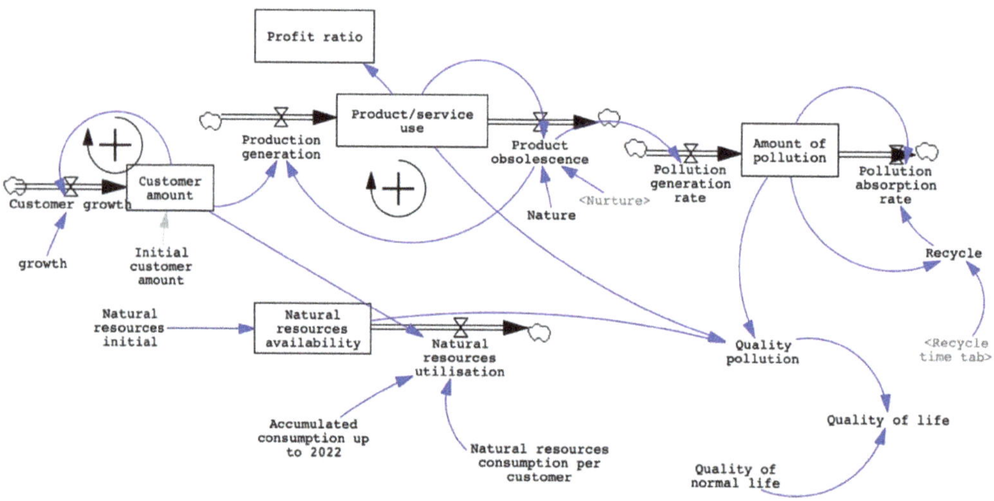

Figure 2. The integrated model of PSS car-sharing.

During the process of model integration, it was discovered that all dimensions are interconnected. For instance, the rate of product obsolescence in the economic dimension is an indicator that influences the rate of pollution generation, which in turn affects the amount of pollution. Increasing customer demand influences the availability of natural resources through the natural resource utilisation rate in the environmental dimension. Likewise, the availability of natural resources and the level of pollution impact the social dimension of life quality. Vensim software was utilised for this case study since it enables the integration of variables from other view-named shadow variables.

4. Simulation Results

Conceptually, to simulate the model, stocks describe the material or other accumulations; they are the system's states. The quantitative model's notation is dependent on the software. In terms of PSS sustainability, each sustainability indicator can be viewed as a stock variable that reflects the status of the PSS, although this is not always the case. The figure depicts the relationships between the variables using quantitative equations. This stock and flow diagram was simulated using the initial values of several variables.

The sustainability of each dimension was measured based on the long-term behaviour of each indicator, which may be displayed as a stock variable illustrating the state change of PSS from a sustainability perspective. However, a comprehensive view should be assumed on the interpretation of these data as the support they offer goes beyond just continuing with PSS, if sustainability is measured as good from all perspectives of sustainability and not continuing if not. The strength of this strategy resides in the simulation technique's capacity for intensive study. This provides strategic insights for rethinking the PSS concept: the attained levels of sustainability for the three dimensions may be set differently depending on the intended use.

This section presents a case study of the car-sharing system to demonstrate the assessment of PSS sustainability attainment. As energy and the environment have been highlighted as global challenges, a green transportation system has attracted considerable interest. The car-sharing system has been widely embraced as an example of a green transportation system in many places around the world. The business model for car-sharing is essentially the same everywhere; however, the following details are based on situations in Indonesia.

Moreover, this scenario is appropriate for demonstrating the operation of the proposed approach since it considers the environmental objective and the economic and social repercussions due to its public nature. Despite the fact that the case study was based on a real business that reflects the context of a car-sharing system, in order to achieve a simple and understandable illustration that focuses on the purpose of the case study, certain assumptions were made; the number of measured indicators and the scope of the presented model were reduced. In addition, for the data of some contextual variables, behaviour patterns were hypothesised based on indications from the literature and previous cases. The specifics of the presumed environment are described in each pertinent step.

The generated model in the stock and flow diagram was prepared for execution. However, verifying the model to prevent simulation errors would be preferable. Errors in the model will result in inaccurate simulation results, or fatalities will prevent the model from being simulated. The variables associated with policy and circumstance were quantified. However, for factors that cannot be designed by the PSS structure, some assumptions were made based on empirical evidence from the literature review [59,65,67]. The quality of life approach of Fuller et al. [59] showed, for instance, that pollution is responsible for one in every six deaths worldwide. The statement was then quantitatively translated for simulation purposes. As demonstrated by Lee et al. [46], this type of hypothetical approach has been widely employed to maintain the model's integrity in various simulation forecasts.

Figure 3 depicts the results of the simulation. The simulation results are extremely intriguing for future discussion to obtain an objective evaluation of the sustainability of PSS car-sharing and to undertake additional research on the formation of the rebound effect. The complete simulation results are presented in Table A1 (Appendix A).

The results of the simulation showed that there was an exponential increase in profit, which was influenced by the increase in product/service use. Furthermore, the increase in product/service was affected by two main factors, namely, the increase in product obsolescence and product generation. This increase in profit area is certainly a good achievement for the company if viewed from one dimension alone. Unfortunately, this increase in profit area was not accompanied by an increase in other dimensions of the sustainability dimension. The results of this study show that the amount of pollution increased significantly following the increase in the profit area. Although recycling strategies have been adopted for the waste generated, the rate of increase in product/service usage and the faster rate of obsolescence influenced by nurture was greater than the ability to recover, resulting in an increase in the amount of pollution. In 2023, the accumulated amount of pollution is 881,178 kg CO_2 eq. However, by 2027, the forecast value of the resulting amount of pollution will reach 3,129,180 kg CO_2 eq. This shows that various reactive strategies such as recycling are currently not sufficient to reduce the rate of increase in the amount of pollution generated because the amount of pollution in the system is influenced by the pollution absorption rate and the pollution generation rate. The increase in the amount of pollution shows that the pollution absorption rate is not faster than the pollution generation rate, which is directly affected by product obsolescence. Meanwhile, product obsolescence in PSS is strongly influenced by nature and nurture, which can accelerate the rate of product obsolescence. In other environmental simulations, namely, natural resources, there is a decrease in the availability of natural resources. Although this decrease is not captured clearly through numerical calculations due to the large reserves of natural resources, the high consumption of natural resources in the long run is very influential for the next generation. The availability of natural resources, in addition to being influenced by the reserves available on Earth, is also influenced by the increase in its use. Based on the simulation results, the consumption of natural resources needed for 5 years alone will reach 1.3 million metric tons, which will greatly affect the availability of natural resources. Furthermore, in an effort to assess the social dimension, this research evaluated the quality of life outcomes. Quality of life is affected by "quality pollution", which refers to the extent to which pollution affects the quality of life, while quality pollution is affected by the amount of pollution, product/service use, and the availability of natural resources. The simulation results for this social dimension are quite interesting, as the business model run with the PSS looks very promising for the first few years. The graph of the quality of life improvement continues to increase since being implemented, but soon the curve showed that social performance will decrease in the following years. The cause of this decline is certainly influenced by various structures in the system that are interrelated with each other, either due to the increase in pollution that cannot be counteracted by the recovery quota or the unstoppable increase in product obsolescence caused by nurture.

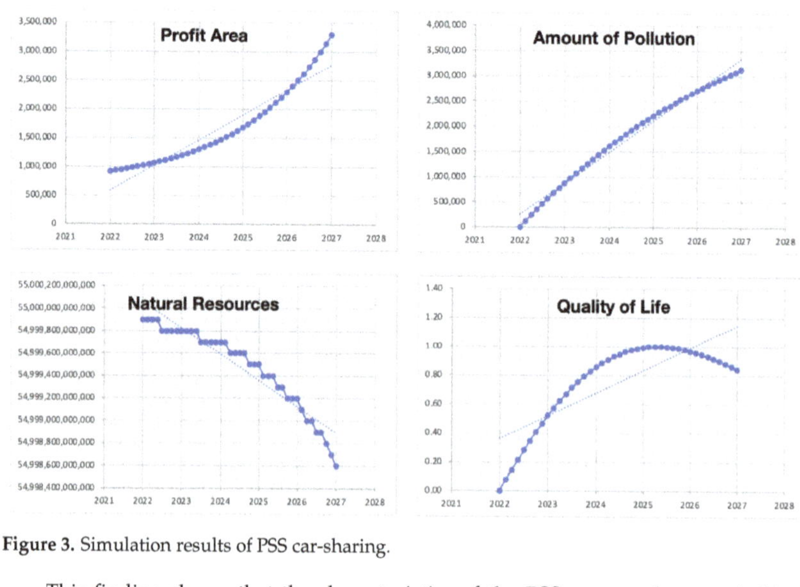

Figure 3. Simulation results of PSS car-sharing.

This finding shows that the characteristics of the PSS are not always suitable for implementation in various situations and conditions. In this case, ownership decreased the emotional attachment of users, causing the problems captured through nurture in this study. Furthermore, this study clearly showed that there was a rebound effect in the implementation of the PSS, where positive improvements only occurred in the area of increasing profits, but the achievement of sustainability targets decreased in all other aspects. The results of this study show that a systemic approach in the implementation of PSS is needed because each variable in the system is interrelated with each other and influences each other so that they cannot be separated.

5. Discussion

5.1. The Influence of Nurture on the PSS

Decades before this study's inception, Nicola asserted that PSS had the potential to cause ecological damage, as characterised by second-order effects. Several additional scholars began to recognise the emergence of rebound effects, utilising distinct terminology. However, Herring and Sorrel [21] clearly described the distinction between the rebound effect and side effect in terms of terminology. The side effect, which was utilised by previous scholars, is an element of the rebound effect. Examining the rebound effect on PSS implementation, this study took Herring and Sorrel's perspective. However, no study has currently assessed the rebound effect in PSS implementation, which can result in trade-offs. By incorporating nurture as a new variable, this study was the first to simulate the formation of the rebound effect in PSS implementation, according to the authors.

The nurture variable has emerged as a new variable where it has become a logical consequence of PSS where there is no transfer of ownership; hence, consumer behaviour will unquestionably alter. After incorporating these consumption-phase behaviour patterns into the simulation, the results indicate a significant change in system performance. Figure 4 illustrates the comparison of sustainability attainment. When nurture is incorporated into the calculation, a fundamental difference is recognised and shown by the formation of the rebound effect. The positive achievement in the economic dimension is indicated by the variable in the profit ratio that has increased over the years. The increase in the profit ratio is attributable to a rise in the product/service use. At this point, it is evident that the production generation and product obsolescence variables are responsible for increased product/service use (see Figure A1 in Appendix A). In detail, product obsolescence is influenced by the primary factor, nurture, which accelerates the pace of obsolescence as well as the nature and product/service use, which constitute a feedback loop. While production generation is driven by the increase in the number of consumers and product obsolescence, as illustrated in Figure A2, the number of items that must be generated increases with the rate of obsolescence and the increase in customer amount. However, this growth has the effect of increasing the amount of pollution produced, thus harming the environmental dimension. This study demonstrated that the increase in pollution was due to two factors: the pollution generation and pollution absorption rates. The faster a product is consumed, the quicker it must be replaced in order to continue providing services to consumers, resulting in greater pollution. As depicted in Figure A3, pollution generation is directly caused by product obsolescence, whereas pollution absorption is affected by the recycling capacity and the amount of pollution itself, creating a feedback loop. In the first two years of the social dimension, it appears promising to improve the quality of life by lowering the number of items, thereby reducing the consumption of natural resources and the amount of pollution. Nevertheless, the gain in the social dimension was short-lived. Considered characteristics that influence the quality of life include the availability of natural resources [46], pollution [68], and population/consumers [46], as illustrated in Figure A4. Unfortunately, the results of natural resource availability could not be accurately recognised in the comparative simulation. The ample supply of iron in the earth may be a contributing factor as well as the absence of a direct relationship between nurture and natural resources in the simulation as the loop cycles along the economic path (see Figure A5). This should be the focus of future studies. However, according to the authors, this event is consistent with Vezzoli's [11] definition of unwanted side effects.

Figure 4. The comparative of sustainability attainment.

This study finally demonstrates that the PSS business model is improving the economic dimension reflected by the indicator for the profit area, where growth is exponential. In contrast, the social dimension, as measured by the quality of life indicator, appears promising in the early years and peaks in March 2025. The following year, however, the performances will decrease. In the context of the study of system dynamics, this pattern parallels the overshoot and collapse pattern. An essential premise of growth is that the carrying capacity of the environment is fixed. In reality, one of the factors that determines the quality of life is the support of natural resources and the generation of products, which is speeding up due to the obsolescence affected by the change in customer behaviour due to non-ownership. In the environmental dimension, the amount of pollution increases linearly with the rate of product generation as a result of either a growth in the number of existing customers or the number of new customers.

5.2. Comparative Analysis

In the environmental dimension, the amount of pollution was the assessed indicator. Figure 4 demonstrates that the curve was nearly linear when nature was included. To comprehend the underlying structure of this expansion, the ecological idea of carrying capacity is useful. In accordance with the ecological notion of carrying capacity, the high output rate due to nurture and the growing number of customers has not been compensated by the extremely restricted recycling capacity. Increasing the recycling capacity could provide balancing feedback to increase the amount of pollution for future mitigation strategies. In contrast, when nurturing was excluded from the evaluation, the difference in the amount of emissions created was very significant. The growth of the profit area when nurture was included in the simulation followed an exponential pattern. According to Sterman [69], exponential growth is the outcome of positive (self-reinforcing) feedback. The higher the number, the greater its net increase, which will lead the number to rise exponentially. This case study indicates that the increase in profit is linked to the rise in the number of customers and product/service used. The rise in car usage is also affected by production generation, which is hastened by nurture. Positive feedback loops promote growth, amplify deviations, and encourage change, whereas the activities of negative feedback loops do not appear to be able to control a decline that is moving further away from the goal in the absence of nurturing. This study demonstrated that nurturing contributed positively to the future expansion of the economic profit area. When nurture was considered in the social dimension, which is based on the quality of life indicator, it was discovered that the pattern approached an S-shaped pattern. According to Sterman [69], a fundamental assumption behind S-shaped growth is that the environment's carrying capacity is fixed. In the simulation of this car-sharing case study, the quality of life was determined by considering the fixed availability of natural resources while the product/service use increases, resulting in a decrease in the capacity of natural availability. The amount of pollution also increased, resulting in a significant straightening of the car-sharing curve pattern in the quality of life indicator over time. In contrast, when nurture was omitted from the simulation, the quality of life increased exponentially and appeared to be quite high. However, this condition does not reflect the actual real condition because it disregards the true condition of the nurture variable.

In the same simulation, when the nurture variable was removed and a value of 5 years was only assigned to nature, the quality of life performance increased dramatically until it was comparable to that of a normal life. This is because the lower obsolescence rate was matched with a recovery capacity that could absorb low-outmoded products, allowing for optimal product recovery and a reduction in pollution. According to Buberger et al. [57], the production of a single vehicle consumes plenty of natural resources and generates a great deal of pollution. Consequently, if the rate of obsolescence is completely out of control as a result of consumer behaviour, the environmental and societal consequences will be extremely serious. In addition, simulations have been conducted to forecast the climate over the following century. The results are identical, and the loss in quality of life performance is becoming more apparent as it approaches zero, indicating a departure from normal quality of life. The complete results of the comparative simulations are presented in Table A2 (Appendix A).

5.3. Limitations

Despite its contributions and benefits, there were limitations to this study. The first was that the input data did not entirely use the available data from the case studies. The service provider's data related to economic factors such as the number of clients, the damage rate, and the number of products/services utilised. In the meantime, secondary data were utilised for information on other dimensions such as the capacity to recycle, the number of emissions during the life cycle, and the availability of natural resources. However, the procedures for using secondary data in this study adhered to the pedigree matrix approach utilised by ecoinvent to ensure data integrity and quality declaration. Some of the points considered for data collection in this study included reliability, completeness, temporal correlation, geographical correlation, and further technological correlation.

In addition, this study examined the rebound effect from the perspective of an improvement in the technical efficiency. In contrast, Walnum et al. [70] explained that there were more perspectives on the rebound effect such as psychological, evolutionary, and socio-planning, and that interdisciplinary collaboration is required to build and develop a very complex system. However, this research limited the case study to the rebound effect perspective of technical efficiency development since, as demonstrated by Santarius and Soland's [71] investigation, this perspective strongly influences customer behaviour.

5.4. Future Direction/Policy Recommendations

This study demonstrated that in the implementation of PSS, nature can no longer be utilised as a variable for determining the life cycle of products/services, as the PSS differs substantially from the conventional business model, in which product ownership is shifted. Therefore, consumer behaviour during the consumption time of the product or service fluctuates dramatically due to oscillations in the propensity to retain. Consequently, future research should investigate how to manage aftermarket behaviour. Some studies such as that by Fargnoli et al. [72] on use-oriented manufacturers have demonstrated the importance of aftermarket services for optimizing the product life cycle. Some actions such as technical assistance network ownership can contribute to the achievement of a circular economy.

Other research has highlighted the significance of implementing policies to slow the rate of change in consumer behaviour. Vivanco et al. [31] and Maxwell [30] proposed a bonus-malus scheme policy, also known as feebates or taxes, which is a variant of an environmental tax in which subsidies are used to incentivise environmentally conscious decisions and attitudes, whereas Scheepens et al. [73] stated in their research that the PSS had a "double objective" consisting of reduced environmental costs and increased value. Scheepens et al. [73] proposed an eco-efficient value creation policy to assist in avoiding a number of the risks associated with a circular business model design (e.g., having positive outcomes at the product level, but having negative effects at the social level; having positive effects on the environment, but not having enough customer-perceived value to overcome intense market competition). In contrast, Sassaneli et al. [74] showed that the high level of abstraction of PSS concepts and a lack of attention to knowledge management could be a problem during the life cycle phase and proposed a method called GuRuMeth, which utilises a circular economy approach to detail the design stages and identify its impact at various phases of its life cycle, so that it can be used as a new approach to prevent this problem during the design phase. This research was founded on the concept of design for X (DfX) in an in-depth study by Sassanelli et al. [75] that resulted in design for product service supportability.

Product–service systems (PSSs) are complex systems that necessitate a cautious policy selection strategy. Although the policies proposed by previous researchers are sound, for future direction, it is necessary to consider the complexities of the PSS. Due to the unique characteristics of a PSS, it is essential to consider potential policy restrictions and unintended consequences. Therefore, careful consideration should be given to the selection of policies that can be adapted to particular circumstances. Simulations can help identify prospective problems and provide insights into the optimal policies that can be adapted to specific contexts. In general, while the proposed policies provide a useful framework, they must be carefully selected, implemented, and evaluated to ensure that they achieve their intended goals.

6. Conclusions

This study proposed a dynamic and multidimensional approach to the measurement of PSS sustainability by using a creative combination of SD and sustainability dimensions, with the primary objective of identifying sustainability achievements and detecting potential rebound effects caused by the policy itself. Consequently, this approach may be used to determine the long-term sustainability behaviour of a PSS that considers the interdependencies among the three pillars of PSS sustainability, to identify trade-offs between dimensions, and lead to rebound effects. Furthermore, as a dynamic and multidimensional assessment indicator that considers the sustainability characteristics of a PSS, this method can be used effectively to evaluate numerous PSS solutions or to analyse the concept of PSS including the potential rebound effect in the PSS, which can negate the benefits of PSS due to greater negative impacts during the implementation phase. This study indicates a growth in profit area from 923,080 in 2022 to 3,300,000 in 2027. In addition, the forecasted value of the quantity of pollution generated in 2027 is 3,129,180 kg CO_2 equivalent, and the required consumption of natural resources for five years is 1.3 million metric tons, which has a significant impact on the availability of natural resources. The results of the simulation for the social dimension are quite intriguing, with the PSS business model appearing to be very promising in the first few years. The graph of improved quality of life since implementation continues to rise, but in the subsequent years reveals a decline in social performance. This study confirms that nurture should be a prominent issue in a PSS since certain PSS features can significantly affect the achievement of set targets.

This study identified a discernible knowledge gap in the earlier study on nurture in a PSS. Therefore, the rebound effect could not be accurately identified since previous research did not consider nurture in the PSS evaluation. Consequently, the novelty of this study resides in its capacity to comprehensively identify rebound effects that account for nurture and quantify the influence of all behaviours on the system structure through sustainability parameters. In terms of knowledge and the theoretical foundation, this research contributes to closing the knowledge gap on the elements that significantly contribute to the rebound effect and trade-off of sustainability outcomes during PSS implementation since, to date, no PSS research has focused on environmental variables that cause product obsolescence. While the contribution of this research to the industry is its ability to detect the impact of every measure on the rebound effect so that policymakers must be more cautious, for instance, during the period of product use by customers, this study is additionally applicable to forecast future PSS attainments and the potential emergence of rebound effects of the business. In addition, the results of this study serve as a warning to industries that have previously believed that implementing a PSS will inherently result in sustainability. The industry reacted positively to this result, particularly its ability to predict the conditions for attaining sustainability in the coming years. The adopted systemic approach demonstrates convincingly that every decision has an effect on all aspects of the system. In addition, the industry regards the results of this study as a starting point for developing new policies to control the pace of customer behaviour change.

It should be noted, however, that this research was conducted in the context of car-sharing, where emissions and the consumption of natural resources are high enough to have a significant impact on the entire system if the life cycle is shortened; thus, this simulation cannot be generalised to all cases. Furthermore, the case study of the public car-sharing system is important as an illustration of comprehensible settings, and the underlying assumptions limit the applicability of the conclusions. In light of this, it is worthwhile for future studies to focus on enriching and systematising the indicators for each dimension and for the interrelationships between dimensions to enhance the data collection based on the literature and the case study used in this study. Additional development and validation by established practises for system dynamics modelling is required to yield more definitive results.

Author Contributions: Conceptualization, S.A.; Methodology, S.A.; Software, S.A.; Validation, S.A., Y.M.; Formal analysis, S.A.; Investigation, S.A.; Resources, S.A.; Data curation, S.A.; Writing—original draft preparation, S.A.; Writing—review and editing, S.A., Y.M., Y.T., H.W. and Y.S.; Visualization, S.A.; Supervision, Y.S.; Funding acquisition, Y.S. All authors have read and agreed to the published version of the manuscript.

Funding: This research received no external funding.

Institutional Review Board Statement: Not applicable.

Data Availability Statement: Not applicable.

Conflicts of Interest: The authors declare no conflict of interest.

Appendix A

Table A1. Simulation results.

Time (Year)	Amount of Pollution	Natural Resources Availability	Quality of Life	Profit Ratio
2022	0	54,999,900,000,000	0.0000	923,080
2022.12	119,639	54,999,900,000,000	0.0742	937,503
2022.25	236,428	54,999,900,000,000	0.1460	952,828
2022.38	350,438	54,999,900,000,000	0.2153	969,110
2022.5	461,739	54,999,800,000,000	0.2821	986,410

Table A1. *Cont.*

Time (Year)	Amount of Pollution	Natural Resources Availability	Quality of Life	Profit Ratio
2022.62	570,398	54,999,800,000,000	0.3461	1,000,000
2022.75	676,482	54,999,800,000,000	0.4074	1,020,000
2022.88	780,054	54,999,800,000,000	0.4659	1,050,000
2023	881,178	54,999,800,000,000	0.5215	1,070,000
2023.12	979,914	54,999,800,000,000	0.5740	1,090,000
2023.25	1,076,320	54,999,800,000,000	0.6236	1,120,000
2023.38	1,170,460	54,999,800,000,000	0.6701	1,140,000
2023.5	1,262,390	54,999,700,000,000	0.7134	1,170,000
2023.62	1,352,160	54,999,700,000,000	0.7536	1,200,000
2023.75	1,439,830	54,999,700,000,000	0.7907	1,230,000
2023.88	1,525,440	54,999,700,000,000	0.8246	1,270,000
2024	1,609,060	54,999,700,000,000	0.8554	1,300,000
2024.12	1,690,720	54,999,700,000,000	0.8830	1,340,000
2024.25	1,770,490	54,999,600,000,000	0.9076	1,380,000
2024.38	1,848,400	54,999,600,000,000	0.9291	1,420,000
2024.5	1,924,510	54,999,600,000,000	0.9476	1,470,000
2024.62	1,998,850	54,999,600,000,000	0.9631	1,520,000
2024.75	2,071,480	54,999,500,000,000	0.9759	1,570,000
2024.88	2,142,430	54,999,500,000,000	0.9858	1,620,000
2025	2,211,750	54,999,500,000,000	0.9931	1,680,000
2025.12	2,279,480	54,999,400,000,000	0.9978	1,740,000
2025.25	2,345,650	54,999,400,000,000	1.0000	1,810,000
2025.38	2,410,310	54,999,400,000,000	0.9999	1,880,000
2025.5	2,473,500	54,999,300,000,000	0.9976	1,950,000
2025.62	2,535,240	54,999,300,000,000	0.9931	2,030,000
2025.75	2,595,590	54,999,200,000,000	0.9867	2,110,000
2025.88	2,654,560	54,999,200,000,000	0.9785	2,200,000
2026	2,712,210	54,999,200,000,000	0.9685	2,300,000
2026.12	2,768,550	54,999,100,000,000	0.9570	2,400,000
2026.25	2,823,620	54,999,000,000,000	0.9440	2,510,000
2026.38	2,877,460	54,999,000,000,000	0.9297	2,620,000
2026.5	2,930,100	54,998,900,000,000	0.9143	2,740,000
2026.62	2,981,560	54,998,900,000,000	0.8977	2,870,000
2026.75	3,031,870	54,998,800,000,000	0.8803	3,000,000
2026.88	3,081,070	54,998,700,000,000	0.8620	3,150,000
2027	3,129,180	54,998,600,000,000	0.8430	3,300,000

Table A2. Comparative results.

Time (Year)	Nurture Included			Nurture Excluded		
	Amount of Pollution	Quality of Life (Normalised)	Profit Ratio	Amount of Pollution	Quality of Life (Normalised)	Profit Ratio
2022	0	0.00	923,080	0	0.0000	923,080
2022.12	119,639	0.00	937,503	5640	0.0142	900,003
2022.25	236,428	0.00	952,828	11,279	0.0289	877,503
2022.38	350,438	0.00	969,110	16,919	0.0439	855,566
2022.5	461,739	0.00	986,410	22,559	0.0593	834,177
2022.62	570,398	0.00	1,000,000	28,198	0.0752	813,322
2022.75	676,482	0.00	1,020,000	33,838	0.0914	792,989
2022.88	780,054	0.00	1,050,000	39,478	0.1081	773,165
2023	881,178	0.00	1,070,000	45,117	0.1253	753,836
2023.12	979,914	0.00	1,090,000	50,757	0.1429	734,990
2023.25	1,076,320	0.01	1,120,000	56,397	0.1609	716,615
2023.38	1,170,460	0.01	1,140,000	62,036	0.1795	698,700
2023.5	1,262,390	0.01	1,170,000	67,676	0.1986	681,233
2023.62	1,352,160	0.01	1,200,000	73,316	0.2181	664,202
2023.75	1,439,830	0.01	1,230,000	78,955	0.2382	647,597
2023.88	1,525,440	0.01	1,270,000	84,595	0.2589	631,408
2024	1,609,060	0.01	1,300,000	90,234	0.2801	615,623
2024.12	1,690,720	0.01	1,340,000	95,874	0.3018	600,232
2024.25	1,770,490	0.01	1,380,000	101,514	0.3242	585,227
2024.38	1,848,400	0.01	1,420,000	107,153	0.3471	570,597
2024.5	1,924,510	0.01	1,470,000	112,793	0.3707	556,332
2024.62	1,998,850	0.01	1,520,000	118,432	0.3949	542,424
2024.75	2,071,480	0.01	1,570,000	124,072	0.4197	528,864
2024.88	2,142,430	0.01	1,620,000	129,712	0.4452	515,643
2025	2,211,750	0.01	1,680,000	135,351	0.4714	502,752
2025.12	2,279,480	0.01	1,740,000	140,991	0.4983	490,184
2025.25	2,345,650	0.01	1,810,000	146,630	0.5259	477,930
2025.38	2,410,310	0.01	1,880,000	152,270	0.5543	465,983
2025.5	2,473,500	0.01	1,950,000	157,909	0.5834	454,334
2025.62	2,535,240	0.01	2,030,000	163,549	0.6133	442,976
2025.75	2,595,590	0.01	2,110,000	169,188	0.6440	431,903
2025.88	2,654,560	0.01	2,200,000	174,828	0.6755	421,106
2026	2,712,210	0.01	2,300,000	180,467	0.7079	410,579
2026.12	2,768,550	0.01	2,400,000	186,107	0.7411	400,316
2026.25	2,823,620	0.01	2,510,000	191,746	0.7752	390,309
2026.38	2,877,460	0.01	2,620,000	197,386	0.8102	380,553
2026.5	2,930,100	0.01	2,740,000	203,025	0.8462	371,041
2026.62	2,981,560	0.01	2,870,000	208,665	0.8832	361,766
2026.75	3,031,870	0.01	3,000,000	214,304	0.9211	352,724
2026.88	3,081,070	0.01	3,150,000	219,944	0.9600	343,908
2027	3,129,180	0.01	3,300,000	225,583	1	335,312

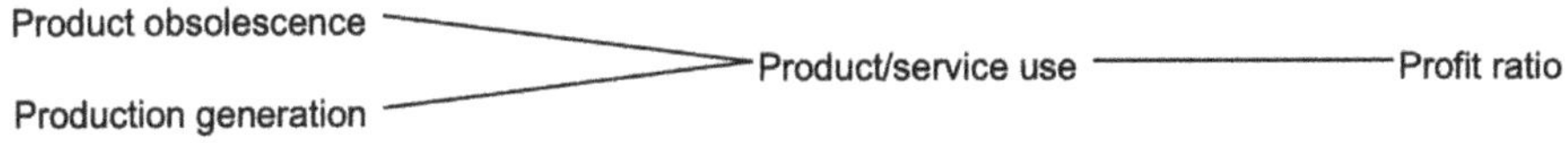

Figure A1. The causes tree of "Profit ratio".

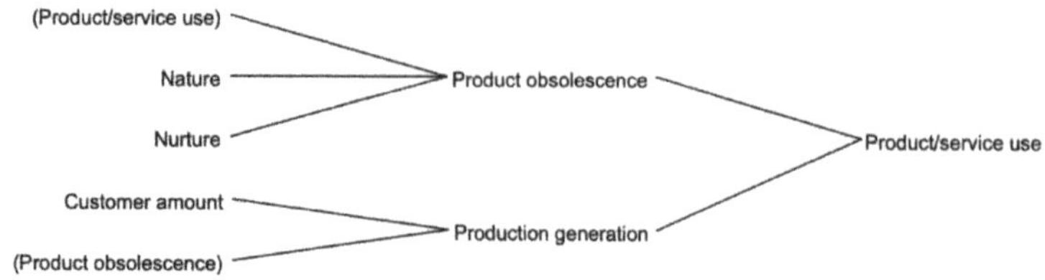

Figure A2. The causes tree of "product/service use".

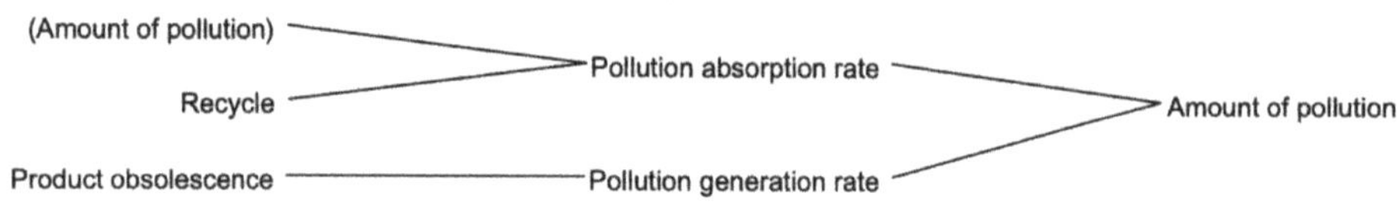

Figure A3. The causes tree of "amount of pollution".

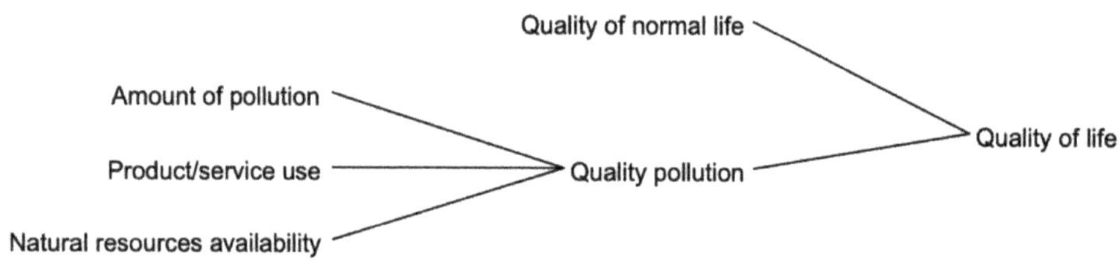

Figure A4. The causes tree of "Quality of life".

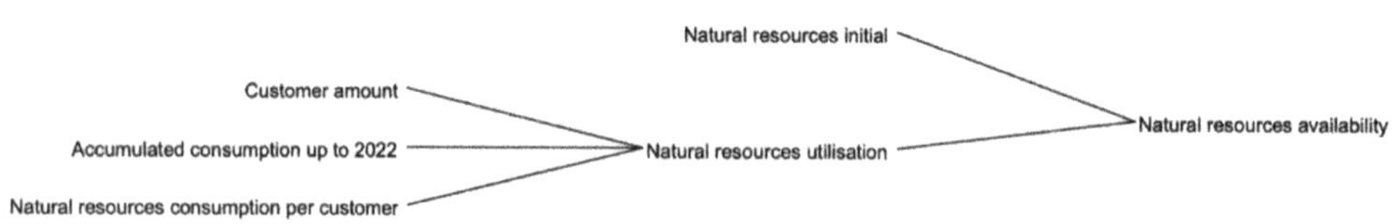

Figure A5. The causes tree of "natural resources availability".

References

1. Mugge, R.; Schoormans, J.P.L.; Schifferstein, H.N.J.; Mugge, R.; Schoormans, J.P.L.; Schifferstein, H.N.J. Design Strategies to Postpone Consumers' Product Replacement: The Value of a Strong Person-Product Relationship. *Des. J.* **2005**, *8*, 38–48. [CrossRef]
2. Bianchi, N.P.; Evans, S.; Revetria, R.; Tonelli, F. Influencing Factors of Successful Transitions towards Product-Service Systems: A Simulation Approach. *Int. J. Math. Comput. Simul.* **2009**, *3*, 30–43.
3. Mont, O. Clarifying the Concept of Product-Service System. *J. Clean. Prod.* **2002**, *10*, 237–245. [CrossRef]
4. Baines, T.S. State-of-the-Art in Product-Service Systems. *Proc. Inst. Mech. Eng. Part B J. Eng. Manuf.* **2007**, *221*, 1543–1552. [CrossRef]
5. Boons, F.; Lüdeke-freund, F. Business Models for Sustainable Innovation: State-of-the-Art and Steps towards a Research Agenda. *J. Clean. Prod.* **2013**, *45*, 9–19. [CrossRef]
6. Ponsioen, T.C.; Vieira, M.D.M.; Goedkoop, M.J. Surplus Cost as a Life Cycle Impact Indicator for Fossil Resource Scarcity. *Int. J. Life Cycle Assess.* **2014**, *19*, 872–881. [CrossRef]
7. Meijkamp, R. Changing Consumer Behaviour through Eco-Efficient Services. An Empirical Study of Car Sharing in the Netherlands. Ph.D. Thesis, Delft University of Technology, Delft, The Netherlands, 2000.
8. Manzini, E. A Strategic Design Approach to Develop Sustainable Product Service Systems: Examples Taken from the "environmentally Friendly Innovation" Italian Prize. *J. Clean. Prod.* **2003**, *11*, 851–857. [CrossRef]
9. Tukker, A. Product-Services as a Research Field: Past, Present and Future. Reflections from a Decade of Research. *J. Clean. Prod.* **2006**, *14*, 1552–1556. [CrossRef]
10. Nicola, W.; Robert, A.; Kai, H. Assessing the Rebound Effect and Other Macro Impacts of S-PSS Strategies. Presented at INSEAD-CMER, Fontainebleau, France, 9 May 2002.
11. Vezzoli, C.; Kohtala, C.; Srinivasan, A.; Diehl, J.C.; Fusakul, S.M.; Xin, L.; Sateesh, D. *Product-Service System Design for Sustainability*; Greenleaf Publishing: Sheffield, UK, 2017; pp. 1–502. ISBN 1-909493-69-4.

12. Alfarisi, S.; Mitake, Y.; Tsutsui, Y.; Wang, H.; Shimomura, Y. A Study of the Rebound Effect on the Product-Service System: Why Should It Be a Top Product-Service Priority? *Procedia CIRP* **2022**, *109*, 257–262. [CrossRef]
13. Tukker, A. Product Services for a Resource-Ef Fi Cient and Circular Economy e a Review. *J. Clean. Prod.* **2015**, *97*, 76–91. [CrossRef]
14. Cherry, C.E. Why Is Ownership an Issue? Exploring Factors That Determine Public Acceptance of Product-Service Systems. *Sustainability* **2018**, *10*, 2289. [CrossRef]
15. Cox, J.; Griffith, S.; Giorgi, S.; King, G. Consumer Understanding of Product Lifetimes. *Resour. Conserv. Recycl.* **2013**, *79*, 21–29. [CrossRef]
16. Yamamoto, H.; Murakami, S. Product Obsolescence and Its Relationship with Product Lifetime: An Empirical Case Study of Consumer Appliances in Japan. *Resour. Conserv. Recycl.* **2021**, *174*, 105798. [CrossRef]
17. Vezzoli, C. Why Have "Sustainable Product-Service Systems" Not Been Widely Implemented? Meeting New Design Challenges to Achieve Societal Sustainability. *J. Clean. Prod.* **2012**, *35*, 288–290. [CrossRef]
18. Annarelli, A.; Battistella, C.; Nonino, F. Product Service System: A Conceptual Framework from a Systematic Review. *J. Clean. Prod.* **2016**, *139*, 1011–1032. [CrossRef]
19. Demyttenaere, K. The Influence of Ownership on the Sustainable Use of Product-Service Systems—A Literature Review. *Procedia CIRP* **2016**, *47*, 180–185. [CrossRef]
20. Thiesen, J.; Christensen, T.S.; Kristensen, T.G.; Andersen, R.D.; Brunoe, B.; Gregersen, T.K.; Thrane, M.; Weidema, B.P. LCA Case Studies Rebound Effects of Price Differences. *Int. J. Life Cycle Assess.* **2008**, *13*, 104–114. [CrossRef]
21. Herring, H. Technological Innovation, Energy Efficient Design and the Rebound Effect. *Technovation* **2007**, *27*, 194–203. [CrossRef]
22. Sarmiento, P.A.Q.; Mayancela, R.; Suárez, L. Análisis de La Relación Entre Gestión de Calidad, Gestión Del Conocimiento e Innovación En Las Pymes. *Iber. Conf. Inf. Syst. Technol. (CISTI)* **2019**, *6*, 18867025.
23. Morecroft, J.D. *Strategic and Business*; John Wiley & Sons: New York, NY, USA, 2015; ISBN 978-1-118-84468-7.
24. Haase, R. Product/Service-System Origins and Trajectories: A Systematic Literature Review of PSS Definitions and Their Characteristics. *Procedia CIRP* **2017**, *64*, 157–162. [CrossRef]
25. Geng, X. An Integrated Approach for Rating Engineering Characteristics' Final Importance in Product-Service System Development. *Comput. Ind. Eng.* **2010**, *59*, 585–594. [CrossRef]
26. Kang, M. Product Service Systems as Systemic Cures for Obese Consumption and Production. *J. Clean. Prod.* **2008**, *16*, 1146–1152. [CrossRef]
27. Tukker, A.; Tischner, U. (Eds.) *New Business for Old Europe: Product-Service Development, Competitiveness and Sustainability*; Greenleaf Publishing: Sheffield, UK, 2006.
28. Berkhout, P.H.G.; Muskens, J.C.; Velthuijsen, J.W. Defining the rebound effect. *Energy Pol.* **2000**, *28*, 425–432. [CrossRef]
29. Kjaer, L.L. Guidelines for Evaluating the Environmental Performance of Product/Service-Systems through Life Cycle Assessment. *J. Clean. Prod.* **2018**, *190*, 666–678. [CrossRef]
30. Maxwell, D.; Owen, P.; McAndrew, L.; Muehmel, K.; Neubauer, A. *Addressing the Rebound Effect: A Report for The European Commission DG Environment*; European Commission: Brussels, Belgium, 2011.
31. Vivanco, D.F.; VAn der Voer, E.; Font Vivanco, D.; Kemp, R.; van der Voet, E. How to Deal with the Rebound Effect? A Policy-Oriented Approach. *Energy Policy* **2016**, *94*, 114–125. [CrossRef]
32. Laurenti, R. Unintended Environmental Consequences of Improvement Actions: A Qualitative Analysis of Systems' Structure and Behavior. *Syst. Res. Behav. Sci.* **2016**, *33*, 381–399. [CrossRef]
33. Liedtke, C.; Baedeker, C.; Hasselkuß, M.; Rohn, H.; Grinewitschus, V. User-Integrated Innovation in Sustainable LivingLabs: An Experimental Infrastructure for Researching and Developing Sustainable Product Service Systems. *J. Clean. Prod.* **2015**, *97*, 106–116. [CrossRef]
34. Kuo, T. Simulation of Purchase or Rental Decision-Making Based on Product Service System. *Int. J. Adv. Manuf. Technol.* **2011**, *52*, 1239–1249. [CrossRef]
35. Gottberg, A.; Longhurst, P.J.; Cook, M.B. Exploring the Potential of Product Service Systems to Achieve Household Waste Prevention on New Housing Developments in the UK. *Waste Manag. Res.* **2010**, *28*, 228–235. [CrossRef]
36. Kjaer, L.L. Challenges When Evaluating Product/Service-Systems through Life Cycle Assessment. *J. Clean. Prod.* **2016**, *120*, 95–104. [CrossRef]
37. Mylan, J. Understanding the Diffusion of Sustainable Product-Service Systems: Insights from the Sociology of Consumption and Practice Theory. *J. Clean. Prod.* **2015**, *97*, 13–20. [CrossRef]
38. Nes, N.V.; Cramer, J. Product Lifetime Optimization: A Challenging Strategy towards More Sustainable Consumption Patterns. *J. Clean. Prod.* **2006**, *14*, 1307–1318. [CrossRef]
39. Park, H. A Chance Discovery-Based Approach for New Product–Service System (PSS) Concepts. *Serv. Bus.* **2015**, *9*, 115–135. [CrossRef]
40. Schifferstein, H.N.J.; Zwartkruis-Pelgrim, E.P.H. Consumer-Product Attachment: Measurement and Design Implications. *Int. J. Des.* **2015**, *2*, 1–13.
41. Laurenti, R.; Sinha, R.; Singh, J. Some Pervasive Challenges to Sustainability by Design of Electronic Products—A Conceptual Discussion. *J. Clean. Prod.* **2015**, *108*, 281–288. [CrossRef]
42. Sassanelli, C.; Rosa, P.; Rocca, R.; Terzi, S. Circular Economy Performance Assessment Methods: A Systematic Literature Review. *J. Clean. Prod.* **2019**, *229*, 440–453. [CrossRef]

43. Weidmann, D.; Maisenbacher, S.; Kasperek, D.; Maurer, M. Product-Service System Development with Discrete Event Simulation Modeling Dynamic Behavior in Product-Service Systems. In Proceedings of the 2015 Annual IEEE Systems Conference (SysCon) Proceedings, Vancouver, BC, Canada, 13–16 April 2015; IEEE: Vancouver, BC, Canada, 2015; pp. 133–138.
44. Grüneisen, P.; Stahl, B.; Kasperek, D.; Maurer, M.; Lohmann, B. Qualitative System Dynamics Cycle Network of the Innovation Process of Product Service Systems. *Procedia CIRP* **2015**, *30*, 120–125. [CrossRef]
45. Lee, S. Dynamic and Multidimensional Measurement of Product-Service System (PSS) Sustainability: A Triple Bottom Line (TBL)-Based System Dynamics Approach. *J. Clean. Prod.* **2012**, *32*, 173–182. [CrossRef]
46. Lee, J. Dominant Innovation Design for Smart Products-Service Systems (PSS): Strategies and Case Studies. In Proceedings of the Annual SRII Global Conference, San Jose, CA, USA, 23–25 April 2014; pp. 305–310.
47. Richardson, G.P. Problems with Causal-Loop Diagrams. *Syst. Dyn. Rev.* **1986**, *2*, 158–170. [CrossRef]
48. Forrester, J.W. Lessons from System Dynamics Modeling. *Syst. Dyn. Rev.* **1987**, *3*, 136–149. [CrossRef]
49. Statistics Indonesia. Percentage of Population by Province and Gender. Available online: https://www.bps.go.id/dynamictable/2018/03/20/1288/persentase-penduduk-menurut-provinsi-dan-jenis-kelamin-2009-2018.html (accessed on 11 December 2022).
50. Belgiawan, P.F.; Schmöcker, J.; Fujii, S. Understanding Car Ownership Motivations among Indonesian Students. *Int. J. Sustain. Transp.* **2014**, *10*, 37–41. [CrossRef]
51. Statista. Car-Sharing Indonesia. Available online: https://www.statista.com/outlook/mmo/shared-mobility/shared-rides/car-sharing/indonesia (accessed on 7 January 2023).
52. Acerbi, F.; Sassanelli, C.; Taisch, M. A Conceptual Data Model Promoting Data-Driven Circular Manufacturing. *Oper. Manag. Res.* **2022**, *15*, 838–857. [CrossRef]
53. Statista. Reserves of Iron Ore Worldwide from 2010 to 2021. 2022. Available online: https://www.statista.com/statistics/1168572/global-reserves-of-iron-ore/ (accessed on 16 November 2022).
54. Drive Sustainability Iron Ore Value Chain. 2022.
55. Statista. Production Volume of Usable Iron Ore Worldwide from 2006 to 2021. Available online: https://www.statista.com/statistics/589945/iron-ore-production-gross-weight-worldwide/ (accessed on 17 November 2022).
56. Verwertung, H.N. Klimabilanz von E-Fahrzeugen & Life Cycle Engineering. 2019. Available online: https://www.hanswernersinn.de/sites/default/files/VW_Klimabilanz_von_E-Fahrzeugen_Life_Cycle_Engineering_0.pdf (accessed on 29 November 2022).
57. Buberger, J.; Kersten, A.; Kuder, M.; Eckerle, R.; Weyh, T. Total CO_2 Equivalent Life-Cycle Emissions from Commercially Available Passenger Cars. *Renew. Sustain. Energy Rev.* **2022**, *159*, 112158. [CrossRef]
58. Kojima, M. Vehicle Recycling in the ASEAN and Other ASIAN Countries. ERIA Research Project Report No. 16. 2017. Available online: https://think-asia.org/handle/11540/9394 (accessed on 29 November 2022).
59. Fuller, R.; Landrigan, P.J.; Balakrishnan, K.; Bathan, G.; Bose-O'Reilly, S.; Brauer, M.; Caravanos, J.; Chiles, T.; Cohen, A.; Corra, L.; et al. Pollution and Health: A Progress Update. *Lancet Planet. Health* **2022**, *6*, e535–e547. [CrossRef]
60. Lozano, R.; Huisingh, D. Inter-Linking Issues and Dimensions in Sustainability Reporting. *J. Clean. Prod.* **2011**, *19*, 99–107. [CrossRef]
61. Labuschagne, C.; Brent, A.C.; van Erck, R.P. Assessing the Sustainability Performances of Industries. *J. Clean. Prod.* **2005**, *13*, 373–385. [CrossRef]
62. Assefa, G.Ã.; Frostell, B. Social Sustainability and Social Acceptance in Technology Assessment: A Case Study of Energy Technologies. *Technol. Soc.* **2007**, *29*, 63–78. [CrossRef]
63. Roy, M. Introduction to Sustainable Development. In *Sustainable Development Strategies*; Elsevier: Amsterdam, The Netherlands, 2021; pp. 1–25. ISBN 978-0-12-818920-7.
64. WCED. *Report of the World Commission on Environment and Development: Our Common Future*; Oxford University Press: Oxford, UK; New York, NY, USA, 1987.
65. Badiru, A.B.; Omitaomu, O.A. *Handbook of Industrial Engineering Equations, Formulas, and Calculations*; CRC Press: Boca Raton, FL, USA, 2011; ISBN 978-1-4200-7627-1.
66. ESCAP UN. Integrating the Three Dimensions of Sustainable Development: A Framework and Tools. Greening of Economic Growth Series. Bangkok. 2015. Available online: https://repository.unescap.org/handle/20.500.12870/3161 (accessed on 1 December 2022).
67. Corsini, L.; Moultrie, J. What Is Design for Social Sustainability? A Systematic Literature Review for Designers of Product-Service Systems. *Sustainability* **2021**, *13*, 5963. [CrossRef]
68. Ceschin, F.; Gaziulusoy, I. Evolution of Design for Sustainability: From Product Design to Design for System Innovations and Transitions. *Des. Stud.* **2016**, *47*, 118–163. [CrossRef]
69. Sterman, J.D. *Business Dynamics, System Thinking and Modeling for Complex World*; MCGraw Hill: New York, NY, USA, 2002; ISBN 0-07-231135-5.
70. Walnum, H.J.; Aall, C.; Løkke, S. Can Rebound Effects Explain Why Sustainable Mobility Has Not Been Achieved. *Sustainability* **2014**, *6*, 9510–9537. [CrossRef]
71. Santarius, T.; Soland, M. How Technological E Ffi Ciency Improvements Change Consumer Preferences: Towards a Psychological Theory of Rebound Effects. *Ecol. Econ.* **2018**, *146*, 414–424. [CrossRef]

72. Fargnoli, M.; Haber, N.; Tronci, M. Case Study Research to Foster the Optimization of Supply Chain Management through the PSS Approach. *Sustainability* **2022**, *14*, 2235. [CrossRef]
73. Scheepens, A.E.; Vogtländer, J.G.; Brezet, J.C. Two Life Cycle Assessment (LCA) Based Methods to Analyse and Design Complex (Regional) Circular Economy Systems. Case: Making Water Tourism More Sustainable. *J. Clean. Prod.* **2016**, *114*, 257–268. [CrossRef]
74. Sassanelli, C.; Da Costa Fernandes, S.; Rozenfeld, H.; Mascarenhas, J.; Terzi, S. Enhancing Knowledge Management in the PSS Detailed Design: A Case Study in a Food and Bakery Machinery Company. *Concurr. Eng. Res. Appl.* **2021**, *29*, 295–308. [CrossRef]
75. Sassanelli, C.; Pezzotta, G.; Pirola, F.; Terzi, S.; Rossi, M. Design for Product Service Supportability (DfPSS) Approach: A State of the Art to Foster Product Service System (PSS) Design. *Procedia CIRP* **2016**, *47*, 192–197. [CrossRef]

sustainability

MDPI

Article

Measuring the Sustainable Entrepreneurial Performance of Textile-Based Small–Medium Enterprises: A Mediation–Moderation Model

Sidney Mangenda Tshiaba [1,*], Nianxin Wang [1,*], Sheikh Farhan Ashraf [2], Mehrab Nazir [1] and Nausheen Syed [3]

1 School of Economics and Management, Jiangsu University of Science and Technology, Zhenjiang 212003, China; Mehrabnazir9@gmail.com
2 School of Management, Jiangsu University, Zhenjiang 212100, China; fhsheikh08@gmail.com
3 Department of Business Administration, Government College Women University, Faisalabad 38000, Pakistan; nausheen.dr@gmail.com
* Correspondence: sidneytshiaba@outlook.com (S.M.T.); nianxin.wang@gmail.com (N.W.)

Abstract: This research aimed to examine the role of knowledge management practices in sustainable entrepreneurship performance. This study also investigated the relationships between six concepts: knowledge sharing behavior, innovative capacity, absorptive capacity, dynamic capability, opportunity recognition, and sustainable entrepreneurship. A self-administered questionnaire was used for data collection from 486 entrepreneurs randomly selected from textile-based SMEs in the Democratic Republic of the Congo (DRC). The findings show that knowledge management practices positively and significantly impact sustainable entrepreneurship performance and SMEs' dynamic capabilities. Moreover, opportunity recognition strengthens the relationship between SMEs' dynamic capabilities and sustainable entrepreneurship performance. This study offers valuable insights and directions for researchers and practitioners interested in the field of entrepreneurship.

Keywords: sustainable entrepreneurial performance; dynamic capabilities; opportunity recognition; knowledge sharing behavior; innovative capacity; absorptive capacity

Citation: Mangenda Tshiaba, S.; Wang, N.; Ashraf, S.F.; Nazir, M.; Syed, N. Measuring the Sustainable Entrepreneurial Performance of Textile-Based Small–Medium Enterprises: A Mediation–Moderation Model. *Sustainability* **2021**, *13*, 11050. https://doi.org/10.3390/su131911050

Academic Editors: Claudio Sassanelli, Sergio Terzi and Marc A. Rosen

Received: 8 August 2021
Accepted: 29 September 2021
Published: 6 October 2021

1. Introduction

Across the world, the importance of sustainable entrepreneurship has increased as a potential solution to different problems. Practitioners claim that entrepreneurs may tackle issues caused by natural degradation by inventing new sustainable practices [1,2]. Sustainable entrepreneurship performance, in addition to success, is based on knowledge and represents an element of competitive advantage [3]. Knowledge management practices such as knowledge sharing behaviors, innovations, and absorption capacities build a relationship between entrepreneur capabilities and sustainable entrepreneurship performance [4]. According to Antunes and Pinheiro [5], knowledge management practices can help small and medium enterprises (SMEs) develop and prosper with better business performance in the long term.

Prior studies have examined the impact of knowledge management practices on sustainable entrepreneurship [6] and the relationship between leadership skills and sustainability. Moreover, knowledge-based theory (KBT) can efficiently manage and create unique dynamic capabilities which contribute to sustainable entrepreneurship performance [4]. Therefore, entrepreneurs with knowledge management practices will likely achieve good sustainable entrepreneurship performance [5]. Li [3] explained that the sharing, acquisition, and application of knowledge contribute to innovation and performance [7]. The knowledge management practices of entrepreneurs have progressively become of interest to researchers, especially in business studies, as a means of enhancing entrepreneurship performance [8,9].

When knowledge management engages in sustainability, the organization's attitude evolves, and social responsibility is valued equally to economic viability [10]. The significance of KMPs consolidating at the worldwide level demonstrates the convergence in the utilization of KMPs in organizations [11]. Knowledge management (KM) practices can be broken down into three sets, namely, knowledge sharing behavior, innovative capacity, and absorptive capacity [12–15]. Sustainable development techniques can be built on the foundation of knowledge management. Knowledge sharing behavior, innovative capacity, and absorptive capacity are considered vital attributes of knowledge management practices in sustainability criteria, especially in an entrepreneurial context [15]. As a result, businesses must rely more heavily on their knowledge-generating resources. In sustainable development, knowledge management practices are considered a new philosophy of development that attempts to improve engagement with social, economic, and environmental values [16,17]. The KMPs utilized in sustainability explain a transformation in the organization's stance when social and environmental responsibilities are held as equal to commercial viability. Knowledge management (hereafter known as KM) practices have become an essential source of enhancing the sustainable performance of organizations.

In this regard, KM can play a crucial role by facilitating the sharing of information among various time zones and geographical regions [18]. There is a growing need for approaches to improve KM processes and procedures throughout the assessment of environmental, social, and economic consequences, given the increased demand for sustainable entrepreneurial performance [19]. This study focused on the knowledge management practices contributing to many dimensions of sustainability that are best described in the literature.

Recent studies have emerged that integrate entrepreneurship with sustainability and encompass the broad concept of sustainable entrepreneurship, which includes financial, environmental, and moral ideals [20]. Due to the rapid changes in the environment, SME entrepreneurs confront a number of challenges in finding opportunities that might help them solve problems and improve their performance [3]. Small–medium enterprises (SMEs) can play a significant role in developing a country—they are generally considered a key pillar of economic development in developing countries [21]. Therefore, several entrepreneurial capabilities are essential to enhance sustainable entrepreneurial performance (SEP) and achieve organizational development goals [12]. Good SEP depends not only on the willingness and commitment to become an entrepreneur [22,23] but also on the knowledge and capabilities essential to becoming a sustainable entrepreneur [24].

Entrepreneurs have many opportunities to utilize the available resources for higher profitability and ensure their sustainable organizational performance [25]. Several studies have proven the potential role of SMEs in enhancing economic growth, wealth creation, and employment, particularly in emerging countries [26]. Therefore, it is necessary to explore the integrated relationship among several sustainable entrepreneurial success factors, particularly in developing countries such as the Democratic Republic of the Congo (DRC) [27]. SMEs have played a major part in contributing to this country's gross domestic product (GDP). Furthermore, the SME sector of the DRC represents 25% of the country's exports, 35% of manufacturing, and 53% of the hotel, restaurant, wholesaler, and retail trade sectors. Additionally, 20% of SMEs are active in the industrial sector, and 22% are engaged in the service sector.

SEP is directly associated with knowledge management practices: both SEP and knowledge management play a positive role in ensuring business growth [28,29]. Numerous studies suggested that improvement in SEP supports and sustains an organization's market value [30]. However, there are several fundamentals which are involved in SEP such as knowledge sharing behavior (KSB), innovative capacity (IC), and absorptive capacity (AC), which are directly related to the success of an entrepreneur [31]. The exchange of skills and experiences within an organization is known as KSB [32]. Information regarding organizational schedules, depositories, and repositories and across organizational boundaries is practiced and eventually relies on members' KSB for performance [33]. The sharing of knowledge in an organization depends on the organizational atmosphere and entrepreneurs' behavior, which is beneficial for sustainable performance [12]. When KSB is restricted, the gaps ascend, which creates hurdles in performances [34].

Prior studies illustrated that innovation and innovative capability (IC) are also related to SEP [35]. The linkage of inner capacity with abilities that comes with something new is known as IC—IC is directly correlated with the nature of SEP [35]. The IC of an individual comes in the form of entrepreneurship. Furthermore, entrepreneurs' strategic planning and absorptive and innovative capacities enhance SEP [36,37]. AC categorizes abilities and the assimilation and utilization of knowledge for SME performance. Entrepreneurs with AC can absorb knowledge from competitors and apply knowledge within their organization to enhance performance [36]. Researchers indicated that AC combines three necessary abilities: peripheral knowledge, understanding of knowledge, and integration of innovative knowledge for SEP [9]. These practices may help to manage the knowledge that can be used for achieving organizational goals [38]. Therefore, it is important to measure the impact of such knowledge management practices on the EP of SMEs.

However, studies have proved that an entrepreneur's dynamic capability (DC) has a vital role in increasing performance, which can be further availed by using organizational resources to create, design, and modify an organization according to market conditions and challenges [39,40]. DC replicates valuable resources such as innovative and absorptive capacities for competitive advantages and sustainable performances [41]. The main goal of this research was to examine the relationship and impact of knowledge management practices on sustainable entrepreneurship performance using dynamic capabilities as mediators and opportunity recognition as a moderator [42]. It has also been proven that the KSB of an entrepreneur significantly contributes to improving dynamic capacities [13]. Prior studies mostly explored this in the context of knowledge management strategies connected to DC or SEP in various industries [4], but not specifically in textile-based SMEs. A variety of textile-related SMEs, including weaving, ginning, knitting, power looms, and manual dying units, contribute significantly to the entire textile sector and economic development, particularly in developing nations. The lack of focus on this particular issue motivated the researchers to evaluate the EP of this sector through a holistic research model grounded in recourse-based theory.

Furthermore, the concept of entrepreneurial opportunity recognition (OR) observes the position, demand, and market value for a new product, and it deeply affects SEP [43,44]. Numerous researchers claimed that "an opportunity may be the chance to meet a market need through a creative combination of resources to deliver superior value" [45]. The researchers argue that opportunity means recognizing a market need with the available capabilities, which improves performance. Entrepreneurs employ opportunity sources to discover, evaluate, and exploit opportunities [46], which enhances SEP and organizational performance by increasing the capabilities of entrepreneurs [47]. This study proposes an integrated research framework ensuring the moderating role of OR to strengthen the relationship of DC and SEP in SMEs, which has not been studied yet in a similar context (as per our best knowledge). The rest of this paper is divided into several sections covering the theoretical justification, hypothesis development, methodology, results, discussion, conclusion, and study implications.

Sustainable entrepreneurship is regarded as a creative, market-oriented personality—a style of value generation that provides new start-ups using environmental management techniques or cleaner manufacturing procedures [48]. The primary aim of the current study was to develop a business venue using KSB, innovation capacity, and absorptive capacity to create sustainable entrepreneurship [49]. This research is based on knowledge sharing behaviors (KSB), innovative capability (IC), and absorptive capacity (AC) to enhance sustainable entrepreneurial performance [50], and entrepreneurial behavior results from sustainable performance [51]. In existing research, intentional models involve understanding sustainable entrepreneurial performance [52]. The primary motivation for designing the current study is that the literature on the textile sector has barely addressed the relationship between the variables mentioned above. Achieving holistic business performance is associated with considering all aspects of sustainable development, particularly in the textile sector [53]. The primary objective of this study was to explore the relationship and impact of knowledge management practices on sustainable entrepreneurship performance through dynamic entrepreneurial capabilities.

Thus, investigating a complementary perspective would fill a research gap, and this study covers the existing gap in the literature of knowledge management practices towards sustainable performance. There have been no formal studies that examine the impact of combining the concepts of knowledge sharing behavior, innovative capacity, and absorptive capacity to achieve sustainable performance. Secondly, this study measures the sustainable entrepreneurship performance of SME entrepreneurs by using dynamic capability as a mediator because the significance of the SME sector is increasing gradually. Thirdly, most previous studies focused on other industries and examined the role of knowledge management practices in business performance [3,52]. The relevance of opportunity recognition in the relationship between dynamic capability and long-term entrepreneurial performance has also been overlooked in prior studies. As a result, we employed opportunity recognition to mediate the relationship between dynamic capability and sustainable entrepreneurship performance. This dynamic capability is well suited to a specific target market in order to improve sustainable entrepreneurial performance. Thus, this study considered opportunity recognition to be a moderating element in the association between dynamic capabilities and sustainable entrepreneurial performance.

2. Theoretical Justification and Hypothesis Development

The conceptual framework is based on empirical studies and fundamental theories. This study looked at the significance of antecedents in relation to [54] Schumpeter's entrepreneurship theory, which is based on entrepreneurs and SEP. The concept of entrepreneurship theory (ET) supports SEP based on organizational support and resources. Moreover, the resource-based theory presented introduced "resource-based theory", emphasized difficulties in imitating the organization's features for a greater performance and viable advantage, and concluded that AC and IC are directly linked with performance [55]. Resource-based theory is applied to analyze and deduce a company's internal assets, highlighting resources, capabilities, and capacities in a framing strategy to achieve performance stability [56,57]. OR provides an entrepreneur a chance to create a new notion for a product and SEP [58,59]. In addition, knowledge-based theory (KBT) indicates that if knowledge management practice is applied efficiently, it creates a unique skill that leads to better sustainable entrepreneurship performance [60]. Therefore, businesses with more robust knowledge management practices are likely to accomplish business sustainability [12,42]. Li et al. [3] stated that knowledge management processes such as sharing, acquiring, and implementing knowledge constantly improve innovation capacity, which contributes to improved sustainable entrepreneurial performance. Therefore, we also integrated the above theories to develop a holistic theoretical framework to meet the objectives of this research.

2.1. *Knowledge Sharing Behavior, Dynamic Capability, and Sustainable Entrepreneurial Performance*

The interaction of social culture and sharing and exchanging knowledge with technical skills in an organization is known as KSB [61]. KSB is always voluntary; sharing and exploring any information in the organization or with the entrepreneur cannot be forced [62]. Bartol and Srivastava [61] described KSB as spreading important information within the organization, which becomes a valuable asset for performance [63,64]. KSB increases the tendency to understand organizational domestic and economic challenges an entrepreneur faces in sustainable performance [63]. The employee starts sharing knowledge in an organization with the entrepreneur and believes in intrinsic benefits, monetary benefits, self-satisfaction, promotion, and social recognition in the organization [65], which causes a negative influence on SEP. The external information shared in an organization, through socialism or initialization, becomes significant knowledge in performance [66].

Many researchers argued that organizational performance and SEP performance move in parallel, and that entrepreneurs' DC is critical to both [67]. An entrepreneur's DC considers KSB a significant asset in the organization and a major source for enhancing dynamic entrepreneurial capabilities in achieving the maximum competitive advantage in SEP [68]. An entrepreneur's planning and DC enhance and assist in directing, acting, and decision making for competitive organizational advantages and SEP [12,69].

Hypothesis 1 (H1). *KSB has a positive influence on DC.*

Hypothesis 2 (H2). *KSB has a positive influence on EP.*

Hypothesis 3 (H3). *DC mediates the relationship between KSB and SEP.*

2.2. *Innovative Capacity, Dynamic Capability, and Sustainable Entrepreneurial Performance*

Villa introduced the concept of IC, which is used to examine the level of innovation and invention, including potential ideas for economic activity [70]; meanwhile, the researcher also argued that "borrowing" brings innovation rather than "invention." The combination of an entrepreneur's capabilities, power, and abilities, which create something different, is known as an innovation [71,72]. IC is directly associated with the nature of entrepreneurs, and it comes in the form of entrepreneurship [64,73]. Several studies have also observed that an entrepreneur's IC plays an important role in improving SEP [74]. When entrepreneurs face certain uncertainties, IC assists in gaining, creating, and utilizing inner qualities. IC also improves decision making power and leadership skills, serves as a financial adviser in the organization, is vigilant of organization, awareness, and allocation of better opportunities with better substitutes, and becomes more beneficial for SEP [75].

Meanwhile, researchers argued that absorbing external knowledge leads the entrepreneur towards IC and SEP [76]. Furthermore, DC enhances the IC of an entrepreneur in developing a new product for the market and SEP [77]. Therefore, it can be concluded that the DC of an entrepreneur always creates a value chain with IC and performance.

Hypothesis 4 (H4). *IC has a positive influence on DC.*

Hypothesis 5 (H5). *IC has a positive influence on EP.*

Hypothesis 6 (H6). *DC mediates the relationship between IC and SEP.*

2.3. *Absorptive Capacity, Dynamic Capability, and Entrepreneurial Performance*

AC is defined as the ability to recognize and assimilate new and external knowledge, which is applied for commercial purposes [78]. The AC of an entrepreneur is to absorb innovation for change and better performance [79]. Entrepreneurial AC is to focus and absorb cognitive features in learning, evaluating, and formatting outside knowledge on a large scale for SEP [80]. Here, the researchers considered AC as a potential mechanism for SEP. Sulistyo and Siyamtinah [81] stated that AC affects assimilating and acquisition, which brings a change in EP. The role of AC supports strategic planning, and creating, absorbing, building, and utilizing available opportunities [82]. Meanwhile, identifying and configuring the core competencies of entrepreneurs through dynamic capabilities enhance SEP. Absorptive capacity potential is realized when the level of realized AC rises, and the entrepreneur can use the potential AC for SEP [83].

The DC of entrepreneurs is to adapt, abandon, reconfigure, and increase valuable resources, which help in the creation and development of new values for SEP [78], arguing three types of dynamic capabilities: possession, deployment, and upgrading capabilities, which are enhanced through wisdom, and through creating, adapting, integrating, and developing resources to obtain the maximum competitive advantages [81]. AC contributes to the understanding and utilization of valuable information with dynamic capabilities, in order to generate the optimal marketing strategies for long-term financial profit and SEP [84]. Prior studies explained that AC enhances the process of evaluation and adaption in SEP [85]. The combination of AC and DC has a significant influence on SEP in an organization [86]. The DC of an entrepreneur emphasizes the mechanism of IC in developing, creating, and managing, which helps entrepreneurs in performance [87]. Therefore, AC and DC are necessary to gain ideas and implications for SEP.

Hypothesis 7 (H7): *AC has a positive influence on DC.*

Hypothesis 8 (H8): *AC has a positive influence on EP.*

Hypothesis 9 (H9) : *DC mediates the relationship between AC and EP.*

2.4. Dynamic Capability and Sustainable Entrepreneurial Performance

Many economists deny the role of entrepreneurs as primary, while in the real world, entrepreneurs are considered the primary decision makers [88] and rulers of the economy [89]. Entrepreneurs are recognized as the backbone for organizational and economic growth. Entrepreneurial capabilities are implemented for a sustainable business model, organizational change, and SEP [71]. According to resource-based theory, DC plays a vital role in SEP [71]. DC is to peruse and observe opportunities at the right time and place to acquire the market with business strategies, available resources, capacities, and capabilities for SEP [90]. Earlier studies suggested that the DC of entrepreneurs is the primary source for a rapid and better change in organizational culture [91].

The DC of entrepreneurs restructures and changes the organizational environment directly associated with SEP [71]. The DC of entrepreneurs is the most reliable and sound source for taking competitive advantages and plays a mediating role between entrepreneurial resources and SEP [70]. As per resource-based theory, the DC of an entrepreneur contributes to accepting, maintaining, developing, and accomplishing new challenges with opportunities in the market and SEP [92]. DC is to understand, investigate, and analyze the entrepreneurial competency level and enhance an entrepreneur's resource capacity for SEP in an organization [92].

Hypothesis 10 (H10). *DC has a positive influence on SEP.*

2.5. Opportunity Recognition, Dynamic Capability, and Entrepreneurial Performance

The concept of recognition of opportunities is closely linked to entrepreneurship. Enterprise opportunities are acknowledged by conditions in which the presentation and commercialization of new goods, services, raw materials, and arrangement practices are of maximum priority compared to the cost of production [93]. While identifying entrepreneurship as a subjective issue, an opportunity itself is an objective phenomenon identified by a given person at one time [93]. Numerous entrepreneurial opportunities are generated in developing nations because faster and more efficient countries generate many possibilities for innovative participants and often distribute these to the local market [94].

In prior research, different researchers argued that entrepreneurs are different while perceiving OR [95]. The theory of OR also proposes that entrepreneurs' cognition makes entrepreneurial processes and performance more sustainable [96]. Meanwhile, Akkaya and Üstgörül's [97] study also discussed the mediating role of OR in association with entrepreneurial performance and found it to be a critical factor in enhancing SEP. In addition, several researchers have indicated that entrepreneurs' self-made tactics are essential in the OR process [98]. This study integrated OR to test its effect on the link between DC and SEP due to the lack of research attention on this essential factor. We propose the following:

Hypothesis 11 (H11). *OR has a moderating effect on the relationship between DC and SEP (an increase in OR will strengthen the relationship between DC and SEP.*

2.6. Conceptual Framework

Figure 1 shows the conceptual model for the studied variables, and the purpose of this study is to explore the impact of KSB, innovation, and AC on SEP. Furthermore, this study describes the role of DC as a mediator between KSB, IC, and AC with SEP. This study also explored the role of OR as a moderator between DC and SEP.

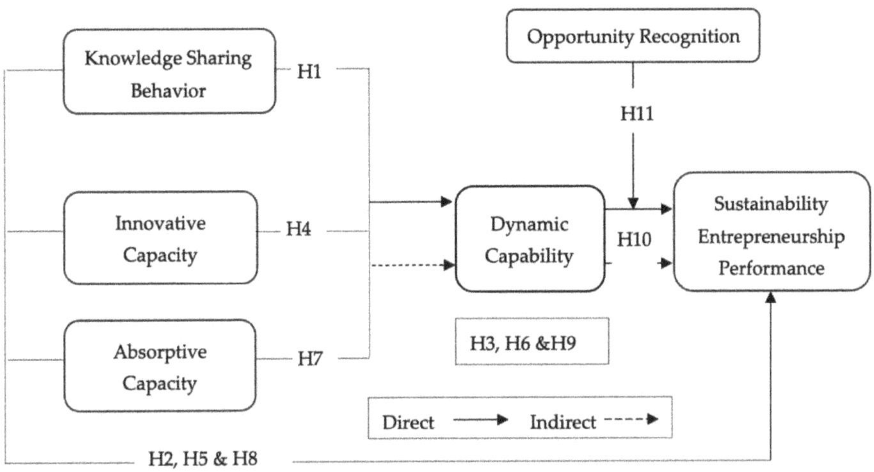

Figure 1. Research framework.

3. Materials and Methods

Creswell et al. [99] identified that a quantitative research technique is the best way to check the statistical relationship between variables. This study is based on a deductive method of research because it focuses on testing hypotheses arising from a current theory [100]. Therefore, we employed the survey method to test our hypotheses. We applied a cross-sectional study with a convenience sampling technique on textile-based SMEs from Kinshasa. The researcher used a time lag approach and collected data in three rounds [101]. A total of 500 digital and paper-pencil questionnaires were distributed and emailed to the target population, and 486 respondents responded fairly. There were different steps in collecting data from the respondents, and we collected data for knowledge management practices and dynamic capability measures. However, there are no significant data for registered SMEs in the chamber of commerce of Congo. Therefore, we approached respondents through emails and physically for different cities' listed SMEs [90].

Congo is also regarded as one of the least innovative countries in the textile industry. As a result, the purpose of this research is to see how highly certified firms think about KM practices in terms of achieving long-term entrepreneurial success. Because the fundamental source of knowledge is the acquisition and application of information, which leads to sustainable performance, it may help to understand the specific status of green practices and provide practical consequences to other non-certified businesses in Congo. The data for this study were collected from respondents using non-probability convenience sampling. Furthermore, they were better equipped with appropriate information and, at the same time, played an essential role in knowledge transmission among diverse departments [102]. We approached middle and senior managers with formal approval and requested that they participate in data collection, as did the previous researcher. The organization as a whole is represented by these responders.

In addition, the researcher also assured the respondents that their information is confidential and that the research is purely for academic purposes. The questionnaire was initially drafted in English, but it was intended for use in French, the official language in

Congo. Therefore, the English questionnaire was translated into French and four other national languages (Kikongo (Kituba), Lingala, Swahili, and Tshiluba) by the researchers and professional native translators, working independently to ensure consistency and to make it easier for the respondents to comprehend [101].

Moreover, we also tried to encourage some female entrepreneurs to participate; however, most refused to participate. Therefore, our sample is only based on male entrepreneurs. The partial least square (PLS)-structural equation modeling (SEM) technique analyzed the proposed research model using Smart-PLS v3. Smart-PLS is a powerful tool used to test mediation–moderation models and works with multivariate and normal distributions simultaneously [103,104]. Additionally, it is helpful for measuring the validity and reliability of studies.

3.1. Demographics of Respondents

Table 1 describes the sample statistic frequency distribution of the targeted respondents. The sample statistics include age, qualification, the business sector of an entrepreneur, and the business tenure. The results show that most of the respondents fall in the age group of 33–39 years old (31.48%), while 16.25% of the respondents belong to the age group 26–32, and only 9.87% are mostly young entrepreneurs below 25 years. A total of 17.9% of the respondents are 40–46, and the remaining 24.48% are senior entrepreneurs above 47 years. Most of the respondents are highly qualified, and only 25.92% have attained a middle school certificate. During the data collection phase, we discovered that most senior entrepreneurs do not have higher education and yet are running a successful enterprise. This may be due to their leadership abilities, financial support, or many other reasons. The textile industry in Kinshasa consists of several sub-units such as knitting, weaving, seizing, power looms, and manual drying units. Therefore, we considered all these units for data collection, and the percentages are presented in Table 1. Furthermore, a question related to their experience is also described in the same table.

Table 1. The sample statistics of the respondents.

Particulars	Description	Frequency	Percentage
Gender	Male	486	100%
Age	18–25	48	9.87%
	26–32	79	16.25%
	33–39	153	31.48%
	40–46	87	17.90%
	47 Above	119	24.48%
Educational Qualification	Middle School	126	25.92%
	High School	159	32.71%
	Graduation Level	117	24.07%
	University Level	67	13.78%
	Professional Education	17	3.49%
Business Sector	Knitting	147	30.25%
	Weaving	84	17.28%
	Seizing	79	16.26%
	Power Looms	93	19.14%
	Manual Drying Units	83	17.08%
Business Tenure	1–5 years	74	15.22%
	6–10 years	127	26.13%
	11–15 years	126	25.92%
	16–20 years	87	17.90%
	21–25 years	34	6.99%
	25 years above	38	7.81%

3.2. The Measures

A structured questionnaire was developed to ensure that the content of the research model was practical and realistic. All variables were constructed and operationalized using the existing literature on sustainable entrepreneurial performance, knowledge sharing capacity, absorptive capacity, dynamic capability, innovative capacity, and opportunity recognition. A 5-point Likert scale ranging from 1 (strongly disagree) to 5 (strongly agree) was employed to measure all constructs. Five items were adapted from the study by Hsu [105] to assess information sharing abilities. We used five measuring constructs from Hurley's [106] study to evaluate inventive capability as an exemplary item, "I frequently participate in knowledge sharing activities." To test absorptive capacity, four items were used [107]. A representative item was "risk-taking is encouraged in our firm." The item "our firm regularly considers the consequences of changing market demand in terms of new ways to provide services" was used to test the dynamic capability measured using two dimensions: exploration and exploitation, with three items each. This scale was adapted from the study by [108,109].

3.3. Measurement Model

Table 2 shows the results of the convergent validity and reliability analysis of the data collected from the respondents. To confirm the convergent validity, we used Smart-PLS3 to conduct confirmatory factor analysis (CFA), composite reliability (CR), and average variance extracted (AVE). Cronbach's alpha values were also checked to ensure reliability. The overall values of Cronbach's alpha range from 0.936 to 0.953, which is higher than the threshold value, according to Table 2. The levels of CR and AVE are also higher than the prescribed values, validating the study's reliability and convergent validity [104,110]. This study also looked at discriminant validity, which is the degree to which components differ experimentally from one another [111]. The criterion for discriminant validity analysis is shown in Table 3. The results suggest that discriminant validity is not a problem because the diagonal values (square root of AVE) are higher than the inter-construct correlations, as advised by [112].

Table 2. Convergent validity and reliability.

Constructs		Factor Loading	Alpha	CR	AVE
KNOWLEDGE SHARING BEHAVIOR	KSB1	0.942	0.953	0.953	0.804
	KSB2	0.914			
	KSB3	0.899			
	KSB4	0.804			
	KSB5	0.917			
INNOVATIVE CAPACITY	IC1	0.922	0.936	0.935	0.743
	IC2	0.872			
	IC3	0.852			
	IC4	0.866			
	IC5	0.793			
ABSORPTIVE CAPACITY	AC1	0.965	0.936	0.936	0.785
	AC2	0.826			
	AC3	0.855			
	AC4	0.893			
DYNAMIC CAPABILITY	DC1	0.822	0.945	0.946	0.746
	DC2	0.759			
	DC3	0.900			
	DC4	0.915			
	DC5	0.880			

Table 2. *Cont.*

Constructs		Factor Loading	Alpha	CR	AVE
OPPORTUNITY RECOGNITION	OR1	0.774	0.942	0.940	0.725
	OR2	0.902			
	OR3	0.929			
	OR4	0.897			
	OR5	0.819			
	OR6	0.773			
ENTREPRENEURIAL PERFORMANCE	EP1	0.726	0.950	0.949	0.630
	EP2	0.765			
	EP3	0.825			
	EP4	0.812			
	EP5	0.812			
	EP6	0.955			
	EP7	0.786			
	EP8	0.768			
	EP9	0.747			
	EP10	0.767			
	EP11	0.747			

Table 3. Fornell–Lacker criterion discriminant validity.

	AC	DC	SEP	IC	KSB	OR
AC	0.886					
DC	0.427	0.864				
SEP	0.435	0.415	0.794			
IC	0.339	0.371	0.447	0.862		
KSB	0.553	0.427	0.453	0.453	0.897	
OR	0.237	0.366	0.379	0.186	0.247	0.851

Note: Diagonal values are the square root of the average variance extracted from each construct.

Furthermore, heterotrait–monotrait ratio (HTMT) analysis for discriminant validity was also applied [113]. A value of the HTMT ratio closer to one indicates a lack of discriminant validity in the path analysis [112]. To clearly distinguish the two factors, the HTMT ratio should be less than one [112]. The current study results shown in Table 4 show that the values are in accordance with the threshold values. Therefore, we can conclude that there is no issue of discriminant validity at all.

Table 4. Heterotrait–monotrait (HTMT) ratios.

	AC	DC	EP	IC	KSB
DC	0.428				
SEP	0.432	0.410			
IC	0.337	0.370	0.446		
KSB	0.551	0.427	0.453	0.450	
OR	0.233	0.366	0.378	0.186	0.245

3.4. Structural Model

To test the hypotheses, we applied PLS-SEM in the current study. Figure 2 shows the results of the path analysis, which are also described in Table 5. The value of the adjusted R-square of the dependent variable is 0.402, showing that these selected variables explain a total of 40% of the variation. Nevertheless, this study considered DC as a mediator, showing a 26.5% variation. The consistent bootstrapping test was applied for confirming the significance of the structural model [97].

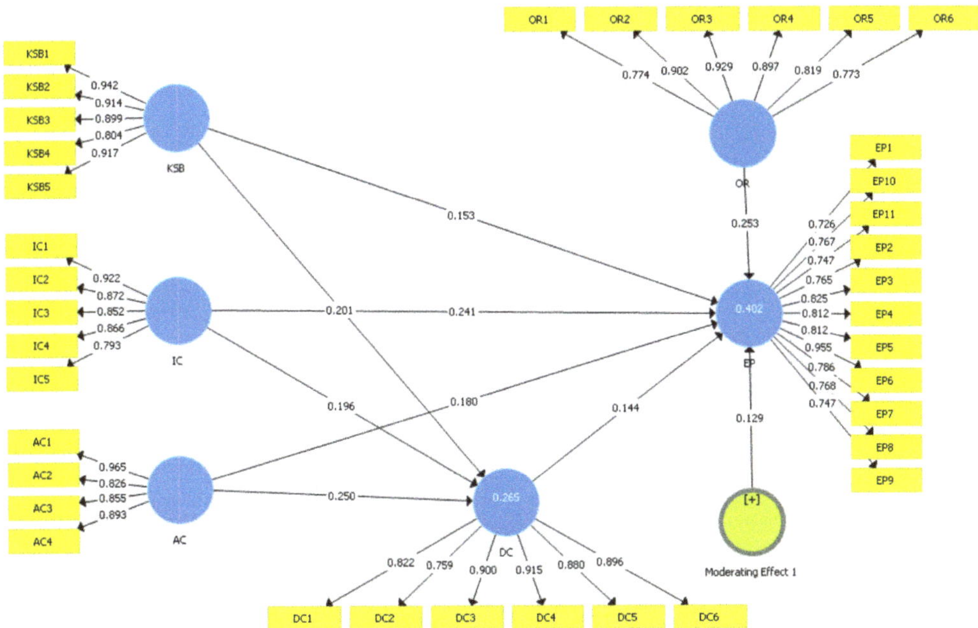

Figure 2. Results of path analysis.

Table 5. SEM results with bootstrapping (total direct effect).

Hypothesis	Relationship	β	S.D	t-Values	p-Values	Decision
H1	KSB → DC	0.201	0.048	4.194	0.000	Supported
H2	KSB → SEP	0.153	0.060	2.541	0.011	Supported
H4	IC → DC	0.196	0.045	4.378	0.000	Supported
H5	IC → EP	0.241	0.049	4.925	0.000	Supported
H7	AC → DC	0.250	0.053	4.696	0.000	Supported
H8	AC → SEP	0.180	0.054	3.339	0.001	Supported
H10	DC → SEP	0.144	0.051	2.814	0.005	Supported

According to the results of Table 5, H1 showed a direct positive effect of KSB on DC, and therefore H1 is supported ($\beta = 0.201$; $t = 4.194$; $p < 0.000$) with the direct positive and significant relationship between KSB and DC. H2 demonstrated a direct positive effect of KSB on SEP, and therefore H2 is supported ($\beta = 0.153$; $t = 2.541$; $p < 0.011$), indicating that KSB has a positive and significant impact on SEP. H4 explained a direct positive effect of IC on DC, and therefore H4 is supported ($\beta = 0.196$; $t = 4.378$; $p < 0.000$), indicating that IC has a positive and significant effect on DC. Meanwhile, H5 also showed a direct and positive effect of IC on EP, and therefore H5 is supported ($\beta = 0.241$; $t = 4.925$; $p < 0.000$), indicating that IC has a positive and significant effect on EP. H7 also explored a direct positive effect of AC on DC, and therefore H7 is supported ($\beta = 0.250$; $t = 4.696$; $p < 0.000$), indicating that AC has a direct and significant effect on DC. At the same time, H8 also showed a direct positive effect of AC on SEP; therefore, H8 is supported ($\beta = 0.180$; $t = 3.339$; $p < 0.001$), showing a positive and significant effect between AC and SEP. The last direct effect of H10 showed a positive direction of DC on SEP; therefore, H10 is supported ($\beta = 0.144$; $t = 2.814$; $p < 0.005$), showing that DC has a positive and significant impact on SEP.

Table 6 shows the indirect effects of KSB, IC, and AC on sustainable entrepreneurial performance through DC and the moderating effect of OR on the relationship between DC and SEP. This study also measured the mediating and moderating role of DC and

OR. Table 6 represents the values of the SEM results for the specific indirect effects. The results for H3 confirm that DC mediates the relationship between KSB and SEP; therefore, H3 ($\beta = 0.029$; $t = 2.204$; $p < 0.028$) is supported, showing that there is partial mediation between KSB and SEP through DC. H6 showed that DC mediates the relationship between IC and SEP; therefore, H6 is supported with values of ($\beta = 0.028$; t = 2.270; $p < 0.023$), showing partial mediation between IC and SEP.

Table 6. SEM results with bootstrapping (specific indirect effect).

Hypotheses	Constructs	β	(SD)	*t*-Values	*p*-Values	Decision
H3	KSB → DC → SEP	0.029	0.013	2.204	0.028	Partially mediates
H6	IC → DC → SEP	0.028	0.012	2.270	0.023	Partially mediates
H9	AC → DC → SEP	0.036	0.015	2.395	0.017	Partially mediates
H11	ORxDC → SEP	0.129	0.030	4.269	0.000	Moderation proved

Meanwhile, H9 confirms that DC mediates the relationship between AC and SEP; therefore, H3 ($\beta = 0.036$; $t = 2.395$; $p < 0.017$) is supported, showing partial mediation between AC and SEP through DC. Moreover, this study considers the moderating effect of OR on the relationship between the DC and SEP of textile-based SMEs in Kinshasa, Congo. The results are presented in Table 6, showing that OR significantly and positively moderates the relationship between DC and SEP ($\beta = 0.129$; $t = 4.269$; $p < 0.000$). Figure 3 represents the moderation effect of OR on SEP, showing that OR significantly strengthened the positive relationship of DC and SEP.

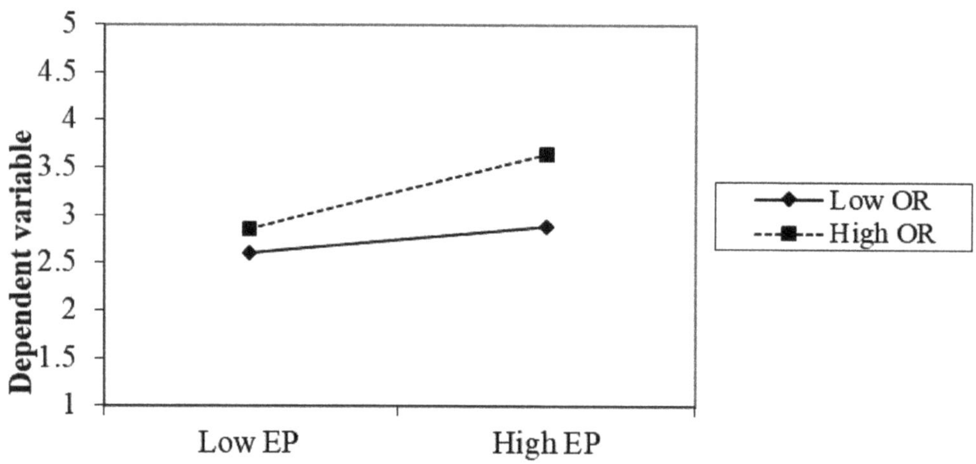

Figure 3. Moderation result.

4. Discussion

This research explored the impact of knowledge management practices on sustainable entrepreneurship performance, with mediating and moderating effects of dynamic capabilities and opportunity recognition. The path coefficient supports the provided hypotheses empirically and identifies significant findings with *p*-value < 0.05 and *t*-value > 2.

Based on the overall statistical results of our study, H1 for knowledge sharing behavior provides a significant effect on DC, which is in line with [114]. This means that KSB can spread important information within the organization, which becomes a valuable asset for sustainable performance [107,115]. KSB can increase the tendency to understand the organizational domestic and economic challenges an entrepreneur faces in sustainable performance. Knowledge sharing behavior demonstrates the effect of dynamic capabilities, which help to determine, integrate, and reconfigure internal and external capabilities for better sustainable performance. The result of H2 shows that knowledge sharing behavior has a good association with sustainable entrepreneurship performance. The findings are also consistent with those of earlier investigations by [116]. H3 showed an indirect effect of dynamic capabilities on the relationship between knowledge sharing behavior and sustainable entrepreneurial performance. Knowledge sharing behavior enhances the sustainable entrepreneurial performance through an indirect relation with dynamic capabilities. This finding is similar to that of previous studies by [105].

The finding of H4 suggests that innovative capacity influences dynamic capabilities, and this conclusion is similar to that of previous investigations [114]. The conclusion of H5 shows that the effect of innovative capacity on sustainable entrepreneurial performance is favorable, and the findings are similar to those of a previous study by Jantunen et al. [116]. The results of H6 suggest that innovative capacity positively influences sustainable entrepreneurial performance through dynamic capabilities, which is linked to Furman's [102] previous study. When entrepreneurs face adversity, innovative capacity contributes to acquiring, creating, and applying inner values. Innovative capacity boosts decision making power and leadership abilities, assists as a financial adviser in the organization, is attentive to organization, awareness, and allocation of better opportunities with better substitutes, and benefits sustainable entrepreneurial performance [75]. Furthermore, H6 confirms that dynamic capacity positively mediates the association between innovative capacity and sustainable entrepreneurial performance. These results are consistent with those of previous studies by [42].

Additionally, the study of H7 revealed that absorptive capacity has positive influences on dynamic capabilities, and these findings are consistent with those of previous research [117]. Dynamic capability reorganizes and modifies the organizational environment which is closely linked to sustainable entrepreneurial performance [71]. Dynamic capability is the most dependable and sound source for gaining competitive advantages, and it acts as a link between entrepreneurial resources and SEP [13,70]. Nonetheless, H8 claims that absorption capacity positively impacts long-term entrepreneurship success, and the findings are linked to those of [31]. According to H9, dynamic capability favorably and significantly mediates between absorptive capacity and sustainable entrepreneurial performance [105,118].

The H10 dynamic capabilities positively impact sustainable entrepreneurial performance [119,120]. Finally, H11's findings reveal that opportunity recognition moderates the relationship between dynamic capacity and sustainable entrepreneurial performance. The data show a significant and favorable moderation effect on the relationship between dynamic talents and sustainable entrepreneurial success [121]. Several studies have found that self-made approaches used by entrepreneurs are critical in the opportunity recognition process [85]. Because there has been less research on the link between DC and SEP, this study combined OR to see how it affects the link.

5. Implications

This study contributes to the growing body of knowledge about SME performance, dynamic capabilities, potential opportunities, and knowledge management capabilities. The results of the research mentioned above help lower the failure rate of firms, which is better understood by the government and non-government textile sectors. Secondly, this study reveals that textile-based SMEs with a low performance can benefit significantly from this study's findings. Furthermore, this research aids SMEs in developing more effective methods of knowledge transfer to foster a strong organizational climate that can better compete against competitors. The lack of internal and external information that the SME faces can impact the company's long-term performance. Using the dynamic capacities of the organization, SMEs can also develop their organizational and entrepreneurial potential. Furthermore, this research has broader implications for an industrial practitioner in the field of small–medium performance toward substantial firm and entrepreneurial performance.

6. Conclusions

This study extends the existing literature by exploring the importance of knowledge management practices with sustainable entrepreneurship performance, opportunity recognition, and dynamic capabilities of textile-based SMEs. Several studies have looked at how knowledge management techniques affect SME sustainability, but little attention has been paid to dynamic capabilities' vital role. As a result, the impact of opportunity recognition and SMEs' dynamic capabilities connected to knowledge management methods on sustainable entrepreneurship performance is important.

This study's findings indicate that knowledge management methods had a considerable impact on SME entrepreneurial performance, as evidenced by the significant beta coefficient, t-values, and p-values. Furthermore, the findings reveal that dynamic capabilities are critical to SME performance and that opportunity recognition moderates the relationship between dynamic capabilities and entrepreneurial performance. These arguments show how knowledge management strategies help entrepreneurs perform better, which impacts unemployment and economic growth. In addition, other industries have underlined the need to understand existing attitudes toward green products. As a result, the textile sector should look at these rapidly growing consumer content strategies. Textile management teams must fulfill the demands of increasing customer satisfaction and loyalty. As a result, research into this new type of textile consumer is needed.

7. Limitations and Future Research

The present study has limitations concerning the data source and sampling, which affect the internal and external validity of the study. The data consist of a single source. This research used cross-sectional data; however, longitudinal data should be used in the future. This approach would improve the study in order to promote the success of Congo's SME industries. To obtain a higher performance in SMEs, a detailed and better conclusion for the research could include government policies and demographics as a control variable. Another limitation of this research is that, due to the chosen region, the sample population was confined to males. Gender biases may have influenced the findings of this study. Any future study should take both males and females into account.

Lastly, this study focused on male entrepreneurs in the Congolese textile industry. However, to be more inclusive, the study may include additional and different industries, such as enterprises with more male and female entrepreneurs. Furthermore, future research can be conducted on a similar pattern in a different time zone. It has also been stated that knowledge and innovation capacities are not constant, which may be enhanced with the situation developed. Hence, people can vary in their knowledge and learning abilities throughout time. Future researchers should undertake a longitudinal study on the spectrum that is outlined in this research.

Author Contributions: Conceptualization, S.M.T. and N.W.; methodology, S.M.T. and M.N.; validation, N.W. and S.M.T.; formal analysis, M.N.; investigation, S.M.T.; resources, N.W.; data curation, S.M.T.; writing original draft preparation, S.M.T.; writing review and editing, S.F.A. and N.S.; visualization, S.M.T.; supervision, N.W.; project administration, N.W. and S.M.T.; funding acquisition, N.W. All authors have read and agreed to the published version of the manuscript.

Funding: This research was funded by The National Natural Science Foundation of China, grant numbers 71971101 and 71972090, and the key project of philosophy and social science research in colleges and universities of Jiangsu Province, grant number 2019SJZDA032.

Institutional Review Board Statement: Not Applicable.

Informed Consent Statement: Not Applicable.

Data Availability Statement: The datasets used in this study are available in the text and cited in the Reference section.

Acknowledgments: We are grateful to Vivian Andoh for her immense support to this paper.

Conflicts of Interest: The authors declare that the research was conducted in the absence of any commercial or financial relationships that could be constructed as a potential conflict of interest.

Abbreviations

Knowledge sharing behavior (KSB)	KS behavior is defined as "a set of individual behaviors involving sharing one's work-related knowledge and expertise with other members within one's organization" [15].
Innovation capacity (IC)	Innovation capacity is defined as continually improving firms' capabilities and resources to discover opportunities to develop new products [4,15].
Absorptive capacities (AC)	The ability of firms to recognize, assimilate and apply new knowledge for the benefit of their business performance [41].
Dynamic capabilities (DC)	Dynamic capabilities are defined as "the firm's ability to integrate, build, and reconfigure internal and external competencies to address rapidly changing environments" [42].
Opportunity recognition (OR)	Opportunity recognition (OR) is how entrepreneurs identify potential ways to identify new business based on their opportunities. Opportunity recognition (OR) is how entrepreneurs identify potential ways towards identifying new businesses based on the opportunities they identify [43,44].
Sustainability entrepreneurship performance (SEP)	Sustainable performance of an organization refers to its ability to meet the needs and expectations of customers and other stakeholders in the long-term, balanced by an effective management organization by organization staff awareness by learning and applying appropriate improvements and innovation [20].

References

1. Middermann, L.H.; Kratzer, J.; Perner, S. The impact of environmental risk exposure on the determinants of sustainable entrepreneurship. *Sustainability* **2020**, *12*, 1534. [CrossRef]
2. Abubakar, A.M.; Elrehail, H.; Alatailat, M.A.; Elçi, A. Knowledge management, decision-making style and organizational performance. *J. Innov. Knowl.* **2019**, *4*, 104–114. [CrossRef]
3. Li, C.; Ashraf, S.F.; Shahzad, F.; Bashir, I.; Murad, M.; Syed, N.; Riaz, M. Influence of knowledge management practices on entrepreneurial and organizational performance: A mediated-moderation model. *Front. Psychol.* **2020**, *11*, 2862. [CrossRef]
4. Makhloufi, L.; Laghouag, A.; Sahli, A.A.; Belaid, F. Impact of entrepreneurial orientation on innovation capability: The mediating role of absorptive capability and organizational learning capabilities. *Sustainability* **2021**, *13*, 5399. [CrossRef]
5. Antunes, H.D.J.G.; Pinheiro, P.G. Linking knowledge management, organizational learning and memory. *J. Innov. Know.* **2020**, *5*, 140–149. [CrossRef]
6. Cerchione, R.; Esposito, E. A systematic review of supply chain knowledge management research: State of the art and research opportunities. *Int. J. Prod. Econ.* **2016**, *182*, 276–292. [CrossRef]
7. Zheng, Y.; Han, W.; Yang, R. Does government behaviour or enterprise investment improve regional innovation performance-Evidence from China. *Int. J. Technol. Manag.* **2021**, *85*, 274. [CrossRef]
8. Lotfi, M.; Yousefi, A.; Jafari, S. The effect of emerging green market on green entrepreneurship and sustainable development in knowledge-based companies. *Sustainability* **2018**, *10*, 2308. [CrossRef]
9. Schaltegger, S.; Wagner, M. Sustainable entrepreneurship and sustainability innovation: Categories and interactions. *Bus. Strat. Environ.* **2011**, *20*, 222–237. [CrossRef]

10. Martins, V.; Rampasso, I.; Anholon, R.; Quelhas, O.; Filho, W.L. Knowledge management in the context of sustainability: Literature review and opportunities for future research. *J. Clean. Prod.* **2019**, *229*, 489–500. [CrossRef]
11. Vickers, N.J. Animal communication: When i'm calling you, will you answer too? *Curr. Biol.* **2017**, *27*, R713–R7152017. [CrossRef]
12. Lüdeke-Freund, F. Sustainable entrepreneurship, innovation, and business models: Integrative framework and propositions for future research. *Bus. Strat. Environ.* **2020**, *29*, 665–681. [CrossRef]
13. Pai, F.-Y.; Chang, H.-F. The effects of knowledge sharing and absorption on organizational innovation performance–A dynamic capabilities perspective. *Interdiscip. J. Info. Know. Manag.* **2013**, *8*, 083–097. [CrossRef]
14. Liu, S.M.; Hu, R.; Kang, T.W. The effects of absorptive capability and innovative culture on innovation performance: Evidence from Chinese high-tech firms. *J. Asian Financ. Eco. Bus.* **2021**, *8*, 1153–1162.
15. Kurniawan, P.; Hartati, W.; Qodriah, S.L.; Badawi, B. From knowledge sharing to quality performance: The role of absorptive capacity, ambidexterity and innovation capability in creative industry. *Manag. Sci. Lett.* **2020**, 433–442. [CrossRef]
16. Chang, D.L.; Marques, J.S.; Da Costa, E.M.; Selig, P.M.; Yigitcanlar, T. Knowledge-based, smart and sustainable cities: A provocation for a conceptual framework. *J. Open Innov. Technol. Mark. Complex.* **2018**, *4*, 1–17. [CrossRef]
17. Van Reijsen, J.; Helms, R.; Batenburg, R.; Foorthuis, R. The impact of knowledge management and social capital on dynamic capability in organizations. *Know. Manag. Res. Pract.* **2009**, *13*, 401–417. [CrossRef]
18. Mohamed, M.; Stankosky, M.; Mohamed, M. An empirical assessment of knowledge management criticality for sustainable development. *J. Know. Manag.* **2009**, *13*, 271–286. [CrossRef]
19. Bucci, M.; El-Diraby, T.E. The functions of knowledge management processes in urban impact assessment: The case of Ontario. *Impact Asses. Proj. Apprais.* **2018**, *36*, 265–280. [CrossRef]
20. Youssef, A.B.; Boubaker, S.; Omri, A. Entrepreneurship and sustainability: The need for innovative and institutional solutions. *Technol. Forecast. Soc. Chang.* **2018**, *129*, 232–241.
21. Acs, Z.J.; Estrin, S.; Mickiewicz, T.; Szerb, L. Entrepreneurship, institutional economics, and economic growth: An ecosystem perspective. *Small Bus. Econ.* **2018**, *51*, 501–514. [CrossRef]
22. De Clercq, D.; Honig, B.; Martin, B. The roles of learning orientation and passion for work in the formation of entrepreneurial intention. *Int. Small Bus. J. Res. Entrep.* **2012**, *31*, 652–676. [CrossRef]
23. González-Serrano, M.H.; Sanz, V.A.; González-García, R.J. Sustainable sport entrepreneurship and innovation: A bibliometric analysis of this emerging field of research. *Sustainability* **2020**, *12*, 5209. [CrossRef]
24. Terán-Yépez, E.; Marín-Carrillo, G.M.; Casado-Belmonte, M.D.P.; Capobianco-Uriarte, M.D.L.M. Sustainable entrepreneurship: Review of its evolution and new trends. *J. Clean. Prod.* **2019**, *252*, 119742. [CrossRef]
25. Shane, S.; Locke, E.A.; Collins, C.J. Entrepreneurial motivation. *Hum. Resour. Manag. Rev.* **2003**, *13*, 257–279. [CrossRef]
26. Veronica, S.; Manlio, d.G.; Shlomo, T.; Antonio, M.P.; Victor, C. International social SMEs in emerging countries: Do governments support their international growth? *J. World Bus.* **2019**, *55*, 100995. [CrossRef]
27. Urban, B.; Kongo, M. The relevance of human capital to firm performance: A focus on the retail industry in Kinshasa, Democratic Republic of Congo. *Acta Commer.* **2015**, *15*, 1–9. [CrossRef]
28. Ngassa, T.C. The influence of entrepreneurship training on profit gaps amongst young male and female entrepreneurs in congo. *Mod. Econ.* **2021**, *12*, 1092–1104. [CrossRef]
29. Awan, U.; Khattak, A.; Rabbani, S.; Dhir, A. Buyer-driven knowledge transfer activities to enhance organizational sustainability of suppliers. *Sustainability* **2020**, *12*, 2993. [CrossRef]
30. Hasan, F.S.; Almubarak, M.M.S. Factors influencing women entrepreneurs' performance in SMEs. *World J. Entrep. Manag. Sustain. Develop.* **2016**, *12*. [CrossRef]
31. Gebauer, H.; Worch, H.; Truffer, B. Absorptive capacity, learning processes and combinative capabilities as determinants of strategic innovation. *Eur. Manag. J.* **2012**, *30*, 57–73. [CrossRef]
32. Al-Husseini, S.; Elbeltagi, I. Evaluating the effect of transformational leadership on knowledge sharing using structural equation modelling: The case of Iraqi higher education. *Int. J. Leader. Educ.* **2018**, *21*, 506–517. [CrossRef]
33. Azaizah, N.; Reychav, I.; Raban, D.R.; Simon, T.; McHaney, R. Impact of ESN implementation on communication and knowledge-sharing in a multi-national organization. *Int. J. Inf. Manag.* **2018**, *43*, 284–294. [CrossRef]
34. Shao, Z.; Feng, Y.; Liu, L. The mediating effect of organizational culture and knowledge sharing on transformational leadership and Enterprise Resource Planning systems success: An empirical study in China. *Comput. Hum. Behav.* **2012**, *28*, 2400–2413. [CrossRef]
35. Vargas, N.; Lloria, M.B.; Salazar, A.; Vergara, L. Effect of exploitation and exploration on the innovative as outcomes in entrepreneurial firms. *Int. Entrep. Manag. J.* **2018**, *14*, 1053–1069. [CrossRef]
36. Rahmani, B.; Morid, S.P.; Shahed, S.H. Tourism capacity in the development of sustainable entrepreneurship in rural areas (Case Study: Central District of Hamadan). *J. Urban. Manag.* **2018**, *17*, 65–97.
37. Hakuzimanaa, J.; Masasib, B. Performance evaluation of irrigation schemes in Rugeramigozi marshland, Rwanda. *Water Conserv. Manag.* **2020**, *4*, 15–19. [CrossRef]
38. Lane, P.; Salk, J.; Lyles, M.J.S.M.J. Knowledge acquisition and performance in transitional economy international joint ventures. *Strateg. Manag. J.* **2001**, *22*, 97–102.
39. Feng, T.; Cai, D.; Wang, D.; Zhang, X. Environmental management systems and financial performance: The joint effect of switching cost and competitive intensity. *J. Clean. Prod.* **2016**, *113*, 781–791. [CrossRef]

40. Revuelto-Taboada, L.; Simón-Moya, V. A comprehensive understanding of social and sustainable entrepreneurship. *Manag. Decis.* **2012**, *50*, 744–748.
41. Giniuniene, J.; Jurksiene, L. Dynamic capabilities, innovation and organizational learning: Interrelations and impact on firm performance. *Procedia Soc. Behav. Sci.* **2015**, *213*, 985–991. [CrossRef]
42. Protogerou, A.; Caloghirou, Y.; Lioukas, S. Dynamic capabilities and their indirect impact on firm performance. *Ind. Corp. Chang.* **2011**, *21*, 615–647. [CrossRef]
43. Jiang, W.; Chai, H.; Shao, J.; Feng, T. Green entrepreneurial orientation for enhancing firm performance: A dynamic capability perspective. *J. Clean. Prod.* **2018**, *198*, 1311–1323. [CrossRef]
44. Cohen, B.; Winn, M.I. Market imperfections, opportunity and sustainable entrepreneurship. *J. Bus. Vent.* **2007**, *22*, 29–49. [CrossRef]
45. Clark, K.; Ramachandran, I. Subsidiary entrepreneurship and entrepreneurial opportunity: An institutional perspective. *J. Int. Manag.* **2019**, *25*, 37–50. [CrossRef]
46. Chung, L.H.; Gibbons, P.T. Management, Corporate entrepreneurship: The roles of ideology and social capital. *Group Organ. Manag.* **1997**, *22*, 10–30. [CrossRef]
47. Yi, G. From green entrepreneurial intentions to green entrepreneurial behaviors: The role of university entrepreneurial support and external institutional support. *Int. Entrepreneur. Manag. J.* **2021**, *17*, 963–979. [CrossRef]
48. Kloepfer, K.; Castrogiovanni, G.J. Entrepreneurship: Venture creation subprocesses, subdomains, and interfaces. *Int. Entrep. Manag. J.* **2018**, *14*, 681–696. [CrossRef]
49. Gast, J.; Gundolf, K.; Cesinger, B. Doing business in a green way: A systematic review of the ecological sustainability entrepreneurship literature and future research directions. *J. Clean. Prod.* **2017**, *147*, 44–56. [CrossRef]
50. Domańska, A.; Żukowska, B.; Zajkowski, R.J.P.E. Green entrepreneurship as a connector among social, environmental and economic pillars of sustainable development. Why some countries are more agile? *Probl. Ekorozw.* **2018**, *13*, 67–76.
51. Bae, T.J.; Qian, S.; Miao, C.; Fiet, J.O. The relationship between entrepreneurship education and entrepreneurial intentions: A meta–analytic review. *Entrep. Theory Pract.* **2014**, *38*, 217–254. [CrossRef]
52. Fayolle, A.; Liñán, F.; Moriano, J.A. Beyond entrepreneurial intentions: Values and motivations in entrepreneurship. *Int. Entrep. Manag. J.* **2014**, *10*, 679–689. [CrossRef]
53. Muñoz, P.; Kibler, E.; Mandakovic, V.; Amorós, J.E. Local entrepreneurial ecosystems as configural narratives: A new way of seeing and evaluating antecedents and outcomes. *Res. Policy* **2020**, 104065. [CrossRef]
54. Schumpeter, J.A. *The Economics and Sociology of Capitalism*; Princeton University Press: Princeton, NJ, USA, 1991.
55. Grant, R. The Resource-Based Theory of Competitive Advantage: Implications for Strategy Formulation. *Calif. Manag. Rev.* **1999**, *10*, 3–23. [CrossRef]
56. Alvarez, S.A.; Busenitz, L.W. The entrepreneurship of resource-based theory. *J. Manag.* **2001**, *27*, 755–775. [CrossRef]
57. Smirat, I.M.H.; Mohd Shariff, M.N. Research, Strategy, structure, and family firm performance: The relationships of the resource-base view and the contingency approach. *Aust. J. Bus. Manag. Res.* **2014**, *4*, 1.
58. Schmitt, A.; Rosing, K.; Zhang, S.X.; Leatherbee, M. A dynamic model of entrepreneurial uncertainty and business opportunity identification: Exploration as a mediator and entrepreneurial self-efficacy as a moderator. *Entrep. Theory Pract.* **2017**, *42*, 835–859. [CrossRef]
59. Gerlach, A. Management, Sustainable entrepreneurship and innovation. *Soc. Responsib. Environ. Manag.* **2003**, *29*.
60. Kane, G.C.; Palmer, D.; Phillips, A.N.; Kiron, D. Winning the digital war for talent. *MIT Sloan Manag. Rev.* **2017**, *58*, 17.
61. Srivastava, A.; Bartol, K.M.; Locke, E.A. Empowering leadership in management teams: Effects on knowledge sharing, efficacy, and performance. *Acad. Manag. J.* **2006**, *49*, 1239–1251. [CrossRef]
62. Käser, P.A.; Miles, R.E. Understanding knowledge activists' successes and failures. *Long Range Plan.* **2002**, *35*, 9–28. [CrossRef]
63. Jilani, M.M.A.K.; Fan, L.; Islam, M.T.; Uddin, A. The influence of knowledge sharing on sustainable performance: A moderated mediation study. *Sustainability* **2020**, *12*, 908. [CrossRef]
64. Yan, J.; Meng, Y.; Yang, X.; Luo, X.; Guan, X. Privacy-preserving localization for underwater sensor networks via deep reinforcement learning. *IEEE Trans. Inf. Forensics Secur.* **2020**, *16*, 1880–1895. [CrossRef]
65. Abbas, J.; Hussain, I.; Hussain, S.; Akram, S.; Shaheen, I.; Niu, B. The impact of knowledge sharing and innovation upon sustainable performance in islamic banks: A mediation analysis through an SEM approach. *Sustainability* **2019**, *11*, 4049. [CrossRef]
66. Bock, G.W.; Kim, Y.G. Breaking the myths of rewards: An exploratory study of attitudes about knowledge sharing. *Info. Resour. Manag. J.* **2002**, *15*, 14–21. [CrossRef]
67. Setini, M.; Yasa, N.N.K.; Supartha, I.W.G.; Giantari, I.G.A.K.; Rajiani, I. The passway of women entrepreneurship: Starting from social capital with open innovation, through to knowledge sharing and innovative performance. *J. Open Innov. Technol. Mark. Complex.* **2020**, *6*, 25. [CrossRef]
68. Pezeshkan, A.; Fainshmidt, S.; Nair, A.; Frazier, M.L.; Markowski, E. An empirical assessment of the dynamic capabilities–performance relationship. *J. Bus. Res.* **2016**, *69*, 2950–2956. [CrossRef]
69. Schmitt, U. Supporting the sustainable growth of SMEs with content-and collaboration-based personal knowledge management systems. *J. Entrep. Innov. Emerg. Econ.* **2018**, *4*, 1–21. [CrossRef]

70. Mudalige, D.; Ismail, N.A.; Malek, M.A. Exploring the role of individual level and firm level dynamic capabilities in SMES' internationalization. *J. Int. Entrep.* **2018**, *17*, 41–74. [CrossRef]
71. Eikelenboom, M.; de Jong, G. The impact of dynamic capabilities on the sustainability performance of SMEs. *J. Clean. Prod.* **2019**, *235*, 1360–1370. [CrossRef]
72. Gao, H.; Hsu, P.-H.; Li, K.; Zhang, J. The real effect of smoking bans: Evidence from corporate innovation. *J. Financ. Quant. Anal.* **2018**, *55*, 387–427. [CrossRef]
73. Gonzalez, R.V.D. Effects of learning culture and teamwork context on team performance mediated by dynamic capability. *J. Knowl. Manag.* **2021**. [CrossRef]
74. Hameed, K.; Arshed, N.; Yazdani, N.; Munir, M. Motivating business towards innovation: A panel data study using dynamic capability framework. *Technol. Soc.* **2021**, *65*, 101581. [CrossRef]
75. Beltramino, N.S.; García-Perez-de-Lema, D.; Valdez-Juárez, L.E. The structural capital, the innovation and the performance of the industrial SMES. *J. Intel. Cap.* **2020**, *21*, 913–945. [CrossRef]
76. Sunny, S.A.; Shu, C. Investments, incentives, and innovation: Geographical clustering dynamics as drivers of sustainable entrepreneurship. *Small Bus. Econ.* **2017**, *52*, 905–927. [CrossRef]
77. Fernandes, C.I.I.; Veiga, P.M.; Peris-Ortiz, M.; Rueda-Armengot, C. What impact does innovation and sustainable entrepreneurship have on competitiveness? *Int. J. Soc. Ecol. Sustain. Dev.* **2017**, *8*, 56–66. [CrossRef]
78. Flatten, T.C.; Engelen, A.; Zahra, S.A.; Brettel, M. A measure of absorptive capacity: Scale development and validation. *Eur. Manag. J.* **2011**, *29*, 98–116. [CrossRef]
79. Kostopoulos, K.; Papalexandris, A.; Papachroni, M.; Ioannou, G. Absorptive capacity, innovation, and financial performance. *J. Bus. Res.* **2011**, *64*, 1335–1343. [CrossRef]
80. Sciascia, S.; D'Oria, L.; Bruni, M.; Larrañeta, B. Entrepreneurial Orientation in low- and medium-tech industries: The need for Absorptive Capacity to increase performance. *Eur. Manag. J.* **2014**, *32*, 761–769. [CrossRef]
81. Sulistyo, H. Siyamtinah Innovation capability of SMEs through entrepreneurship, marketing capability, relational capital and empowerment. *Asia Pac. Manag. Rev.* **2016**, *21*, 196–203. [CrossRef]
82. Augier, M.; Teece, D.J. Dynamic Capabilities and the Role of Managers in Business Strategy and Economic Performance. *Organ. Sci.* **2009**, *20*, 410–421. [CrossRef]
83. Albort-Morant, G.; Henseler, J.; Cepeda-Carrión, G.; Leal-Rodríguez, A.L. Potential and Realized Absorptive Capacity as Complementary Drivers of Green Product and Process Innovation Performance. *Sustainability* **2018**, *10*, 381. [CrossRef]
84. Fang, E.; Zou, S. Antecedents and consequences of marketing dynamic capabilities in international joint ventures. *J. Int. Bus. Stud.* **2009**, *40*, 742–761. [CrossRef]
85. Dzhengiz, T.; Niesten, E.; Dzhengiz, T. Competences for Environmental Sustainability: A Systematic Review on the Impact of Absorptive Capacity and Capabilities. *J. Bus. Ethics* **2019**, *162*, 881–906. [CrossRef]
86. Aboelmaged, M.; Hashem, G. Absorptive capacity and green innovation adoption in SMEs: The mediating effects of sustainable organisational capabilities. *J. Clean. Prod.* **2019**, *220*, 853–863. [CrossRef]
87. Zhai, Y.-M.; Sun, W.-Q.; Tsai, S.-B.; Wang, Z.; Zhao, Y.; Chen, Q. An Empirical Study on Entrepreneurial Orientation, Absorptive Capacity, and SMEs' Innovation Performance: A Sustainable Perspective. *Sustainability* **2018**, *10*, 314. [CrossRef]
88. Amui, L.B.L.; Jabbour, C.J.C.; Jabbour, A.B.L.D.S.; Kannan, D. Sustainability as a dynamic organizational capability: A systematic review and a future agenda toward a sustainable transition. *J. Clean. Prod.* **2017**, *142*, 308–322. [CrossRef]
89. Saunila, M. Innovation capability in SMEs: A systematic review of the literature. *J. Innov. Know.* **2020**, *5*, 260–265. [CrossRef]
90. Akter, S.; Jamal, N.; Ashraf, M.; McCarthy, G.; Varsha, P. The rise of the social business in emerging economies: A new paradigm of development. *J. Soc. Entrep.* **2019**, *11*, 282–299. [CrossRef]
91. Nave, A.; Franco, M. University-Firm cooperation as a way to promote sustainability practices: A sustainable entrepreneurship perspective. *J. Clean. Prod.* **2019**, *230*, 1188–1196. [CrossRef]
92. Mousavi, S.; Bossink, B.; van Vliet, M. Dynamic capabilities and organizational routines for managing innovation towards sustainability. *J. Clean. Prod.* **2018**, *203*, 224–239. [CrossRef]
93. Sibindi, N. Dynamic managerial capabilities as antecedent of corporate entrepreneurship: A conceptual model. *J. Entrep. Innov.* **2021**, 2. [CrossRef]
94. Rebs, T.; Thiel, D.; Brandenburg, M.; Seuring, S. Impacts of stakeholder influences and dynamic capabilities on the sustainability performance of supply chains: A system dynamics model. *J. Bus. Econ.* **2019**, *89*, 893–926. [CrossRef]
95. Inigo, E.A.; Albareda, L.; Ritala, P. Business model innovation for sustainability: Exploring evolutionary and radical approaches through dynamic capabilities. *Ind. Innov.* **2017**, *24*, 515–542. [CrossRef]
96. Hashim, N.A.B.; Raza, S.; Minai, M.S. Relationship between entrepreneurial competencies and small firm performance: Are dynamic capabilities the missing link? *Acad. Strat. Manag. J.* **2018**, *17*, 1–10.
97. Akkaya, B.; Üstgörül, S. Sustainability of SMEs and health sector in a dynamic capabilities perspective. *Chall. Oppor. SMEs Ind.* **2020**, 43–64. [CrossRef]
98. Asante, E.A.; Affum-Osei, E. Entrepreneurship as a career choice: The impact of locus of control on aspiring entrepreneurs' opportunity recognition. *J. Bus. Res.* **2019**, *98*, 227–235. [CrossRef]

99. Creswell, J.W.; Plano Clark, V.L.; Gutmann, M.L.; Hanson, W.E. An Expanded Typology for Classifying Mixed Methods Research into Designs. Available online: https://us.corwin.com/sites/default/files/upm-binaries/19291_Chapter_7.pdf (accessed on 8 August 2021).
100. Bryman, A. Barriers to integrating quantitative and qualitative research. *J. Mix. Meth. Res.* **2007**, *1*, 8–22. [CrossRef]
101. Brislin, R.W. Back-translation for cross-cultural research. *J. Cross. Psych.* **1970**, *1*, 185–216. [CrossRef]
102. Yusr, M.; Mokhtar, S.S.M.; Perumal, S.; Abdullateef, A.O.; Fathilah, R.; Yunus, N.K.Y. Managing absorptive capacity to enhance the influence of TQM practices on product innovation performance. *Int. J. Econ. Res.* **2017**, *14*, 325–332.
103. Hair, J.F.; Ringle, C.M.; Sarstedt, M. PLS-SEM: Indeed a silver bullet. *J. Market. Theory Pract.* **2011**, *19*, 139–152. [CrossRef]
104. Hair, J.F., Jr.; Sarstedt, M.; Hopkins, L.; Kuppelwieser, V.G. Partial least squares structural equation modeling (PLS-SEM): An emerging tool in business research. *Euro. Bus. Rev* **2014**, *26*, 106–121. [CrossRef]
105. Hsu, M.-H.; Ju, T.L.; Yen, C.-H.; Chang, C.-M. Knowledge sharing behavior in virtual communities: The relationship between trust, self-efficacy, and outcome expectations. *Int. J. Hum. Comput. Stud.* **2007**, *65*, 153–169. [CrossRef]
106. Hurley, R.F.; Hult, G.T.M. Innovation, market orientation, and organizational learning: An integration and empirical examination. *J. Mark.* **1998**, *62*, 42–54. [CrossRef]
107. Leal-Rodriguez, A.L.; Ariza-Montes, A.; Roldán, J.L.; Leal-Millan, A. Absorptive capacity, innovation and cultural barriers: A conditional mediation model. *J. Bus. Res.* **2014**, *67*, 763–768. [CrossRef]
108. Atuahene-Gima, K. Resolving the capability–rigidity paradox in new product innovation. *J. Market.* **2005**, *69*, 61–83. [CrossRef]
109. Kuckertz, A.; Kollmann, T.; Krell, P.; Stöckmann, C. Understanding, differentiating, and measuring opportunity recognition and opportunity exploitation. *Int. J. Entrep. Behav. Res.* **2017**, *23*, 78–97. [CrossRef]
110. Ab Hamid, M.R.; Sami, W.; Sidek, M.H.M. Discriminant Validity Assessment: Use of Fornell & Larcker criterion versus HTMT Criterion. *J. Phys. Conf. Ser.* **2017**, *890*, 012163. [CrossRef]
111. Fornell, C.; Larcker, D.F. Evaluating structural equation models with unobservable variables and measurement error. *J. Market. Res.* **1981**, *18*, 39–50. [CrossRef]
112. Henseler, J.; Hubona, G.S.; Ray, P.A. Using PLS path modeling in new technology research: Updated guidelines. *Ind. Manag. Data Syst.* **2016**, *116*, 2–20. [CrossRef]
113. Fornell, C.; Bookstein, F.L. Two structural equation models: LISREL and PLS applied to consumer exit-voice theory. *J. Market. Res.* **1982**, *19*, 440–452. [CrossRef]
114. Ferreira, J.; Coelho, A. Dynamic capabilities, innovation and branding capabilities and their impact on competitive advantage and SME's performance in Portugal: The moderating effects of entrepreneurial orientation. *Int. J. Innov. Sci.* **2020**, *12*, 255–286. [CrossRef]
115. Yeganerad, M.; Tahery, M. The mediating effect of organizational culture and knowledge sharing on transformational leadership and Enterprise Resource Planning (ERP) systems success at the University. *Res. Edu, Lead. Mana.* **2014**, *1*, 137–163.
116. Jantunen, A.; Puumalainen, K.; Saarenketo, S.; Kyläheiko, K. Entrepreneurial orientation, dynamic capabilities and international performance. *J. Int. Entrep.* **2005**, *3*, 223–243. [CrossRef]
117. Chaudhary, S.; Batra, S. Absorptive capacity and small family firm performance: Exploring the mediation processes. *J. Knowl. Manag.* **2018**, *22*, 1201–1216. [CrossRef]
118. Han, J.; Won, E.-J.; Kang, H.-M.; Lee, M.-C.; Jeong, C.-B.; Kim, H.-S.; Hwang, D.-S.; Lee, J.-S. Marine copepod cytochrome P450 genes and their applications for molecular ecotoxicological studies in response to oil pollution. *Mar. Pollut. Bull.* **2017**, *124*, 953–961. [CrossRef] [PubMed]
119. Pezeshkan, A.; Smith, A.; Fainshmidt, S.; Sedeh, A.A. National business systems and firm innovation: A study of developing economies. *J. Bus. Res.* **2016**, *69*, 5413–5418. [CrossRef]
120. Bamel, U.K.; Bamel, N. Organizational resources, KM process capability and strategic flexibility: A dynamic resource-capability perspective. *J. Knowl. Manag.* **2018**, *22*, 1555–1572. [CrossRef]
121. Sanz-Velasco, S.A. Opportunity development as a learning process for entrepreneurs. *Int. J. Entrep. Behav. Res* **2006**, *12*, 251–271. [CrossRef]

sustainability

Article

Integration of the Circular Economy Paradigm in Companies from the Northwest of the Iberian Peninsula

Maria Manuel Sá [1,2,3,*], Carla Oliveira-Silva [1,4], Manuel Paulo Cunha [1], Artur Gonçalves [5], Jesús Diez [6], Ines Méndez-Tovar [6] and Eva Curto Izquierdo [7]

1 Unidade de Investigação em Ciências Empresariais (UNICES), University of Maia, Av. Carlos Oliveira Campos, 4475-690 Maia, Portugal; cosilva@ismai.pt (C.O.-S.); mpcunha@ismai.pt (M.P.C.)
2 Geo-Space Sciences Research Centre (CICGE), 4169-007 Porto, Portugal
3 Research Centre for Health and Environment (CISA), 4200-072 Porto, Portugal
4 Research Centre for Business Sciences (NECE-UBI), 6200-209 Covilhã, Portugal
5 Centro de Investigação de Montanha (CIMO), Instituto Politécnico de Bragança, Campus de Santa Apolónia, 5300-253 Bragança, Portugal; ajg@ipb.pt
6 Fundación del Patrimonio Natural de Castilla y León, 47008 Valladolid, Spain; jesus.diez@patrimonionatural.org (J.D.); ines.mendez@patrimonionatural.org (I.M.-T.)
7 Escuela de Organización Industrial, Área de Innovación Emprendimiento y Pymes, 28040 Madrid, Spain; evacurto@eoi.es
* Correspondence: maria.sa@ismai.pt

Abstract: Over recent decades, Circular Economy (CE) has become a major topic when organizations try to develop their business amid the constrains of resource limitation and the desire to reduce their environmental impact. This study's main purpose is to assess the integration of CE practices in public and private organizations in the northwest of the Iberian Peninsula. Through an online survey distributed to 294 companies from the cited region, we assessed their perceptions on CE, including such aspects as the area(s) it was integrated in, why, with what difficulties or what was necessary to accomplish it, and how the impact of the implemented CE practices was measured. Results showed that companies associate CE mostly with "resource optimization". "Entity's vision and mission" was the main strategic area where CE was implemented. The main motivation why entities/organizations embraced CE was "environmental reasons", while "lack of information and guidance" and "lack of financial resources" represented the main obstacles to CE implementation. Non-parametrical statistical tests were used to compare the answers of three groups of people with different positions within the company/entity (manager, executive, and technician), as well as to compare the answers of two activity sectors (industry and services).

Keywords: circular economy; industry/services strategies; manager/executive/technicians' perceptions

Citation: Sá, M.M.; Oliveira-Silva, C.; Cunha, M.P.; Gonçalves, A.; Diez, J.; Méndez-Tovar, I.; Izquierdo, E.C. Integration of the Circular Economy Paradigm in Companies from the Northwest of the Iberian Peninsula. *Sustainability* **2022**, *14*, 7940. https://doi.org/10.3390/su14137940

Academic Editors: Sergio Terzi and Claudio Sassanelli

Received: 20 May 2022
Accepted: 27 June 2022
Published: 29 June 2022

1. Introduction

The concept of Circular Economy (CE) was widely disseminated and promoted by British sailor Ellen MacArthur, when in 2005, in a lonely circum-navigation trip around the globe, the activist had to reuse what few resources there were available in the boat, intensifying her consciousness of the importance of their preservation [1,2]. This awareness propelled her to create a Foundation, named after herself, in 2010 [3], emerging soon after as a strong advocate for and the face of Circular Economy.

Initially, CE as a concept dealt more with how materials were kept, restored, and reintroduced cyclically in the value chain, so as to create economic advantages to suppliers and users, as well as benefitting the environment, thanks to the decreased extraction and importation of components and source-materials [4]. CE was seen as a key element to promote the decoupling between economic growth and increased consumption of resources, since, according to the circular approach, economic value is extracted from the materials already in circulation in the economy.

According to the European Commission [5], "in circular economy the value of products and materials is maintained for as long as possible; waste and resource use are minimized, and resources are kept within the economy when a product has reached the end of its life, to be used again and again to create further value". With CE, any economic activity should operate in a closed-loop process of resource-production-consumption-regenerated resource.

Today, CE is seen as a much wider endeavor, associated with many other concepts, such as eco-design, innovation, industrial symbiosis, reverse logistics, responsible production, optimization of resource use, bioeconomy, blue economy, new business models, etc. [6]; added to an educational character, meaningful to the entire community. CE is a new, more rational way of thinking, anticipating the conception, production, use, and end of the product's life cycle, so as to minimize, dematerialize, remanufacture, and relocate all the materials/components and resources necessary, as well as reducing the environmental and social impact in every stage of the product's life cycle. CE implies the adoption of a cleaner production, greater accountability, and awareness by the producers and consumers, and usage of renewable materials and technologies, in addition to the adoption of suitable policies and tools [7]. CE dictates, in short, a holistic transformation of business [8].

CE involves so many interconnected concepts that many authors have scrambled to find the best definition by reviewing the literature and critically debating various circular economy conceptualizations [6,7,9–14].

Other authors [15,16] propose resources and materials' reduction strategies by submitting R list or R strategies, i.e., strategies that enhance circularity within the production chain. These are normally ranked by decreasing circularity, that is, a lower R means higher circularity: R0—Refuse (making the product redundant, by having it lose its function or having a totally different product offering the same function, as can be seen with digitalization—cable TV services preventing the acquisition of CD, DVD, or Blu-ray products, as well as their respective players); R1—Rethink (making the product more intensive through sharing or with multifunctional products); R2—Reduce (increasing the production or/and usage efficiency or/and consuming fewer resources and natural materials); R3—Reuse (reuse of the discarded product by a different consumer or even the original one if it is still in good conditions and can carry on fulfilling its original function); R4—Repair (repairing and maintaining a product); R5—Refurbish (restoring an old product and updating it); R6—Remanufacture (using the discarded product's parts/components in a new product with the same function); R7—Repurpose (using the discarded product's parts/components in a new product with a different function); R8—Recycle (processing materials in order to obtain the same material with equal or lower quality); R9—Recover (recovering the materials' energy).

Despite all the talk about CE, it is still in its embryo phase [16,17]. The European Commission has proposed a series of programs [5,18–22] to accelerate the transition to CE, creating a cleaner way to promote a more competitive Europe, in cooperation with its economic agents, its consumers, its citizens, and its civil society organizations.

Even though CE has become a major slogan for Portuguese [23], Spanish [24], and European authorities, its implications for companies' policies have not yet been fully met.

Currently, there is growing pressure from all of society to create socially, environmentally, and economically sustainable development. Companies need to find the triple bottom line (people, profit, and planet), where commercial interests, society's interests, and a minimization of the environmental impact all overlap [25].

According to Naudé [26], only when companies align sustainable development with business vision and strategy can sustainable development be implemented. CE has a clear impact on sustainable development, as it is related to multiple Sustainable Development Goals (SDGs), namely concerning consumption and production (SDG12), affordable and clean energy (SDG7), industry, innovation, and infrastructure (SDG9), clean water and sanitation (SDG6), decent work and economic growth (SDG8), and sustainable cities and communities (SDG11) [11].

Many authors [7,9] highlight the particular challenges that executives and managers face when trying to implement sustainable development in the business context, in a practical and realistic manner. The same authors agree that the best way to operationalize sustainable development is through CE practices.

Panchal et al. [11] did a literature review where the articles were classified according to the sustainable development goals addressed, the type of industry, and CE performance—looking at R strategies in particular. Authors concluded that the majority of CE performance studies focuses on environmentally sustainable development and overlooks the social aspects. Kirccherr et al. [9] added that quoting good CE practices, as well as obstacles to its implementation, not only helps refine the concept of CE among academics and professionals but can also be instructive to those interested in its rollout. Questioning institutions and companies about the strategic areas CE was integrated in, why did the company embrace it, as well as any difficulties found, would be a source of great help to the whole community.

Insights from case studies of CE business models would also help organizations understand the decisions, management concerns, and challenges they face [27]. Several authors [27–29] use case studies of companies that adopted CE practices and identified relevant managerial practices for CE business models. According to Ünal et al. [28] and Urbinati et al. [29], managers willing to embrace CE principles may benefit from a set of practices that create and capture value, thus leveraging peculiar dimensions of the company's business model: (i) the value network; and (ii) the customer value proposition and interface. This taxonomy, considering two larger circularity dimensions, was initially proposed by Urbinati et al. [30]. Managerial practices that promote the value network include: energy efficiency initiatives (reduction of greenhouse gases' emissions and environmental impact); use of eco-friendly materials (natural, recyclable and easy to separate); engagement of supply chain stakeholders, so they can become aware of CE tools and capabilities; practices related to effective communication with the supply chain stakeholders and upstream partners; design aimed at recycling and/or remanufacturing and/or reusing and/or disassembling practices. Managerial practices that promote customer value proposition and interface dimension include: the direct sale of products, making customers responsible for their use during and at the end of their lifecycle; the offer of complementary services when the product is sold (maintenance or take-back, thus guaranteeing the producer's eventual recovery of the customer's purchases); leasing or renting products; pay-per-use (allowing customers to benefit from short term use without further commitment); promotion on the company's website; advertising and sales personnel present in the store; communication of circularity through all communication channels; and customer involvement in circularity initiatives.

Unal et al. [28] propose a third dimension, managerial commitment, as a moderating factor between the value network and the customer value proposition and the interface dimensions, essential for reaching the intended goals of circular economy businesses.

Customer involvement and interaction are considered key points for the success and design of CE business models [27,29,30], allowing for an improved perception of customer preferences and the rationality of purchasing CEBM-driven products.

However, in the case of manufacturing organizations, their sustainability is highly dependent on the sustainability of their supply chain [31].

Yadav et al. [32] performed a bibliometric study focusing on the sustainable supply chain perspective in the automotive industry, exploring the challenges for both CE and Industry 4.0. Managerial and organizational issues, economic issues, supplier issues, and socio-cultural issues were identified, in this order, as the main challenges in this particular sector.

Nandi et al. [27] also highlight the benefits of supply chain collaboration. These authors give examples of CE business models that demonstrate how supply chains can be transformed to transition into sustainable economy models. However, they warn us that some CE business model practices can harm the social and environmental sustainability dimensions and, therefore, should be carefully evaluated and implemented in thoughtful and inclusive ways.

Bocken et al. [33] and Rosa et al. [34] performed a systematic literature review on existing Circular Business Models (CBMs) and their classification methods, by selecting the most promising ones.

The objectives of this study are to analyze and evaluate the integration of CE in companies from the northwest of Iberian Peninsula, to know the perspectives held by different groups with different positions in the company, regarding this domain, and to compare the integration of CE in different activity sectors.

The structure of this study is as follows: Section 2 describes the sample and the methods used for collecting the data (literature review and questionnaire survey) as well as the methodology followed in the data analysis (procedures, software, and used statistical tests); Section 3 presents the results that emerged from the questionnaire survey and the inferential analysis of the data. Section 4 discusses the main results of the survey (closed and open answers' descriptive analysis and inferential analysis) as well as their significance compared to the existing literature. Section 5 outlines the main conclusions. Finally, Section 6 presents some limits of the study and directions for future research.

2. Methods and Sample

2.1. Research Context

CircularLabs is an INTERREG POCTEP project (0495_CIRCULAR_LABS_6_E), which aims to promote the integration of CE in business models, counting the participation of 11 partners from three Iberian regions: north of Portugal, and Galicia and Castilla y León in Spain. According to the data made available on the Eurostat website, the largest sector in all three of these regions is the services sector, accounting for 66.7%, 68.2%, and 69.3% of all activity, respectively. This is followed by the industry and construction sector with 31.8%, 26.4%, and 25.5%, and finally the agricultural, forestry, and fishing sector with 1.5%, 5.4%, and 5.2% [35]. These regions show similarities to most Polish, Romanian, and Slovakian regions, as well as the western part of France, from Bretagne to the Côte de Azur, regarding both their activity sectors and their GDP per capita [35].

This project includes multiple transnational actions such as training initiatives, workshops, events, expositions, marketplaces, research, and others. Among its first initiatives this project launched an online survey for organizations/companies located in the Iberian northwest regions, focusing on their perception of Circular Economy and the implementation of this complex concept in their organizations' activities. The aim of this survey was to provide an insight on the current panorama concerning Circular Economy and to understand the limitations and potential for its development among organizations. Thus, our first goal was to map organizations' perception of CE, the strategic areas where it was implemented, difficulties and obstacles faced, as well as any necessary resources for the entities' transition to CE. Later, we aimed at comparing points of view from three groups occupying different positions within the companies (executives, managers, and technicians), as well as the two most representative activity sectors (industry and services) as it regards to CE.

2.2. Questionnaire Design

The questions of the online survey were based on the contents of the Circular Economy Action Plan [21], the Circular Economy in Cities and Regions Synthesis Report [36], and the Circular Economy Strategy of Castilla y León 2020–2030 [37]. The questionnaire is composed of five sections: Section 1. General information (regarding the entity in question); Section 2. Concept of Circular Economy; Section 3. Vision for Circular Economy; Section 4. Obstacles and difficulties in the implantation of CE; Section 5. Circular Economy initiatives. Section 5 shall not be covered in this paper. The survey is available in Appendix A.

2.3. Sample

This study's sample was contacted through mailing list distribution addressing organizations in the north of Portugal and the north of Spain. Over five thousand organizations (companies, foundations, local councils, etc.) were reached across the study regions. The survey remained opened between 15 December 2019 and 9 November 2020.

The sample is described in Table 1 and includes 294 respondents. The table shows the distribution of entities/companies per region, activity sector, number of employees, possession of some type of environmental certification, and respondent position within the company.

Table 1. Sample description.

	n	Percentage
Region		
Castilla y León	144	48.9
Galicia	69	23.5
North of Portugal	64	21.8
Others	17	5.8
Sector		
Services	100	34.0
Others	69	23.5
Industry	49	16.7
Public Sector	31	10.5
Agriculture	30	10.2
Tourism	15	5.1
Position		
Manager	105	35.7
Executive	68	23.1
Technician	60	20.4
Other	35	11.9
CEO	16	5.4
Environmental manager	10	3.4
Employees		
Less than 50	232	78.9
Between 50 and 250	30	10.2
More than 250	32	10.9
EMS Certification		
None	222	75.5
Others	30	10.2
ISO 14001	27	9.2
EMAS, ISO 14001	9	3.1
ISO 14001, Others	4	1.4
EMAS	1	0
EMAS, ISO 14001, Others	1	0
Total	294	

2.4. Data Analysis Methodology

All the closed answers, initially in Excel format, as they were multiple answer questions, allowing for more than one answer, were fragmented in *n* variables, with *n* being the number of items for each answer, and codified in 0 and 1 (0 for unselected items and 1 for every selected item). After that, the data were imported and treated with IBM-SPSS Statistics 27. In a first approach, a descriptive analysis of the answers was made. After that analysis, knowing that the respondents were mainly managers, executives, or technicians and that the sample was composed mainly of two sectors—industry and services—non-parametric statistical tests were applied. Chi-square tests of independence and goodness-of-fit were used to study the following hypotheses:

Hypothesis H1. *Proportionately, the answers' distribution to each "Question about CE" is identical;*

Hypothesis H2. *The answers to "Question about CE" are independent from "Position occupied within the company: Manager, Executive and such, or Technician";*

Hypothesis H3. *The answers to "Question about CE" are independent from "Activity sector, be it industry or services".*

The open answers were processed with the qualitative data analysis software NVivo12. The data were analyzed using Qualitative Content Analysis [38]. This method consists of an assessment of the frequency of the same or similar codes throughout a text, followed by a report that highlights the similarities and differences in the data. Thus, the open answers' analysis was conducted as follows: (1) translating the answers from Portuguese and Spanish to English; (2) back-translating the answers to their original language to confirm the translation's quality; (3) answers' codification; (4) code frequencies' analysis.

3. Results

3.1. Descriptive Analysis

3.1.1. Closed Answers

The concepts that the entities most frequently associated with Circular Economy, with the possibility of multiple choice, are shown in Figure 1. The most popular choice was "resources optimization" at 14.4%, followed by "waste reduction" at 13.4%, "responsible production" at 12.8%, and "recycling" and "repair and reuse" both with 11.5%.

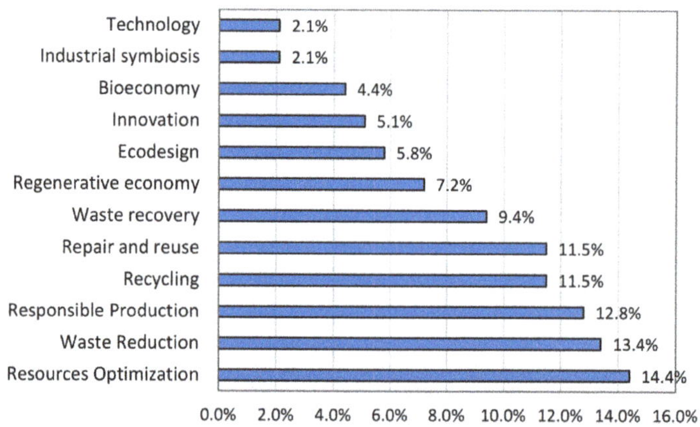

Figure 1. Concepts most frequently associated with CE.

Regarding the strategic areas where the entities/organizations most implemented CE, results can be found in Figure 2. With the option of multiple choice, we could determine that 15.8% chose "entity's mission and vision", 14.7% "environmental policy or environmental management system", and 10.3% "raw material purchasing-supply policy".

As it regards motivations to embrace Circular Economy, "environmental reasons" was the most frequently mentioned answer with 30.3%, followed by "socioeconomic reasons" (16.3%) and "corporate reputation" (16.2%). "Financial reasons", such as the access to subsidies or tax benefits, was the least reported issue, at 3.8% (Figure 3).

Regarding the difficulties faced by the entities in the process of implementing CE, the most mentioned answer was "lack of information and guidance" (21.4%), followed by "lack of financial resources" (19.6%) and "lack of technological solutions" (14.9%) (Figure 4).

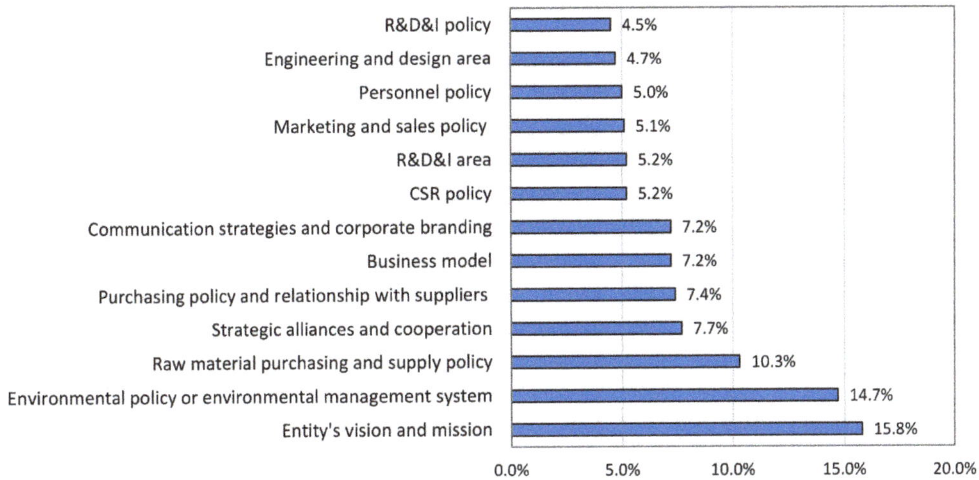

Figure 2. Strategic areas where CE was implemented. R&D&I: Research and Development and Innovation; CSR: Corporate Social Responsibility.

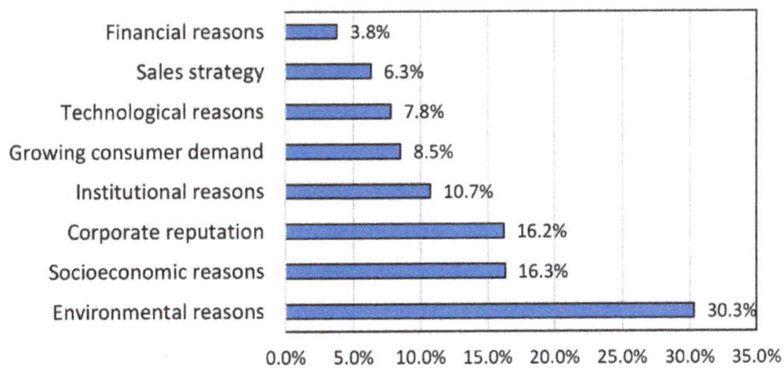

Figure 3. Reasons why the entities embraced CE.

Figure 4. Difficulties faced by the entities in the process of implementing CE.

When asked about the factors that can ease the transition to CE, results show that many organizations reported the need for additional information and guidance, financial resources, and technological solutions. Additional less relevant factors included employees' training and more personnel (Figure 5).

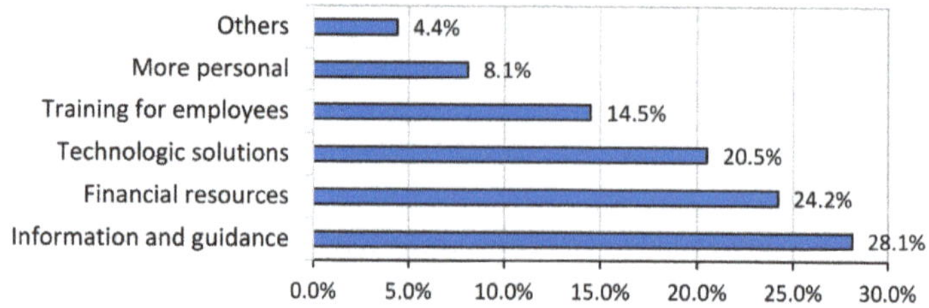

Figure 5. Factors that can ease the transition to CE.

3.1.2. Open Answers

As previously stated, the data from the open responses were analyzed with a text analysis tool, NVivo, to determine most frequently used terms. As it regarded the organizations' understanding of CE, the text analysis found the words "product" and "waste" to be the most frequent, followed by "use", "economy", "resource", and "reuse" as is illustrated in the word cloud below (Figure 6). The top ten most frequently used terms are listed in Table 2.

Figure 6. Word cloud of most frequently used terms in respondents' answer concerning what they understand by "circular economy".

Table 2. Top 10 most frequent terms used by respondents stating what they understand by "circular economy".

Term	Frequency
products	124
waste	111
use	89
economy	81
resources	79
reuse	71
materials	68
reduce	45
recycling	42
generation	40
sustainable	38
economic	38

Regarding the open question "How is circular economy's impact measured?" only 119 answers were registered. Most of the respondents, 39 (32.8%), admitted they did not measure it and four of these (3.4%) added they did not know how to perform such measurement. Other answers provided a miscellaneous perspective, composed of diverse answers such as "sales results", "rough approximation", "according to events' dissemination", "by monitoring", "team meeting", etc. These and the remaining answers are grouped in Table 3.

Table 3. Frequency distribution of the answers to the open-ended question "How is CE's impact measured?".

Answers: "How is CE's Impact Measured?"	Frequency
"It's not measured"	39
"In the implementation phase"	4
Through:	
"indicators … management, KPI [1], GRI [2], sustainability"	10
"consumption reduction"	7
"environmental calculation", "CO_2 emissions decrease" "carbon footprint"	6
"cost decrease"	4
"waste reduction"	4
"energy drop"	4
"% of recycled residue"	4
"client acceptance/appreciation"	4
Diverse answers	33

[1] Key Performance Indicator. [2] Global Reporting Initiative.

We would also like to stress how it is that in the bigger companies (more than 250 employees) that the use of indicators is more prevalent, as open answers suggest, while smaller companies (less than 50 employees) do not measure the impact of CE. Regarding regional differences, Galicia was where there was a greater resource to indicators, followed by Castilla y Leon and, only after that, Portugal.

3.2. Data Inferential Analysis

The Chi-square goodness of fit test was applied to all questions in order to assess whether the proportion of items selected was identical. The p-values were all less than 0.0005, suggesting that the proportion of responses for the different items/answers, in each question, was not the same (Table 4).

Table 4. Chi-square and p-value resulting from the application of the goodness of fit test to each question.

Questions	χ^2	p-Value
Concepts most associated with CE	159.187 ***	<0.0005
Strategic areas where CE was implemented	100.001 ***	<0.0005
Reasons why the entities embraced CE	532.875 ***	<0.0005
Difficulties faced in the implementation of CE	186.358 ***	<0.0005
Needs in order to transition to CE	155.399 ***	<0.0005

*** $p < 0.001$.

The Chi-Square test of independence was used to determine if the two nominal variables: "Questions about CE" and "Position occupied within the company: Manager, Executive, and Technician" were independent (Hypothesis H2). The obtained results of χ^2 and p-value are presented in Table 5.

Table 5. Chi-square and *p*-value resulting from the application of Chi-square test of independence to the nominal variables "Questions" and "Position occupied in company".

Questions	χ^2	*p*-Value
Concepts most associated with CE	55.434 ***	<0.001
Strategic areas where CE was implemented	55.431 ***	<0.001
Reasons why the entities embraced CE	32.379 **	0.009
Difficulties faced in the implementation of CE	34.391 *	0.011
Needs in order to transition to CE	11.450	0.491

* $p < 0.05$. ** $p < 0.01$. *** $p < 0.001$.

The results show that there is a significant relationship between the first four questions and "Position occupied within the company". Thus, for each item within a question we compared the answers' proportions for each position within the company, using Bonferroni's correction. The results showing differences of statistical significance are presented in Table 6.

Table 6. Proportion of answers to each question according to the positions occupied within the company and respective *p*-value, when there is a difference of statistical significance.

Question	Proportion (Number of Answers by Number of Individuals in a Given Position)			
	Manager A	Executive and Such B	Technicians C	*p*-Value
Concepts most frequently associated with CE				
Eco-design	0.035	0.065	0.085	(A,C) * 0.041
Responsible Production	0.117	0.175	0.102	(B,C) * 0.042
Recycling	0.153	0.095	0.080	(A,C) * 0.018
Strategic areas where CE was implemented				
Strategic partnerships and cooperation	0.054	0.107	0.092	(A,B) ** 0.003
CSR policy	0.054	0.064	0.025	(B,C) * 0.017
Purchasing policy and relationship with suppliers	0.078	0.081	0.042	(B,C) * 0.014
Reasons why the entities embraced CE				
Socioeconomic reasons	0.171	0.154	0.103	(A,C) * 0.049
Environmental reasons	0.331	0.291	0.278	(A,C) * 0.033
Institutional reasons	0.069	0.114	0.183	(A,C) ** 0.004
Difficulties faced in the implementation of CE				
Lack of technological solutions	0.164	0.182	0.096	(B,C) * 0.015
Lack of commitment from the top management	0.020	0.057	0.072	(A,C) * 0.035
Lack of training from the employees	0.060	0.082	0.136	(A,C) * 0.022

* $p < 0.05$. ** $p < 0.01$.

The results, based on a two-sided test, show that, proportionately, technicians associate "eco-design" to CE more significantly than other positions, and the difference in answers between technicians and managers is statistically significant (*p*-value = 0.041). Alternatively, managers associate CE with "recycling" proportionately more than the other two positions, and, like before, there is a statistically significant difference in comparison with the technicians' answers (*p*-value = 0.018). Executives give higher importance to "responsible production", with differences of statistical significance (0.042) between the proportion of executives' answers and those of technicians.

Regarding the strategic areas where CE was implemented, a greater proportion of executives' answers include "strategic partnerships and cooperation", "CSR (Corporate and social reputation) policy", and "purchasing policy and relationship with suppliers".

Differences of statistical significance can be found in the executives and managers' answers' proportion for "partnerships and cooperation", as a strategic area of CE implementation, as well as in the executives and technicians' answers' proportion for the other two areas.

Concerning the reasons why entities chose to embrace CE, differences of statistical significance were found between the managers' and technicians' positions: while the first highlight "environmental and socioeconomic reasons", the latter value "institutional reasons".

As it regards to the difficulties faced in the implementation of CE, technicians emphasize "lack of commitment from top management" and "lack of employee training", demonstrating differences of statistical significance when compared to the managers. Executives consider that difficulties come from "lack of technological solutions", showing differences of statistical significance when compared to the technicians as well.

Similarly, the Chi-square test of independence was used to determine if the two nominal variables, "Questions about CE" and "Activity sector: Industry and service" were independent (Hypothesis H3). The results of χ^2 and p-value are shown in Table 7.

Table 7. Chi-square and p-value resulting from the application of Chi-square test of independence to the nominal variables "Questions" and "Activity sector: industry or services".

Question	χ^2	p-Value
Concepts most associated with CE	21.865 *	0.039
Strategic areas where CE was implemented	31.303 **	0.003
Reasons why the entities embraced CE	14.463	0.070
Difficulties faced in the implementation of CE	13.320	0.149
Needs in order to transition to CE	8.327	0.215

* $p < 0.05$. ** $p < 0.01$.

Only the answers to the two first questions depend on the activity sector. Thus, for each item within a question, we compared the answers' proportions between the two activity sectors using Bonferroni's correction. The results with differences of statistical significance are shown in Table 8.

Table 8. Proportion of answers to each question according to sector—industry or services—and respective p-value, when there is a difference of statistical significance.

	Proportion (Number of Answers by Number of Individuals in a Given Position)		
Question	**Industry**	**Services**	**p-Value**
Concepts most associated with CE			
Responsible production	0.161	0.100	0.031 *
Recycling	0.084	0.141	0.048 *
Repair and reuse	0.077	0.138	0.034 *
Strategic areas where CE was implemented			
Environmental policy or environmental management	0.130	0.110	0.033 *
Raw material purchasing and supply policy	0.120	0.073	0.001 **
R&D&I area	0.060	0.031	0.013 *

* $p < 0.05$. ** $p < 0.01$.

Regarding the question about "concepts most associated with CE", we noted that industry makes a stronger connection with the concept of "responsible production", while services are more inclined to "recycling" and "repair and reuse", with differences of statistical significance (p-value < 0.05). As it concerns the strategic areas where CE was implemented, industry shows a greater tendency to introduce CE in "environmental policy or environmental management system", "raw material purchasing and supply policy", and "R&D&I" than the services sector.

4. Discussion

4.1. Descriptive Analysis

Regarding the concepts most frequently associated with CE, we would like to stress how, among the five most mentioned words, we found the 3 R's (reduce, reuse, recycle), the three starting CE principles. CE perception comprises preemptive practical actions (resources optimization, waste reduction) and corrective actions (recycling, repair, and reuse)—which is in line with previous studies [39,40]. According to Darmandieu et al. [40], companies initially go for a combination of preemptive and corrective actions, so as to enhance resource efficiency, and only at a later stage do they progress to more expensive practices, demanding more considerable investments. Maybe this mindset explains why concepts such as "innovation", "eco-design", and other practical preemptive efforts that call for an early investment are quoted to a lesser degree. The industry sector amounts to 16.7% of our sample and just 2.1% of respondents picked "industrial symbiosis". Considering that the majority of industries today belong to some sort of industrial association, and the greater part of industries are located in their own designated site, close to each other, in industrial parks, it would be expected that industrial symbiosis would naturally develop. Nevertheless, this concept was among the least chosen.

On the other hand, companies' awareness of sustainable development is clearly present when "responsible production" is the third most cited concept.

"Entity's vision and mission" was the strategic area where entities most admitted to having integrated CE (15.8%). Considering how sustainable development operationalization is best achieved through CE practices [7,9] and, according to Naudé [26], only when companies align sustainable development with business vision and strategy can it be successfully implemented, our results suggest that entities from the studied regions are on track to achieve sustainable development. The second most quoted strategic area was "environmental policy or environmental management system" (14.7%). According to Garcés-Ayerbe et al. [41], the first CE practices implemented by small and medium sized enterprises, at the time, pertained to pollution control; currently, companies implement strategies that involve proactive environmental strategies of pollution prevention. Banerjee [42] proposes that companies integrate environmental issues in the company's strategy because they have discovered that the more proactive they are in what regards environmental issues, the more sustainable the company becomes in the long term. These environmental concerns are in line with the priority objectives of the eighth General Union Environment Action Program (EAP) to guide the EU's environmental policy to 2030 and align it with the European Green Deal [20,43].

This statement helps us frame our own results which show that the main motivation for companies to embrace CE was "environmental reasons" (30.3% of answers), with "socioeconomic reasons" (16.3%) in second place. Socioeconomic reasons include production costs reduction, material and energetic resources reduction, the creation of new business models, job creation, etc. Darmandieu et al. [40] concluded that circularity in production processes generates a reduction of firms' production costs. Cutting down costs is a great motivation for companies to implement circular practices [44,45]. Coincidentally, product circularity creates what are known as green jobs, which seem to be related to the increased revenues—benefit of environmental innovation in products [46].

"Corporate reputation" was also pointed out as one of the reasons to change to CE, at 16.2%. In a UN Global Compact-Accenture CEO Study [47], where 766 CEOs from all around the globe were interviewed, 75% stated that the main reasons for selecting sustainability strategies were the enhancement of corporate reputation and the potential to decrease costs and increase revenue. Currently, companies face both internal and external social pressures if they do not implement sustainable development strategies and practices, in addition to the permanent need to remain economically competitive and viable [26,48,49]. A study about small and medium Spanish enterprises showed that companies were worried about their company's image [50]. It is expected that prestige and company image improvement come from real sustainability strategies' communication, dissemination and promotion to customers [50], which ends up translating to an increase in market shares.

Regarding any difficulties and obstacles faced in the implementation of CE, "lack of information and guidance" (21.4%), "lack of financial resources" (19.6%), and "technological solutions" (14.9%) rank highest. Funding plays a central role in CE innovation and stimulation. There are several European programs, such as Horizon 2020, Life Programme, COSME, EEAGrants, European Fund for Strategic Investments, Climate-Kic, etc., as well as instruments and initiatives at the national level (Portugal and Spain) that may function as funding opportunities for companies. Nevertheless, information about financial and fiscal support mechanisms to companies who may want to invest in CE do not seem to have been efficiently communicated. This statement is endorsed by the fact that only 4.2% of respondents selected financial reasons (subsidies, tax benefits, etc.) as a motivation to embrace CE.

The lack of information exchange between companies regarding CE benefits is also pointed out by Van Eijk [51] and Van Buren et al. [52] as a barrier to the implementation of CE business models. The transition from a linear economy to a circular one will only be possible if there is a collective effort from all stakeholders, aiming at a knowledge and innovation exchange. If a company's information is considered confidential, it will be harder to develop CE business models [53].

However, information is frequently deemed confidential due to the fierceness of the competition and how economic interests often overshadow social and environmental ones.

"Lack of financial resources" and "lack of technical skills" (technical and technological know-how) were also the most frequently mentioned barriers in literature review papers about small and medium enterprises [50,54]. The authors consider these two barriers to be intimately connected, since without financial means companies cannot invest in employee training, much less in hiring external specialists.

In a literature review by Aranda et al. [55], it was concluded that fund availability, the quality of the company's financial resources, and public subsidies and incentives have a positive effect in stimulating the implementation of CE policies, since they reduce exposition to risk and increase financial feasibility and the profitability potential of the CE investment projects. Ritzén and Sandström [56] claim the barriers for moving towards CE to be financial and technological, as well as structural, operational, and attitudinal: structural due to the lack of information exchange and unclear responsibility distribution, operational due to the lack of infrastructure (responsibilities and task division), and attitudinal due to the great aversion to risk and the business logic of taking small safe steps in the development of the organization.

These difficulties/obstacles described by companies from Spain and the north of Portugal are later requalified as company needs in the following question.

When you ask for a definition of CE in open-ended questions, most answers do not show a comprehensive understanding of the concept, giving only a definition which covers but a fraction of what CE entails. The definitions we received mainly focused on products/materials and waste. Here are some examples from answers we got: "Products which come back into the productive cycle", "a new production, consumption and valorization model in which waste is a resource; the end of single use economics", "It's reusing materials, products and waste", "Processes to limit the use of new materials, rather choosing to lengthen the life cycle of products which are to be thrown away". The top ten most frequently used words in this question confirms the results found by Nobre and Tavares [10].

From the 294 answers we got to the question "How is circular economy impact measured?", 175 (59.5%) were blank. From the 119 answers, 39 (32.8%) said it was not being measured. It is important to stress that if we count our blank answers as "not being measured", that means 72.8% of companies are not measuring the impact of CE. Golinska et al. [57] claimed that the problem with sustainable development application to daily business operations was the lack of sustainability indicators. The same authors did a literature review and came up with a set of criteria/indicators used to assess companies' performance in the economic, environmental, and social areas. In our present study, looking at the number of ambiguous and subjective answers concerning the way CE's impact is being measured, as well as the answers stating clearly that measuring procedures are not known, we seem to be facing a problem of unfamiliarity with the existence of indicators and/or how to apply them. Nevertheless, the use of indicators to calculate economic, social, and environmental payback from the implementation of CE practices in a company is absolutely crucial.

From the ten respondents who used indicators, only two mentioned using those which rated the company's sustainability performance in all three areas (economical, environmental, and social)—using GRI and sustainability indicators. If we consider that "waste reduction", "energy drop", "consumption reduction", and "percentage of recycled waste" contribute to the company's environmental performance, we can say that 25 companies (21% of organizations resorting to them) use environmental indicators—once again showcasing CE origins, originally considered as a tool for projecting environmental solutions.

4.2. Inferential Analysis

The outcome suggests that the focus of managers is on reactive actions or end of the pipe interventions, such as recycling, that allow companies to decrease costs of raw material transportation and the final destination of waste, such as landfills or incineration.

On the other hand, technicians highlight eco-design, an action that precedes the productive process itself, allowing a better management of the product's life cycle, its value chain, and integration in closed loops processes. In these processes, waste is used as an input, thus eliminating the notion of an undesirable by-product within and outside the industrial system [58], highlighting the technicians' greater knowledge and sensitivity to the sustainability issue. Once again, the results suggest that there is a need to raise awareness to the subject, especially among management members, towards whom training should be directed. Finally, executives value "responsible production" more than other company roles. This group seems to be more concerned with the "minimization of waste, energy, and pollution", as well as the company's image to be presented to stakeholders and the community at large. Executives, having a significant influence over the company's strategy, when considering CE implementation, give proportionately more attention to the areas of "strategic partnership and cooperation", "corporate and social responsibility policy", and "purchasing policy and relationship with suppliers", than their managers and technicians counterparts.

Alternatively, "strategic partners and cooperation" is the least relevant area to managers, who seem to be more focused on their own company—at the micro level. The benefits that can emerge from the meso level—an industrial park, for example [59]—are in line with the 17th Sustainable Development Goal of the UN's "Partnerships for the goals" [60]. Companies in the same region and even in the same sector of activity should create symbiosis, such as sharing infrastructure and equipment.

Regarding the reasons why entities chose to embrace the transition to CE, results suggest that there are no great differences between the three groups' perspectives as it regards to their valuing of socioeconomic and environmental reasons, even if managers value this perspective more than executives and technicians. Nevertheless, from an institutional point of view, executives and technicians seem to have a better understanding of CE repercussions.

Langen et al. [61], in a study about the perceptions and awareness of three professional groups—administrators, economists, and researchers—concluded that administrators/managers were also more focused on socioeconomic issues, in particular, in using CE for economic growth and job creation.

Managers' lesser sensibility towards institutional motivations for adopting CE suggests that, once again, this group is focused on a micro-vision of their business, on their own company, and are not as aware of the benefits that may come from listening to stakeholders' opinions and analyzing their close competitors' behavior. We had already seen this when looking at the importance managers gave to strategic partners and cooperation.

Regarding the obstacles faced in CE implementation in companies, it is interesting to see that the lack of technological solutions is mentioned at a greater scale by executives and managers and less so by technicians, which suggests that technicians might be aware of solutions that are not known, valued, or even considered by their hierarchical superiors, maybe due to reasons of a financial nature.

On the other hand, sustainable and environmentally friendly production and consumption technologies depend on technical know-how which can often be found in the innovative solutions offered by the company's suppliers and customers. The company itself frequently has neither the time nor the financial means to spend looking for solutions [52,62]. If we consider this, it is not surprising that executives and managers highlight the absence of technological solutions and consider "purchasing policy and relationship with suppliers" the main area for CE application, significantly more so than their technician counterparts.

Regarding "lack of commitment from the top management", this is relatively devalued by managers, and more significantly stressed by technicians and executives, which suggests that even if top management is in fact committed to CE, that is not the perception of their collaborators. On the other hand, "lack of commitment from the top management" can be a consequence of other barriers such as "lack of CE awareness", "lack of CE knowledge", and "lack of sense of urgency to the implementation of CE", aggravated by lack of know-how [61]. According to Droege et al. [63], "lack of commitment from the top management" is a cultural barrier which leads to structural problems in the company. Managers' aversion to risk can make CE implementation more difficult, even after an assessment of the benefits associated with it [53,64].

Regarding the concepts most associated with CE, the industry sector values "responsible production", while services give precedence to "recycling" and "repair and reuse"—concepts which align themselves with the activity sector.

In fact, "responsible production" has been receiving increasing attention from the industrial sector [65] since it was presented as the 12th objective for sustainable development [60]. Responsible production is based on an Extended Producer Responsibility (EPR) approach [66], a policy which makes the company accountable for the environmental impact of their products manufacture, since its conception, until the end of its life cycle [67]. Since it makes the producer directly responsible for the real impacts of production, it is easy to understand why the industrial sector is so focused on responsible production.

Alternatively, the services sector does not assume any responsibilities of this kind. In a study conducted by the University of Gloucestershire in the UK, it was possible to determine that the hospitality sector dedicates itself mainly to waste management and recycling [68]. The food sector also focuses its CE efforts on recycling, namely waste, cooking oil, and paper, as well as paying some attention to energy and resources minimization [69]. Regarding the strategic areas in which CE was implemented, the industry sector values "environmental policy or environmental management system", "raw material purchasing and supply policy", and "R&D&I" more than the services sector.

The importance attributed to "environmental policies and environmental management" is a consequence of the growing pressure on industrial activity, namely in what regards emission restriction, production of solid waste, and waste dumping in landfills [58]. Concerning the purchase of material resources and stocking policies, we have already discussed Extended Producer Responsibility (EPR) [66]. On the other hand, it is important to highlight that the supply chain and its management are considered to be of great importance to responsible production in the literature [65]. Moreover, there is empirical evidence encouraging the promotion of circular supply chains, aiming for increased operational efficiency, company competition, and economic performance [70,71]. This means that on a micro and meso level, there is evidence that a sustainable supply chain allows for CE performance improvement in eco-industrial parks [59].

Finally, the current focus of research, development, and innovation is lined up with the perception that the industrial sector's focus should be the innovative conception of products [72], eco-innovation [44,73], green technologies [74], responsible production [75], and use technologies from Industry 4.0, such as Artificial Intelligence (AI), Internet of Things (IoT), Blockchain, and Big Data [71,76–81]. The literature further states that innovation is the basis for new ways of producing goods, retaining resources value in the productive processes, and promoting disruption and reformulation, thus contributing to the application of CE [82].

5. Conclusions

This study offers some perspective on the current CE integration panorama in organizations from three regions in the north of the Iberian Peninsula, allowing for the understanding of their motivations, difficulties, and needs.

The concept of CE, as perceived by the majority of respondents, does not really encompass its complexity, particularly if we consider the fact that the most commonly offered definitions entail the concepts of waste valorization/reduction/reuse/recycle. Nevertheless, results indicate that there is a greater awareness from these companies' officials regarding sustainable development and responsible production, in particular. On the other hand, industrial symbiosis seems to be underdeveloped in these regions. In that sense, existing regional industrial associations should take measures to increase industrial symbiosis between their affiliated companies.

Environmental issues continue to be the main reason for CE practices implementation in organizations. However, the image the company transmits to employees, stakeholders, and the community has been gaining importance, with corporate reputation as a secondary factor.

Many companies complain about the lack of information and guidance, not only concerning productive and operational processes that can lead the way for a transition to CE, but also concerning the availability of financial and fiscal support, namely for those companies that are interested in investing in CE. It is noticeable that there is a general unfamiliarity with European and transnational financial guidance, and information regarding support programs.

Moreover, results show that the answers to questions concerning the perception on CE, its integration and difficulties faced in its implementation depend on the respondent's position within the company (manager, executive, and technician). Similarly, we were able to find differences between the two different activity sectors—industry and services, as it regards to their perception of CE and areas of its implementation. Because these opinions are representative of each particular sector/organization and position occupied, these results may be used in future research with due caution, if applied to a different context.

Our results further show that the majority of companies, about 72.8%, does not measure the impact of implemented CE practices. Consequently, there is a need for strengthening the dissemination of information and training indicators regarding CE in general, and particularly regarding monitoring indicators. Larger companies show a greater awareness about the need for assessment indicators when compared to smaller companies. Changing our measuring practices is paramount as it can help demonstrate the benefits of CE implementation, ensuring recognition of its competitive advantages, beyond short term perception.

It is crucial and urgent to stimulate every entity/organization and society at large to raise awareness of the importance of CE as a vital premise for long term sustainability.

Overall, this article offers a new perspective to the current knowledge on Circular Economy practices implemented by organizations in a cross-border context, between Portugal and Spain. Additionally, it can be considered a valuable contribution to the current knowledge on the factors affecting CE implementation by organizations. As its results were already used as a basis for the actions developed under the Circular Labs POCTEP Project, this article offers valuable contributions that should be considered in the future development of CE initiatives, such as awareness campaigns, funding programs, or cooperative actions.

6. Limitations and Future Research

The answer dispersion through each question's items hindered the application of some statistical tests, namely for inter-region comparisons, which was made impossible by low frequencies. As previously mentioned, the data were collected from three regions in the Iberian Peninsula and although some valuable conclusions can be deducted, the study would profit from more data from these regions. Looking at future work, it would also be interesting to replicate the study in different regions and check if the results are the same.

Author Contributions: Conceptualization, J.D. and I.M.-T.; methodology, M.M.S. and C.O.-S.; data curation, M.M.S. and C.O.-S.; writing—original draft; M.M.S., M.P.C. and C.O.-S.; writing—review and editing, M.M.S. and A.G.; visualization, A.G.; supervision, M.P.C.; project administration, E.C.I. and I.M.-T.; funding acquisition, J.D. and E.C.I. All authors have read and agreed to the published version of the manuscript.

Funding: Interreg V-A Spain-Portugal POCTEP 2014-2020, European Commission, under ERDF (European Regional Development Fund). Project code: 0495_CIRCULAR_LABS_6_E. This article was also supported by the FCT/MCTES Project CIMO (UIDB/00690/2020).

Institutional Review Board Statement: The study was conducted in accordance with the Declaration of Helsinki, and approved by Conselho de Ética e Deontologia da Universidade da Maia Approval Code: Parecer NO. 96/2022 Approval Date: 13 June 2022.

Informed Consent Statement: Informed consent was obtained from all subjects involved in the study.

Data Availability Statement: The data presented in this study are available on request from the corresponding author. The data are not publicly available due to the continuation of research and other manuscripts that are in process.

Acknowledgments: The authors would like to thank the Spain–Portugal Cross-border Program (POCTEP), with Financial Support from the European Union under Program ERDF (European Regional Development Fund).

Conflicts of Interest: The authors declare no conflict of interest.

Appendix A Questionnaire

1. General Information

1.1 What position do you occupy within your company?
1—Head of environmental management
2—Manager
3—CEO and such
4—Technical staff
5—Administration and finance
6—Other
If you answered "other" please specify: Your answer
1.2 Which sector does your entity belong to?
1—Industry
2—Services
3—Agrarian
4—Tourism
5—Public sector
6—Other
If you answered "other" please specify: Your answer
1.3 How many employees does your company have?
1—From 0 to 50 employees
2—From 50 to 250 employees

3—Greater than 250 employees
1.4 Does your company have any kind of environmental certification? *
1—None
2—EMAS
3—ISO 14001
4—Other
1.5 In what region is your company located?
1—Castilla y Leon
2—Galicia
3—North of Portugal
4—Other

2. Circular Economy

2.1 State the three concepts you most associate with circular economy.
1—Bioeconomy
2—Eco-design
3—Regenerative economy
4—Innovation
5—Responsible production
6—Waste reduction
7—Recycling
8—Resource optimization
9—Repairing and reusing
10—Industrial symbiosis
11—Technology
12—Waste valorization
What do you understand by circular economy? Your answer

3. Vision for Circular Economy in Your Company

3.1 In what strategic area was circular economy integrated in your company?
1—Entity's mission and vision
2—Environmental policy or environmental management system
3—Raw material purchasing-supply policy
4—Engineering and design
5—Staff policy
6—Business model
7—R&D&I policy
8—Strategic partnerships and cooperation
9—CSR policy
10—Communication strategies and corporate branding
11—Marketing and sales policy
12—Purchasing policy and relationship with suppliers
13—R&D&I area
Give other examples if needed: Your answer
3.2 In your opinion, why has your company embraced a transition to circular economy?

1—Socioeconomic issues (economic crisis, new business models, shortage of material and energetic resources, job creation, production costs economy, ...)

2—Environmental issues (climate emergency, environmental impact, ...)

3—Institutional issues (global pacts, strategies, plans and government programs, ...)

4 -Technological issues (key sectors' technological development, promotion of innovation, ...)

5—Financial issues (subsidies, tax benefits, ...)

6—Growing customer demand

7—Corporate reputation

8—Sales strategy

4. Obstacles and Difficulties

4.1 What are the main obstacles you faced in trying to implement circular economy in your company?

1—Lack of information and guidance
2—Lack of technological solutions
3—Lack of financial resources
4—Lack of staff
5—Lack of a clear legal and/or regulatory framework
6—Lack of customer interest
7—Lack of commitment from top management
8—Lack of employees' training
9—Other
If you answered "other" please specify: Your answer

4.2 What would you need to facilitate your company's transition to a circular economy?

1—Information and guidance
2—Technological solution
3—Financial resources
4—More staff
5—Employees' training
6—Other
If you answered "other" please specify: Your answer

4.3 If circular economy actions are already in place in your company, please state how their impact is being measured: Your answer

5. Circular Economy Practices

If you wish to share or promote any good circular economy practices already in place in your organization, please state so in the following boxes.

5.1 Ecologic design for products or services (aiming at waste prevention, product durability, recycling and repairing potential):

5.2 Purchasing criteria (supply of secondhand product with ecologic tags, remanufactured or updated, made from recycled materials):

5.3 Servitization and new business models (co-operative economics, subscribing to a service, rather than buying a product, etc., …):

5.4 Cooperation or commercial symbiosis (subproducts and material resources swap, resources or equipment sharing, etc.):

5.5 Staff training in environmental issues and/or circular economy.

References

1. The Ellen MacArthur Foundation. History of the Ellen MacArthur Foundation. [WWW Document]. Available online: https://www.ellenmacarthurfoundation.org/our-story/milestones (accessed on 19 June 2022).
2. Sassanelli, C.; Rosa, P.; Rocca, R.; Terzi, S. Circular Economy performance assessment methods: A systematic literature review. *J. Clean. Prod.* **2019**, *229*, 440–453. [CrossRef]
3. Ellen MacArthur Foundation. 2017. Available online: https://archive.ellenmacarthurfoundation.org/pt/fundacao-ellen-macarthur/a-fundacao (accessed on 24 February 2022).
4. Braungart, M.; McDonough, W. *Cradle to Cradle: Remaking the Way We Make Things*; McGraw-Hill: Madrid, Spain, 2005.
5. Communication from the Commission to the European Parliament, the Council, the European Economic and Social Committee and the Committee of the Regions Closing the Loop—An EU Action Plan for the Circular Economy COM/2015/0614 Final. Available online: https://eur-lex.europa.eu/legal-content/PT/TXT/?uri=CELEX:52015DC0614 (accessed on 24 February 2022).
6. Geisendorf, S.; Pietrulla, F. The circular economy and circular economic concepts—A literature analysis and redefinition. *Thunderbird Int. Bus. Rev.* **2018**, *60*, 771–782. [CrossRef]
7. Ghisellini, P.; Cialani, C.; Ulgiati, S. A review on circular economy: The expected transition to a balanced interplay of environmental and economic systems. *J. Clean. Prod.* **2016**, *114*, 11–32. [CrossRef]
8. Jesus, A.; Antunes, P.; Santos, R.; Mendonça, S. Eco-innovation in the transition to a circular economy: An analytical literature review. *J. Clean. Prod.* **2018**, *172*, 2999–3018. [CrossRef]
9. Kirchherr, J.; Reike, D.; Hekkert, M. Conceptualizing the circular economy: An analysis of 114 definitions. *Resour. Conserv. Recycl.* **2017**, *127*, 221–232. [CrossRef]

10. Nobre, G.C.; Tavares, E. The quest for a circular economy final definition: A scientific perspective. *J. Clean. Prod.* **2021**, *314*, 1279. [CrossRef]
11. Panchal, R.; Singh, A.; Diwan, H. Does circular economy performance lead to sustainable development? A systematic literature review. *J. Environ. Manag.* **2021**, *293*, 112811. [CrossRef]
12. Chiappetta Jabbour, C.J.; Fiorini, P.D.C.; Ndubisi, N.O.; Queiroz, M.M.; Piato, É.L. Digitally-enabled sustainable supply chains in the 21st century: A review and a research agenda. *Sci. Total Environ.* **2020**, *725*, 138177. [CrossRef]
13. Rosa, P.; Sassanelli, C.; Urbinati, A.; Chiaroni, D.; Terzi, S. Assessing relations between Circular Economy and Industry 4.0: A systematic literature review. *Int. J. Prod. Res.* **2020**, *58*, 1662–1687. [CrossRef]
14. Sassanelli, C.; Rosa, P.; Terzi, S. Supporting disassembly processes through simulation tools: A systematic literature review with a focus on printed circuit boards. *J. Manuf. Syst.* **2021**, *60*, 429–448. [CrossRef]
15. Potting, J.; Hekkert, M.P.; Worrell, E.; Hanemaaijer, E. *Circular Economy: Measuring Innovation in PRODUCT Chains*; Report 2544; Netherlands Environmental Assessment Agency, PBL Publishers: The Hague, The Netherlands, 2016.
16. Fonseca, L.M.; Domingues, J.P.; Pereira, M.T.; Martins, F.F.; Zimon, D. Assessment of Circular Economy within Portuguese Organizations. *Sustainability* **2018**, *10*, 2521. [CrossRef]
17. Mayer, A.; Haas, W.; Wiedenhofer, D.; Krausmann, F.; Nuss, P.; Blengini, G.A. Measuring progress towards a circular economy. *A monitoring framework for economy-wide material loop closing in the EU28. J. Ind. Ecol.* **2018**, *23*, 62–76. [CrossRef] [PubMed]
18. European Commission 2018. Commission Adopted Circular Economy Package. Available online: https://ec.europa.eu/environment/topics/circular-economy/first-circular-economy-action-plan_pt#ecl-inpage-937 (accessed on 9 March 2022).
19. European Commission 2019. Commission Adopted the Final Circular Economy Package. Available online: https://ec.europa.eu/environment/topics/circular-economy/first-circular-economy-action-plan_pt#ecl-inpage-937 (accessed on 9 March 2022).
20. European Commission 2019. European Commission adopted European Green Deal. Available online: https://ec.europa.eu/environment/topics/circular-economy/first-circular-economy-action-plan_pt#ecl-inpage-937 (accessed on 9 March 2022).
21. European Commission 2019. Circular Economy Action Plan. In Commission Staff Working Document (No. 90; SWD 2019). Available online: https://eur-lex.europa.eu/legal-content/EN/TXT/PDF/?uri=CELEX:52019SC0090&from=EN (accessed on 9 March 2022).
22. European Commission 2020. European Commission Adopted New Circular Economy Action Plan. Available online: https://ec.europa.eu/environment/topics/circular-economy/first-circular-economy-action-plan_pt#ecl-inpage-937 (accessed on 9 March 2022).
23. Diário da República Portuguesa nº 236/2017 de 11 de Dezembro—Resolução do Conselho de Ministros n.º 190-A 2º Suplemento, Série I. Plano de Ação para a Economia Circular em Portugal. Available online: https://files.dre.pt/1s/2017/12/23602/0005400073.pdf (accessed on 10 March 2022).
24. Gobierno, D.E. Estrategia Española de Economía Circular, España Circular 2030 Aprobada por Acuerdo del Consejo de Ministros el 2 de Junio 2020. Available online: https://www.miteco.gob.es/images/es/180206economiacircular_tcm30-440922.pdf (accessed on 10 March 2022).
25. Savitz, A.W.; Weber, K. *The Triple Bottom Line How Today's Best-Run Companies Are Achieving Economic, Social, and Environmental Success—And How You Can too*; Jossey-Bass Publishers: San Francisco CA, USA, 2006.
26. Naudé, M. Sustainable development in companies: Theoretical dream or implementable reality? *Corp. Ownersh. Control* **2011**, *8*, 352–364. [CrossRef]
27. Nandi, S.; Hervani, A.A.; Helms, M.M. Circular Economy Business Models—Supply Chain Perspectives. *IEEE Eng. Manag. Rev.* **2020**, *48*, 193–201. [CrossRef]
28. Ünal, E.; Urbinati, A.; Chiaroni, D. Managerial practices for designing circular economy business models: The case of an Italian SME in the office supply industry. *J. Manuf. Technol. Manag.* **2019**, *30*, 561–589. [CrossRef]
29. Urbinati, A.; Rosa, P.; Sassanelli, C.; Chiaroni, D.; Terzi, S. Circular business models in the European manufacturing industry: A multiple case study analysis. *J. Clean. Prod.* **2020**, *274*, 122964. [CrossRef]
30. Urbinati, A.; Chiaroni, D.; Chiesa, V. Towards a new taxonomy of circular economy business models. *J. Clean. Prod.* **2017**, *168*, 487–498. [CrossRef]
31. Moktadir, A.; Mithun, S.; Rajesh, R.; Kumar, S. Modeling the interrelationships among barriers to sustainable supply chain management in leather industry. *J. Clean. Prod.* **2018**, *181*, 631–651. [CrossRef]
32. Yadav, G.; Luthra, S.; Jakhar, S.K.; Mangla, S.K.; Rai, D.P. A framework to overcome sustainable supply chain challenges through solution measures of industry 4.0 and circular economy: An automotive case. *J. Clean. Prod.* **2020**, *254*, 120112. [CrossRef]
33. Bocken, N.M.P.; Short, S.W.; Rana, P.; Evans, S. A literature and practice review to develop sustainable business model archetypes. *J. Clean. Prod.* **2014**, *65*, 42–56. [CrossRef]
34. Rosa, P.; Sassanelli, C.; Terzi, S. Towards Circular Business Models: A systematic literature review on classification frameworks and archetypes. *J. Clean. Prod.* **2019**, *236*, 117696. [CrossRef]
35. Eurostat 2021. Regions in Europe—2021 Interactive Edition. Available online: https://ec.europa.eu/eurostat/cache/digpub/regions/#total-population (accessed on 19 June 2022).
36. Junta de Castil y Léon. *Thpe Circular Economy in Cities and Regions: Synthesis Report*; OECD Publishing: Paris, France, 2020. [CrossRef]

37. Junta de Castil y León 2020. Estrategia Economía Circular de Castilla y León 2020–2030. Available online: https://patrimonionatural.org/ficheros/5eeb521e74759_ESTRATEGIA-ECONOMiA-CIRCULAR-CASTILLA-Y-LEoN.pdf (accessed on 19 June 2022).
38. Vaismoradi, M.; Snelgrove, S. Theme in qualitative content analysis and thematic analysis. *Forum Qual. Soc. Res.* **2019**, *20*, 23. [CrossRef]
39. Sakai, S.-I.; Yoshida, H.; Hirai, Y.; Asari, M.; Takigami, H.; Takahashi, S.; Tomoda, K.; Peeler, M.V.; Wejchert, J.; Schmid-Unterseh, T.; et al. International comparative study of 3R and waste management policy developments. *J. Mater. Cycles Waste Manag.* **2011**, *13*, 86–102. [CrossRef]
40. Darmandieu, A.; Garcés-Ayerbe, C.; Renucci, A.; Rivera-Torres, P. How does it pay to be circular in production processes? Eco-innovativeness and green jobs as moderators of a cost-efficiency advantage in European small and medium enterprises. *Bus. Strategy Environ.* **2021**, *31*, 1184–1203. [CrossRef]
41. Garcés-Ayerbe, C.; Rivera-Torres, P.; Suárez-Perales, I.; Leyva-de Hiz, D. Is it possible to change from a linear to a circular economy? An overview of opportunities and barriers for European small and medium-sized enterprise companies. *Int. J. Environ. Res. Public Health* **2019**, *16*, 851. [CrossRef]
42. Banerjee, S.B. Organisational Strategies for sustainable development: Developing a research agenda for the new millennium. *Aust. J. Manag.* **2002**, *27*, 105–117. [CrossRef]
43. Parliament Adopts EU Environmental Objectives until 2030. Available online: https://www.europarl.europa.eu/news/pt/press-room/20220304IPR24804/parliament-adopts-eu-environmental-objectives-until-2030 (accessed on 15 June 2022).
44. Bonzanini Bossle, M.; Dutra Barcellos, M.; Marques Vieira, L.; Sauvée, L. The drivers for adoption of eco-innovation. *J. Clean. Prod.* **2016**, *113*, 861–872. [CrossRef]
45. Prieto-Sandoval, V.; Jaca, C.; Ormazabal, M. Towards a consensus on the circular economy. *J. Clean. Prod.* **2018**, *179*, 605–615. [CrossRef]
46. Horbach, J.; Janser, M. The role of innovation and agglomeration for employment growth in the environmental sector. *Ind. Innov.* **2016**, *23*, 488–511. [CrossRef]
47. Lacy, P.; Cooper, T.; Hayward, R.; Neuberger, L. A New Era of Sustainability. A UN Global Compact-Accenture CEO Study. 2010. Available online: https://www.compromisorse.com/upload/estudios/000/53/AccentureUNGCStudy10.pdf (accessed on 22 March 2022).
48. Wilkinson, A.; Hill, M.; Gollan, P. The sustainability debate. *Int. J. Oper. Prod. Manag.* **2001**, *21*, 1492–1502. [CrossRef]
49. Cho, C.H.; Roberts, R.W. Environmental reporting on the internet by America's Toxic 100: Legitimacy and self-presentation. *Int. J. Account. Inf.* **2010**, *11*, 1–16. [CrossRef]
50. Ormazabal, M.; Prieto-Sandoval, V.; Puga-Leal, R.; Jaca, C. Circular Economy in Spanish SMEs: Challenges and opportunities. *J. Clean. Prod.* **2018**, *185*, 157–167. [CrossRef]
51. Van Eijk, F. *Barriers & Drivers towards a Circular Economy. Literature Review*; A-140315-R-Final; Acceleratio: Naarden, The Netherlands, 2015.
52. Van Buren, N.; Demmers, M.; Van der Heijden, R.; Witlox, F. Towards a Circular Economy: The Role of Dutch Logistics Industries and Governments. *Sustainability* **2016**, *8*, 647. [CrossRef]
53. Dekoninck, E.A.; Domingo, L.; O'Hare, J.A.; Pigosso, D.C.A.; Reyes, T.; Troussier, N. Defining the challenges for ecodesign implementation in companies: Development and consolidation of a framework. *J. Clean. Prod.* **2016**, *135*, 410–425. [CrossRef]
54. Rizos, V.; Behrens, A.; Van der Gaast, W.; Hofman, E.; Ioannou, A.; Kafyeke, T.; Flamos, A.; Rinaldi, R.; Papadelis, S.; Hirschnitz-Garbers, M.; et al. Implementation of Circular Economy Business Models by Small and Medium sized Enterprises (SMEs): Barriers and Enablers. *Sustainability* **2016**, *8*, 1212. [CrossRef]
55. Aranda-Usón, A.; Portillo-Tarragona, P.; Marín-Vinuesa, L.M.; Scarpellini, S. Financial Resources for the Circular Economy: A Perspective from Businesses. *Sustainability* **2019**, *11*, 888. [CrossRef]
56. Ritzén, S.; Sandström, G.Ö. Barriers to the circular economy—Integration of perspectives and domains. *Procedia CIRP* **2017**, *64*, 7–12. [CrossRef]
57. Golinska, P.; Kosacka, M.; Mierzwiak, R.; Werner-Lewandowska, K. Grey decision making as a tool for the classification of the sustainability level of remanufacturing companies. *J. Clean. Prod.* **2015**, *105*, 28–40. [CrossRef]
58. Lieder, M.; Rashid, A. Towards circular economy implementation: A comprehensive review in context of manufacturing industry. *J. Clean. Prod.* **2016**, *115*, 36–51. [CrossRef]
59. Zeng, H.; Chen, X.; Xiao, X.; Zhou, Z. Institutional pressures, sustainable supply chain management, and circular economy capability: Empirical evidence from Chinese eco-industrial park firms. *J. Clean. Prod.* **2017**, *155*, 54–65. [CrossRef]
60. United Nations General Assembly. Transforming our World: The 2030 Agenda for Sustainable Development. 2015. Available online: https://www.refworld.org/docid/57b6e3e44.html (accessed on 22 March 2022).
61. Langen, S.K.; Vassillo, C.; Ghisellini, P.; Restaino, D.; Passaro, R.; Ulgiati, S. Promoting circular economy transition: A study about perceptions and awareness by different stakeholders groups. *J. Clean. Prod.* **2021**, *316*, 128166. [CrossRef]
62. Calogirou, C.; Sørensen, S.Y.; Larsen, P.B.; Pedersen, K.; Kristiansen, K.R.; Mogensen, J.; Alexopoulou, S.; Papageorgiou, M. *SMEs and the Environment in the European Union*; PLANET SA: Athens, Greece; Danish Technological Institute: Taastrup, Denmark; DG Enterprise and Industry; European Commission: Brussel, Belgium, 2010.

63. Droege, H.; Raggi, A.; Ramos, T.B. Overcoming current challenges for circular economy assessment implementation in public sector organisations. *Sustainability* **2021**, *13*, 1182. [CrossRef]
64. Liu, Y.; Bai, Y. An exploration of firms' awareness and behavior of developing circular economy: An empirical research in China. *Resour. Conserv. Recycl.* **2014**, *87*, 145–152. [CrossRef]
65. Liu, F.; Lai, K.-H.; Cai, W. Responsible production for sustainability: Concept analysis and bibliometric review. *Sustainability* **2021**, *13*, 1275. [CrossRef]
66. Cai, Y.-J.; Choi, T.-M. Extended producer responsibility: A systematic review and innovative proposals for improving sustainability. *IEEE Trans. Eng. Manag.* **2021**, *68*, 272–288. [CrossRef]
67. OECD. *Extended Producer Responsibility: Update Guidance for Efficient Waste Management*; OECD Publishing: Paris, France, 2016. [CrossRef]
68. Jones, P.; Comfort, D. A circular case: The circular economy and the service industries. *Int. J. Manag. Cases* **2020**, *22*, 13–23.
69. DeMicco, F.; Seferis, J.; Bao, Y.; Scholz, M.E. The Eco-Restaurant of the Future: A Case Study. *J. Foodserv. Bus. Res.* **2014**, *17*, 363–368. [CrossRef]
70. Del Giudice, M.; Chierici, R.; Mazzucchelli, A.; Fiano, F. Supply chain management in the era of circular economy: The moderating effect of big data. *Int. J. Logist. Manag.* **2020**, *32*, 337–356. [CrossRef]
71. Yu, Z.; Khan, S.A.R.; Umar, M. Circular economy practices and industry 4.0 technologies: A strategic move of automobile industry. *Bus. Strategy Environ.* **2022**, *31*, 796–809. [CrossRef]
72. Gue, I.H.V.; Promentilla, M.A.B.; Tan, R.R.; Ubando, A.T. Sector perception of circular economy driver interrelationships. *J. Clean. Prod.* **2020**, *276*, 123204. [CrossRef]
73. Bag, S.; Dhamija, P.; Bryde, D.J.; Singh, R.K. Effect of eco-innovation on green supply chain management, circular economy capability, and performance of small and medium enterprises. *J. Bus. Res.* **2022**, *141*, 60–72. [CrossRef]
74. Fernando, Y.; Wah, W.X.; Shaharudin, M.S. Does a firm's innovation category matter in practising eco-innovation? Evidence from the lens of Malaysia companies practicing green technology. *J. Manuf. Technol. Manag.* **2016**, *27*, 208–233. [CrossRef]
75. Sharma, R.; Jabbour, C.J.C.; Sousa Jabbour, A.B. Sustainable manufacturing and industry 4.0: What we know and what we don't. *J. Enterp. Inf. Manag.* **2020**, *34*, 230–266. [CrossRef]
76. Matthyssens, P. Reconceptualizing value innovation for Industry 4.0 and the Industrial Internet of Things. *J. Bus. Ind. Mark.* **2019**, *34*, 1203–1209. [CrossRef]
77. Nosalska, K.; Piątek, Z.M.; Mazurek, G.; Rządca, R. Industry 4.0: Coherent definition framework with technological and organizational interdependencies. *J. Manuf. Technol. Manag.* **2020**, *31*, 837–862. [CrossRef]
78. Sony, M.; Naik, S. Key ingredients for evaluating Industry 4.0 readiness for organizations: A literature review. *Benchmarking: Int. J.* **2019**, *27*, 2213–2232. [CrossRef]
79. Chen, C.-L. Cross-disciplinary innovations by Taiwanese manufacturing SMEs in the context of Industry 4.0. *J. Manuf. Technol. Manag.* **2020**, *31*, 1145–1168. [CrossRef]
80. Ćwiklicki, M.; Wojnarowska, M. Circular economy and industry 4.0: One-way or two-way relationships? *Eng. Econ.* **2020**, *31*, 387–397. [CrossRef]
81. Pagano, A.; Carloni, E.; Galvani, S.; Bocconcelli, R. The dissemination mechanisms of Industry 4.0 knowledge in traditional industrial districts: Evidence from Italy. *Compet. Rev. Int. Bus. J.* **2021**, *31*, 27–53. [CrossRef]
82. Sehnem, S.; Queiroz, A.A.F.; Pereira, S.C.F.; Santos Correia, G.; Kuzma, E. Circular economy and innovation: A look from the perspective of organizational capabilities. *Bus. Strategy Environ.* **2022**, *31*, 236–250. [CrossRef]

sustainability

MDPI

Article

Corporate Performance, Market-Industry Competition and Enterprise Environmental-Protection Investment

Liu Yang [1], Junting Tan [2], Weiyi Xia [3,4,*], Zhaomei Chi [1], Han Qin [2], Quanxin Gan [5,*] and Qin Yang [1,4]

1 School of International Education, Guangxi University of Finance and Economics, Nanning 530003, China; liuyang@gxufe.edu.cn (L.Y.); czm412@gxufe.edu.cn (Z.C.); gxufegqx@hotmail.com (Q.Y.)
2 MPAcc Center, Guangxi University of Finance and Economics, Nanning 530003, China; junting-tan118@hotmail.com (J.T.); xxsqinhan@hotmail.com (H.Q.)
3 Fangchenggang College, Guangxi University of Finance and Economics, Nanning 530003, China
4 Panyapiwat Institute of Management, Bangkok 10240, Thailand
5 Admissions and Employment Office, Guangxi University of Finance and Economics, Nanning 530003, China
* Correspondence: 2017250054@gxufe.edu.cn (W.X.); quanxin.ga@stu.nida.ac.th (Q.G.)

Abstract: Worldwide, many countries regard green as a keyword related to development, and investments into environmental protection are an important way for enterprises to achieve green development. Therefore, clarifying which factors influence enterprises to invest into environmental protection is very important. Starting from micro-enterprises and using the data from companies listed in China's A-share manufacturing industry from 2008 to 2019, in this study, we empirically analyze the relationship between corporate performance (CP) and the scale of investments by enterprises into environmental protection (EI) and analyze the moderating effect of industry competition on the relationship between CP and EI. The result shows that (1) a positive correlation can be found between CP and EI; (2) fierce industry competition can increase the positive impact of CP on EI; and (3) compared with industries with non-heavy pollution, fierce industrial competition increases the positive impact of CP on EI in industries with heavy pollution. The research results show that performance is a key factor influencing enterprises' decisions about investments into environmental protection, and industry competition can stimulate enterprises to invest into environmental protection. This study explores the internal and external factors influencing an organization to promote active behaviors of investing into environmental protection, provides a reference for enterprises to explore "win–win" paths, and provides a certain theoretical basis for the government to improve relevant regulations.

Keywords: corporate performance; enterprise environmental-protection investment; industry competition

Citation: Yang, L.; Tan, J.; Xia, W.; Chi, Z.; Qin, H.; Gan, Q.; Yang, Q. Corporate Performance, Market-Industry Competition and Enterprise Environmental-Protection Investment. *Sustainability* **2022**, *14*, 5459. https://doi.org/10.3390/su14095459

Academic Editors: Sergio Terzi, Claudio Sassanelli and Donato Morea

Received: 11 February 2022
Accepted: 28 April 2022
Published: 1 May 2022

1. Introduction

In 2020, the Chinese government clearly put forward the goals of reaching their "carbon peak" in 2023 and "carbon neutrality" in 2060. At present, from the perspective of industrial structures, the proportion of industries with heavy pollution in China compared with the overall economy ranks the highest among developed economies, and the amount of air pollution produced by one unit output of the industries with heavy pollution is nine times that of the service industry. Changing this polluting economic structure is one way to achieve the carbon peak and carbon neutrality goals set by the government. Among these methods, changing the investment structure by increasing investments into environmental protection is a good starting point for changing this economic structure. The Chinese government has continued to increase their investments into environmental protection. However, China needs to invest about CNY 4 trillion into environmental protection annually, but the Chinese government's environmental-protection investment funds only account for about 10% of this need. Therefore, the Chinese government urgently needs to encourage everyone to invest into environmental protection to meet China's investment needs for environmental protection.

Enterprises are the main consumers of resources and the main manufacturers of pollutants. Therefore, they should bear the responsibility of investing into environmental protection and reducing the impact of business activities on the environment [1]. As environmental externalities cannot be internalized, environmental-protection projects often provide insufficient returns, and the rate of return is lower than the rate of return required by the market. Thus, enterprises investing into environmental protection is more a reflection of their response to external pressure and, therefore, the existing literature pays more attention to the influence of external factors such as government environmental regulation [2–4], the public [5–8], and consumers [9–11] on an enterprise's investment behaviors regarding environmental protection, and is based on institutional theory and stakeholder theory. However, explaining why some companies spend limited resources on environmental projects when they share the same institutional background or the same regional environment from an external perspective, while some enterprises focus on other projects, may be difficult.

Focusing on the factors influencing enterprises to invest into environmental protection, some studies in the literature empirically studied the internal factors of an enterprise, such as ownership concentration [12,13], environmental management system [14–16], managers' private income [17], political connection [18], internal control quality [19,20], and other factors related to their corporate governance structure and corporate social capital, but no consistent conclusions have been drawn. Therefore, what internal factors affect decision making by enterprises regarding investments into environmental protection is still worth exploring.

The existing literature discusses the pressure-formation process for high-performance enterprises and its impact on investment behaviors from the perspective of enterprise behavior theory and prospect theory. He et al. [21] found that the main way for high-performance enterprises to turn high expectations into reality and to alleviate the pressures of playing catch up is to adopt short-term investment behaviors such as hollowed-out, related-party transactions. When the actual performance of an enterprise is higher than the expected performance target, the greater is the possibility of negative behaviors being adopted by the enterprise [22,23]. Guo and Chen [24] found that, when an enterprise is in a state of excellent performance, redundant resources and successful paths that are accumulated during the early stages of operation will help the enterprise to implement M&A and to obtain a higher level of performance.

On the issue of environmental protection, a small amount of literature combines corporate performance with an enterprise's investments into environmental protection. Some scholars believe that high-performance enterprises have more disposable funds, are more likely to obtain external financing, and are more optimistic about future expectations. Therefore, high-performance enterprises will be more willing to increase their investments into environmental protection [25]. However, this view has not been unanimously recognized. Hitchens et al. [26] found no significant correlation between corporate performance and the scale of environmental investment by European SMEs. Additionally, some research results on the relationship between corporate performance and enterprise environmental investment have not reached a consistent conclusion.

In contrast to other investment decisions made by enterprises, investments into environmental protection are strongly affected by external factors. According to the characteristics of investments into environmental protection, the initial cost of an investment into environmental protection is often greater than its resultant income, which undoubtedly increases the business risk of an enterprise, resulting in some enterprises being unwilling to actively invest in environmental protection. Will the business risk be lessened with an increase in corporate performance? On the one hand, the level of an enterprise's performance represents their developmental abilities, profitability, and disposable capital, which affects the point of reference for decision making, thus affecting their decision making regarding investments into environmental protection. On the other hand, as the standard for evaluating the salary, promotion, and dismissal of managers, performance plays a part in the incentive mechanism for managers, which affects the enterprise's preferences towards taking risks, thus affecting decisions made regarding investments into environmental protection.

In addition, the existing literature mostly studies the moderating effect of internal governance factors on CP and EI from the perspective of an internal organization. A small amount of literature considers that, in different competitive environments, enterprise managers may adopt different competitive methods and strategies, thus affecting investment decisions for environmental protection. To explore the above problems, in this paper, we study the moderating effect of industry competition on the relationship between CP and EI from the perspective of risk aversion. Through the study of this moderating effect, we can strengthen the correlation between the internal and external factors of an organization and an enterprise's investments into environmental protection, and help provide reasons to encourage enterprises to actively invest in environmental protection. Finally, considering the different sensitivities of enterprises within industries with heavy pollution and non-heavy pollution to environmental 'legitimacy', we find that there are great differences in enterprises' decisions regarding investments into environmental protection. This paper further studies the differences in the moderating effect of industry competitiveness on the relationship between CP and EI for different industry attributes. The above research results will help China and other countries committed to green development determine what are the main factors that encourage enterprises to bear the responsibility of environmental protection and provides a theoretical and practical basis for realizing the goals of green development.

The rest of this paper is arranged as follows: Section 2 provides the theoretical background and research hypothesis. Section 3 introduces the sample, data, model, and variables used in the empirical analysis. Section 4 analyzes the empirical results. Finally, Section 5 gives the research conclusions, the significance of this study, the limitations of this study, and possible future research directions.

2. Theoretical Background and Research Hypothesis

2.1. Corporate Performance and Investments by Enterprises into Environmental Protection

In the literature, theoretical research on investment behaviors focuses on financial variables and research on the impact of profit on investment is usually called the profit theory of investment [27]. The establishment of the profit theory of investment was based on the following two views. The first view holds that the profit achieved determines the expected level of profit, while the level of investment is determined by the expected level of profit. Profit is thus used as a proxy variable of expected capital in the investment model; while, in the profit theory of investment, profit directly determines the level of investment. The second view is that investment ratios are limited by the supply of funds. In incomplete markets with information asymmetry and agency costs, investment decision making is more sensitive to the availability of internal funds because they have more cost advantages than external funds. Jensen [28] believed that the better the cash flow, the more enterprises tend to increase their investments.

Investments by enterprises into environmental protection are characterized by general investment projects and affect environmental performance at the same time, so they are characterized by the comprehensive income from economic performance and environmental performance. However, investment projects regarding environmental protection are generally focused on environmental-protection-technology R&D, pollution end treatment, environmental-protection-equipment purchase, cleaner production, and other aspects that makes producing direct economic benefits based on investments into environmental protection difficult in the short term [29]. Moreover, these large investment expenditures put pressure on profits and increase operating risks [30]. Risk-averse managers tend to reduce investments into environmental protection to reduce operating costs and to improve revenue. As the primary goal of enterprises is "survival" [31], this pressure to balance profit with risk is an important factor barring many enterprises from investing into environmental protection.

For high-performing enterprises, first, enterprises with outstanding business performances tend to show an upward trend. With improvements in the actual performance of enterprises, managers will improve their point of reference for decision making [32]. A higher existing profit level means a higher expected profit level. High-performing enterprises will thus not take "survival" but rather "seeking development" as their reference point for decision making. Therefore, high-performance enterprises will give more consideration to the environmental demands of stakeholders, and national strategies for continuously strengthening environmental governance will be met with more rapid environmental investments [24].

Second, the personal risk preferences of managers change with a change in the reference point for decision making. When taking "survival" as the reference point for decision making, managers show risk aversion. When taking "seeking development" as the reference point for decision making, managers show a risk preference, which affects the decision making of an enterprise regarding investments into environmental protection. Managers of high-performing enterprises will overestimate the positive impact of investments into environmental protection on the economic and social benefits awarded to enterprises in the short term, such as enhancing market competitiveness and improving corporate image, while underestimating the negative impact of investments into environmental protection on business risks when increasing the scale of investments into environmental protection. For enterprises that do not have outstanding business performance, they will be more inclined toward risk aversion. The threat rigidity hypothesis of Staw et al. [33] holds that decision makers in crisis prefer to rely on past experiences and existing knowledge to strengthen their control over existing resources, such as reducing costs and improving efficiency. Therefore, their decision-making behavior tends toward being stable rather than taking risks [34]. Compared with daily production and operational investments, investments into environmental protection do not produce an "immediate" effect, their explicit expected economic return is low, and their acquisition time is unknown. Enterprise managers tend to reduce the level of investment into environmental protection for the sake of business stability.

Third, high-performing enterprises not only have more internal capital but also are more likely to obtain external capital support [25,35], which can make up for the net loss from investments into environmental protection in the initial stages, to a certain extent, and can help stabilize investor confidence [36]. The trade-off between the costs and benefits affects the decision-making process of an enterprise regarding investments into environmental protection. When enterprises bear the cost of investments into environmental protection, they will also obtain a certain amount of income. The cost of environmental-protection investment is often large in the early stages and small in the later stages, while the income generated from investments into environmental protection follows an opposite trend. The net profit and loss of investments into environmental protection is the difference between the income from investments into environmental protection minus the cost. In the initial stages of investments into environmental protection, the cost is often greater than the income, that is, the net loss, which is one of the main reasons why some enterprises are unwilling to invest in environmental protection. The high profit level of high-performing enterprises provides support for enterprises regarding their resources. This will reduce the risk of market fluctuation caused by a net loss in environmental-protection investment and stabilize investors' confidence in the future performance of enterprises. Based on the above analysis, we propose the following research hypothesis:

Hypothesis 1 (H1). *Corporate performance is positively correlated with the scale of an enterprise's investment into environmental protection.*

2.2. The Moderating Role of Industry Competition

As an important external governance mechanism, industry competition can encourage managers to improve business efficiency, to a certain extent. Under different competitive environments, the role of the joint constraints of incentive mechanisms within the enterprise and external governance mechanisms in agency problems will change, further affecting the level of investment by an enterprise into environmental protection. Corporate performance is an evaluation standard that affects the salary, promotion, and dismissal of managers. In a market with fierce competition, the comparability between corporate performance levels increases [22], and it is easier to evaluate the abilities of managers [37]. On the contrary, in a market with weak competition, the role of corporate performance in evaluations will be significantly reduced. Therefore, fierce competition makes the relationship between corporate performance and actual environmental behavior more intuitive. In particular, when the design and production of similar products are the same, the difference in environmental impact mainly comes from investments into environmental protection [38]. At this time, the market becomes more sensitive to the proportion of corporate performance invested in green-technology R&D, pollution control, and other investments. As the pressure of environmental protection increases, managers' perception of environmental-pollution risks and losses will be amplified accordingly. In this situation, the pursuit of risk by high-performance managers is further strengthened based on the psychological decision-making process of loss aversion. When determining what investment projects to engage in, managers will overestimate the possible benefits of investments into environmental protection and underestimate the risks of investments into environmental protection, which will then promote an increase in the scale of investments into environmental protection by enterprises.

In addition, fierce competition facilitates the antagonistic relationship between enterprises within the same industry, triggering the pressures of "innovation competition" [39]. Therefore, the more intense the market competition, the more enterprises need to find new competitive advantages to alleviate this competitive pressure. Among them, a new competitive advantage can be realized through an enterprise's investment into environmental protection, especially an enterprise's investment into environmental-product innovation and environmental-technology R&D to meet the increasing demand for green standards, to increase the differentiated comparative advantages of products, to help an enterprise establish a good green reputation and brand image, and to improve its brand effect [40] to, therefore, alleviate the competitive pressure of the industry. As Zhang et al. [41] found, ecological labels and environmental values will have a positive impact on the purchase intentions of green products.

When the market competition is low, competitive elimination in the market is weak. Even if enterprises invest more into environmental protection and improve their corporate image, the impact on stimulating consumers to buy green products and technologies will be small. On the contrary, it may expose the problems of environmental pollution for enterprises, attract the attention of regulatory authorities, and affect the normal production and operation of enterprises. This is the phenomenon called "whip the fast and hard working—unfair punishment" [42]. To maintain corporate image and interests, high-performing enterprises are more likely to invest resources into other projects, rather than environmental-protection projects. Based on this, we propose the following research hypothesis:

Hypothesis 2 (H2). *Fierce competition will intensify the positive impact of corporate performance on the scale of investments into environmental protection.*

2.3. *Influence of Industry Differences on Moderating Effect*

In order to achieve the goals of green development, most countries implement different environmental policies based on the actual situation of different regions and industries. It can be seen that different industry attributes and different environmental regulations and standards inevitably affect the strategic decision making of enterprises. Compared with enterprises in industries with non-heavy pollution, enterprises from industries with heavy pollution produce a greater degree of environmental pollution and environmental damage. Therefore, enterprises in industries with heavy pollution are increasingly facing stricter and higher emission-reduction standards. For example, China has clearly begun to implement a differentiated qualification certification, government support and other measures for classified enterprises. These policies undoubtedly make heavy-polluting enterprises bear more environmental-protection responsibilities and increase their willingness to reach environmental legitimacy and to avoid environmental-pollution risks.

When faced with fierce competition, enterprises will inevitably make investment decisions in accordance with governmental regulations about the environment. Compared with industries with non-heavy pollution, enterprises in industries with heavy pollution are obviously facing more stringent environmental-regulation pressures. The dual external pressure of market competition and environmental regulation will encourage high-performing enterprises to invest in environmental-protection projects to alleviate this external pressure and to thus achieve both the goals of legitimacy and acquiring a green reputation. Based on this, we propose the following, research Hypothesis 3:

Hypothesis 3 (H3). *Compared with industries with non-heavy pollution, fierce market competition will intensify the positive impact of CP on EI in industries with heavy pollution.*

Therefore, we propose the following theoretical model, shown in Figure 1 below.

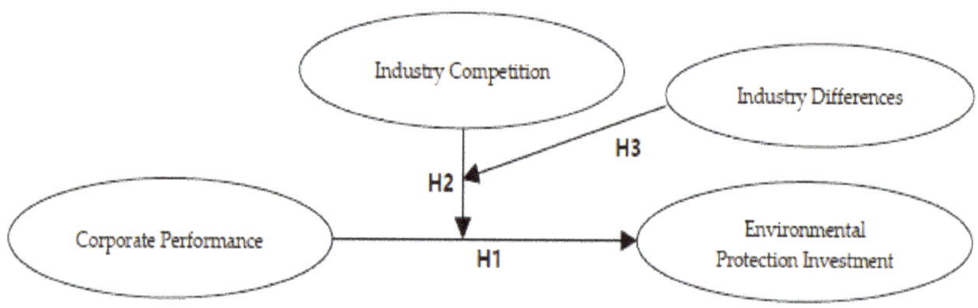

Figure 1. Theoretical model.

3. Research Design

3.1. Research Samples and Data Sources

3.1.1. Research Samples

On 22 November 2007, the State Council of China issued the *"11th Five-Year Plan for National Environmental Protection"*. Strictly speaking, 2008 was the opening year of the 11th five-year plan of environmental protection, and the environmental-protection investment data for enterprises were only updated to 2019, so we chose 2008–2019 as the research object of A-share manufacturing listed companies in China. In order to ensure the reliability of the data, the data for missing data and data of ST (specially treatment) and PT (particular transfer) enterprises were deleted, 16,145 research samples were finally identified.

This study further divided the whole sample into industries with heavy pollution and with non-heavy pollution. It was used to test the difference in the moderating effects of industry competition on the relationship between CP and EI under different industry attributes. Following Li and Lu [43], and Bai and Zhang [20], the industries with heavy pollution were determined according to the "List of Classification Management of Environmental Verification Industries of Listed Companies" issued by China's Ministry of Environmental Protection in 2008 and the "Guidelines for Industrial Classification of Listed Companies" issued by the China Securities Regulatory Commission in 2012, including 16 categories such as metallurgy, chemical, petrochemical, coal and thermal power. If an enterprise was from an industry with heavy pollution, we set the value to 1; otherwise, it was 0. After the classification, there were 7384 enterprises in the heavy-pollution industries group and 8761 enterprises in the non-heavy-pollution industries group.

3.1.2. Data Sources

The data on enterprise's investments into environmental protection were obtained by manually screening and sorting based on the detailed data from a list of "construction projects" in the CSMAR database, which includes data on corporate performance, industry competition, and other financial indicators. To avoid the influence of extreme values, all continuous variables were "Winsorize" tailed at the upper and lower 1% levels.

3.2. Empirical Model

Mixed data from multiple years and multiple enterprises were used in this study. The fixed effect model and cluster-robust standard error (robust) were used in the regression analysis. To test Hypothesis 1, the regression model used was as follows:

$$EI_{i,t} = \alpha + \beta_1 CP_{i,t} + \beta_2 CONTROL_{i,t} + YEAR_t + \xi_{i,t} \quad (1)$$

In model (1), EI represents the scale of enterprise environmental-protection investment; CP represents corporate performance; CONTROL represents a collection of control variables; ξ represents the disturbance term; α represents the constant term; and β_1 represents the coefficient of explanatory variables, with its direction being the focus of this study, indicating the impact of corporate performance on the scale of an enterprise's investment into environmental protection.

In industrial competition, we asked whether the effect of corporate performance on the scale of investments into environmental protection for manufacturing enterprises changes. Therefore, according to the moderating effect model and based on model (1), we added variable HHI and cross-product term CP $\times$ HHI, and model (2) was constructed to test Hypothesis 2:

$$EI_{i,t} = \alpha + \beta_1 CP_{i,t} + \beta_2 HHI_{i,t} + \beta_3 CP_{i,t} \times HHI_{i,t} + \beta_4 CONTROL_{i,t} + YEAR_t + \xi_{i,t} \quad (2)$$

In model (2), HHI represents the industry competition intensity and β_3 represents the key coefficient of the moderating effect.

3.3. Variable Definitions

(1) Dependent Variable

The scale of investments by an enterprise into environmental protection (EI): this article followed the practices of Tang and Li [44], and Zhai and Liu [45], calculating EI using total investment/capital stock, in which the total investment is the total amount of new investments into environmental protection in the current year and the capital stock is the average total assets. That is, EI = the newly added investments into environmental protection in the current year/the average total assets.

(2) Independent Variable

Corporate performance (CP): following Zeng et al. [46], this study selected the net interest rate of total assets (ROA) to measure corporate performance, which refers to the ratio of the net profit realized by the enterprise in a certain operating period to the average total assets.

(3) Moderating Variable

Industry competition intensity (HHI): according to industrial theory, competition is more often generated within industrial enterprises. This study used the Herfindahl–Hirschman index to measure this variable with reference to the practices of Gu [47] and Zhang et al. [39]. This indicator measures the amount of competition in an industry. The higher the HHI, the greater the degree of monopoly. On the contrary, the smaller the HHI, the greater the degree of market competition.

$HHI = \sum (X_i/X)^2$, where $X = \sum X_i$, X_i is the book value of the owner's equity in a single enterprise.

(4) Control Variables

According to the previous research, an enterprise's investments into environmental protection are affected by many other factors, mainly including three aspects: the basic characteristics, financial performance, and governance characteristics of the corporation [16,43,48]. Therefore, based on the above research, this study set the following control variables:

(1) The size of enterprise (SIZE). With the increase in enterprise scale, enterprises will have more opportunities to invest in environmental protection [49]. Following Duan and Xu [50], this index was measured using the natural logarithm of total assets at the end of the year.

(2) The age of enterprise (AGE). The longer an enterprise has been established, the more technology and knowledge advantages it has. In order to reduce uncertain losses, enterprises tend to invest into environmental protection. Following Guo and Zhang [51], the difference between the year of establishment and the year of observation was selected to measure the age of enterprises.

(3) Property right (SOE). According to property right theory, state-owned holding enterprises will increase their investments into environmental protection in response to the national economy and people′s livelihoods. Following Cui et al. [52], it was set as a virtual variable, with 1 for state-owned enterprises and 0 otherwise.

(4) Financial risk (LEV). The higher the financial risk, the greater the debt-repayment pressure, the more serious the financing constraint, and the less investment into environmental protection. Following Tan and Yang [53], Jiang and Huang [54], and Gebhardt et al. [55], we measured financial risk using the asset–liability ratio.

(5) Operational risk (VOL). The greater the operating risk, the less likely a long-term investment will be made, which is negatively correlated with the enterprise′s investments into environmental protection. Following Yang et al. [56], and Giuli and Kostovetsky [57], the annual stock price volatility of the enterprise was used to measure operating risk.

(6) Financial slack (SLACK). The more slack resources managers have, the more complacent managers will be, which makes them optimistic and unlikely to try alternative strategies [58]. It reduces the motivation of managers to invest in environmental protection. Following Yang et al. [16] and Zou et al. [59], we used the ratio of cash and cash equivalents to total assets at the end of the period to measure this index.

(7) Shareholding ratio of institutional investors (INSHARE). As an external force of the corporate governance mechanism, institutional investors have strong abilities to perform information analyses and mining. Institutional investors′ shareholding can reduce the degree of information asymmetry and the financial constraints of enterprises, improve the level of corporate governance, and allow enterprises to fulfill their environmental-protection responsibilities. Following Zhu et al. [60], this index was measured by the ratio of the number of shares held by institutional investors to the total share of capital owned by the enterprise.

(8) Proportion of independent directors (IDD). Independent directors influence an enterprise's investments into environmental protection by encouraging and restricting directors and by supervising and balancing the board of directors. In a sense, the scale of independent directors represents the influence of independent directors when expressing independent opinions to the board of directors or at a general meeting of shareholders. The larger the scale, the greater the possibility that the environmental-protection investment issues proposed by independent directors will be discussed and adopted by the board meeting [36]. Following Liu and Li [61], this index was measured by the proportion of independent directors to the number of the board of directors.

(9) CEO change (TURN). After a change in CEO, the new CEO will have short-term expectations and be unwilling to bear too many investment risks. Therefore, a new CEO will try to avoid investing resources into projects with limited maturity and lacking effectiveness, which will affect the enterprise's environmental-protection investment behaviors.

Following Huang [62], we set it as a virtual variable. If the CEO changed, we set it to 1; otherwise, it is 0.

(10) Agency cost (COST). The agency cost is caused by a conflict of interest between management and shareholders. On the agency issue, in order to obtain private benefits from investment, an enterprise's management will be committed to building an empire and attempt to maximize assets, which will affect the enterprise's decisions to invest in environmental protection. Following Jensen and Meckling [63], and Guo [64], this indicator was measured based on the proportion of administrative expenses in operating revenue.

In addition, this study also controlled the annual fixed effect (YEAR) to exclude the heterogeneity from different years.

Details of the variable symbols and specific settings are shown in Table 1 below.

Table 1. Research variable description.

Type	Name	Code	Indicators' Description
Dependent variable	The scale of an enterprise's investment into environmental protection	EI	Newly added investments into environmental protection in the current year/average total assets
Independent variable	Corporate performance	CP	Net profit/average total assets
Moderating variable	The industry competitive intensity	HHI	Herfindahl–Hirschman index calculated using the recorded owners' equity
Control variables	The size of the enterprise	SIZE	Natural log of total assets
	The age of the enterprise	AGE	Difference between establishment and observation year
	Property right	SOE	1 for state-owned enterprises; otherwise, 0
	Financial risk	LEV	Total liabilities/total assets
	Financial slack	SLACK	Cash and cash equivalents/total assets
	Operating risk	VOL	Annual stock-price volatility in the current year
	CEO change	TURN	If the CEO changed, set to 1 in the current year; otherwise, 0
	Agency Cost	COST	Management expenses/operating income
	Proportion of independent directors	IDD	Number of independent directors/total number of directors
	Shareholding ratio of institutional investors	INSHARE	Institutional investor holdings/total shares
	Annual dummy variables	YEAR	Setting 11 dummy variables in 12 years

4. Empirical Results and Analysis

4.1. Descriptive Statistics

The descriptive statistics of variables in this study are shown in Table 2. Among them, the maximum of EI is 5.9010, the minimum is 0, the mean is 0.3042, and the median is 0, indicating that more than half of the sample enterprises do not reach the average amount of investments into environmental protection. The standard deviation of 1.0271 is greater than the mean and median, indicating that there are obvious individual differences in the amount of investment into environmental protection by the sample enterprises. The difference between the maximum (0.2398) and the minimum (−0.2221) CP is large, and the standard deviation is greater than the mean and median, indicating that there are significant individual differences in the performances of the sample enterprises. The mean industry competitiveness (HHI) is 27.2311, which is higher than the median, 23.0590, indicating that most of the sample enterprises are in an industry with a high degree of competition. The mean SOE is 0.3052, indicating that 30.52% of the sample enterprises are state-owned enterprises. The mean TURN is 0.1096, indicating that 10.96% of the sample enterprises had a CEO turnover during 2008–2019. From the descriptive statistics of other variables in Table 2, it can be found that there is a large difference between the maxima and the minima, indicating that there are large difference between the financial risk, operational risk, and internal governance environments of the sample enterprises.

Table 2. Descriptive statistics.

Variables	N	Mean	Std	Minimum	Maximum	Median
EI	16,145	0.3042	1.0271	0	5.9010	0
CP	16,145	0.0489	0.0620	−0.2221	0.2398	0.0450
HHI	16,145	27.2311	18.8159	3.0736	67.0646	23.0590
SIZE	16,145	21.9174	1.1685	18.4887	26.8076	21.7518
AGE	16,145	20.3595	5.0092	7	37	20
SOE	16,145	0.3052	0.4605	0	1	0
LEV	16,145	38.9713	19.6609	5.4152	92.4247	37.8942
SLACK	16,145	0.7136	3.1169	0.0060	26.8284	0.1332
VOL	16,145	48.9663	23.0038	17.9081	157.6656	43.1473
TURN	16,145	0.1096	0.3124	0	1	0
COST	16,145	0.0905	0.0650	0.0061	0.7891	0.0780
IDD	16,145	37.1087	5.2373	0	57.1429	33.3333
INSHARE	16,145	35.2940	23.5946	0.0775	88.2443	34.8098

4.2. Correlation Analysis

The correlation coefficient analysis of the independent variable, dependent variable and moderating variable in this study is shown in Table 3. It can be seen that the correlation coefficient between EI and CP (r = 0.026) is positive and significant at the 10% level, indicating that an improvement in corporate performance will help to improve investments into environmental protection.

The correlation coefficient between HHI and EI (r = 0.014) is positive but not significant, and the correlation coefficient between HHI and CP (r = 0.080) is positive and significant at the 1% level. The following sections will further test our hypotheses based on a regression analysis.

4.3. Multiple Regression Analysis

To determine whether to use the fixed-effect model or the random-effect model, we conducted the Hausman test. According to the Hausman test results, the p value is <0.01, so the fixed-effect model should be used instead of the random-effect model. The following hypothesis tests adopt the two-way fixed-effect model regression.

4.3.1. Regression Analysis of the Impact of CP on EI

The regression results between CP and EI are shown in Table 4. According to model (1) in Table 4, the coefficient between CP and EI ($\beta_1 = 0.398$) is positive and significant at the 5% level, indicating that, with an improvement in corporate performance, the anti-risk ability is improved. Since "seeking development" is the expected goal of high-performance enterprises, which encourages enterprises to invest more actively into environmental protection, Hypothesis 1 is verified.

Table 3. Correlation test results of main variables.

Variables	EI	CP	HHI	SIZE	AGE	SOE	LEV	SLACK	VOL	TURN	COST	IDD	INSHARE
EI	1												
CP	0.026 *	1											
HHI	0.014	0.080 ***	1										
SIZE	−0.034 **	−0.045 ***	−0.051 ***	1									
AGE	−0.084 ***	−0.099 ***	−0.015 **	0.207 ***	1								
SOE	−0.122 ***	−0.161 ***	−0.083 ***	0.302 ***	0.188 ***	1							
LEV	−0.002	−0.419 ***	−0.159 ***	0.337 ***	0.124 ***	0.330 ***	1						
SLACK	−0.012	0.028 ***	−0.041 ***	−0.131 ***	0.026 ***	0.086 ***	0.031 ***	1					
VOL	0.039 ***	0.022 ***	0.006	−0.248 ***	−0.111 ***	−0.083 ***	−0.069 ***	0.167 ***	1				
TURN	−0.066 ***	−0.112 ***	−0.023 ***	−0.023 ***	0.042 ***	0.157 ***	0.096 ***	0.015 **	0.005	1			
COST	0.045 ***	−0.175 ***	0.101 ***	−0.305 ***	0	−0.091 ***	−0.115 ***	−0.037 ***	0.090 ***	0.046 ***	1		
IDD	−0.008	−0.011	0.003	0.093 ***	0.022 ***	−0.058 ***	−0.039 ***	−0.060 ***	0.011	−0.073 ***	0.032 ***	1	
INSHARE	−0.056 ***	0.060 ***	−0.036***	0.415 ***	0.108 ***	0.327 ***	0.199 ***	−0.053 ***	−0.236 ***	0.052 ***	−0.120 ***	−0.037 ***	1

Note: *, **, *** represent the significance level of 1%, 5%, and 10%, respectively.

Table 4. Regression results.

Variables	EI					
	Full Samples		**Heavy-Polluting Industry**		**Non−Heavy Polluting Industry**	
	Model (1)	**Model (2)**	**Model (1)**	**Model (2)**	**Model (1)**	**Model (2)**
CP	0.398 **	0.932 ***	0.545 *	1.225 ***	0.010	0.564
	(2.20)	(3.24)	(1.94)	(3.28)	(0.04)	(1.16)
HHI		0.002		0.001		0.002
		(1.21)		(0.54)		(0.70)
CP×HHI		−0.021 **		−0.024 **		−0.023
		(−2.04)		(−1.99)		(−1.31)
SIZE	0.151 ***	0.150 ***	0.132 ***	0.131 ***	0.149 ***	0.148 ***
	(4.53)	(4.50)	(2.81)	(2.78)	(3.19)	(3.19)
AGE	0.090	0.092	0.195 *	0.196 **	−0.444	−0.440
	(0.56)	(0.57)	(1.94)	(1.96)	(−1.41)	(−1.40)
SOE	0.045	0.045	0.012	0.012	0.080	0.082
	(0.66)	(0.66)	(0.14)	(0.13)	(0.75)	(0.76)
LEV	0.001	0.001	−0.001	−0.001	0.004 **	0.004 **
	(1.01)	(1.02)	(−0.88)	(−0.91)	(2.33)	(2.33)
SLACK	−0.004	−0.004	0.001	0.001	−0.002	−0.002
	(−0.83)	(−0.86)	(0.10)	(0.11)	(−0.32)	(−0.38)
VOL	−0.000	−0.000	−0.001	−0.001	−0.000	−0.000
	(−0.45)	(−0.41)	(−0.81)	(−0.83)	(−0.29)	(−0.23)
TURN	−0.049 **	−0.049 **	−0.015	−0.015	−0.076 **	−0.076 **
	(−2.15)	(−2.16)	(−0.44)	(−0.44)	(−2.55)	(−2.56)
COST	0.662 ***	0.654 ***	0.590	0.563	0.507 **	0.507 **
	(3.38)	(3.34)	(1.60)	(1.53)	(2.15)	(2.16)
IDD	−0.002	−0.002	−0.004	−0.003	−0.002	−0.002
	(−0.82)	(−0.83)	(−0.76)	(−0.73)	(−0.57)	(−0.59)
INSHARE	−0.000	−0.000	−0.001	−0.001	−0.000	−0.000
	(−0.76)	(−0.74)	(−0.61)	(−0.60)	(−0.46)	(−0.45)
YEAR	Control	Control	Control	Control	Control	Control
Constant	−4.688	−4.745	−6.252 ***	−6.282 ***	5.632	5.511
	(−1.42)	(−1.45)	(−2.74)	(−2.77)	(0.90)	(0.88)
N	16,145	16,145	7,384	7384	8761	8761
R^2	0.013	0.014	0.024	0.025	0.017	0.018
Adj−R^2	0.0119	0.0122	0.0214	0.0219	0.0149	0.0150
F	4.969 ***	4.776 ***	4.358 ***	4.427 ***	2.690 ***	2.468 ***

Note: *, **, *** represent the significance level of 1%, 5%, and 10%, respectively.

From the results of the relationship between the control variables and EI in model (1) from Table 4, we can see that the correlation between the coefficient of firm size (SIZE) and EI ($\beta = 0.151$) is positive and significant at the 1% level, the correlation between the coefficient of CEO turnover (TURN) and EI ($\beta = -0.049$) is negative and significant at the 5% level, and the correlation between the coefficient of agency cost (COST) and EI ($\beta = 0.662$) is positive and significant at the 1% level.

4.3.2. Regression Analysis of the Moderating Effect of HHI on CP and EI

According to model (2) in Table 4, the coefficient of CP and EI ($\beta_1 = 0.932$) is positive and significant at the 1% level, indicating that the higher the performance level of manufacturing enterprises, the larger the scale of investments into environmental protection. Thus, Hypothesis 1 is further verified.

In addition, the coefficient CP × HHI and EI ($\beta_3 = -0.021$) is negative and significant at the 5% level. The smaller the HHI value, the higher the industry competition. These results show that higher industry competition triggers "green competition" among manufacturing enterprises, resists the long-term environmental impact and business risks brought on by fierce competition, and encourages high-performance enterprises to invest more resources into environmental protection projects. Hypothesis 2 is thus validated.

From the relationship between the control variables and EI in model (2), it can be seen that it is consistent with the relationship between the control variables and EI in model (1).

4.3.3. Regression Analysis on Influence of Industry Differences on Moderating Effect

According to the grouping regression results in Table 4, compared with non-heavy-pollution industries, the coefficient ($\beta_1 = 0.545$) of CP and EI in heavy-pollution industries is positive and significant at the 10% level. After adding HHI and CP × HHI, the coefficient of CP and EI in heavy-pollution industries increases from 0.545 to 1.225 and is significant at the 1% level. Additionally, the coefficient ($\beta_3 = -0.024$) of CP × HHI is negative and significant at the 5% level, indicating that fierce industry competition increases the positive impact of CP on EI in heavy-pollution industries.

4.4. Robustness Tests

This paper used the methods of index substitution, adding control variables and reducing samples to test the robustness, and continued to use the fixed-effect model for the regression analysis, as follows:

First, replace the explanatory variable. Replace net profit/total assets with net profit/average total assets. It can be seen from Table 5 that the coefficient of CP is still significantly positive, and that the coefficient of CP × HHI is negative and significant at the level of 5%. The coefficient of CP × HHI in heavy-pollution industries is negative and significant at the 5% level.

Table 5. Robustness Test 1.

Variables	CP (Alternative Measures)				Added Control Variables			
	Full Samples		Heavy Polluting Industry	Non-Heavy Polluting Industry	Full Samples		Heavy Polluting Industry	Non-Heavy Polluting Industry
	Model (1)	Model (2)	Model (2)	Model (2)	Model (1)	Model (2)	Model (2)	Model (2)
CP	0.004 **	0.009 ***	0.012 ***	0.007	0.432 **	0.970 ***	1.247 ***	0.553
	(2.29)	(3.35)	(3.23)	(1.41)	(2.37)	(3.36)	(3.33)	(1.14)
HHI		0.002	0.001	0.002		0.002	0.001	0.002
		(1.24)	(0.57)	(0.77)		(1.27)	(0.68)	(0.73)
CP × HHI		−0.000 **	−0.000 **	−0.000		−0.021 **	−0.023 *	−0.023
		(−2.14)	(−2.06)	(−1.52)		(−2.06)	(−1.94)	(−1.31)
SIZE	0.151 ***	0.149 ***	0.131 ***	0.147 ***	0.148 ***	0.147 ***	0.125 ***	0.148 ***
	(4.52)	(4.49)	(2.76)	(3.19)	(4.31)	(4.27)	(2.68)	(3.02)
AGE	0.090	0.092	0.196 **	−0.439	0.088	0.090	0.198**	−0.446
	(0.56)	(0.57)	(1.97)	(−1.40)	(0.54)	(0.56)	(1.99)	(−1.42)
SOE	0.046	0.045	0.012	0.081	0.040	0.040	−0.011	0.092
	(0.67)	(0.66)	(0.14)	(0.76)	(0.58)	(0.58)	(−0.12)	(0.85)
LEV	0.001	0.001	−0.002	0.004 **	0.001	0.001	−0.001	0.004 **
	(1.03)	(1.02)	(−0.94)	(2.36)	(1.08)	(1.09)	(−0.83)	(2.42)
SLACK	−0.004	−0.004	0.001	−0.002	−0.004	−0.004	0.001	−0.003
	(−0.85)	(−0.87)	(0.08)	(−0.38)	(−0.85)	(−0.88)	(0.13)	(−0.46)
VOL	−0.000	−0.000	−0.001	−0.000	−0.000	−0.000	−0.001	−0.000
	(−0.45)	(−0.39)	(−0.81)	(−0.21)	(−0.53)	(−0.49)	(−0.86)	(−0.23)
TURN	−0.049 **	−0.049 **	−0.016	−0.076 **	−0.049 **	−0.050 **	−0.015	−0.076 **
	(−2.16)	(−2.17)	(−0.46)	(−2.56)	(−2.18)	(−2.19)	(−0.46)	(−2.57)
COST	0.665 ***	0.657 ***	0.561	0.515 **	0.652 ***	0.644 ***	0.511	0.515 **
	(3.38)	(3.35)	(1.51)	(2.19)	(3.37)	(3.33)	(1.41)	(2.18)
IDD	−0.002	−0.002	−0.003	−0.002	−0.002	−0.002	−0.003	−0.002
	(−0.82)	(−0.84)	(−0.74)	(−0.59)	(−0.77)	(−0.78)	(−0.67)	(−0.56)
INSHARE	−0.000	−0.000	−0.001	−0.000	−0.001	−0.001	−0.001	−0.000
	(−0.76)	(−0.72)	(−0.59)	(−0.45)	(−0.79)	(−0.77)	(−0.64)	(−0.45)
MSHARE					−0.028	−0.027	−0.402	0.236
					(−0.15)	(−0.15)	(−1.46)	(0.99)
TOP1					−0.220	−0.220	−0.292	−0.108
					(−0.83)	(−0.83)	(−0.70)	(−0.32)
BALANCE					0.018	0.020	0.059	−0.003
					(0.39)	(0.43)	(0.96)	(−0.04)
YEAR	Control	Control	Control	Control	Control	Control	Control	Control
Constant	−4.676	−4.733	−6.272 ***	5.517	−4.517	−4.572	−6.075 ***	5.625
	(−1.41)	(−1.44)	(−2.77)	(0.88)	(−1.35)	(−1.38)	(−2.67)	(0.90)
N	16,145	16,145	7384	8761	16,145	16,145	7384	8761
R^2	0.013	0.014	0.025	0.018	0.014	0.014	0.027	0.018
Adj-R^2	0.0120	0.0123	0.0219	0.0151	0.0121	0.0124	0.0235	0.0152
F	4.977 ***	4.759 ***	4.438 ***	2.475 ***	4.549 ***	4.422 ***	4.166 ***	2.376 ***

Note: *, **, *** represent the significance level of 1%, 5%, and 10% respectively.

Second, increase the control variables. Referring to Zhang et al. [39], the control variables (the shareholding ratio of managers, the shareholding ratio of the largest shareholder, and the degree of equity balance) are added to the model. The shareholding ratio of managers = the number of shares held by management/total share capital; the shareholding ratio of the first largest shareholder = number of shares held by the first largest shareholder/total shares; and the degree of equity balance = total number of shares held by the second to fifth largest shareholder/number of shares held by the first largest shareholder. The results of Table 5 show that the coefficient of CP is still significantly positive after adding these control variables and that the coefficient of CP × HHI is still significantly negative in the full samples. The coefficient of CP × HHI in heavy-pollution industries is negative and significant at the 10% level.

Third, in view of the impact of the implementation of the new Environmental Protection Law of the People's Republic of China in 2015 on the results of this study, we selected A-share manufacturing listed companies from 2015 to 2019 as the research object for the robustness test. It can be seen from the results in Table 6 that, except for the significance level and directional change in a small number of control variables, the coefficient of CP is significantly positive and the coefficient of CP × HHI is significantly negative in all samples. The coefficient of CP × HHI in heavy-pollution industries is negative and significant at the 1% level.

Table 6. Robustness test two.

Variables	Used the 2015–2019 Samples			
	Full Samples		Heavy-Polluting Industry	Non−Heavy Polluting Industry
	Model (1)	Model (2)	Model (2)	Model (2)
CP	0.469 **	1.585 ***	1.806 ***	0.447
	(2.21)	(3.98)	(3.38)	(0.67)
HHI		0.005	0.004	−0.003
		(1.34)	(0.98)	(−0.60)
CP × HHI		−0.042 ***	−0.048 ***	−0.007
		(−3.26)	(−3.14)	(−0.29)
SIZE	0.152 ***	0.149 ***	0.163	0.157 **
	(2.71)	(2.71)	(1.53)	(2.42)
AGE	0.022	0.024	0.126	−0.465
	(0.15)	(0.17)	(1.45)	(−1.46)
SOE	−0.204 *	−0.204 *	−0.245 *	−0.106
	(−1.93)	(−1.95)	(−1.78)	(−0.70)
LEV	0.003 *	0.003 *	0.003	0.004 **
	(1.69)	(1.81)	(1.06)	(2.19)
SLACK	0.148	0.151	0.334	−0.036
	(1.15)	(1.17)	(1.59)	(−0.22)
VOL	−0.001	−0.001	−0.000	−0.001 *
	(−1.54)	(−1.54)	(−0.59)	(−1.74)
TURN	−0.041	−0.040	0.007	−0.087 **
	(−1.39)	(−1.37)	(0.15)	(−2.56)
COST	0.680 ***	0.649 ***	0.474	0.429
	(2.80)	(2.73)	(0.80)	(1.58)
IDD	−0.003	−0.003	−0.000	−0.006
	(−0.66)	(−0.65)	(−0.05)	(−1.12)
INSHARE	−0.001	−0.001	−0.001	−0.002
	(−1.40)	(−1.49)	(−0.81)	(−1.43)
MSHARE	−0.163	−0.170	−0.573	0.071
	(−0.58)	(−0.60)	(−1.06)	(0.24)
TOP1	−0.883 *	−0.893 *	−2.054 **	0.148
	(−1.72)	(−1.75)	(−2.44)	(0.31)

Table 6. *Cont.*

Variables	Used the 2015–2019 Samples			
	Full Samples		Heavy-Polluting Industry	Non−Heavy Polluting Industry
	Model (1)	Model (2)	Model (2)	Model (2)
BALANCE	−0.096	−0.094	−0.244 *	0.041
	(−1.32)	(−1.30)	(−1.82)	(0.55)
YEAR	Control	Control	Control	Control
Constant	−3.041	−3.142	−5.077 *	5.861
	(−0.98)	(−1.03)	(−1.77)	(0.96)
N	9049	9049	3953	5096
R^2	0.022	0.024	0.052	0.019
Adj−R^2	0.0200	0.0217	0.0476	0.0153
F	4.568 ***	4.410 ***	4.359 ***	1.930 ***

Note: *, **, *** represent the significance level of 1%, 5%, and 10%, respectively.

Therefore, it can be considered that Hypothesis 1, Hypothesis 2, and Hypothesis 3 have good robustness.

5. Discussions and Suggestions

5.1. Discussions

Based on the data from a list of China' s A-share manufacturing companies from 2008 to 2019, this paper finds the following: First, corporate performance is positively correlated with the scale of investments into environmental protection, indicating that high-performance enterprises show higher risk preferences and are more willing to invest into environmental protection projects. Second, fierce industrial competition has increased the positive effect of corporate performance on the scale of enterprises' investments into environmental protection, indicating that industry competition brings about an external pressure to enterprises. To resist the long-term environmental impact and operational risks caused by fierce competition, high-performance enterprises are encouraged to invest more resources into environmental protection projects. Third, compared with non-heavy-polluting industries, fierce industry competition has increased the positive impact of corporate performance on the scale of investments into environmental protection in heavy-polluting industries, indicating that industry attributes are important factors affecting an enterprise's decision making about investments into environmental protection.

In addition, we find that firm size (SIZE) is significantly positively correlated with EI at the 1% level. From the perspective of management behavior, large-scale enterprises are more willing to invest into environmental protection to show their ability to "seek development" and to establish a good green image to meet financing needs and stakeholders' environmental demands [65,66]. CEO turnover (TURN) has a significantly negative correlation with EI, indicating that CEO turnover affects the risk-taking level of enterprises. CEO successors tend to invest into projects with obvious short-term benefits due to a tendency towards "loss aversion" [67,68]. Therefore, for CEO successors, environmental investment is a suboptimal choice when resources are limited. Agency cost (COST) is significantly positive correlated with EI at the 1% level, indicating that agency cost is caused by a conflict of interest between management and shareholders. On the agency problem, the existing literature mainly focuses on the mechanism behind the impacts of manager's private income and private costs on investment decision making. Jensen [69] and other scholars believe that, in order to obtain private benefits from investment, enterprise managers need to be committed to building an empire and attempt to maximize assets [70]. Investments into environmental protection are characterized by large amounts of investment and long periods of investment, which are conducive to motivating managers to control income by increasing investments into environmental protection.

5.2. Suggestions

From the perspective of the enterprise: First, improve their image as a green business. Enterprises are profit-making organizations. The motive of increasing the profits of an enterprise drive managers to make investment decisions and owners of the enterprise to raise awareness of environmental protection by managers and to improve their green business philosophy, which is helpful in realizing sustainable development. Second, pay attention to improving internal governance mechanisms. Managers of high-performance enterprises may have a risk preference, so it is necessary to strengthen the internal governance mechanisms of high-performance enterprises to prevent managers from insufficient or excessive environmental investment and, thus, to find a balance between efficient environmental investment and promoting sustainable and healthy development and environmental governance.

From the perspective of the government, first, give full play to the pressures of the market on enterprises to fulfill their environmental responsibility. On the one hand, the government should appropriately relax the threshold for market access in industries with a monopoly. If the government allows more enterprises to join a monopolized industry, industry competition will form and encourage competition among enterprises to develop low-carbon alternatives. On the other hand, the government should cultivate public awareness of environmental protection and encourage the public to practice green consumption and behaviors. Public demand for environmental protection encourages enterprises to consciously invest into environmental protection.

Second, the government should adjust these measures to local conditions, verify our system according to multiple levels and angles, and encourage enterprises to assume environmental responsibility. Environmental information disclosure between the government and enterprises should be improved to effectively strengthen the responsibility of enterprises as environmental-protection investors. In addition, the government should improve market-based environmental-management policy tools. For example, helping enterprises raise environmental protection funds through various ways; reducing problems with costs in the financing process; improving the efficiency of green financing; and giving preferential treatment regarding loan amount, loan interest rate, loan term, loan conditions, and other aspects for environmental investment.

5.3. Limitations and Further Research

This study has certain limitations. First, this study focused only on the A-stock manufacturing industry in China, meaning that generalizing the findings of the study to dissimilar business contexts is challenging. For that reason, replicating this study in different industries and countries can be attempted in future studies. Second, although the data in this study are sufficient to support the research conclusions, the latest data can still be added to future research to further enhance the representativeness of these conclusions.

Author Contributions: Conceptualization, Q.G. and Z.C.; methodology, L.Y. and J.T.; software, L.Y.; formal analysis, Q.Y. and W.X.; investigation, Q.Y. and J.T.; resources, Z.C.; data curation, L.Y. and H.Q.; writing—original draft preparation, Q.G. and W.X.; writing—review and editing, J.T. and Q.Y.; supervision, L.Y. and W.X. All authors have read and agreed to the published version of the manuscript.

Funding: This research is supported by following programs: The National Social Science Foundation of China (20XGL014), (18BJY015) and (19BGL085), National Natural Science Foundation of China (72103051), Guangxi Philosophy and Social Science Project of China (20BGL001, 21FKS040) and (20AJY002).

Institutional Review Board Statement: Not applicable.

Informed Consent Statement: Not applicable.

Data Availability Statement: The data presented in this study are available on request from the corresponding author.

Conflicts of Interest: The authors declare no conflict of interest.

References

1. Sun, Z.L. The development and Characteristics of Environmental Accounting Information Disclosure in Japan. *Account. Townsh. Enterp. China* **2014**, *3*, 18–20.
2. Yang, S. Research on Environmental Regulation, Social Responsibility and Corporate Environmental Investment. Master's Thesis, Shanxi Finance and Economics University, Taiyuan, China, 8 May 2015.
3. Wang, Y.H. Environmental Regulation, Government Subsidy and Corporate Environmental Investment. Master's Thesis, Anhui University of Finance & Economics, Bengbu, China, 1 December 2017.
4. Liao, Y.L.; Lei, S.Y.; Song, M.J.; Zeng, L.N.; Deng, W.K.; Hu, D.F. Corporate Environmental Investment Behavior and Cooperation Evolution Considering Centralized Sanction and Reference Dependence. *Chin. J. Manag. Sci.* **2022**, *3*, 1–11. [CrossRef]
5. Konar, S.; Cohen, M.A. Information as Regulation: The Effect of Community Right to Know Laws on Toxic Emissions. *J. Environ. Econ. Manag.* **1997**, *321*, 109–124. [CrossRef]
6. Xiong, Y.; Xu, X. The Impact of Environmental Regulation on Foreign Direct Investment in China: An Empirical Analysis based on Panel Data Model. *Econ. Rev.* **2007**, *2*, 122–160.
7. Zheng, S.Q.; Wan, G.H.; Sun, W.Z.; Luo, D.L. Public Demands and Urban Environmental Governance. *Manag. World* **2013**, *6*, 72–84.
8. Yang, L.; Gan, Q.X.; Ma, D.S. Public Environmental Attention and Corporate Environmental Protection Investment-Based on the Regulatory Perspective of Green Image. *Financ. Mon.* **2020**, *8*, 33–40.
9. Kesidou, E.; Demirel, P. On the Drivers of Eco-innovations: Empirical Evidence from the UK. *Res. Policy* **2012**, *415*, 862–870. [CrossRef]
10. An, Z.R. Research on Environmental Investment Mechanism of Enterprises: From the Perspective of Environmental Sustainability. Ph.D. Thesis, Beijing Jiaotong University, Beijing, China, 31 March 2017.
11. Yi, S.; Xue, Q.Z. Green Supply Chain Management and Green Innovation: Empirical Research based on Chinese Manufacturing Enterprises. *Sci. Res. Manag.* **2016**, *37006*, 103–110.
12. Mackenzie, C.; Rees, W.; Rodionova, T. Do Responsible Investment Indices Improve Corporate Social Responsibility? FTSE4Good's Impact on Environmental Management. *Corp. Gov. Int. Rev.* **2013**, *215*, 495–512. [CrossRef]
13. Tang, G.P.; Li, L.H.; Wu, D.J. Environmental Regulation, Industry Attributes and Enterprises Environmental Protection Investment. *Account. Res.* **2013**, *6*, 83–89.
14. Inoue, E.; Arimura, T.H.; Nakano, M. A New Insight into Environmental Innovation: Does the Maturity of Environmental Management Systems Matter? *Ecol. Econ.* **2013**, *941*, 156–163. [CrossRef]
15. Jabbour, C.J.C. Business School Environmental Management: Creating State-of-the-art Maps. *Contemp. Thought Mag. Adm.* **2014**, *84*, 1–22.
16. Yang, L.; Qin, H.; Xia, W.Y.; Gan, Q.; Li, L.; Su, J.; Yu, X. Resource slack, Environmental Management Maturity and Enterprise Environmental Protection Investment: An Enterprise Life Cycle Adjustment Perspective. *J. Clean. Prod.* **2021**, *309*, 127339. [CrossRef]
17. Tian, S.S.; Li, Q. Private Income, Property Right and Environmental Investment of Enterprises: Considering the Influence of Institutional Pressure. *Financ. Account. Mon.* **2016**, *21*, 21–26.
18. Hu, L.X.; Gao, J. Political Connection, Financing Constraints and Corporate Environmental Investment Research. *Value Eng.* **2020**, *3909*, 92–94.
19. Yang, L.; Qin, H.; Gan, Q.; Su, J. Internal Control Quality, Enterprise Environmental Protection Investment and Finance Performance: An Empirical Study of China's Share Heavy Pollution Industry. *Int. J. Environ. Res. Public Health* **2020**, *17*, 6082. [CrossRef]
20. Bai, S.X.; Zhang, Z.Z. Can Internal Control Execution Enhance Corporate Environmental Investment? *Res. Financ. Econ. Issues* **2022**, *02*, 104–111.
21. He, X.G.; Li, J.; Lv, F.F.; Deng, H. Analysis on the Speculative Behavior of High Performing Enterprises: A Data Test from Chinese Listed Companies. *China Ind. Econ.* **2015**, *5*, 110–121.
22. Xu, X.Q.; Wang, J.; Ma, J. High Aspirations on the Incidence of Corporate Negative Behavior: Evidence from the Listed Manufacturing Companies in China. *Nankai Bus. Rev.* **2016**, *1902*, 137–144.
23. Mishina, Y.; Dykes, B.J.; Block, E.S. Why "Good" Firms Do Bad Things: The Effects of High Aspirations, High Expectations, and Prominence on the Incidence of Corporate Illegality. *Acad. Manag. J.* **2010**, *534*, 701–722. [CrossRef]
24. Guo, R.; Chen, X.R. Resource and Pressure Constraint Framework Choice of Strategic Adjustment Direction of Excellent Enterprises. *Financ. Mon.* **2021**, *12*, 129–137.
25. Murovec, N.; Erker, R.S.; Prodan, I. Determinants of Environ-mental Investments: Testing the Structural Model. *J. Clean. Prod.* **2012**, *37*, 265–277. [CrossRef]
26. Hitchens, D.; Clausen, J.; Trainor, M.; Keil, M.; Thankappan, S. Competitiveness, Environmental Performance and Management of SMEs. *Greener Manag. Int.* **2003**, *44*, 45–57. [CrossRef]
27. Lu, Z.F.; Han, X.; Chang, Q. Research on the Relationship Between Corporate Long-term Debt and Investment Behaviors: An Empirical Analysis Based on Chinese Listed Companies. *J. Manag. Word* **2006**, *1*, 120–128.
28. Jensen, M.C.; Fama, E.F. Organization Forms and Investment Decisions. *J. Financ. Econ.* **1985**, *141*, 101–119.
29. Liu, Y.Y.; Huang, Z.Y.; Liu, X.X. Environmental Regulation, Executive Compensation Incentives and Corporate Environmental Investment: Evidence from the Implementation of the 2015 Environmental Protection Law. *Account. Res.* **2021**, *5*, 175–192.
30. Arouri, M.E.H.; Caporale, G.M.; Rault, C.; Sova, R.; Sova, A. Environ-mental Regulation and Competitiveness: Evidence from Romania. *Ecol. Econ.* **2012**, *81*, 130–139. [CrossRef]

31. Song, Y.C.; Ren, H.F.; Zhang, Y.Y. Does Equity Refinancing Promote Innovation in Manufacturing Firms? Explanation from the Perspective of Competition. *Nankai Manag. Rev.* **2022**, 3, 1–27. Available online: https://124.222.168.6/kcms/detail/12.1288.F.20210922.1328.002.html (accessed on 23 September 2021).
32. Song, T.B.; Zhong, X.; Chen, W.H. Will the better the Performance of the Enterprise, the more Environmental Investment? Empirical Evidence from Chinese Manufacturing Listed Companies. *East China Econ. Manag.* **2017**, *315*, 126–133.
33. Staw, B.M.; Sandelands, L.E.; Dutton, J.E. Threat Rigidity Effects in Organizational Behavior: A Multilevel Analysis. *Adm. Sci. Q.* **1981**, *264*, 501–524. [CrossRef]
34. Li, G.; Lv, W.X.; Zhou, C.H. Performance Feedback and Organizational Responses: A Literature Review and Prospects. *Foreign Econ. Manag.* **2019**, *41*, 86–108.
35. Hall, J. Environmental supply chain dynamics. *J. Clean. Prod.* **2000**, *86*, 455–471. [CrossRef]
36. Xie, D.M.; Wang, P. Tax-reducing Incentives, the Scale of Independent Directors and Environmental Protection Investment of Heavily Polluting Enterprises. *Account. Res.* **2021**, *8*, 137–152.
37. Nalebuff, B.; Stiglitz, J. Information, Competition, and Markets. *Am. Econ.* **1983**, *73*, 78–83.
38. Pagell, M.; Wiengarten, F.; Fynes, B. Institutional Effects and the Decision to Make Environmental Investments. *Int. J. Prod. Res.* **2013**, *512*, 427–446. [CrossRef]
39. Zhang, Y.M.; Xing, C.; Zhang, Y. The Impact of Media Coverage on Green Technology Innovation of High-Polluting Enterprises. *Chin. J. Manag.* **2021**, *1804*, 557–568.
40. Wang, M.Y.; Li, Y.M.; Wang, Z.T. U-shaped Relationship between Enterprises' Environmental Performance and Economic Performance of Green Technology Innovation and the Moderation Effect of Government Competition Regulation. *Sci. Manag. Res.* **2021**, *5*, 107–116.
41. Zhang, X.M.; Wang, X.N. Research on the Influence of Eco-Label on the Purchase Intention of Green Products with the Mediating Role of Consumer Perceived Value. *Ecol. Econ.* **2019**, *35*, 63–68.
42. Li, Q.; Feng, B. Does Companies Quietly Disclose Environmental Information? A Study of the Relationship between Environmental Investment and Environmental Information Disclosure Quality under the Pressure of Competition. *J. Cent. South Univ. Financ. Law* **2015**, *4*, 58–66.
43. Li, W.J.; Lu, X.A. Do Institutional Investors Care Firm Environmental Performance? Evidence from the Most Polluting Chinese Listed Firms. *Financ. Res.* **2015**, *12*, 97–112.
44. Tang, G.P.; Li, L.H. Ownership Structure, Property Right and Corporate Environmental Investment: Empirical Evidence from China's A-share Listed Companies. *Res. Financ. Econ. Issues* **2013**, *3*, 93–100.
45. Zhai, H.Y.; Liu, Y.W. Does Environmental Justice Specialization Promote Corporate Environmental Government. *China Popul. Resour. Environ.* **2019**, *29*, 138–147.
46. Zeng, A.M.; Zhang, C.; Wei, Z.H. Impact of Financial Crisis, Flexible Financial Reserve and Firm Investment Behavior: Empirical Evidence from Chinese Listed Companies. *Manag. World* **2013**, *4*, 107–120.
47. Gu, L. Product market competition, R&D investment, and stock returns. *J. Financ. Econ.* **2016**, *1192*, 441–455.
48. Cai, W.; LI, G. The Drivers of Eco-innovation and its Impact on Performance: Evidence from China. *J. Clean. Prod.* **2018**, *176*, 110–118. [CrossRef]
49. Bansal, P. Evolving Sustainably: A Longitudinal study of Corporate Sustainable Development. *Strateg. Manag. J.* **2005**, *263*, 197–218. [CrossRef]
50. Duan, Y.Q.; Xu, S.L. Ordered Environmental Regulation and Heavy Pollution Enterprise Investment: Incentive or Inhibition? Quasi-natural Experiments Based on the New 'Environmental Law'. *Financ. Dev. Stud.* **2021**, *7*, 54–61.
51. Guo, M.Q.; Zhang, C.P. Empirical Research on Firm Age, R & D Investment and Performance: Based on Empirical Data of A-share Listed Companies in Zhejiang Province. *Account. Newsl.* **2017**, *36*, 52–56.
52. Cui, X.M.; Wang, J.Y.; Wang, M. Environmental Investment, CEO Overseas Experience and Corporate Value: Value Added or Impaired?—Analysis from the Perspective of Branding Theory. *Audit. Econ. Res.* **2021**, *36*, 86–94.
53. Tan, H.L.; Yang, J. Empirical Study on the Impact of Corporate Social Responsibility on Corporate Governance and its Performance. *Ind. Technol. Econ.* **2009**, *28*, 152–155.
54. Jiang, F.X.; Huang, J.C. Marketization Process and Dynamic Adjustment of Capital Structure. *Manag. World* **2011**, *3*, 124–134.
55. Gebhardt, W.R.; Lee, C.M.C.; Swaminathan, B. Toward an Implied Cost of Capital. *J. Account. Res.* **2001**, *391*, 135–176. [CrossRef]
56. Yang, H.Y.; Li, C.Y. Empirical Study on the Correlation Between Executive Compensation Incentives and Shareholder Wealth—Taking listed Companies in Guangxi from 2001 to 2006 as Samples. *Friends Account.* **2010**, *5*, 84–87.
57. Giuli, A.D.; Kostovetsky, L. Are Red or Blue Companies more Likely to Go Green? Politics and Corporate Social Responsibility. *J. Financ. Econ.* **2014**, *111*, 158–180. [CrossRef]
58. Kim, H.; Kim, H.; Lee, P.M. Ownership Structure and the Relationship Between Financial Slack and R&D Investments: Evidence from Korean Firms. *Organ. Sci.* **2008**, *19*, 404–418.
59. Zou, H.L.; Zeng, S.X.; Lin, H.; Zhai, Y.M. Board Characteristics, Resource Slack and Environmental Performance: An Empirical Analysis of Manufacturing Listed Companies. *J. Syst. Manag.* **2016**, *2*, 193–202.
60. Zhu, M.; Ning, J.H.; Yuan, Z.M. Institutional Investor Heterogeneity, Environmental Regulation and Corporate Environmental Investment. *Financ. Dev. Res.* **2019**, *7*, 12–20.

61. Liu, X.; Li, X.R. Government Decentralization and Corporate Fraud: Empirical Evidence of State-owned Listed Companies. *Account. Res.* **2016**, *4*, 34–41.
62. Huang, Q.H.; Zhang, F.F.; Chen, X.D. Innovation Incentive Effect of Executive Short-term Pay. *Sci. Res. Manag.* **2019**, *40*, 257–265.
63. Jensen, M.; Meckling, W.H. Theory of the Firm: Managerial Behavior, Agency Costs and Ownership Structure. *J. Financ. Econ.* **1976**, *34*, 305–360. [CrossRef]
64. Guo, X.Y. Environmental Regulation, Agency Costs and Corporate Environmental Investment. Master's Thesis, Yangzhou University, Yangzhou, China, 25 June 2019.
65. Du, X. Research on the Correlation between Company Characteristics and Its Environmental Investment: Based on Petrochemical Listed Companies. *J. Xi'an Shiyou Univ.* **2015**, *2402*, 16–21.
66. Chi, Z.; Wang, J.Y. Environmental Quality, Executive Hometown Identification and Investment in Corporate Environmental Assets. *Macroeconomics* **2021**, *9*, 149–160.
67. Liu, X.; Xue, Y.Z. Research on CEO Successor Selection Mechanism and Corporate Risk-taking after CEO Change—Based on CEO Successor Age Perspective. *Manag. Rev.* **2016**, *5*, 137–149.
68. Li, X.S. The Impact of CEO Change on Company Innovation AND R&D Investment: "Cautious" or "aggressive". *Friends Account.* **2020**, *21*, 72–79.
69. Jensen, M.C. Agency Cost of Free Cash Flow, Corporate Finance and Takeovers. *Am. Econ. Rev.* **1986**, *762*, 323–329.
70. Hart, O.; Moore, J. Debt and Seniority: Analysis of the Role of Hard Claims in Constraining Management. *Am. Econ. Rev.* **1995**, *85*, 567–585.

sustainability

Article

Treatment of Industrial Wastewater in a Floating Treatment Wetland: A Case Study of Sialkot Tannery

Adeel Younas [1], Love Kumar [2,*], Matthew J. Deitch [2], Sundus Saeed Qureshi [3], Jawad Shafiq [4], Sohail Ali Naqvi [1], Avinash Kumar [5], Arjmand Qayyum Amjad [1] and Sabzoi Nizamuddin [6]

1 Freshwater Programme, World Wide Fund for Nature-Pakistan (WWF-Pakistan), WWF-Pakistan Office, Lahore 54600, Punjab, Pakistan
2 Soil, Water and Ecosystem Sciences Department, IFAS West Florida Research and Education Center, University of Florida, Gainesville, FL 32603, USA
3 Institute of Environmental Engineering and Management, Mehran University of Engineering and Technology, Jamshoro 76062, Sindh, Pakistan
4 Department of Local Government & Community Development, Lahore 54000, Punjab, Pakistan
5 Department of Chemistry and Biochemistry, Florida Atlantic University, Boca Raton, FL 33431, USA
6 School of Engineering, RMIT University, Melbourne 3000, Australia
* Correspondence: lovekumar@ufl.edu

Abstract: The city of Sialkot in Pakistan is a hub of leather tanneries, with approximately 260 tanneries in operation and, while producing millions of leather products per day, the city discharges millions of gallons of untreated effluent into drains each day. In order to devise a cost-effective system for the treatment of tannery wastewater, a floating treatment wetland (FTW) was established to treat the effluent using local plant species through phytoremediation. The efficiency of the FTW was tested with three different plant species, each grown separately and operating for three months in the FTW tank. Two of the plant species introduced, water hyacinth and water lettuce, were floating and vascular; the third plant species *Typha latifolia* was vegetated on a floating mat of styrofoam while the roots extended down to the contaminated water. Wastewater from a tannery drain was pumped into the FTW tank with a flow of 0.5 L per minute and was given a retention time of six days. The influent and effluent from the FTW were periodically tested to determine the percentage removal of contaminants, primarily the total suspended solids, biochemical oxygen demand, chemical oxygen demand, and chromium. After two months with each species, a significant change in the quality of wastewater was measured: chromium was removed by up to 95 percent by the water hyacinth and water lettuce and 33 percent by the *Typha latifolia.* The pilot model indicates that FTWs are an effective system to treat effluent from tanneries in a cost-effective way as an alternative to establishing an expensive treatment system with high associated operational costs. It can help in achieving the circular economy concept of conventional wastewater schemes towards more sustainable ones. Moreover, to achieve the principles of circular economy and environmentally friendly development, it is crucial that the substances used for a wetland foundation have the capacity to be recycled, are available at a cheap price, and are locally available.

Keywords: circular economy; circular bioeconomy; floating treatment wetland; phytoremediation; tannery wastewater; tropical wetlands; *Typha latifolia*

Citation: Younas, A.; Kumar, L.; Deitch, M.J.; Qureshi, S.S.; Shafiq, J.; Ali Naqvi, S.; Kumar, A.; Amjad, A.Q.; Nizamuddin, S. Treatment of Industrial Wastewater in a Floating Treatment Wetland: A Case Study of Sialkot Tannery. *Sustainability* **2022**, *14*, 12854. https://doi.org/10.3390/su141912854

Academic Editors: Claudio Sassanelli and Sergio Terzi

Received: 14 September 2022
Accepted: 29 September 2022
Published: 9 October 2022

1. Introduction

Water availability and quality are global challenges [1]. Worldwide, organic and inorganic pollutants from commercial, agricultural, and domestic sectors have polluted around 70% of the available freshwater resources [2]. Contamination by heavy metals [3] and other pollutants through industrialization and urbanization poses a serious threat to aquatic and human life [4–6]. Industrial processes in industries such as textiles, mining, electroplating, and leather processing release many inorganic contaminants into the environment including lead, chromium, copper, and cadmium [7–10]. While processes such as filtration, complexation, coagulation, flocculation, chemical precipitation, chemical oxidation/reduction, membrane technology, electrochemical treatment, biosorption, and complexation are commonly used for the removal of wastewater contaminants [1,11–14], many of these methods require wastewater treatment plants with chemicals, ions, and other materials [1]. These methods are expensive, do not sufficiently remove these contaminants, and require high energy [15,16].

In many countries, such as Pakistan, India, Argentina, China, Mexico, Brazil, and South Korea, leather tanneries and related industries are a main driver of the economy [17]. The leather value chain begins with the husbandry of animals and ends with the leather goods, with the intermediate steps including the collection of skins and hides from the slaughtered animals and processing of hides, which is often conducted in small labor-intensive workshops in larger capital-intensive factories [18]. Growing demand for leather products worldwide has placed pressures on manufacturers of all sizes to produce garments more quickly. To meet these demands, manufacturers may turn to different methods that allow more efficient production [19–21], often without considering the cost; for example, in recent years, most tanneries changed their production from traditional vegetable tanning to chromium tanning technology [17].

The leather industry is also dependent on sub-industries, such as meat production, livestock rearing, and tanning, as part of leather production, which raises extensive environmental concerns [22]. According to the INETI (Instituto Nacional de Engenharia, Tecnologia e Inovação) organization, tanneries cause many environmental problems such as wastewater discharge from tannery industries; this wastewater has high inorganic and organic chemicals including chromium and trace organic and synthetic chemicals such as the dyes, pesticides, processing chemical solvents, and finishing agents [23]. The contamination of groundwater occurs when wastewater from these industries seeps through the soil from the unlined pipes, drains, and ponds, or from spills and dumps [23,24]. In large leather-producing cities such as Sialkot in Pakistan (the largest leather producing city in the country), the discharge of effluents from the Sialkot tanneries adds up to approximately 1.1 million liters per day, which causes the contamination of the surface and groundwater of Sialkot [23]. Given the threat it poses to human health and aquatic life, there is an urgent need to deal properly and scientifically with the large amount of effluent from the leather industry.

Conventional wastewater treatment methods are costly and environmentally intrusive, and they require engineering expertise, labor management, and process activities [14]. They also require a lot of time and money [25]. In contrast, methods for improving water quality based on natural mechanisms can avoid many of these challenges [26]. In the field of environmental engineering, one such modern nature-based solution for water treatment is floating treatment wetlands (FTWs). This soil-less planting technique has been developed to treat different types of wastewater to the point where it can be discharged and used again for purposes such as fisheries and agriculture without a measurable effect on life and the environment [27–29]. FTWs are a reliable method to treat wastewater and surface water runoff [30,31]. To enhance the purification reactions in FTWs, hydrophytes are often used to accumulate pollutants in the tissues, usually in the rhizosphere, of plants. For the removal of contaminants, some plants prompt additional physicochemical and biological processes leading to the additional removal of contaminants [32]. Some hydrophytes such as water hyacinth [32] are well-known for their capacity to remove contaminants in polluted water, owing to its ability to absorb pollutants. According to previous studies, regional hydrophyte species should be selected when designing the plant community of FTWs so that the selected plants are adapted to the climate conditions [32]. The floating mat underside the plant roots supplies a significant contact area for the formation of connected biofilms and the trapping of dispersed particles. The plants must obtain their energy straight from the water column because they are not rooted in the soil like in subsurface flow-built wetlands, which may increase the rates of nutrient and component absorption into the biomass. Due to their stability, they can withstand large variations in water depth and have the potential to improve treatment efficiency by retaining more water during flow periods, therefore extending the period that the wastewater is held in the wetland. Due to the oxygen carried by the roots, the roots of these species can survive in hypoxic or anoxic conditions. The microbial transformation of aquatic contaminants is supported by the radial oxygen loss to the rhizosphere [33,34]. Despite the fact that FTWs and conventional CWs are considered as processes that are straightforward to build and operate, many complicated processes take place that have a direct impact on system performance and the efficacy of the removal of pollutants, including inlet contaminant concentrations, hydraulic loadings, pH, the presence of micro-organisms in the rhizosphere, redox conditions, and temperature [35]. Similar to how CW uses plants that float freely, in FTWs, the roots of the plants are in a constant physical relationship with the water. Macrophytes in these environments take up nutrients directly from the water. Through a hanging network of roots, rhizomes, and the connected biofilm, which is in charge of both biochemical activities and significant physical functions such as filtration and particulate capture, the plants offer a biologically active surface area [36]. The root growth of FTWs offers a bigger surface area than other conventional methods of treatment, such as built wetlands, for the establishment of a biofilm [37]. While reducing wastewater flow and turbulence, increasing sedimentation, and trapping/filtering suspended material, this biofilm serves as an ecosystem for several bacterial communities and becomes essential for the sequestration of nutrients from the water through nitrification, nitrification for nitrogen and adsorption for retaining phosphorous [38]. Sedimentation, in addition to vegetation, is crucial to the function of FTWs. One or more of the key methods for lowering oxygen, nitrogen, total phosphorus, and orthophosphate is the settling of suspended materials and their trapping in plant roots. However, it is likely that phosphorus from sediments is transferred back into the water section during a lengthy hydraulic retention time, leading to an increase in P levels [39]. Studies have demonstrated that FTWs can be a productive, low-cost [40,41], and low-maintenance method of improving water quality across a wide range of uses; for example, there has recently been interest in evaluating the effectiveness of FTWs in the treatment of industrial wastewater and stormwater runoff [38,42,43]. Floating wetlands can be built in any lagoon, existing water body, or built structure without the need for any digging or earth shifting to remove contaminants from the water and without the need for further land acquisition [38]. The present study aims to assess the ability of different plant species to treat wastewater from a tannery production plant using an FTW system. This study serves as a reference and guide for academics and policymakers in the design and use of FTWs to remediate water contamination from diverse sources.

2. Current Problems and Prospects of Using Plants for Wastewater Treatment

Plants and media have a key part in the removal of manmade wetland technologies. Phytoremediation allows plants in FTWs to lower pollutant levels [38]. Plant growth material supplies physical support for vegetation formation. Moreover, extra surfaces for biofilm growth and nutrient adsorption may enhance the sedimentation and pollutant filtration [42,43]. Gravel is the media mostly used in constructed wetlands [44], though Priya et al. (2013) found that sand was a more effective treatment than gravel. Sirianuntapiboon et al. (2006) reported that constructed wetlands with media including both sand and soil in combination results in the maximum contaminant removal efficiency. Various studies have used different types of media (e.g., vermiculite, zeolite, and lime) for extraction of certain compounds from waste effluent [45–47]. To compare how well two parallel hybrid wetlands remove pollutants after being fed industrial wastewater (with limited biodegradability), Saeed et al. (2019) studied a system that had two stages of vertical flow (VF) wetlands, following the final surface flow (SF) wetland round of treatment. They found that in both systems, the concentration-based mean overall reduction rates for all of the different types of waste (specifically, NH4-N, TN, P, BOD, and COD) were removed by at least 90% [48]. Cristina et al. used a light expanded clay substrate composed of plants and without plants. They found 41–58% BOD removal, and lower nutrient removal. Moreover, they found no changes in the removal over the 17 months of study [49]. In FTWs, the influence of the plant type depends on several factors such as plant production, the physical effects of root systems, micro-organisms, evapotranspiration, the uptake by plants, and the weather [50–53]. Multiple studies have indicated that adsorption and sedimentation are the primary reduction tools for metals in FTWs [37,40,42]. According to the research, pH values from 6 to 8 and higher temperatures (from 6 to 26 °C) preferred positively charged adsorption on small particles and organic materials. Metals are generally linked with sulfides in anaerobic soils, forming insoluble sulfides in water. Moreover, the physicochemical parameters of water in FTWs, such as pH, temperature, DO, and organic material levels might alter heavy metal removal effectiveness and subsequent release [54–56].

Priya and Selvan [57] found that contaminants are stored in the plant root system, then moved to the shoots and other parts of the plant. When the plant is harvested, these contaminants are removed from the system. A hyperaccumulator or accumulator can remove pollutants from water and soil. This green technology is seen as a long-term and promising alternative to conventional water and soil treatment methods for developing countries. More than 500 species have been found to be capable of storing metals from contaminated soils in their roots [58,59]. It is difficult to determine how the metal (Cu, Ni, and Pb) is distributed between the roots and the plant leaves and stems [60]. Due to their carcinogenic and bioaccumulative properties, metal ions are among the most dangerous water pollutants. Wastewaters containing metal ions are produced in a variety of industries, including metal finishing, mining, cement, leather, textiles, and paints. Various hazardous metal ions, including chromium, nickel, zinc, lead, arsenic [3], cobalt, cadmium, and copper are found in industrial effluents and pose a threat to ecosystems [1].

Climate change, particularly temperature changes, may impact the development of microalgae. Overall, increased temperatures can encourage the formation of organic carbon in manmade wetlands, which, in turn, can help the expansion of microalgae in such areas [61]. In addition, increasing the temperature and irradiance of microalgae to a proper range could promote enzyme activity and metabolic activity and improve nutrient removal efficiency and biomass production, all of which would be beneficial for pollutant removal [62]. While numerous methods have demonstrated good results in terms of pollutant removal, its application in cold climates has been a neglected topic. Rarely has research been undertaken to create treatment methods that are effective at low temperatures. The performance of treatment processes is reduced in cold climates where the average temperature is around 10 degrees Celsius or less and the average temperature in winter is less than 3 degrees Celsius. Micro-organisms and plants cease to function efficiently in cold climates, resulting in the reduced efficiency of treatment methods [63,64].

Other factors may also limit the capacity of FTWs to reduce pollutant concentrations. While BOD, COD, and chromium are efficiently handled by some of the plants used in this study, these plants may not be as efficient at reducing concentrations of other metals or pollutants. Higher pollutant concentrations may also constrain efficiency: the net change may be the same, but if the concentrations of pollutants are higher in the FTW inflow, the percentage of reduction would be lower. Finally, in our study, FTW size and inflow were constant; further research would be beneficial to assess whether and how FTWs could scale up in size to treat higher inflow volumes or higher concentrations of pollutants [65,66]. Any inability, loss, or unknown reaction may deviate as a contributing factor to plant species, and it may also produce increased disadvantages of phytoremediation. Phytoremediation has certain drawbacks according to the literature. These include the concentration, toxicity, and bioavailability of contaminants as well as the capacity of plants to withstand pressure. Low effectiveness or wake of phytomanagement, lack of sufficient macro/micronutrients in contaminated media, and finally, the physiological qualities and limitations of plant species are some of the potential drawbacks of phytoremediation. Multicontaminant interactions, adaptation of plant species to climate change and pollution in urban environments, and genetic traits and specifications of plant species classify them into groups such as "indicators", "low tolerance", "high tolerance", and "hyperaccumulator". Hyperaccumulator plant species often serve as an extractor of rare metals [65]. These are all important topics for further research on FTWs in developing countries. This study provides important information on how small and medium enterprises (SMEs) can employ low-cost tools for effective wastewater treatment. The specifications for design, construction, and operation provided here are from an actual SME producer in Sialkot; if applied broadly, the cumulative benefit of several SMEs adopting this type of treatment tool can help to achieve local water quality goals. In places such as Sialkot where SMEs are a predominant type of production facility, small-scale treatment through FTWs may be the most feasible type of wastewater treatment SMEs can employ. Like many developing regions, Pakistan is promoting the incorporation of cleaner production concepts into its industries [67,68], wherein cost-effective strategies are employed to increase overall efficiencies and reduce the risks to human health and ecosystems [69]. FTWs align with these goals. More research on the use of floating treatment wetlands as a low-cost, energy-efficient wastewater treatment process in developing countries can help to highlight the value of such tools as well as explore the parameters and limitations that make them successful in reducing pollutants from wastewater discharge.

3. Methodology

3.1. Study Area

The city of Sialkot, located in the northeastern Punjab province in Pakistan, is known as one of Pakistan's most industrialized regions. Major industrial sectors include the production of leather products, surgical instruments, diesel engines, beverages, iron and steel, and pharmaceuticals. There are approximately 264 tanneries, 244 leather garment manufacturing units, and 900 leather sports product manufacturing units present in Sialkot [70]. Leather tanning facilities generate large quantities of wastewater from their processes. Therefore, one leather tannery was selected to participate in a pilot FTW system to treat its wastewater using cost effective and environmentally friendly techniques.

The weather in Sialkot changes a lot over the course of the year and even on a single day. The region usually has hot, dry summers and cold monsoon rains. In January, the average temperature is 5 °C, and in June, it is 40 °C. Sialkot gets most precipitation during a monsoon season (July to September), with less intense precipitation continuing through cold winters (December and February), and spring (March and April), before a relatively dry premonsoon season (summer between May and June) [71].The average amount of rain that Sialkot gets each year is 934.7 mm [72].

3.2. Sample Collection and Analysis

A total of 18 composite influent wastewater samples were collected twice every 15 days from the inlet of the FTW tank, which was located on the drain of a tannery that collected the effluent from all the operations from raw hide to finished leather. An additional 18 composite effluent wastewater samples were collected twice after every 15 days at the outlet of the FTW tank to assess contamination and removal efficiency (Figure 1). The samples were collected at a rate of 0.5 L per minute, approximately the rate at which the equalized wastewater flowed into the pilot system. The hydraulic retention time of the FTW tank was 14 days. This study ran from September to April. The color of the collected wastewater samples was visually analyzed. The temperature of the samples was assessed using industrial thermometer with a range from 0 to 100 °C [73].

The total suspended solids (TSS), biological oxygen demand (BOD), and chemical oxygen demand (COD) were determined by AWWA/APHA Standard Methods for the Examination of Water and Wastewater, 2340 D, 5210 D, and 5220 D, respectively. The heavy metal chromium (Cr) was qualified in water samples using atomic absorption as outlined by AWWA/APHA method 3111/3120B. The pilot scale FTW system was designed and established at a drain of a tannery in Sialkot; the pilot unit consisted of 3 chambers designed at the flow rate of 0.72 m^3/day (0.5 L per minute, as described above). The design and selection of model was decided reviewing some local and international studies. Ayaz et al. (2020) used four chamber wetland sizes of 70, 145, 50 and 50 liters. They used four different species for the treatment of industrial wastewater [74]. Moreover, a similar type of design was used in serval studies for other pilot projects [75–78]. The unit was made of steel with the design characteristics shown in Table 1.

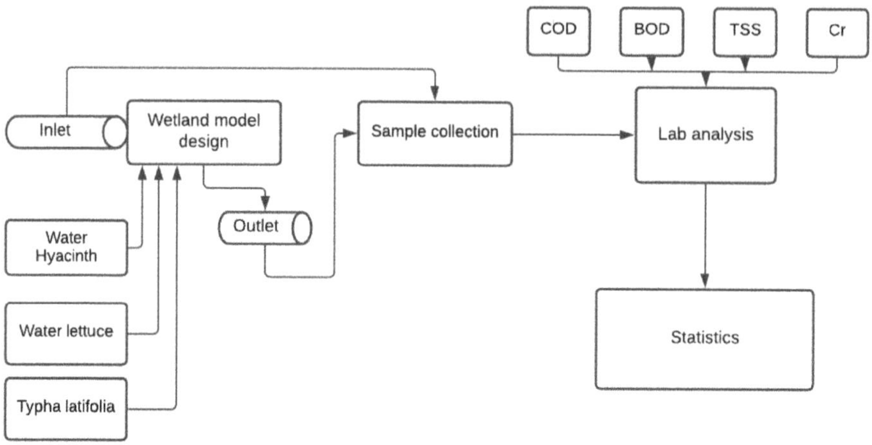

Figure 1. Methodology flow diagram.

Table 1. Design aspects of the floating treatment wetland used to treatment tannery wastewater.

Design	Measurement	Unit
Design flow	0.72	m^3/day
Detention time	6	days
Width	1.37	M
Length	3.6	m
Depth	0.84	m
Area	5.01	m^2
Volume	4.31	m^3

The selection of the plants used in pilot FTW system was based on vascular species that locally originate in Sialkot. Three vascular plant species (water hyacinth, water lettuce, and *Typha latifolia*) were introduced in the tanks as shown in Figures 2–4, respectively. These plants were resettled from the local areas of Sialkot to the pilot FTW system. The vegetation was planted by hand. The plants were sited at the depth equal to the introduced water level inflow. The number of plants was monitored through the operation. After the selection of plants and installation of the FTW system, the tannery wastewater flowed from the inlet into the FTW. The plants were grown during different periods of time to check the intake of contamination from the plant and the treatment of wastewater.

Water hyacinth (*Eichhornia crassipes*) was the first species introduced in the FTW tank (Figure 2), followed by water lettuce (*Pistia stratiotes*) (Figure 3), and then *Typha latifolia* (Figure 4). These plants were primarily selected because they are locally available, and secondly, these are widely used in the region by researchers in domestic and industrial wastewater treatment [74,75,79–82]. Each plant species was trialed for two months at a time, mid-September 2018 to mid-November 2018, mid-November 2018 to mid-January 2019, and mid-January 2019 to mid-April 2019, respectively. The water was pumped with the help of small pump at a rate of 0.72 m^3/day, and the treated wastewater was discharged back to the drain from the outlet of the FTW. Water sampling and testing was carried out fortnightly (with six sampling events for each plant type) at the inlet (In-1 through In-6) and outlet (Ef-1 through Ef-6) of the floating treatment wetland to obtain effective results.

Figure 2. Water hyacinth in the Sialkot floating treatment wetland (photo: A. Youns and J. Shafiq).

Figure 3. Water lettuce in the Sialkot floating treatment wetland (photo: A. Youns and J. Shafiq).

Figure 4. *Typha latifolia* in the Sialkot floating treatment wetland (photo: A. Youns and J. Shafiq).

Sustainable crop production includes methods of raising vegetables, fruits, grains and other food, and fiber crops in ecologically mindful ways that focus on soil health and biodiversity instead of chemical fertilizers and pesticides. These growing practices require more labor and specialized expertise than chemical-based production, which means that the resulting products are often more expensive. Water lettuce is a floating plant and is popular for those with water gardens because its presence inhibits the growth of algae and cleans the water. For our FTW, water lettuce was harvested on a specified designed area of FTW which gets its nutrients from wastewater. Water lettuce grows best in the early part of the summer when temperatures reach at least 15.5 °C and does not require external thermal devices [81]. Water hyacinth can form thick mats that cover the entire surface of ponds, choking out native species and oxygen in the water [43]. Fish cannot survive in water without oxygen, so it is important to control the growth of the plant. Thus, harvesting water hyacinth in designed wetlands can help with preserving ecosystems. The benefits for the environment are equally important reasons to consider growing *Typha*. *Typha* was grown in peat with a higher water level; this means the peat was functioning as it should for pollution reduction, reducing the carbon emissions from the peat, and the established plants actively stored the carbon. Both actions contribute towards the CO_2 emission reduction targets [83,84]. FTWs were inspected on a weekly basis to check and examine the overall functioning. Key attention was given to the inlet flow of the system, which was checked twice a week, as blockages in the pump and pipe could occur due to suspended solids in the wastewater.

4. Results and Discussion

Tanning industries are a major wastewater source, and tanning industry wastewater has a high organic load due to the process used to turn raw skins or hide into leather [85,86]. In this study, tannery wastewater flowed through floating treatment wetlands to remove pollutants. In the following sections, we describe the results of the treatment of wastewater using three different species.

4.1. Water Hyacinth

Water hyacinth (*Eichhornia Crassipes*) grows abundantly in tropical and subtropical areas of the world. Water hyacinth is a floating water plant that has been named one of the world's 100 most invasive global plants [87]. In recent years, water hyacinth has received attention because of its potential for the abstraction of pollutants when it is used in biological treatment systems [88]. The thick growth of water hyacinth can reduce the quantity of sunlight that reaches the water, thus impeding many photosynthetic species and disturbing the ecological equilibrium. Furthermore, its extensive coating on the surface of bodies of water reduces oxygen transmission into the water [88]. Water hyacinth has lengthy roots floating in water which are believed to provide a medium for aerobic microbes in sewage treatment systems. These microbes eat organic materials and nutrients in wastewater and convert them into inorganic compounds that plants may absorb. Water hyacinth often develops swiftly in non-native countries due to a lack of natural enemies or consumers [87,89].

Water treated by water hyacinth had low amounts of COD, BOD, TSS, and chromium (Cr) after two months compared to the flow into the floating treatment wetland (Figures 5 and 6). The influent value of COD was more than 1500 mg/L, which decreased to less than 10 mg/L after treatment. The BOD influent value was found to be 1000 mg/L, which was reduced to 10 mg/L after treatment. The TSS values were noted to be above 150 mg/L, which reduced to under 50 mg/L after the treatment. The chromium inflow values were noted as above 14 mg/L, whereas the effluent values were reduced to below 2 mg/L. Additional processes such as the gravitational settling of solids and coprecipitation with insoluble compounds may have contributed to the high chromium removal in the wetlands [90]. Overall, the average reduction in the pollution in tannery wastewater by the water hyacinth was more than 90 percent for COD, BOD, and Cr, and 85% for TSS (Table 2). The results for Cr were similar to those reported in a previous study, which found that the water hyacinth removed 87.52% of Cr from the contaminated water source [91].

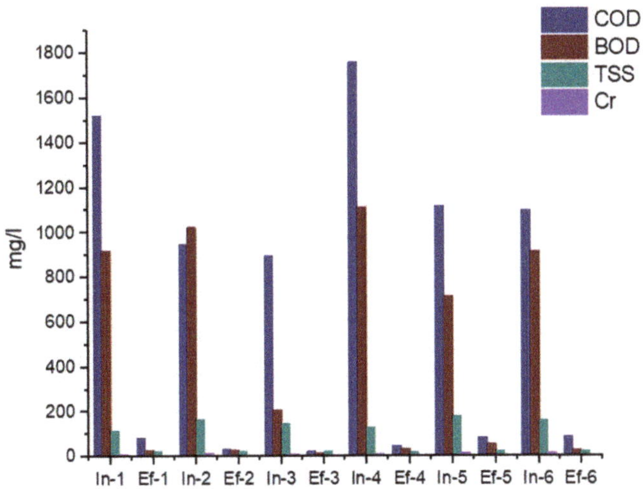

Figure 5. Pollutant concentrations in the FTW inflow (In) and effluent (Ef) over six biweekly sampling periods (1 through 6) using water hyacinth.

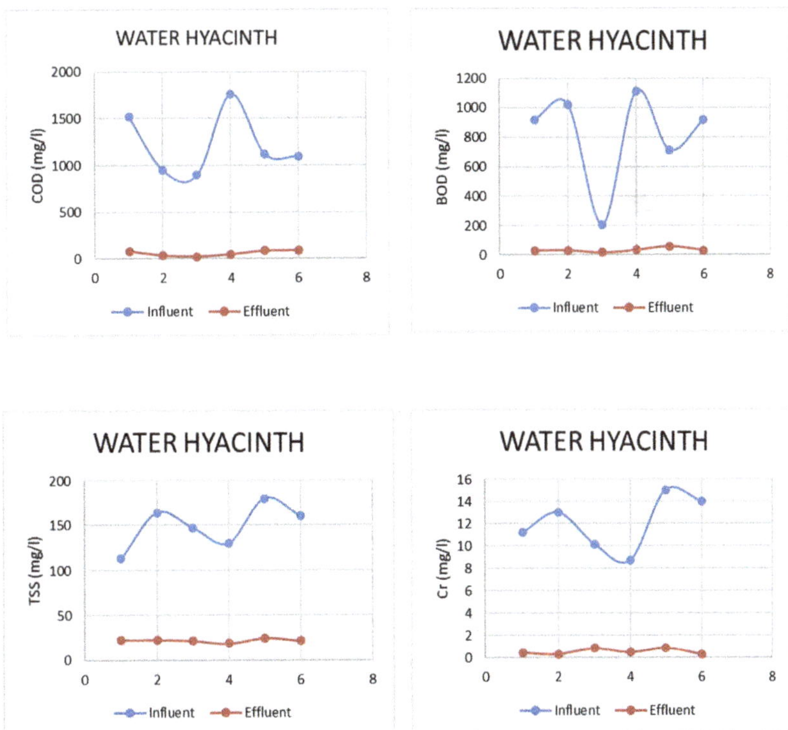

Figure 6. Variations in COD, BOD, TSS, and Cr concentrations into and out of the water hyacinth FTW at each of the six biweekly sampling periods.

Table 2. Averaged effluent concentrations and removal efficiencies for COD, BOD, TSS, and Cr with the hyacinth species.

Parameters	Average Inlet Concentration (mg/L)	Average Outlet Concentration (mg/L)	Removal Efficiency (%)
COD	1222.16	59.83	95
BOD	814.7	31.76	96
TSS	149	22.41	85
Cr	12	0.67	94

4.2. Water Lettuce

Water lettuce, (*Pistia stratiotes*) from the Araceae family, floats on the water surface and its roots hang submerged beneath floating leaves [92]. It is used for medicine and fodder in different countries across the world. Water lettuce is larger in yield compared to other small aquatic weeds such as *Lemna* spp. and may be used in the phytoremediation of a variety of contaminants found in industrial effluent [93]. It has a high capacity for the absorption of pollutants including hazardous heavy metals as well as a high level of cellular proliferation [94]. As a result, it can grow in stressful circumstances and absorb a variety of pollutants inside its plant components.

During the two-month period when water lettuce was in the FTW system, BOD and COD in the tannery effluent ranged from 400 to 1200 mg/L; these values decreased after treatment using water lettuce (Figure 7). However, water lettuce was not as efficient at reducing BOD or COD as water hyacinth, reducing COD, on average, by only 27% and BOD by 41% (Figures 7 and 8; Table 3). The removal of TSS (80%) and Cr (more than 90%) was similar to water hyacinth. Our results are in agreement with other studies which found that the removal of metals such as zinc, cadmium, and nickel from wastewater by water lettuce is extensively efficient [93]. Other research observed that water lettuce can efficiently remove Cr from water at different concentrations of 1, 2, 4, and 6 mg Cr/L [95,96]. Other research on treatment wetland (in vertical and horizontal design) found a higher removal percentage of BOD (82%), along with efficient removal of phosphate (95.4%) and chloride (51%). Additionally, the fecal coliform removed by water lettuce (over 98%) suggested the ability of the plant to uptake nutrients and release toxins for pathogen disinfection. It is also suggested that vertical design could be a better option for wetlands using the species of water lettuce [97].

4.3. Typha Latifolia

Typha latifolia is a well-known emergent hyperaccumulator plant. It can collect metals such as copper, mercury, chrome, copper, and lead up to 0.1 percent of the plant dry weight and iron and zinc up to 1% [98]. In recent times, *P. australis* and *T. latifolia* have been used to remove heavy metals [99]. Over the two months when *Typha latifolia* was grown in the FTW, the analysis of the inlet and outlet samples indicated a reduction in COD of 278 mg/L with 48% removal efficiency, BOD 123 mg/L with 31% efficiency, and chromium of 3.36 mg/L with 33% efficiency (Table 4; Figures 9 and 10). The TSS reduction was 72% with an average inlet and outlet concentration of 220.33 mg/L and 61.68 mg/L. Another study reported that *Typha* species removed 96.2% of cadmium, 83.6% of copper, and 95.9% of lead, respectively [100]. In the *Typha latifolia* study, the BOD and COD results varied slightly, whereas the TSS and chromium results varied greatly between the FTW inflow and effluent. The input COD was found to be 800 mg/L, and effluent COD was reduced to 200 mg/L after treatment. The input BOD was as high as 500 mg/L; after treatment, it was reduced to as low as 200 mg/L (Figure 10). The TSS value in the FTW inflow varied from 60 to 400 mg/L and from 20 to 100 mg/L in the effluent. As result of treatment, the inflow chromium levels varied from 5 to 16 mg/L and effluent from 0 to 15 mg/L (Figure 10). It has been shown that the *Typha* species are more tolerant of metal toxicity than other plant species [101,102].

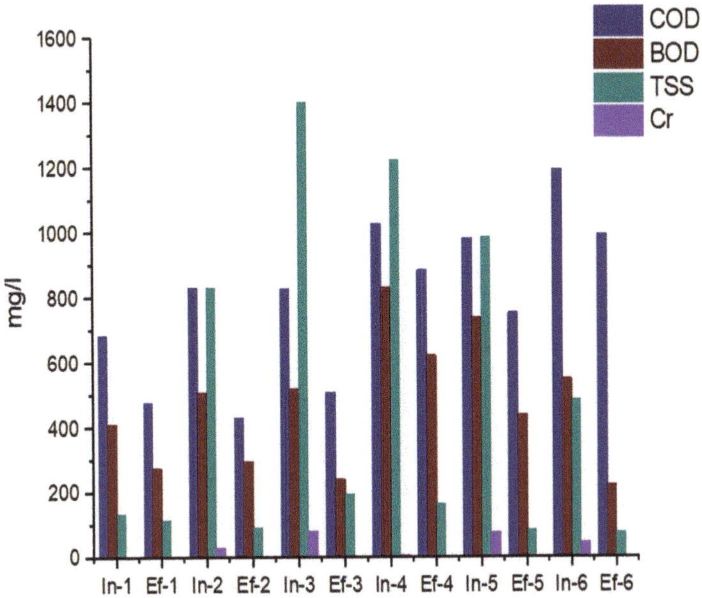

Figure 7. Concentrations of pollutants in the FTW inflow (In) and the effluent (Ef) over the six biweekly sampling periods (1 through 6) using water lettuce.

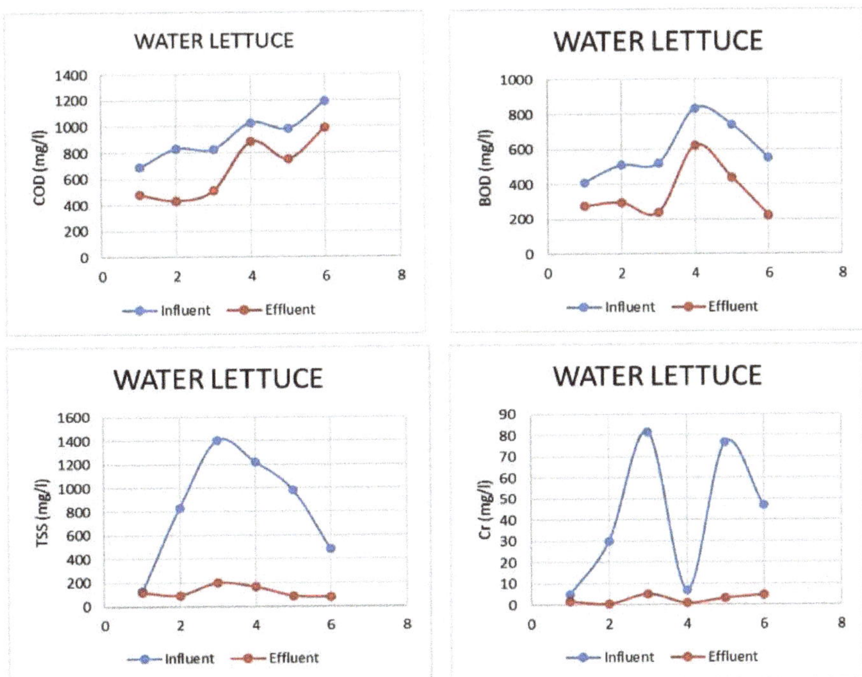

Figure 8. Variations in COD, BOD, TSS, and Cr mass concentration into and out of the water lettuce FTW at each of the six biweekly sampling periods.

Table 3. Averaged effluent concentrations and removal efficiencies for water lettuce.

Parameters	Average Inlet Concentration	Average Outlet Concentration	Removal (%)
COD	924.66	675.16	27
BOD	594.33	349.66	41
TSS	842.66	123.16	85
Cr	41.19	2.5	94

Table 4. Averaged effluent concentrations and removal efficiencies for *Typha latifolia*.

Parameters	Average Inlet Concentration	Average Outlet Concentration	Removal (%)
COD	583.83	305	48
BOD	399.33	276.83	31
TSS	220.33	61.82	72
Cr	10.04	6.68	33

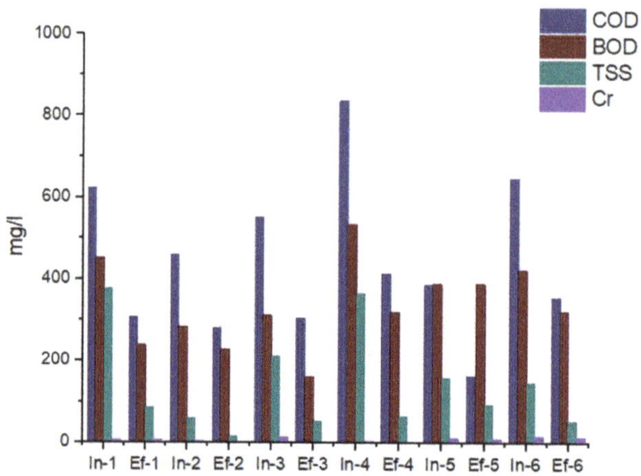

Figure 9. Pollutant concentrations in the FTW inflow (In) and effluent (Ef) over the six biweekly sampling periods (1 through 6) using *Typha latifolia*.

The metal uptake by these wetland plants, precipitation and coprecipitation as insoluble salts, and metal binding to the substrate are processes that are attributed to heavy metal reduction in *T. latifolia* [103]. *T. latifolia* provided substrate and sustaining media for the growth of micro-organisms, which are key players in heavy metal immobilization and uptake by plants. The rhizosphere of these plants may be naturally reduced, which would further augment the cell wall capability to absorb metals through immobilization, which could explain why the overall effect of metal uptake by *T. latifolia* is reduced. Another reason for the decrease in overall heavy metal removal could be that these plants accumulate phytosiderophores [97,104].

Floating treatment wetlands are a less expensive wastewater treatment technique with minimal construction, operation, and maintenance costs [76,82,105] partly because construction materials can be obtained locally and commercially. Moreover, FTWs do not need any complicated technical mechanisms for their installation or manufactured chemical input to sustain functional processes or for maintenance. Hence, low construction and operating costs make this technology a particularly reasonable and practical method in developing countries [56]. According to the research, a 201 m^3 constructed wetland costs approximately 445 USD (1 USD = 176 PKR), a 158 m vertical flow constructed wetland (VFCW) requires 574 USD, a 251 m^3 horizontal flow constructed wetland (HFCW) requires 1425 USD, a 272 m^3 reed system costs 1040 USD, and a secondary constructed wetland costs

1874 USD (it always varies area to area) [106]. The cost of a constructed wetland is around 50%–90% less than the cost of other traditional wastewater treatment technologies [107]. The COD, TSS, TN, and TP removal rates for the HFCW and VFW systems are 61%, 75%, 31%, and 26%, respectively. Another comparison made by the US-EPA is that the life cycle costs for a swamp are lower than the cost of a conventional treatment system designed for the same flow and effluent water quality [104].

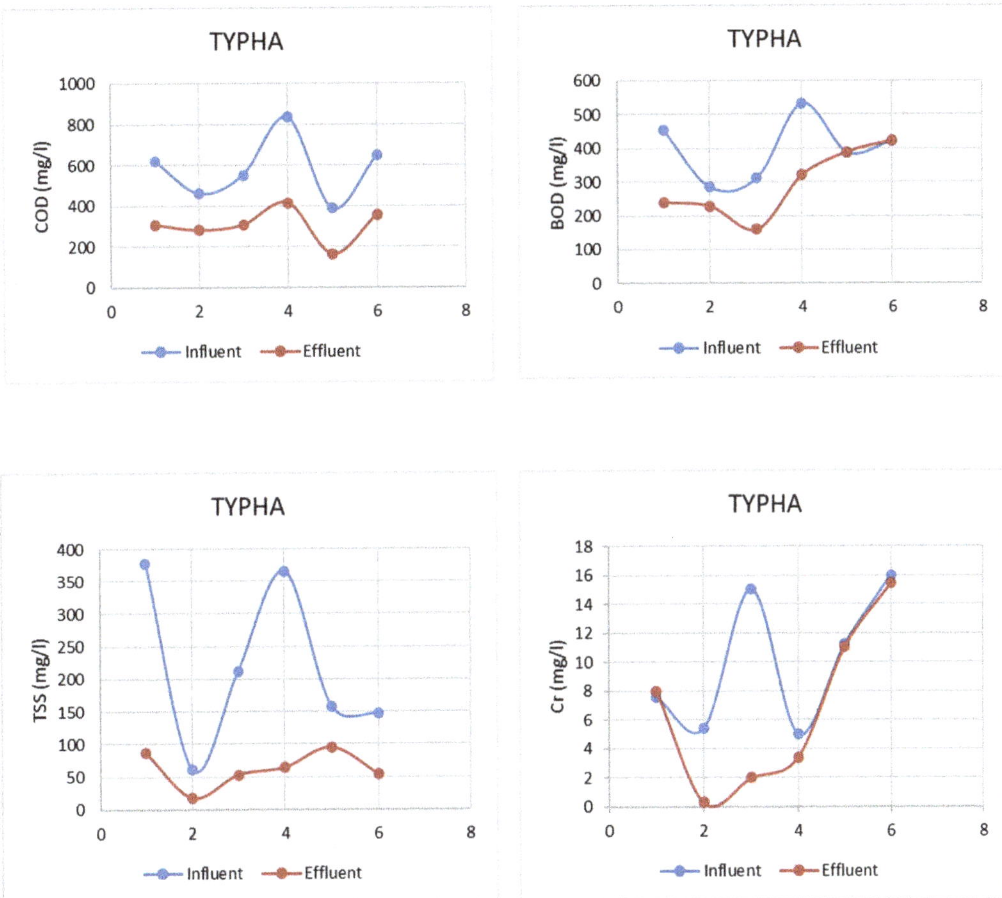

Figure 10. Variations in COD, BOD, TSS and Cr concentration into and out of the *T. latifolia* FTW at each of the six biweekly sampling periods.

A designed floating treatment wetland may be used to treat a wide range of wastewater types (industrial, kitchen, washbasin, etc.). The efficiency of a built wetland is determined by the kind of wetland, species, delivered hydraulic stress, and bed material [105,108]. Water hyacinth (*Eichhornia crassipes*), a free-floating aquatic vegetation that grows quickly, may absorb huge amounts of trace minerals and pollutants through processes that are dependent on root growth. When plants are cultivated in industrial wastewaters with high quantities of macronutrients, their absorption of elements such as heavy metals are frequently boosted. Other factors may limit the capacity for plants used in this study to reduce pollutants; for example, water lettuce is based in hot, humid climates and is not robust in cold areas.

To maintain the optimal plant densities (0.2–0.7 kg dry biomass per m^2), metal- or nutrient-loaded plant biomass must be removed from water bodies for efficient water treatment. If not collected, the bulk of the components would be returned to the water through plant breakdown processes. It has been demonstrated that more intensive management, including more regular and timely harvesting of plant biomass, might result in a greater pollutant reduction. For example, for optimal pollutant reduction, plants in a semitropical environment should be picked every other week during the wet season, when temperatures are also optimal for water lettuce development, to maintain around a three-fourths covering of the surface of the water [91]. The *Typha* genus is commonly applied in wetlands. Most *Typha* research focuses on removal efficiency responses to contaminants. In addition to affecting the ability of a plant to absorb pollutants from water, root structure and root diameter variations are strongly linked to the ecological requirements of plants and removal efficiency [109–111]. Most of the components were gathered in the sediment, which aided in the deposition. The pollutant transfer to above-ground structures was not always considered. Therefore, plants in this study had only a limited effect on reducing pollutants (Table 5). However, macrophytes can operate as phytostabilizers and improve contaminant sequestration, especially when the root biomass grows rapidly. As a whole, the wetlands which were investigated here for the first time in Pakistan have shown a positive performance. This study will be a guiding principle for small and medium enterprises. It will help SMEs to design, construct, assess, and achieve the local wastewater treatment guidelines. Furthermore, detailed research and use of the floating treatment wetland as a low-cost, energy-efficient wastewater treatment process in developing countries are needed. Future research should focus on the efficiency and sustainability of FTWs to improve the environment and operations and to find new ways to improve dead or contaminated plant management and substrates in real-world field trials. Nutrients and other pollutants absorbed by wetlands plants have been found to be released into water when species die and decompose during the cold winter, potentially resulting in poor removal performance. As a result, research and development on optimal plant harvesting methodologies as well as the restoration and regeneration of plant resources in FTWs is critical. It is recommended that floating treatment wetlands be integrated to treat wastewater from tanneries. Many constructed treatment wetlands were designed and considered only for wastewater treatment. However, in addition to their high removal efficiencies, manmade treatment wetlands have lately been demonstrated to offer a significant possibility in the new sustainable and circular economy in industrial and urban environments. As proposed by the "sponge city" idea, treatment wetlands can successfully treat, collect, and recycle nutrients and water for future use. Floating treatment wetlands are a new technique that has proven useful in a variety of applications, including wastewater treatment, bioremediation, and stormwater treatment. The efficiency of their construction and operation as well as the reduced area needed for procedures make them an appealing alternative for integration with treatment ponds and traditional artificial wetlands. However, just one research paper integrating FTW with a mixed CW mechanism was discovered in this review; hence, combining FTW with other technologies, such as advanced oxidation processes, should be investigated further, particularly when the focus is on water reuse.

Table 5. Removal efficiency comparison of plants.

	COD Removal Efficiency %	BOD Removal Efficiency %	TSS Removal Efficiency %	Cr Removal Efficiency %
Water Hyacinth	95	96	84	94
Water lettuce	27	41	85	94
Typha latifolia	48	31	72	33

5. Conclusions

Floating treatment wetland systems are an innovative field that has been demonstrated to be adaptable to multiple tasks, such as domestic wastewater treatment, bioremediation, and industrial wastewater treatment. Their comparatively simple development, operation, and low cost make them an attractive alternative for integration with treatments ponds and conventionally created wetlands. The sustainability of the tannery sector is important for the GDP of Pakistan. The tannery sector utilizes a large amount of water and discharges polluted water. In order to devise a cost-effective system for the treatment of tannery wastewater, a floating treatment wetland model was set up to treat effluent using local plant species through phytoremediation as a pilot. In this study, we found that floating treatment wetland systems can be successfully established for the treatment of the contaminants of tannery wastewater. Our research shows that the water treated by water hyacinth had low amounts of COD, BOD, TSS, and chromium (Cr) after two months compared to the flow into the floating treatment wetland. The influent value of COD was decreased to 10 mg/L from 1500 mg/L, the TSS reduced to 50 mg/L from 150 mg/L, and chromium from 14 mg/L to 2 mg/L. Water lettuce was not as efficient at reducing BOD or COD as water hyacinth, reducing COD, on average, by only 27% and BOD by 41%. The removal of TSS (80%) and Cr (more than 90%) was similar to the water hyacinth. Furthermore, *Typha latifolia* is a well-known emergent hyperaccumulator plant. It can collect metals such as copper, mercury, chrome, copper, and lead up to 0.1 percent of the plant dry weight and iron and zinc up to 1%. Over the two months when *Typha latifolia* was grown in the FTW system, the analysis of inlet and outlet samples indicated a reduction in COD of 278 mg/L with 48% removal efficiency, BOD 123 mg/L with 31% efficiency, and chromium of 3.36 mg/L with 33% efficiency. The TSS reduction was 72%, with an average inlet and outlet concentration of 220.33 mg/L and 61.68 mg/L. The tested pilot-scale FTW was demonstrated to be a successful treatment solution for tannery effluents, and it is a low-cost wastewater treatment method with low development, operating, and maintenance costs.

Author Contributions: Conceptualization, L.K.; Data curation, A.Y. and J.S.; Formal analysis, L.K. and A.K.; Project administration, S.A.N., A.Y. and A.Q.A.; Software, S.S.Q.; Supervision, A.Q.A.; Validation, S.N.; Writing—original draft, L.K.; Writing—review & editing, L.K. and M.J.D. All authors have read and agreed to the published version of the manuscript.

Funding: This research received no external funding.

Institutional Review Board Statement: No applicable.

Informed Consent Statement: No applicable.

Data Availability Statement: More information about study can be found on https://www.wwfpak.org/our_work_/water_/iles/, accessed on 26 September 2022.

Acknowledgments: The publication of this study was supported by the project "International Labor and Environmental Standards Application in Pakistan's SMEs (ILES)" by WWF-Pakistan which was also funded by the European Union and ILO. The research findings, recommendations, and conclusions drawn from this publication are part of the ILES project activities. The contents of this publication are the sole responsibility of the authors and can in no way be taken to reflect the views of WWF-Pakistan or the European Union and ILO. We thank the two anonymous reviewers and editor for constructive suggestions and comments on earlier drafts. We also want to extend our thanks to tannery representatives who participated in this study.

Conflicts of Interest: The authors declare that they have no known competing financial interest or personal relationships that could have appeared to influence the work reported in this paper.

References

1. Montaño-Medina, C.U.; Lopéz-Martínez, L.M.; Ochoa-Terán, A.; López-Maldonado, E.A.; Salazar-Gastelum, M.I.; Trujillo-Navarrete, B.; Pérez-Sicairos, S.; Cornejo-Bravo, J.M. New pyridyl and aniline-functionalized carbamoylcarboxylic acids for removal of metal ions from water by coagulation-flocculation process. *Chem. Eng. J.* **2023**, *451*, 138396. [CrossRef]
2. Shahid, M.; Al-Surhanee, A.; Kouadri, F.; Ali, S.; Nawaz, N.; Afzal, M.; Rizwan, M.; Ali, B.; Soliman, M. Role of Microorganisms in the Remediation of Wastewater in Floating Treatment Wetlands: A Review. *Sustainability* **2020**, *12*, 5559. [CrossRef]
3. Eljamal, O.; Sasaki, K.; Hirajima, T. Sorption Kinetic of Arsenate as Water Contaminant on Zero Valent Iron. *J. Water Resour. Prot.* **2013**, *5*, 563–567. [CrossRef]

4. D'Adamo, I.; Gastaldi, M.; Morone, P.; Rosa, P.; Sassanelli, C.; Settembre-Blundo, D.; Shen, Y. Bioeconomy of Sustainability: Drivers, Opportunities and Policy Implications. *Sustainability* **2021**, *14*, 200. [CrossRef]
5. Sikander, M.; Kumar, L.; Naqvi, S.A.; Arshad, M.; Jabeen, S. Sustainable practices for reduction of environmental footprint in tanneries of Pakistan. *Case Stud. Chem. Environ. Eng.* **2021**, *4*, 100161. [CrossRef]
6. Cheng, Q.; Huang, Q.; Khan, S.; Liu, Y.; Liao, Z.; Li, G.; Ok, Y.S. Adsorption of Cd by peanut husks and peanut husk biochar from aqueous solutions. *Ecol. Eng.* **2016**, *87*, 240–245. [CrossRef]
7. Sarode, S.; Upadhyay, P.; Khosa, M.; Mak, T.; Shakir, A.; Song, S.; Ullah, A. Overview of wastewater treatment methods with special focus on biopolymer chitin-chitosan. *Int. J. Biol. Macromol.* **2018**, *121*, 1086–1100. [CrossRef]
8. Genet, M.; Stokes, A.; Fourcaud, T.; Norris, J.E. The influence of plant diversity on slope stability in a moist evergreen deciduous forest. *Ecol. Eng.* **2010**, *36*, 265–275. [CrossRef]
9. Khan, S.; Waqas, M.; Ding, F.; Shamshad, I.; Arp, H.P.H.; Li, G. The influence of various biochars on the bioaccessibility and bioaccumulation of PAHs and potentially toxic elements to turnips (*Brassica rapa* L.). *J. Hazard. Mater.* **2015**, *300*, 243–253. [CrossRef]
10. D'Adamo, I.; Sassanelli, C. A mini-review of biomethane valorization: Managerial and policy implications for a circular resource. *Waste Manag. Res.* **2022**, 1–12. [CrossRef]
11. Fu, F.; Wang, Q. Removal of heavy metal ions from wastewaters: A review. *J. Environ. Manag.* **2011**, *92*, 407–418. [CrossRef] [PubMed]
12. Kurniawan, T.A.; Chan, G.Y.; Lo, W.-H.; Babel, S. Physico–chemical treatment techniques for wastewater laden with heavy metals. *Chem. Eng. J.* **2006**, *118*, 83–98. [CrossRef]
13. Marinho, B.A.; Cristóvão, R.O.; Boaventura, R.A.R.; Vilar, V.J.P. As(III) and Cr(VI) oxyanion removal from water by advanced oxidation/reduction processes—A review. *Environ. Sci. Pollut. Res.* **2018**, *26*, 2203–2227. [CrossRef] [PubMed]
14. Carolin, C.F.; Kumar, P.S.; Saravanan, A.; Joshiba, G.J.; Naushad, M. Efficient techniques for the removal of toxic heavy metals from aquatic environment: A review. *J. Environ. Chem. Eng.* **2017**, *5*, 2782–2799. [CrossRef]
15. Bolisetty, S.; Peydayesh, M.; Mezzenga, R. Sustainable technologies for water purification from heavy metals: Review and analysis. *Chem. Soc. Rev.* **2019**, *48*, 463–487. [CrossRef]
16. Weng, C.-H.; Lin, Y.-T.; Hong, D.-Y.; Sharma, Y.C.; Chen, S.-C.; Tripathi, K. Effective removal of copper ions from aqueous solution using base treated black tea waste. *Ecol. Eng.* **2014**, *67*, 127–133. [CrossRef]
17. Calheiros, C.; Rangel, A.; Castro, P. Constructed Wetlands for Tannery Wastewater Treatment in Portugal: Ten Years of Experience. *Int. J. Phytoremediation* **2014**, *16*, 859–870. [CrossRef]
18. Kumar, L.; Nadeem, F.; Sloan, M.; Restle-Steinert, J.; Deitch, M.J.; Naqvi, S.A.; Kumar, A.; Sassanelli, C. Fostering Green Finance for Sustainable Development: A Focus on Textile and Leather Small Medium Enterprises in Pakistan. *Sustainability* **2022**, *14*, 11908. [CrossRef]
19. Sassanelli, C.; Rosa, P.; Rocca, R.; Terzi, S. Circular economy performance assessment methods: A systematic literature review. *J. Clean. Prod.* **2019**, *229*, 440–453. [CrossRef]
20. Acerbi, F.; Sassanelli, C.; Terzi, S.; Taisch, M. A Systematic Literature Review on Data and Information Required for Circular Manufacturing Strategies Adoption. *Sustainability* **2021**, *13*, 2047. [CrossRef]
21. Acerbi, F.; Sassanelli, C.; Taisch, M. A conceptual data model promoting data-driven circular manufacturing. *Oper. Manag. Res.* **2022**, 1–20. [CrossRef]
22. Memedovic, O.; Mattila, H. The global leather value chain: The industries, the main actors and prospects for upgrading in LDCs. *Int. J. Technol. Learn. Innov. Dev.* **2008**, *1*, 482. [CrossRef]
23. Memon, Y.I.; Qureshi, S.S.; Kandhar, I.A.; Qureshi, N.A.; Saeed, S.; Mubarak, N.; Khan, S.U.; Saleh, T.A. Statistical analysis and physicochemical characteristics of groundwater quality parameters: A case study. *Int. J. Environ. Anal. Chem.* **2021**, 1–22. [CrossRef]
24. Jiang, W.; Yuan, Z.; Bi, J.; Sun, L. Conserving water by optimizing production schedules in the dyeing industry. *J. Clean. Prod.* **2010**, *18*, 1696–1702. [CrossRef]
25. Rottle, N.; Bowles, M.; Andrews, L.; Engelke, J. Constructed floating wetlands: A "s afe-to-fail" study with m ulti-sector participation. *Restor. Ecol.* **2022**, e13672. [CrossRef]
26. Castellar, J.A.; Torrens, A.; Buttiglieri, G.; Monclús, H.; Arias, C.A.; Carvalho, P.N.; Galvao, A.; Comas, J. Nature-based solutions coupled with advanced technologies: An opportunity for decentralized water reuse in cities. *J. Clean. Prod.* **2022**, *340*, 130660. [CrossRef]
27. Unnithan, M.R.; Vinod, V.P.; Anirudhan, T.S. Synthesis, Characterization, and Application as a Chromium(VI) Adsorbent of Amine-Modified Polyacrylamide-Grafted Coconut Coir Pith. *Ind. Eng. Chem. Res.* **2004**, *43*, 2247–2255. [CrossRef]
28. Faulwetter, J.L.; Burr, M.D.; Cunningham, A.B.; Stewart, F.M.; Camper, A.K.; Stein, O.R. Floating treatment wetlands for domestic wastewater treatment. *Water Sci. Technol.* **2011**, *64*, 2089–2095. [CrossRef]
29. Somprasert, S.; Mungkung, S.; Kreetachat, N.; Imman, S.; Homklin, S. Implementation of an Integrated Floating Wetland and Biofilter for Water Treatment in Nile Tilapia Aquaculture. *J. Ecol. Eng.* **2021**, *22*, 146–152. [CrossRef]
30. Uysal, Y. Removal of chromium ions from wastewater by duckweed, *Lemna minor* L. by using a pilot system with continuous flow. *J. Hazard. Mater.* **2013**, *263*, 486–492. [CrossRef] [PubMed]

31. Pascual, A.; De La Varga, D.; Arias, C.A.; Van Oirschot, D.; Kilian, R.; Álvarez, J.A.; Soto, M. Hydrolytic anaerobic reactor and aerated constructed wetland systems for municipal wastewater treatment—Highwet project. *Environ. Technol.* **2016**, *38*, 209–219. [CrossRef] [PubMed]
32. DalCorso, G.; Fasani, E.; Manara, A.; Visioli, G.; Furini, A. Heavy Metal Pollutions: State of the Art and Innovation in Phytoremediation. *Int. J. Mol. Sci.* **2019**, *20*, 3412. [CrossRef]
33. Afzal, M.; Arslan, M.; Müller, J.A.; Shabir, G.; Islam, E.; Tahseen, R.; Anwar-Ul-Haq, M.; Hashmat, A.J.; Iqbal, S.; Khan, Q.M. Floating treatment wetlands as a suitable option for large-scale wastewater treatment. *Nat. Sustain.* **2019**, *2*, 863–871. [CrossRef]
34. Weragoda, S.K.; Jinadasa, K.B.S.N.; Zhang, D.Q.; Gersberg, R.M.; Tan, S.K.; Tanaka, N.; Jern, N.W. Tropical Application of Floating Treatment Wetlands. *Wetlands* **2012**, *32*, 955–961. [CrossRef]
35. Guerrero, C.M.; Travis, G. *Assessing Decentralized Wastewater Treatment Options in SantaBarbara Count*; Taylor & Francis Group: Oxfordshire, UK, 2009.
36. Dotro, G.; Langergraber, G.; Molle, P.; Nivala, J.; Puigagut, J.; Stein, O.; von Sperling, M. Treatment Wetlands. 2017. Available online: https://library.oapen.org/handle/20.500.12657/31049 (accessed on 22 September 2022).
37. Walker, C.; Tondera, K.; Lucke, T. Stormwater Treatment Evaluation of a Constructed Floating Wetland after Two Years Operation in an Urban Catchment. *Sustainability* **2017**, *9*, 1687. [CrossRef]
38. Lucke, T.; Walker, C.; Beecham, S. Experimental designs of field-based constructed floating wetland studies: A review. *Sci. Total Environ.* **2019**, *660*, 199–208. [CrossRef]
39. Abed, S.N.; Almuktar, S.A.; Scholz, M. Remediation of synthetic greywater in mesocosm—Scale floating treatment wetlands. *Ecol. Eng.* **2017**, *102*, 303–319. [CrossRef]
40. Shahid, M.J.; Arslan, M.; Ali, S.; Siddique, M.; Afzal, M. Floating Wetlands: A Sustainable Tool for Wastewater Treatment. *CLEAN—Soil Air Water* **2018**, *46*, 1800120. [CrossRef]
41. Johnson, S. Literature Review: Pollutant Removal Efficacy of Floating Treatment Wetlands across Water Bodies. Bachelor's Thesis, Portland State University, Portland, OR, USA, 2021. [CrossRef]
42. Winston, R.J.; Hunt, W.F.; Kennedy, S.G.; Merriman, L.S.; Chandler, J.; Brown, D. Evaluation of floating treatment wetlands as retrofits to existing stormwater retention ponds. *Ecol. Eng.* **2013**, *54*, 254–265. [CrossRef]
43. Pavlidis, G.; Zotou, I.; Karasali, H.; Marousopoulou, A.; Bariamis, G.; Tsihrintzis, V.A.; Nalbantis, I. Performance of Pilot-scale Constructed Floating Wetlands in the Removal of Nutrients and Pesticides. *Water Resour. Manag.* **2021**, *36*, 399–416. [CrossRef]
44. Abdelhakeem, S.G.; Aboulroos, S.A.; Kamel, M.M. Performance of a vertical subsurface flow constructed wetland under different operational conditions. *J. Adv. Res.* **2016**, *7*, 803–814. [CrossRef]
45. Eljamal, O.; Thompson, I.P.; Maamoun, I.; Shubair, T.; Eljamal, K.; Lueangwattanapong, K.; Sugihara, Y. Investigating the design parameters for a permeable reactive barrier consisting of nanoscale zero-valent iron and bimetallic iron/copper for phosphate removal. *J. Mol. Liq.* **2019**, *299*, 112144. [CrossRef]
46. Rahmadyanti, E. Integrated System of Biofilter and Constructed Wetland for Sustainable Batik Industry. *Int. J. GEOMATE* **2020**, *18*, 138–148. [CrossRef]
47. García-Valero, A.; Martínez-Martínez, S.; Faz, Á.; Terrero, M.A.; Muñoz, M.; Gómez-López, M.D.; Acosta, J.A. Treatment of WASTEWATER from the Tannery Industry in a Constructed Wetland Planted with *Phragmites australis*. *Agronomy* **2020**, *10*, 176. [CrossRef]
48. Saeed, T.; Khan, T. Constructed wetlands for industrial wastewater treatment: Alternative media, input biodegradation ratio and unstable loading. *J. Environ. Chem. Eng.* **2019**, *7*, 103042. [CrossRef]
49. Calheiros, C.S.; Rangel, A.O.; Castro, P.M. Constructed wetland systems vegetated with different plants applied to the treatment of tannery wastewater. *Water Res.* **2007**, *41*, 1790–1798. [CrossRef]
50. Salt, D.E.; Blaylock, M.; Kumar, N.P.B.A.; Dushenkov, V.; Ensley, B.D.; Chet, I.; Raskin, I. Phytoremediation: A Novel Strategy for the Removal of Toxic Metals from the Environment Using Plants. *Nat. Biotechnol.* **1995**, *13*, 468–474. [CrossRef] [PubMed]
51. Shelef, O.; Gross, A.; Rachmilevitch, S. Role of Plants in a Constructed Wetland: Current and New Perspectives. *Water* **2013**, *5*, 405–419. [CrossRef]
52. Calheiros, C.S.C.; Pereira, S.I.A.; Franco, A.R.; Castro, P.M.L. Diverse Arbuscular Mycorrhizal Fungi (AMF) Communities Colonize Plants Inhabiting a Constructed Wetland for Wastewater Treatment. *Water* **2019**, *11*, 1535. [CrossRef]
53. Arliyani, I.; Tangahu, B.V.; Mangkoedihardjo, S. Plant Diversity in a Constructed Wetland for Pollutant Parameter Processing on Leachate: A Review. *J. Ecol. Eng.* **2021**, *22*, 240–255. [CrossRef]
54. Rijkenberg, M.J.; Depree, C.V. Heavy metal stabilization in contaminated road-derived sediments. *Sci. Total Environ.* **2010**, *408*, 1212–1220. [CrossRef] [PubMed]
55. Boelaert, F.; Amore, G.; van der Stede, Y.; Stoicescu, A.; Nagy, K.; Riolo, F.; Kleine, J.; Messens, W.; Lima, E.; Watts, M.; et al. The European Union summary report on trends and sources of zoonoses, zoonotic agents and food-borne outbreaks in 2014. *EFSA J.* **2015**, *13*, 4329. [CrossRef]
56. Shen, S.; Li, X.; Lu, X. Recent developments and applications of floating treatment wetlands for treating different source waters: A review. *Environ. Sci. Pollut. Res.* **2021**, *28*, 62061–62084. [CrossRef]
57. Priya, E.S.; Selvan, P.S. Water hyacinth (*Eichhornia crassipes*)—An efficient and economic adsorbent for textile effluent treatment—A review. *Arab. J. Chem.* **2017**, *10*, S3548–S3558. [CrossRef]
58. Krämer, U. Metal Hyperaccumulation in Plants. *Annu. Rev. Plant Biol.* **2010**, *61*, 517–534. [CrossRef]

59. Yu, C.; Peng, X.; Yan, H.; Li, X.; Zhou, Z.; Yan, T. Phytoremediation Ability of *Solanum nigrum* L. to Cd-Contaminated Soils with High Levels of Cu, Zn, and Pb. *Water Air Soil Pollut.* **2015**, *226*, 157. [CrossRef]
60. Liang, Y.; Zhu, H.; Bañuelos, G.; Shutes, B.; Yan, B.; Cheng, X. Removal of sulfamethoxazole from salt-laden wastewater in constructed wetlands affected by plant species, salinity levels and co-existing contaminants. *Chem. Eng. J.* **2018**, *341*, 462–470. [CrossRef]
61. Li, W.; Xu, X.; Fujibayashi, M.; Niu, Q.; Tanaka, N.; Nishimura, O. Response of microalgae to elevated CO_2 and temperature: Impact of climate change on freshwater ecosystems. *Environ. Sci. Pollut. Res.* **2016**, *23*, 19847–19860. [CrossRef]
62. Gacheva, G.; Gigova, L. Biological activity of microalgae can be enhanced by manipulating the cultivation temperature and irradiance. *Open Life Sci.* **2014**, *9*, 1168–1181. [CrossRef]
63. Jafarinejad, S.; Jiang, S.C. Current technologies and future directions for treating petroleum refineries and petrochemical plants (PRPP) wastewaters. *J. Environ. Chem. Eng.* **2019**, *7*, 103326. [CrossRef]
64. Tian, X.; Song, Y.; Shen, Z.; Zhou, Y.; Wang, K.; Jin, X.; Han, Z.; Liu, T. A comprehensive review on toxic petrochemical wastewater pretreatment and advanced treatment. *J. Clean. Prod.* **2020**, *245*, 118692. [CrossRef]
65. Farraji, H.; Zaman, N.Q.; Tajuddin, R.M.; Faraji, H. Advantages and Disadvantages of Phytoremediation: A Concise Review. *Int. J. Env. Tech. Sci.* **2016**, *2*, 69–75. Available online: www.journalijets.org (accessed on 23 September 2022).
66. Verhoeven, J.T.A.; Arheimer, B.; Yin, C.; Hefting, M.; Verhoeven, J.T.A.; Arheimer, B.; Yin, C.; Hefting, M.; Verhoeven, J.T.A.; Arheimer, B.; et al. Regional and global concerns over wetlands and water quality. *Trends Ecol. Evol.* **2006**, *21*, 96–103. [CrossRef] [PubMed]
67. Ortolano, L.; Sanchez-Triana, E.; Afzal, J.; Ali, C.L.; Rebellón, S.A. Cleaner production in Pakistan's leather and textile sectors. *J. Clean. Prod.* **2014**, *68*, 121–129. [CrossRef]
68. Kumar, L.; Kamil, I.; Ahmad, M.; Naqvi, S.A.; Deitch, M.J.; Amjad, A.Q.; Kumar, A.; Basheer, S.; Arshad, M.; Sassanelli, C. In-house resource efficiency improvements supplementing the end of pipe treatments in textile SMEs under a circular economy fashion. *Front. Environ. Sci.* **2022**, *10*. [CrossRef]
69. Khalili, N.R.; Duecker, S.; Ashton, W.; Chavez, F. From cleaner production to sustainable development: The role of academia. *J. Clean. Prod.* **2015**, *96*, 30–43. [CrossRef]
70. Rizwan, U.; Malik, R.N.; Abdul, Q. Assessment of groundwater contamination in an industrial city, Sialkot, Pakistan. *Afr. J. Environ. Sci. Technol.* **2009**, *3*, 429–446. Available online: https://www.ajol.info/index.php/ajest/article/view/56273 (accessed on 2 September 2022).
71. Qadir, A.; Malik, R.N. Assessment of an index of biological integrity (IBI) to quantify the quality of two tributaries of river Chenab, Sialkot, Pakistan. *Hydrobiologia* **2008**, *621*, 127–153. [CrossRef]
72. Ghani, A.; Maalik, S. Assessment of diversity and relative abundance of insect fauna associated with Triticum aestivum from district Sialkot, Pakistan. *J. King Saud Univ.-Sci.* **2019**, *32*, 986–995. [CrossRef]
73. Tüfekci, N.; Sivri, N.; Toroz, İ. Pollutants of Textile Industry Wastewater and Assessment of its Discharge Limits by Water Quality Standards. *Turk. J. Fish. Aquat. Sci.* **2007**, *7*, 97–103.
74. Ayaz, T.; Khan, S.; Khan, A.Z.; Lei, M.; Alam, M. Remediation of industrial wastewater using four hydrophyte species: A comparison of individual (pot experiments) and mix plants (constructed wetland). *J. Environ. Manag.* **2019**, *255*, 109833. [CrossRef] [PubMed]
75. Khan, S.; Ahmad, I.; Shah, M.T.; Rehman, S.; Khaliq, A. Use of constructed wetland for the removal of heavy metals from industrial wastewater. *J. Environ. Manag.* **2009**, *90*, 3451–3457. [CrossRef] [PubMed]
76. Shehzadi, M.; Afzal, M.; Khan, M.U.; Islam, E.; Mobin, A.; Anwar, S.; Khan, Q.M. Enhanced degradation of textile effluent in constructed wetland system using Typha domingensis and textile effluent-degrading endophytic bacteria. *Water Res.* **2014**, *58*, 152–159. [CrossRef] [PubMed]
77. Gholipour, A.; Zahabi, H.; Stefanakis, A.I. A novel pilot and full-scale constructed wetland study for glass industry wastewater treatment. *Chemosphere* **2020**, *247*, 125966. [CrossRef]
78. Calheiros, C.S.; Quitério, P.V.; Silva, G.; Crispim, L.F.; Brix, H.; Moura, S.C.; Castro, P.M. Use of constructed wetland systems with Arundo and Sarcocornia for polishing high salinity tannery wastewater. *J. Environ. Manag.* **2012**, *95*, 66–71. [CrossRef]
79. Mant, C.; Costa, S.; Williams, J.; Tambourgi, E. Phytoremediation of chromium by model constructed wetland. *Bioresour. Technol.* **2006**, *97*, 1767–1772. [CrossRef]
80. Naseer, K.; Hashmi, I.; Arshad, M.; Gabriel, H.F. Performance Efficiency of a Large-Scale Integrated Constructed Wetland. *J. Environ. Treat. Tech.* **2021**, *9*, 629–635.
81. Saadi, S.T.A.; Arshad, M.; Qammar, M.U.; Haq, M.A.U. Designing an efficient wetland by decision support system using experimental and modling approach. *Pak. J. Agric. Sci.* **2020**, *57*, 837–847. [CrossRef]
82. Agarry, S.E.; Oghenejoboh, K.M.; Latinwo, G.K.; Owabor, C.N. Biotreatment of petroleum refinery wastewater in vertical surface-flow constructed wetland vegetated with *Eichhornia crassipes*: Lab-scale experimental and kinetic modelling. *Environ. Technol.* **2018**, *41*, 1793–1813. [CrossRef]
83. Wichtmann, W.; Syndrom, C. The CINDERELLA Project: Paludiculture for GHG Emissions Mitigation in Peatlands Aim of the Project. Progressing Paludicultures after Centuries of Peatland Destruction and Neglect. 2017. Available online: http://www.imcg.net/media/2017/imcg_bulletin_1709.pdf (accessed on 2 September 2022).

84. Geurts, J.J.M.; Fritz, C. Paludiculture Pilots and Experiments with Focus on Cattail and Reed in The Netherlands. Technical Report CINDERELLA Project FACCE-JPI ERA-NET Plus on Climate Smart Agriculture. 2018, pp. 1–71. Available online: https://repository.ubn.ru.nl/bitstream/handle/2066/192628/192628pub.pdf (accessed on 2 September 2022).
85. Kantawanichkul, S.; Kladprasert, S.; Brix, H. Treatment of high-strength wastewater in tropical vertical flow constructed wetlands planted with *Typha angustifolia* and *Cyperus involucratus*. *Ecol. Eng.* **2009**, *35*, 238–247. [CrossRef]
86. Dan, T.H.; Quang, L.N.; Chiem, N.H.; Brix, H. Treatment of high-strength wastewater in tropical constructed wetlands planted with Sesbania sesban: Horizontal subsurface flow versus vertical downflow. *Ecol. Eng.* **2011**, *37*, 711–720. [CrossRef]
87. Hashem, M.A.; Hasan, M.; Momen, M.A.; Payel, S.; Nur-A-Tomal, M.S. Water hyacinth biochar for trivalent chromium adsorption from tannery wastewater. *Environ. Sustain. Indic.* **2020**, *5*, 100022. [CrossRef]
88. Elbasiouny, H.; Darwesh, M.; Elbeltagy, H.; Abo-Alhamd, F.G.; Amer, A.A.; Elsegaiy, M.A.; Khattab, I.A.; Elsharawy, E.A.; Ebehiry, F.; El-Ramady, H.; et al. Ecofriendly remediation technologies for wastewater contaminated with heavy metals with special focus on using water hyacinth and black tea wastes: A review. *Environ. Monit. Assess.* **2021**, *193*, 1–19. [CrossRef] [PubMed]
89. Radu, V.M.; Ionescu, P.; Diacu, E.; Ivanov, A.A. Removal of Heavy Metals from Aquatic Environments Using Water Hyacinth and Water Lettuce. *Rev. Chim.* **2018**, *68*, 2765–2767. [CrossRef]
90. Song, Z.; Williams, C.; Edyvean, R. Sedimentation of tannery wastewater. *Water Res.* **2000**, *34*, 2171–2176. [CrossRef]
91. Emerhi, E.A. Physical and combustion properties of briquettes produced from sawdust of three hardwood species and different organic binders. *Adv. Appl. Sci. Res.* **2011**, *2*, 236–246.
92. Dipu, S.; Kumar, A.A.; Thanga, V.S.G. Phytoremediation of dairy effluent by constructed wetland technology. *Environmentalist* **2011**, *31*, 263–278. [CrossRef]
93. Reddy, K.R.; Sutton, D.L.; Bowes, G. Freshwater aquatic plant biomass production in Florida. *Proc. Soil Crop Sci. Soc. Fla.* **1983**, *42*, 28–40.
94. Wickramasinghe, S.; Jayawardana, C.K. Potential of Aquatic Macrophytes *Eichhornia crassipes*, *Pistia stratiotes* and *Salvinia molesta* in Phytoremediation of Textile Wastewater. *J. Water Secur.* **2018**, *4*, 1–8. [CrossRef]
95. Mufarrege, M.M.; Hadad, H.R.; Di Luca, G.A.; Sanchez, G.C.; Maine, M.A.; Caffaratti, S.E.; Pedro, M.C. Organic Matter Effects on the Cr(VI) Removal Efficiency and Tolerance of Typha domingensis. *Water Air Soil Pollut.* **2018**, *229*, 384. [CrossRef]
96. Gupta, P.; Roy, S.; Mahindrakar, A.B. Treatment of Water Using Water Hyacinth, Water Lettuce and Vetiver Grass—A Review. *Resour. Environ.* **2012**, *2*, 202–215. [CrossRef]
97. Yasar, A.; Zaheer, A.; Tabinda, A.B.; Khan, M.; Mahfooz, Y.; Rani, S.; Siddiqua, A.; Mahfooz, M.K.Y. Comparison of Reed and Water Lettuce in Constructed Wetlands for Wastewater Treatment. *Water Environ. Res.* **2018**, *90*, 129–135. [CrossRef]
98. Abou-Elela, S.I.; Hellal, M.S. Municipal wastewater treatment using vertical flow constructed wetlands planted with *Canna*, *Phragmites* and *Cyprus*. *Ecol. Eng.* **2012**, *47*, 209–213. [CrossRef]
99. Kumari, M.; Tripathi, B. Efficiency of Phragmites australis and *Typha latifolia* for heavy metal removal from wastewater. *Ecotoxicol. Environ. Saf.* **2015**, *112*, 80–86. [CrossRef]
100. Ingrao, C.; Vesce, E.; Evola, R.S.; Rebba, E.; Arcidiacono, C.; Martra, G.; Beltramo, R. Chemistry behind leather: Life Cycle Assessment of nano-hydroxyapatite preparation on the lab-scale for fireproofing applications. *J. Clean. Prod.* **2020**, *279*, 123837. [CrossRef]
101. Cardwell, A.; Hawker, D.; Greenway, M. Metal accumulation in aquatic macrophytes from southeast Queensland, Australia. *Chemosphere* **2002**, *48*, 653–663. [CrossRef]
102. Maine, M.; Suñe, N.; Hadad, H.; Sánchez, G.; Bonetto, C. Removal efficiency of a constructed wetland for wastewater treatment according to vegetation dominance. *Chemosphere* **2007**, *68*, 1105–1113. [CrossRef]
103. Hadad, H.; Maine, M.A.; Bonetto, C. Macrophyte growth in a pilot-scale constructed wetland for industrial wastewater treatment. *Chemosphere* **2006**, *63*, 1744–1753. [CrossRef]
104. US EPA. Wastewater Technology Fact Sheet Wetlands: Subsurface Flow. 2000. Available online: https://www3.epa.gov/npdes/pubs/wetlands-subsurface_flow.pdf (accessed on 26 August 2022).
105. Saeed, T.; Muntaha, S.; Rashid, M.; Sun, G.; Hasnat, A. Industrial wastewater treatment in constructed wetlands packed with construction materials and agricultural by-products. *J. Clean. Prod.* **2018**, *189*, 442–453. [CrossRef]
106. Parde, D.; Patwa, A.; Shukla, A.; Vijay, R.; Killedar, D.J.; Kumar, R. A review of constructed wetland on type, treatment and technology of wastewater. *Environ. Technol. Innov.* **2020**, *21*, 101261. [CrossRef]
107. Hassan, I.; Chowdhury, S.R.; Prihartato, P.K.; Razzak, S.A. Wastewater Treatment Using Constructed Wetland: Current Trends and Future Potential. *Processes* **2021**, *9*, 1917. [CrossRef]
108. Yazdani, V.; Golestani, H.A. Advanced treatment of dairy industrial wastewater using vertical flow constructed wetlands. *Desalination Water Treat.* **2019**, *162*, 149–155. [CrossRef]
109. Morari, F.; Ferro, N.D.; Cocco, E. Municipal Wastewater Treatment with *Phragmites australis* L. and *Typha latifolia* L. for Irrigation Reuse. Boron and Heavy Metals. *Water Air Soil Pollut.* **2015**, *226*, 1–14. [CrossRef]

110. Abbas, N.; Butt, M.T.; Ahmad, M.M.; Deeba, F. Phytoremediation potential of *Typha latifolia* and water hyacinth for removal of heavy metals from industrial wastewater. *Chem. Int.* **2021**, *7*, 103–111.
111. Githuku, C.R.; Ndambuki, J.M.; Salim, R.W.; Badejo, A.A. Treatment Potential of *Typha latifolia* in Removal of Heavy Metals from Wastewater Using Constructed Wetlands, I. 2018. Available online: https://repository.lboro.ac.uk/articles/conference_contribution/Treatment_potential_of_Typha_latifolia_in_removal_of_heavy_metals_from_wastewater_using_constructed_wetlands/9593411 (accessed on 2 September 2022).

sustainability MDPI

Article

Analyzing the Leading Role of High-Performance Work System towards Strategic Business Performance

Abdelmohsen A. Nassani [1], Hadi Hussain [2], Joanna Rosak-Szyrocka [3], László Vasa [4,*], Zahid Yousaf [5,*] and Mohamed Haffar [6]

1 Department of Management, College of Business Administration, King Saud University, P.O. Box 71115, Riyadh 11587, Saudi Arabia; nassani@ksu.edu.sa
2 School of Economics and Finance, Xian Jiaotong University, Xi'an 710049, China; bangash0012@hotmail.com
3 Department of Production Engineering and Safety, Faculty of Management, Czestochowa University of Technology, 42-200 Czestochowa, Poland; joanna.rosak-szyrocka@wz.pcz.pl
4 Faculty of Economics, Széchenyi István University, 9026 Győr, Hungary
5 Higher Education Department, Government College of Management Sciences, Mansehra 21300, Pakistan
6 Department of Management, Birmingham Business School, University of Birmingham, Birmingham B15 2TY, UK
* Correspondence: vasa.laszlo@sze.hu (L.V.); muhammadzahid.yusuf@gmail.com (Z.Y.)

Abstract: HPWS is currently perceived as potential source of the competitive skills, capabilities and knowledge of human resources. This study aim to illustrate how high-performance work systems (HPWS) offer the foundation for strategic business performance (SBP) through the mediating function of organizational flexibility and contextualizing manufacturing firms of developing countries by providing an empirically tested framework for analyzing SBP. The current study is based on a quantitative research design. Data were gathered from manufacturing firms from the top, middle and operational management firms. SEM was used to analyze our 589 samples. Findings revealed that HPWS is the only component aiding manufacturing firms' growth. The results illustrate that HPWS will take a long time to achieve SBP if organisational flexibility does not mediate the relationship between HPWS and SBP. Utilizing actual data, this study reveals practical strategies for enhancing the mechanism of business development performance among manufacturing organizations. Furthermore, this research helps to understand the relationship between HPWS and organizational flexibility in attaining SBP.

Keywords: strategic business performance; organizational flexibility; high performance work system; manufacturing organizations

Citation: Nassani, A.A.; Hussain, H.; Rosak-Szyrocka, J.; Vasa, L.; Yousaf, Z.; Haffar, M. Analyzing the Leading Role of High-Performance Work System towards Strategic Business Performance. *Sustainability* **2023**, *15*, 5697. https://doi.org/10.3390/su15075697

Academic Editors: Sergio Terzi and Claudio Sassanelli

Received: 3 February 2023
Revised: 16 March 2023
Accepted: 21 March 2023
Published: 24 March 2023

1. Introduction

Recently, it is acknowledged that human resources are a significant potential source to improve a firm's sustainable competitiveness. This need forces an organization to develop a system proficient in facilitating the best development of its employees, and boost its competitive benefits to make linkage between HRM and strategic performance. A high-performance work system (HPWS) is a promising tool addressing the majestic challenge of a firm's success [1] and is exceptionally popular around the globe for improving competitiveness [2]. The term "flexible work practices", which describes how HPWS may help an organization become more adaptable to change, has been coined to describe the new paradigm of performance excellence that HPWS represents [3]. However, human resource literature does not elaborate on how HPWS provides the groundwork for adaptability in big organizations. The components of HPWS, selective staffing, training, employee commitment and participation, timely performance appraisal and clear job description are crucial for improving the overall organization's performance [4,5]. These components of HPWS still need to deepen understanding of how larger enterprises can perform strategically.

There is a considerable amount of literature linking HPWS to higher performance [1], productivity [6], organizational ambidexterity [7], workforce productivity [8], employee attitudes [9], organizational effectiveness, social capital [10], occupational safety [11], employees' competencies [9], employees' discretionary behavior [12], etc. However, beyond these outcomes of HPWS, detailed research is still required to show how HPWS fosters organizational flexibility to achieve strategic business performance (SBP) contextualizing larger organizations. This research provides a deeper understanding of HPWS by concentrating on the role of organizational flexibility as a mediator in the relationship between HPWS and SBP.

More incredible organizational flexibility results from HPWS deployment in larger organizations [10]. When a company is under pressure from competitors, it may be flexible and react to change by reorganizing its resources to meet market demands [12]. HPWS deployment enables such organizational flexibility [3]. Organizational flexibility encourages an enterprise's development and success and plays a significant part in elucidating SBP [13].

Organizational flexibility sets a strategic action to achieve the organization's objectives and cater the reliable basis for SBP [14]. SBP enables organizations to gain a foothold in the industry by maximizing higher market share and profitability, achieving marketing strategies and accomplishing overall strategic objectives [15]. There is hardly any evidence in the existing literature showing that organizational flexibility drives HPWS to SBP. This study closes this gap by exploring how HPWS influences organizational flexibility to foster SBP.

The research aims to develop a theoretical model showing the impact of HPWS on SBP in the presence of organizational flexibility (See Figure 1). The paper is arranged as follows. Firstly, HPWS, organizational flexibility and SBP are explored in the light of relevant literature and hypotheses development. Secondly, methods and research design are discussed. Third, the analysis and results are presented. Fourth, discussion and conclusions and theoretical and practical implications are presented. Finally, limitations and future research directions are given.

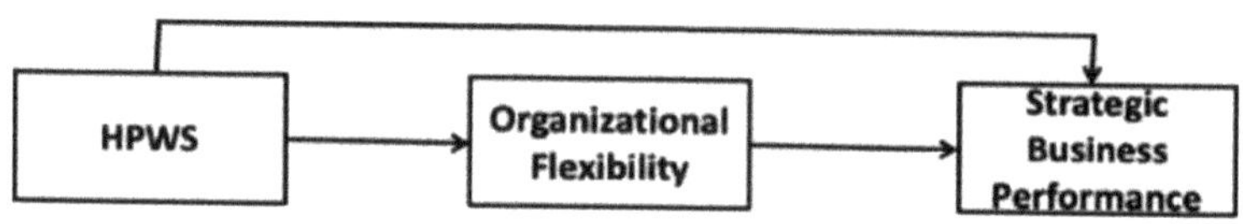

Figure 1. Theoretical Framework.

2. Research Context

2.1. High-Performance Work System (HPWS) and Strategic Business Performance (SBP)

SBP is an imperative consequence of explorative and exploitative learning, initiated with HPWS in terms of selective staffing, training, participation, performance appraisal and clear job description [16]. Selective staffing refers to finding appropriate workers with job-related knowledge, experience and expertise [1]. This process helps identify those who are well matched with the organization and have values similar to what other employees hold [17]. HPWS create knowledge, improves employee discretionary efforts through motivating and empowering them and operates in accordance with organizational structures to confirm SBP [13]. Such new experienced individuals foster heterogeneous but employee's related knowledge, skills and specialties required for enhancing long-term prosperity or sustainability in performance [18].

HPWS practices are integrated and coherent for achieving desired organizational performance through positive employee responses [13]. The study employs the concise framework of [4] a summary of five leading HPWS practices, i.e., selective staffing, training, participation, performance appraisal and clear job description. Training is improving one's knowledge and cognitive abilities through implementing improvements in practical pedagogy by studying sciences and technology [19]. A high-performance work system has recently emerged [1] and is labelled as high-involvement practices [2] and high commitment practices [18]. It creates a higher involvement and higher commitment of the organization's organizational system by creating such HR practices, which enables a high level of strategic performance [20]. SBP performance measures how well a company performs and how it stacks up against its main rivals in critical aspects, including entering a new market, building up its reputation and brand recognition, and responding to threats posed by competitors [21]. HPWS motivates workers to complete their tasks strategically; empowered employees can focus on their tasks and sense greater self-competence during decision-making, influencing SBP [22]. Larger enterprises which implement HPWS make their organizations more meaningful and may perform strategically by searching talented pool of human capital, offering career developments via training, putting forward their participation for improved skills, encouraging employees via timely appraisals and controlling operational mechanisms through offering clear job description [5,23]. SBP may be achievable if firms have robust merchandising, distribution and marketing strategies [16].

Furthermore, there is a difference between performance and SBP as strategic performance measures using modern indicators, methods, and concepts, and the idea of SBP is beyond average performance [24]. Organizations following such standardized HR practices (HPWS) can achieve the targets of SBP [14]. Therefore, it is argued that each component of HPWS determines the basis for larger enterprises to perform strategically. Our discussion leads to the following hypothesis:

H1: *The high-performance work system is positively associated with strategic business performance.*

2.2. High-Performance Work System (HPWS) and Organizational Flexibility

HPWS, in term of selective staffing, introduces new knowledge and experience into the organization, which shape a flexible infrastructure in both organization's strategies and structures [3]. The capability to manage and modify internal operations is called organizational flexibility [5]. As the newcomers share their experiences and success stories with existing employees and remove all the hurdles of organizational rigidity for establishing organizational flexibility [25]. Organizational flexibility is a powerful ability to modify, respond and adjust more incredible job strategies as necessary [13]. Training, a significant component of HPWS, is crucial to organizational flexibility because it initiates novelty and advancement in existing techniques [20]. These improved procedures and processes help determine the basis of an organization's flexibility [26]. In addition, flexible organizations have fewer restrictions on how they might seek support for their HR practices. HPWS, in terms of participation and employees' involvement in the decision-making process, encourages flexible strategies and removes hurdles of strict procedures [3]. Decentralization allows employees to participate in decisions and other operational mechanisms, which set directions for organizational flexibility. In the case of more prominent companies, we argue that HPWS significantly impacts organizational flexibility [25]. Another primary dimension of HPWS is performance appraisal, essential in developing flexible organizations because performance appraisals promote deserving employees and encourage flexible organizational structures [17]. Based on these arguments, we propose that all the dimensions of HPWS, i.e., selective staffing, training, participation and performance appraisal, positively influence organizational flexibility, and developed the following hypothesis:

H2: *A high-performance work system is positively associated with organizational flexibility.*

2.3. Organizational Adaptability and Strategical Business Outcomes

Flexible structures and strategies provide a tactical orientation for a method to bring about a change for the better [27]. This modification enables the company to meet its long-term goals and guarantee SBP [28]. Organizational flexibility boosts business competitiveness into a cutting-edge paradigm, which results in superior strategic performance [29]. To achieve SBP, flexible operations are more critical than inflexible operational mechanisms [30]. When reacting to large rivals, organizational flexibility brings about significant changes, which aid an organization in carrying out tactical activities for forming SBP [5].

Moreover, organizational adaptability allows for the necessary modifications to advance knowledge and skill to accomplish strategic goals [25,31]. Additionally, businesses with a flexible character can handle uncertainty and function strategically [31]. We contend that more organizational flexibility may help more significant enterprises run more effectively and efficiently to achieve SBP [3]. As a result, we suggest the following:

H3: *Strategic business effectiveness is positively related to an organization's adaptability.*

2.4. Organizational Adaptability Has a Mediating Function between HPWS-SBP

HPWS shapes organizational flexibility, a mechanism to transform strategies and structures for achieving sustainability in overall performance [10]. Even though several academics have acknowledged the positive relationship between organizational flexibility and effectiveness [25], the literature scarcely has any data demonstrating how adaptability plays a mediating role in the influence of HPWS on SBP. HPWS, in terms of selective staffing, enhances organizational flexibility through particular practices for increasing strategic business performance, e.g., [32] acknowledging that the involvement of existing team members in the selection process allows them to choose their future colleagues to enhance a flexible and collaborative infrastructure for achieving SBP. Larger enterprises focus on HPWS to set strategic directions through their adaptable aptitude [22].

Furthermore, implementing HPWS in larger organizations enhances an organization's ability to engage and empower employees to focus on less hierarchical structures by generating flexibility [14]. Such organizational flexibility helps to develop and maintain competitiveness by capturing key opportunities for achieving SBP [13]. The relationship between HPWS (selective staffing, training, participation, performance appraisal, clear job description) and SBP is mediated by organizational flexibility.

The selective staffing process introduces newcomers with advanced knowledge, capabilities and experience in related jobs with different personalities or interpersonal skills. Such a mix-up of newly individual staff and older ones leads an organization toward strategic business performance and flexibility [31].

Training emphasizes employees' mindset to motivate them to perform their tasks more flexibly (i.e., extra hours of work, teamwork etc.), which initiates overall organizational flexibility for superior strategic performance [32]. Training educates and fosters employees' experiences and career development by improving existing potential and skills, which enhances overall organizational flexibility for achieving SBP [33].

The participation process allows employees to participate in decision-making, which is crucial to organizational flexibility for improved strategic business performance [34]. Therefore, HPWS, in terms of participation, fosters organizational flexibility through synergies and results in high SBP [3].

Performance appraisal counterbalances the employees' working attitude by appreciating their efforts and highlighting their mistakes [35]. Such checks and balances on employees' performance creates a flexible environment by offering a strategic option for their career planning [25]. Thus, HPWS, in term of performance appraisals, encourages organizational flexibility and ensures SBP [28].

A clear job description improves employees' performance by understanding their work requirements and provides the basis for a flexible attitude [36]. Such clarity about their organizational work makes them more adaptable to serving strategically [27]. Based on the above mentioned literature, we proposed that HPWS, in terms of selective staffing, training, participation, performance appraisal and clear job description, positively influences organizational flexibility, which turns into strategic business performance. Organizational flexibility works as a bridge between HPWS and SBP. Hence, we formulated the following hypothesis.

H4: *The mediation between HPWS and SBP can be seen through the lens of organizational adaptability.*

3. Methodology

3.1. Data and Empirical Analysis

This study used a quantitative research design and random sampling technique. Data were collected through questionnaires as a survey tool within time-lagged (multiple rounds, 90 days spaced) from 583 managers at 76 Pakistani enterprises in the industrial and service sectors. Among the companies listed on the Pakistan Stock Exchange, 80 were chosen randomly. In 76 businesses, access was controlled by connections on a personal and professional level. These firms belonged to different industries, such as marketing, textile manufacturing, insurance, banking, health and electronics. Initially, 2000 senior managers—human resource managers (general manager human resources), the heads of different divisions and other senior managers involved in strategic decisions of the respective firms—were identified and contacted. A total of 1362 senior managers consented to participate in the three data collection rounds. They were given sealed return envelopes containing the confidentiality pledge, the survey questions and an information page outlining the main concepts and study goals.

Before distributing the questionnaire, this study questionnaire was checked by five experts, researchers and three members of academia to measure the reliability and validity of constructs. In the first phase, information was gathered on HPWS about age, sexuality, employment history and level of education. In the third and final phase, we gathered information regarding the mediator (organizational flexibility) and the result (strategic business performance). Within the 1st, 2nd and 3rd waves of data collection, we obtained 732, 669 and 621 replies, respectively. After eliminating invalid and missing values, we were left with 589 viable responses to employ in our tests of the predicted correlations. SPSS 25.0 and AMOS 25.0 were used for statistical analysis, namely structural equation modelling (SEM).

The sample had 408 men (70%) and 175 females (30%). When asked about their level of education, we found that 88% of respondents had a master's degree and 12% held a bachelor's degree. Respondents had an average age and employment history of 51.93 and 19.65 years, respectively. Participants came from a wide variety of backgrounds and occupations. The generalization of our results was improved by collecting data from a wide range of participants [37].

Further, a time-lagged design was used to lessen the variability in commonly used methods [11]. Herman's one-factor analysis was also computed to check the data for systematic variation [33]. A single component explained 23.24% of the total variation. (A number that is far below the threshold of 50%; hence, we can conclude that common technique bias is not a potential issue in our data.)

3.2. Variables and Measures

All factors were scored using a five-point Likert scale ranging from 1 (strongly disagree) to 5 (strongly agree).

3.2.1. HPWS

To quantify HPWS, we used a 16-item scale based on research by [6]. CFI = 0.95, IFI = 0.95, RMSEA = 0.05, TLI = 0.94, GFI = 0.93 and 2 (99) = 260.97 indicated a good match.

3.2.2. Organizational Flexibility

Organizational adaptability was assessed using a scale of eight items from [37]. We obtained the following values for the fit indices: RMSEA = 0.05, CFI = 0.95, IFI = 0.95, TLI = 0.96, GFI = 0.95, and 2 (121) = 313.32.

3.2.3. Strategic Business Performance

The 17-item scale used to assess SBP was developed based on previous research by [14]. The results of the fit indices were as follows: CFI = 0.96, IFI = 0.96, RMSEA = 0.05, TLI = 0.95, GFI = 0.94, and 2 (116) = 337.19.

4. Results

4.1. Non-Independence of the Data

Participants in the present research were recruited from 62 companies, so we had to check for any signs of data manipulation. Therefore, the intraclass correlation coefficient (ICC) (1) was determined for strategic business success. By using the [38] approach, we decided that there was no evidence of non-independence in our data since the ICC (1) value was 0.01 (ns).

4.2. Means and Correlations

Table 1 shows the results of correstions.

Table 1. Displays mean and correlation values.

Construct	Means	SD	1	2	3	4	5
1. HPWS	3.51	0.87					
2. Organizational flexibility	3.01	1.14	0.22 **				
3. Strategic business performance	2.89	0.88	0.16 **	0.34 **			
4. Gender	1.30	0.46	−0.01	−0.04	−0.05		
5. Age	51.93	4.58	−0.03	−0.08	0.01	0.04	
6. Work experience	19.65	2.94	−0.03	−0.03	0.01	−0.02	0.67 **

Note: ** = significant $p < 0.001$.

4.3. Measurement Model

Our evaluation methodology included measures of increased work systems, organizational flexibility and strategic business success. The measurement model was assessed using confirmatory factor analysis. Each item had a substantial loading on the targeted build. There was a high level of engagement between the measurement model and the data, as shown by the fit indices: 2 (768) = 1399.91, 2/df = 1.82, RMSEA = 0.04, CFI = 0.95, IFI = 0.95, TLI = 0.94, GFI = 0.90.

The convergent and discriminant validity of the scales was analyzed by calculating the maximum shared variance (MSV), the average shared variance (ASV) and the average variance extracted (AVE). Based on the data shown in Table 2, it is clear that AVE is higher than 0.50, ASV is lower than MSV, and MSV and ASV are lower than AVE. The correlations between the several measures of interest were also lower than the square root of the AVE values (bolded values on the diagonal of Table 2). For that reason, the measures had above-average convergent validity and discriminant validity. Furthermore, the scales' internal consistency was also high, with a Cronbach's alpha () >0.70 indicating this (Table 2).

Table 2. Convergent and discriminant validity, as well as reliability.

Construct	1	2	3	4	α	AVE	MSV	ASV
1. HPWS	**0.82**				0.95	0.68	0.06	0.04
2. Organizational flexibility	0.25	**0.71**			0.87	0.51	0.16	0.11
3. Strategic business performance	0.17	0.40	**0.78**		0.95	0.61	0.16	0.09

4.4. Structural Model—Direct and Mediation Results

This structural model was evaluated in three stages. In the first structural model, we looked at whether or not there was a direct correlation between high-performance work style (HPWS) and strategic performance. A strategic business performance correlated positively with HPWS (=0.18, p 0.001). An excellent match between the data and the structural model was shown by the fit indices (2 (486) = 999.04, 2/df = 2.05, RMSEA = 0.05, CFI = 0.94, IFI = 0.94, and TLI = 0.94, GFI = 0.90). As a result, H1 was confirmed.

Organizational adaptability moderated HPWS and strategic business success in Model 2. Structural model 2 has a good fit with the data, as measured by the fit indices 2 (768) = 1399.91,

2/df = 1.82, RMSEA = 0.04, CFI = 0.95, IFI = 0.95, and TLI = 0.94, GFI = 0.90; this suggests that organizational flexibility plays a crucial role as a mediator of the relationship between HPWS and strategic business performance.

Finally, we employed bootstrapping with a sample size of 2000 to evaluate the importance of organizational flexibility in mediating the link between HPWS and strategic business success. Table 3 displays the bootstrapping findings.

Table 3. Measures of plausibility, including confidence intervals of 95% and direct and indirect impacts.

Parameter	β	LL	UP
Standardized direct effects			
HPWS→Strategic business performance	0.08	−0.02	0.19
HPWS→Organizational flexibility	0.26 *	0.16	0.34
Organizational Flexibility→Strategic business performance	0.38 *	0.22	0.51
Standardized indirect effects			
HPWS→Organizational flexibility→Strategic business performance	0.10 *	0.06	0.15

Note: * = significant $p < 0.001$.

Table 3 shows a non-negligible positive correlation between HPWS and organizational flexibility (=0.26, 95% CI > 0). It follows that H2 is correct. Similarly, strategic business success was positively related to organizational adaptability (=0.38, 95% CI > 0). We may then conclude that H3 is correct. Importantly, HPWS was shown to have a positive indirect association with strategic business success (=0.10, 95% confidence interval did not overlap with zero) via organizational flexibility. Furthermore, the direct association between HPWS and strategic business performance became minor once the mediator (organizational flexibility) was included. This means that H4 is correct as well. That is to say, flexibility in the workplace is a necessary but not sufficient condition for the beneficial association between HPWS and strategic business success.

5. Discussion

This study's most crucial challenge was contributing significantly to the research on HPWS, organizational flexibility and SBP. This article proposed a model for the larger counterpart to implement HPWS for achieving the targets of SBP. To accomplish these objectives, four hypotheses were developed. H1 of this study proposed that HPWS can determine the strategic performance of larger companies. The concept of SBP differs from the mere version, i.e., short-term wins, and depends on numerous factors. Results proved that HR is a significant element for determining strategic performance, and it is a tricky goal, and organizations should think more broadly about HPWS to create knowledge, improve employee discretionary efforts through motivating and empowering them, and operate in accordance with organizational structures to confirm SBP [13]. Such new experienced individuals foster heterogeneous results, but employee's related knowledge, skills and specialties are required for enhancing long-term prosperity or sustainability in performance [18]. Researchers in HR strongly recommend HPWS for building solid foundations for long-run organizational performance, i.e., SBP. Our results proved that HPWS predicts SBP.

H2 proposed that HPWS predicts organizational flexibility. Based on the real-world experiences of HR professionals, our findings demonstrated that HPWS, in terms of selective hiring, training, participation, performance review, and defined roles and responsibilities, moves a company toward adaptability. HPWS allows for a more adaptable approach to organizational strategy and structure. Results support the previous studies that HPWS, in term of selective staffing, introduces new knowledge and experience into the organization, which shape a flexible infrastructure in both organizations' strategies and structures [3]. The capability to manage and modify internal operations is called organizational flexibility [5]. The third hypothesis of this research was that adaptability in the workplace correlates favorably with SBP. This study's outcomes are congruent with prior literature findings that organizational flexibility boosts business competitiveness in a cutting-edge paradigm, which results in superior strategic performance [29]. To achieve SBP, flexible operations are more critical than inflexible operational mechanisms [30]. When reacting to large rivals, organizational flexibility brings about significant changes, which aid an organization in carrying out tactical activities for forming SBP [5]. The findings corroborated the connection between the two, showing that SBP is founded on businesses' agility to adjust swiftly to new circumstances.

The final hypothesis of this research is that HPWS and SBP are related to organizational flexibility. Even though HPWS has a beneficial direct effect on SBP, manufacturing firms can only be confident of their ability to successfully implement and sustain this strategy if there is some degree of organizational flexibility between them. Findings are consistent with prior research that HPWS shapes organizational flexibility, a mechanism to transform design and structures for achieving sustainability in overall performance [10]. Several academics have acknowledged the positive relationship between organizational flexibility and effectiveness [25]. Furthermore, implementing HPWS in larger organizations enhances an organization's ability to engage and empower employees and focus on less hierarchical structures via generating flexibility [14]. Such organizational flexibility helps to develop and maintain competitiveness by capturing key opportunities for achieving SBP [13]. In other words, the data support our hypothesis and demonstrate that HPWS helps organizations become more adaptable, which is necessary for attaining SBP.

5.1. Implication for Theory and Practice

There were several ways in which this research aided management theory and practice. This first SBP model illustrates the interplay between high-performance work systems (HPWS), organizational adaptability and SBP in service and industrial organizations.

This research contributes to the existing body of knowledge by illuminating the significance of HPWS and demonstrating how its incorporation into practice helps lead to peak performance. Previous HR studies on HPWS had only looked at its effects on company performance [4,39]. Therefore, the current study filled in some gaps in that research. The phrase "Standards of Business Performance" (SBP) is more all−encompassing than just "financial", "market share", "competitive advantage", or "gaining a footing in the industry," etc. In doing so, the authors have built on the work of others who have shown that factors outside HPWS are essential in determining SBP (See the work of [5,7,40]. Medical professionals pursuing SBP should think about the significance of HPWS. Second, this research expands the existing body of knowledge by introducing the mediating role of organizational flexibility in the connection between HPWS and SBP. This is a new framework for HR researchers to un-taping an essential aspect of the HPWS outcome that previous researchers have skipped. This contribution is pertinent for management in practice because HPWS is a bundle of HR practices that can set directions for SBP. However, organizational flexibility perfectly matches the link between HPWS−SBP through enhancing firm planning and controlling mechanisms. HR managers should concentrate on corrections in hierarchal structure and firms' strategies accordingly for a positive HPWS effect on SBP.

This study also has limitations; data have been collected from Pakistan, a developing nation. Next, research may be conducted in developed countries with the same framework. This is cross-sectional data, and questionnaires were used for data collection. Subsequent analysis might be possible through a longitudinal research design based on episodic interviewing.

5.2. Limitations and Future Directions

Despite valuable contributions, this study has several limitations that provide direction for future studies. Firstly, this study used a quantitative research design and random sampling technique. We suggest that in future research, the qualitative, cross-sectional or longitudinal research design is used for data collection to understand findings better. Secondly, this study is conducted on Pakistan's industrial and service sectors. In the future, we recommend that other studies investigate this empirical model's findings on SMEs, tourism and other industries in developing or developed nations. Thirdly, this research provides a better understanding of how HPWS leads to firm-level outcomes, i.e., strategic performance. In upcoming studies, we suggest that researchers must incorporate constructs at the level of the individual and relationships among individuals similar to variables used in this study, such as organizational flexibility. Finally, this research suggests that future studies should investigate the moderation role of technological constructs between HPWS and SBP.

6. Conclusions

This study provides an improved understanding of the association between HPWS and organizational outcomes such as SBP, which is complicated by the need to deliberate multiple levels of the analysis. HPWS are typically implemented at the corporate level, assuming their impacts will also be felt at the organizational level (e.g., flexibility or strategic performance). This model contends that HPWS are effective in dynamic settings requiring knowledge resources. The alternative implication of knowledge workers supports organizational flexibility to consider more unique and valuable practices than others not directly involved with the strategic core. As such, dynamic environments and reliance on knowledge management may represent boundary conditions to the proposed framework.

Author Contributions: Conceptualization, A.A.N. and H.H.; methodology, J.R.-S.; software, L.V.; validation, M.H. and Z.Y.; formal analysis, Z.Y.; investigation, A.A.N.; resources, H.H.; data curation, Z.Y.; writing—original draft preparation, H.H. and J.R.-S.; writing—review and editing, L.V. and M.H.; visualization, L.V.; supervision and project administration, Z.Y.; funding acquisition, A.A.N. All authors have read and agreed to the published version of the manuscript.

Funding: Researchers Supporting Project number (RSP2023R87), King Saud University, Riyadh, Saudi Arabia.

Institutional Review Board Statement: The study was conducted in accordance with the Declaration of Helsinki and approved by the Institutional Review Board (or Ethics Committee) of GCMS (DN.459/234).

Informed Consent Statement: Written informed consent has been obtained from the patient(s) to publish this paper.

Data Availability Statement: Data is unavailable due to privacy or ethical restrictions.

Acknowledgments: Researchers Supporting Project number (RSP2023R87), King Saud University, Riyadh, Saudi Arabia.

Conflicts of Interest: The authors declare no conflict of interest.

References

1. Shin, D.; Konrad, A.M. Causality between high-performance work systems and organizational performance. *J. Manag.* **2017**, *43*, 973–997. [CrossRef]
2. Wood, S.; De Menezes, L.M. High involvement management, high-performance work systems and well-being. *Int. J. Hum. Resour. Manag.* **2011**, *22*, 1586–1610. [CrossRef]
3. Stirpe, L.; Zárraga-Oberty, C. Are High-Performance Work Systems always a valuable retention tool? The roles of workforce feminization and flexible work arrangements. *Eur. Manag. J.* **2017**, *35*, 128–136. [CrossRef]
4. Evans, W.R.; Davis, W.D. High-performance work systems and organizational performance: The mediating role of internal social structure. *J. Manag.* **2005**, *31*, 758–775. [CrossRef]
5. Katou, A.A. Human resources flexibility as a mediating mechanism between high-performance work systems and organizational performance: A multilevel quasi-longitudinal study. *EuroMed J. Bus.* **2022**, *17*, 174–192. [CrossRef]
6. Michaelis, B.; Wagner, J.D.; Schweizer, L. Knowledge as a key in the relationship between high-performance work systems and workforce productivity. *J. Bus. Res.* **2015**, *68*, 1035–1044. [CrossRef]
7. Úbeda-García, M.; Claver-Cortés, E.; Marco-Lajara, B.; Zaragoza-Sáez, P.; García-Lillo, F. High performance work system and performance: Opening the black box through the organizational ambidexterity and human resource flexibility. *J. Bus. Res.* **2018**, *88*, 397–406. [CrossRef]
8. Koser, M.; Rasool, S.F.; Samma, M. High performance work system is the accelerator of the best fit and integrated HR-practices to achieve the goal of productivity: A case of textile sector in Pakistan. *GMJACS* **2018**, *8*, 12.

9. Miao, R.; Zhou, W.; Li, T. High-performance work system and employee attitudes: A perspective of social exchange. *J. Manag. Sci.* **2013**, *26*, 39–49.
10. Jiang, J.Y.; Liu, C.W. High performance work systems and organizational effectiveness: The mediating role of social capital. *Hum. Resour. Manag. Rev.* **2015**, *25*, 126–137. [CrossRef]
11. Zacharatos, A.; Barling, J.; Iverson, R.D. High-performance work systems and occupational safety. *J. Appl. Psychol.* **2005**, *90*, 77. [CrossRef] [PubMed]
12. Elorza, U.; Harris, C.; Aritzeta, A.; Balluerka, N. The effect of management and employee perspectives of high-performance work systems on employees' discretionary behaviour. *Pers. Rev.* **2016**, *45*, 121–141. [CrossRef]
13. Alatailat, M.; Elrehail, H.; Emeagwali, O.L. High performance work practices, organizational performance and strategic thinking: A moderation perspective. *Int. J. Organ. Anal.* **2019**, *27*, 370–395. [CrossRef]
14. Yousaf, Z.; Majid, A. Organizational network and strategic business performance: Does organizational flexibility and entrepreneurial orientation really matter? *J. Organ. Chang. Manag.* **2018**, *31*, 268–285. [CrossRef]
15. Yousaf, Z.; Sahar, N.; Majid, A.; Rafiq, A. The effects of e-marketing orientation on strategic business performance: Mediating role of e-trust. *World J. Entrep. Manag. Sustain. Dev.* **2018**, *14*, 309–320. [CrossRef]
16. Han, J.H.; Kang, S.; Oh, I.S.; Kehoe, R.R.; Lepak, D.P. The goldilocks effect of strategic human resource management? Optimizing the benefits of a high-performance work system through the dual alignment of vertical and horizontal fit. *Acad. Manag. J.* **2019**, *62*, 1388–1412. [CrossRef]
17. Pahos, N.; Galanaki, E. Staffing practices and employee performance: The role of age. In *Evidence-Based HRM: A Global Forum for Empirical Scholarship*; Emerald Publishing Limited: Bingley, UK, 2019; Volume 7, pp. 93–112.
18. Sarikwal, L.; Gupta, J. The impact of high performance work practices and organizational citizenship behaviour on turnover intentions. *J. Strateg. Hum. Resour. Manag.* **2013**, *2*, 11.
19. Ferraz, F.A.D.; Gallardo-Vazquez, D. Measurement tool to assess the relationship between corporate social responsibility, training practices and business performance. *J. Clean. Prod.* **2016**, *129*, 659–672. [CrossRef]
20. McKenzie, D.; Woodruff, C. What are we learning from business training and entrepreneurship evaluations around the developing world? *World Bank Res. Obs.* **2014**, *29*, 48–82. [CrossRef]
21. Rajnoha, R.; Štefko, R.; Merková, M.; Dobrovič, J. Business intelligence as a key information and knowledge tool for strategic business performance management. *E+ M Ekon. A Manag.* **2016**, *19*, 183–203. [CrossRef]
22. Karatepe, O.M. High-performance work practices and hotel employee performance: The mediation of work engagement. *Int. J. Hosp. Manag.* **2013**, *32*, 132–140. [CrossRef]
23. Shen, J.; Benson, J.; Huang, B. High-performance work systems and teachers' work performance: The mediating role of quality of working life. *Hum. Resour. Manag.* **2014**, *53*, 817–833. [CrossRef]
24. Bromiley, P.; Navarro, P.; Sottile, P. Strategic business cycle management and organizational performance: A great unexplored research stream. *Strateg. Organ.* **2008**, *6*, 207–219. [CrossRef]
25. Wang, Y.; Cao, Y.; Xi, N.; Chen, H. High-Performance Work System, Strategic Flexibility, and Organizational Performance—The Moderating Role of Social Networks. *Front. Psychol.* **2021**, *12*, 670132. [CrossRef]
26. Liu, N.C.; Lin, Y.T. High-performance work systems, management team flexibility, employee flexibility and service-oriented organizational citizenship behaviors. *Int. J. Hum. Resour. Manag.* **2021**, *32*, 3912–3949. [CrossRef]
27. Acharya, S. Beyond learning outcomes: Impact of organizational flexibility on strategic performance measures of commercial e-learning providers. *Glob. J. Flex. Syst. Manag.* **2019**, *20*, 31–41. [CrossRef]
28. Ni, G.; Xu, H.; Cui, Q.; Qiao, Y.; Zhang, Z.; Li, H.; Hickey, P.J. Influence mechanism of organizational flexibility on enterprise competitiveness: The mediating role of organizational innovation. *Sustainability* **2020**, *13*, 176. [CrossRef]
29. Koçyiğit, Y.; Akkaya, B. The role of organizational flexibility in organizational agility: A research on SMEs. *Bus. Manag. Strategy* **2020**, *11*, 110–123. [CrossRef]
30. Saeed, M.A.; Tabassum, H.; Zahid, M.M.; Jiao, Y.; Nauman, S. Organizational flexibility and project portfolio performance: The roles of environmental uncertainty and innovation capability. *Eng. Manag. J.* **2022**, *34*, 249–264. [CrossRef]
31. Jain, N.K.; Panda, A.; Choudhary, P. Institutional pressures and circular economy performance: The role of environmental management system and organizational flexibility in oil and gas sector. *Bus. Strategy Environ.* **2020**, *29*, 3509–3525. [CrossRef]
32. Lin, Y.T.; Liu, N.C. High performance work systems and organizational service performance: The roles of different organizational climates. *Int. J. Hosp. Manag.* **2016**, *55*, 118–128. [CrossRef]
33. Prommarat, P.; Pratoom, K.; Muenthaisong, K. A conceptual model of strategic organizational flexibility capability and business survival. In Proceedings of the Allied Academies International Conference, Academy of Strategic Management Proceedings, Washington, DC, USA, 20–24 July 2015; Volume 14, p. 77.
34. Koçyiğit, Y.; Tabak, A. The Interaction Among Organizational Flexibility, Competitive Strategy and Competitive Advantage: A Path Analytic Study1. In *Agile Business Leadership Methods for Industry 4.0*; Emerald Publishing Limited: Bingley, UK, 2020; pp. 303–326.
35. Akhtar, M.; Mittal, R.K. Strategic Flexibility, Information System Flexibility and Enterprise Performance Management. In *Organisational Flexibility and Competitiveness*; Springer: Berlin/Heidelberg, Germany, 2014; pp. 41–51.
36. Chen, Y.; Gao, L.; Zhang, Y. The Impact of Green Organizational Identity on Green Competitive Advantage: The Role of Green Ambidexterity Innovation and Organizational Flexibility. *Math. Probl. Eng.* **2022**, *2022*, 1–18. [CrossRef]

37. Anser, M.K.; Yousaf, Z.; Usman, M.; Yousaf, S. Towards Strategic Business Performance of the Hospitality Sector: Nexus of ICT, E-marketing and Organizational Readiness. *Sustainability* **2020**, *12*, 1346. [CrossRef]
38. Fornell, C.; Lacker, D.F. Structural equation models with unobservable variables and measurement error: Algebra and Statistics. *J. Mark. Res.* **1981**, *18*, 382–388. [CrossRef]
39. Sassanelli, C.; Rosa, P.; Rocca, R.; Terzi, S. Circular economy performance assessment methods: A systematic literature review. *J. Clean. Prod.* **2019**, *229*, 440–453. [CrossRef]
40. Vinante, C.; Sacco, P.; Orzes, G.; Borgianni, Y. Circular economy metrics: Literature review and company-level classification framework. *J. Clean. Prod.* **2021**, *288*, 125090. [CrossRef]

sustainability

MDPI

Article

Interplay in Circular Economy Innovation, Business Model Innovation, SDGs, and Government Incentives: A Comparative Analysis of Pakistani, Malaysian, and Chinese SMEs

Fazal Ur Rehman [1,*], Basheer M. Al-Ghazali [2,3] and Mohamed Riyazi M. Farook [2]

1 Science and Research Centre, Faculty of Economics and Administration, University of Pardubice, 53210 Pardubice, Czech Republic
2 Department of Business Administration-DCC, King Fahd University of Petroleum & Minerals, Dhahran 31261, Saudi Arabia
3 Center for Finance and Digital Economy, King Fahd University of Petroleum & Minerals, Dhahran 31261, Saudi Arabia
* Correspondence: fazalur.rehman@upce.cz or fazal_marwatpk@yahoo.com

Abstract: This study aims to examine the effects of circular economy innovation and BMI (business model innovation) on SMEs (Small and Medium Enterprises) economic, environmental, and social performance along with the mediating role of government incentives in Pakistan, Malaysia, and China. Data were collected through a structured questionnaire in online survey from the owners, CEO, and senior managers of SMEs in Pakistan, Malaysia, and China, and analyzed using PLS-SEM. The results revealed that circular economy innovation and BMI have positive significant effects on SMEs economic, environmental, and social performance in Pakistan, Malaysia, and China. The study also found that government incentives have mediating effects on the relationship between circular economy innovation, BMI, and SMEs economic, environmental, and social performance in Pakistan, Malaysia, and China. This study provides interesting insights about SMEs economic, environmental, and social performance by evaluating the impacts of circular economy innovation, BMI, and amid mediation of government incentives. These useful insights will enable policy makers and practitioners to develop more effective strategies to enhance the economic, environmental, and social performance of SMEs. By reviewing the literature on circular economy innovation, BMI, and government incentives, the main contribution of this study is the evaluation and analysis of circular economy innovation, BMI, and government incentives as they affect SMEs economic, environmental, and social performance in Pakistan, Malaysia, and China. The theoretical and practical implications for academics and practitioners are displayed at the end of the study.

Keywords: circular economy innovation; business model innovation; government incentives; SMEs performance

Citation: Rehman, F.U.; Al-Ghazali, B.M.; Farook, M.R.M. Interplay in Circular Economy Innovation, Business Model Innovation, SDGs, and Government Incentives: A Comparative Analysis of Pakistani, Malaysian, and Chinese SMEs. *Sustainability* **2022**, *14*, 15586. https://doi.org/10.3390/su142315586

Academic Editors: Sergio Terzi and Claudio Sassanelli

Received: 15 September 2022
Accepted: 4 November 2022
Published: 23 November 2022

1. Introduction

The UN sustainable development agenda 2030 has garnered wide-ranging attention from industries and academia around the world, underscoring the importance of scare resources, avoiding environmental degradation, and preventing hunger and poverty in societies. In this regard, the UN sustainability agenda has launched the concept of Triple Bottom Line (TBL) and emerged as the potential driver for competitive advantage [1]. Due to this reason, many multinational companies have launched practices of sustainable development as per the UN agenda [2]; however, SMEs have widely shown scarce engagement in these practices around the world. For example, the British Chamber of Commerce has noted that only 11 percent of UK SMEs have engagement in sustainability measures, instead of a greater role in and contribution to the world economy [3]. However, sustainability-oriented business practices have forced companies to transform from the traditional to a more innovative business model to ensure energy conservation, reduce pollution and wastage of resources, and gain economic, environmental, and social efficiencies. In the race towards sustainable development, competitive advantage, and environmental efficiencies, firms are widely focusing on sustainability-oriented innovation and circularity to ensure their survival in an intense competitive environment.

Sustainability-oriented innovation has involved firms in the practices of BMI and circular economy innovation to achieve economic, environmental, and social efficiencies among firms [3,4]. These innovative practices are promoting closed-loop production and eco-efficiency in operations, innovating business structure, enhancing production and consumption efficiency, and implementing circular inventions [3,5]. Particularly, circular economy innovation has emerged as a new trend in the business environment to achieve sustainable development goals [6]. The trend of CE-innovation is perceived as a solution-based approach for gaining economics development within environmental boundaries [7], and can be applied as a technique to reduce the loss of resources, create a vital ecosystem using the tactics of eco-design, and repairing, recycling, and refurbishing to retain environmental values. These environmental approaches and ecosystem ensure sustainable production, clean drinking water, and a healthy living environment in the societies [7–9]. Especially, circular economy innovation is perceived as a key driver and contributing factor in sustainability [10]. Due to these reasons, many companies have shown a responsible attitude towards popularizing the practice of circular economy strategies [11] and moving from linear production to a more advanced ecosystem that leads to gaining core competencies and improving efficiencies [11,12].

Circular economy innovation provides a foundation for sustainability-oriented innovation to enhance the efficiencies of resources, innovative solutions, and BMI [13,14]. In this regard, BMI as a tool for sustainability and circularity has been acknowledged as an important part of the gray literature by companies and government agencies [15]. BMI, for sustainability and circularity, is perceived as a fundamental capability of businesses to gain a competitive advantage [15], and has led to the concept of circular/sustainable BMI. Further, firms are forced by various factors to consider environmental behavior, and to integrate sustainability and CE principles in their BMs [15,16]. Recently, there are calls for businesses to address these critical societal issues, specifically the ecosystem in business model research [17], and there is still a lack of understanding of the association between BMI and ecosystem, and there is space for a solid framework in this context.

Implementing the practices of circular economy innovation and BMI to transform the setup of SMEs into more sustainable business practices have brought forward the concept of closed-loop activities as a new trend of innovation [18] and, as an outcome, closed-loop innovative business activities lead SMEs to gain economic, environmental, and social efficiencies [3]. However, the literature is scarce in this regard and needs extensive investigation to explore the role of circular economy innovation and BMI in achieving improved economic, environmental, and social performance among SMEs [3,19–21]. In addition, despite a substantial body of research, it is rare to find data on how firms reinvent their business models for circularity [22–27], and how these practices enhance their economic, environmental, and social efficiencies [28–31], especially in Pakistan, Malaysia, and China. It is also difficult to find data on how government policies and support contribute to the transition process towards achieving sustainable development goals. In the existing scenario, previous studies have widely focused on the larger firms [32,33] but paid less attention to SMEs, especially in a comparative context in Pakistan, Malaysia, and China to understand the approaches towards SDGs. Therefore, this study has observed a gap in research in the relevant literature to evaluate the role of circular economy innovation and BMI in achieving economic, environmental, and social performance among SMEs along with aimed mediation of government incentives in Pakistan, Malaysia, and China. Therefore, this study aims to uncover the following raised research questions:

1. To what degree do circular economy innovation, BMI, and government incentives impact economic, environmental, and social performance among SMEs in Pakistan, Malaysia, and China?
2. What is the relationship between circular economy innovation, BMI, and government incentives among SMEs in Pakistan, Malaysia, and China?
3. Do government incentives mediate the relationship between circular economy innovation, BMI, and economic, environmental, and social performance among SMEs in Pakistan, Malaysia, and China?

To uncover the above research questions, we observed a need to study conduction in the defined context. Therefore, this study contributes to the relevant literature by evaluating the impacts of circular economy innovation, BMI, and government incentives on the economic, environmental, and social performance among SMEs in Pakistan, Malaysia, and China based on the resource-based view theory of Barney 1991 [34]. This study also contributes by assessing the mediating role of government incentives between circular economy innovation, BMI, and the economic, environmental, and social performance among SMEs in Pakistan, Malaysia, and China. As well, this study merged the literature of circular economy innovation, BMI, government incentives, and SDGs based on the findings from Pakistan, Malaysia, and China. The application of PLS-SEM in the defined context is also a novel methodological contribution. However, the remainder of this study is as follows; after the introduction, this study thoroughly reviews the theoretical background of the defined variables and explains the theoretical framework of the study. Methodology is presented in the third section, and results are presented in the fourth section of this study. Discussion, implications, and conclusions are included in the last part of this study.

2. Theoretical Background and Hypothesis Development

2.1. The Resource-Based View Theory

The concept of resource-based view was introduced by Barney in 1991 as a method of viewing of environmental circumstances to gain competitive advantages by using the organizational resources [34]. Barney focused on establishing a connection among heterogeneous resources and the mobility of these resources with the firm's strategic objectives to gain a competitive advantage in the target market. Barney also argued for the use of valuable resources such as physical capital resource, human capital resources, and organization capital resources to enable firms' practices to gain overall efficiency and improvement in performance. It is important for an organization to realize the uniqueness of the available resources (particularly, valuable, rare, imperfect able, and non-substitutable) and utilize them in a way to improve the efficiencies and effectiveness of organization [35]. Valuable, rare, imperfect able, and non-substitutable resources can lead to gaining a competitive advantage in the marketplace and creates barriers for competitors in follow-up imitation. The valuable, rare, imperfect able and non-substitutable resources of SMEs can help to improve the brand image and profitability of firms. It can also lead to create a monopoly in the target market and prevent the entrance of new ventures. In the view of resource-based theory, resources and capabilities can improve production efficiencies among firms, which leads to the enhancement of short-term and long-term profitability of firms [34]. Production efficiencies can also support the elimination of adverse impacts on the environment and gain social efficiencies among communities. Therefore, in the view of resource-based theory, innovating business model, and implementing the practices of circular economy innovation, firms can gain a competitive advantage and economic, environmental, and social efficiencies in the market.

2.2. Circular Economy Innovation and Sustainable Development Goals

In light of the UN sustainability agenda 2030 to transform the world through sustainable development for gaining economic, environmental, and social efficiencies among communities, the notion of circular economy has widely emerged into the academic literature and practices. The agenda of sustainable development was published with the aim of revising economic, environmental, and social policies to ensure prosperity, eliminating poverty and hunger among communities, and to ensure the sustainability of this planet [36]. In this regard, many international organizations and agencies, and the European Union, are trying to boost sustainability through accelerating the transition from linear economy to circular economy innovation [37,38], and launched a roadmap to a "resource efficient Europe" in 2011 to increase sustainability. In 2015, the EU initiated a program called "Close the Circle: An Action Plan of the European Union for the Circular Economy" and enforced its implementation in the member states to achieve the economic, environmental, and social efficiencies.

Kirchherr, Reike, and Hekkert [39] have defined circular economy as "a an economic system that is based on business models which replace the 'end-of-life' concept with reducing, alternatively reusing, recycling and recovering materials in production/distribution and consumption processes, thus operating at the micro level such as products, companies, consumers, eco-industrial parks, and macro level (city, region, nation and beyond), with the aim to accomplish sustainable development, which implies creating environmental quality, economic prosperity and social equity, to the benefit of current and future generations. It is enabled by novel business models and responsible consumers." In this regard, the circular economy approach should be treated as an economic model and as a tool to achieve sustainable development goals among communities [40].

Specifically, the strategy of circular economy is useful to achieve sustainable development goals including economic growth, sustainable production and consumption, climate change and environmental efficiencies, and quality of life and social efficiencies [40]. In this regard, many countries consider circular economy as the prime indicator to achieve the objectives of sustainable development goals and sustainable wellbeing, noted as "This action plan will be instrumental in reaching the Sustainable Development Goals (SDGs) by 2030" [40]. Similarly, many other organizations have noted that circular economy innovation has positive significant effects on sustainable development goals, especially in economic, environmental, and social efficiencies [40].

Predominantly, the transition from the traditional to circular economy model has positive significant impacts on sustainable development goals in direct and indirect ways [41], especially economic, environmental, and social efficiencies, as well as international competitiveness. In this regard, circular economy innovation is a prudent strategy to meet the economic, environmental, and social needs among societies and to serve sustainability goals [42]. Furthermore, Khajuria et al. [43] have outlined five main pillars of the sustainable development goals, which are: people, planet, prosperity, peace, and partnership. Precisely, the concept of circular economy is the main gateway to the future, a key component of sustainability, and a helpful mechanism to transition from the traditional production system to the more advanced circular economy system [40,44]. However, as per the UN 2030 agenda, economics, environmental, and social efficiencies are the top priorities among the sustainable development goals but are still in infancy and require wide-ranging investigation in the context of circular economy innovation to develop a more solid theoretical framework [36,45]. Therefore, we can hypothesize that:

H1a: Circular Economy Innovation has positive effects on economic performance among SMEs in Pakistan, Malaysia, and China

H1b: Circular Economy Innovation has positive effects on environmental performance among SMEs in Pakistan, Malaysia, and China

H1c: Circular Economy Innovation has positive effects on social performance among SMEs in Pakistan, Malaysia, and China

2.3. *Business Model Innovation and Sustainable Development Goals*

BMI is perceived as a main indicator of sustainable development, product and service innovation, technological innovation to achieve the goal of competitive advantage, and income generation [46]. Principally, BMI is the higher-level modification and perfection of the foundation of a business model, production process, services structure, and product features to enhance the sustainable development capabilities among firms. In this regard, prior studies have evaluated the role of BMI in achieving the sustainable development goals among enterprises and observed that BMI has positive association with SDGs [46], and lead firms to achieve economic, environmental, and social efficiencies, as well as a competitive advantage. BMI involves the unique structural and foundational elements of enterprises that work as a mean to achieve core competencies in industrial settings and set them apart from competitors. BMI is an effective strategy to obtain resources, enhance efficiencies, explore new markets to create values, and adopt new methods and logics to acquire values that lead firms to achieve the sustainable development goals [46]. However, the literature has highlighted four categories of BMI, which are: full innovation, partial innovation, expansion innovation, and realization innovation.

BMI reintegrates the internal and external capabilities and resources of firms that can improve the operations, efficiencies, market performance, and core competencies of firms, which can lead them to achieve the sustainable development goals. Therefore, BMI is emerging as a new trend of sustainability in industrial setup and enabling the reconfiguration of business capabilities and resources to achieve economic, environmental, and social efficiencies [47]. Due to its high robustness, researchers are widely attempting BMI for sustainability in various settings to understand its role in achieving economic, environmental, and social efficiencies; however, further investigation is needed to highlight its role in creativity, innovation, and ecosystem efficiencies [48]. Particularly, in prior studies, the research on BMI has widely focused on the value creation, value delivery, and value capturing [48]. Therefore, this study has observed a gap in the literature and identified a need to examine the role of BMI in achieving economic, environmental, and social efficiencies and proposed that:

H2a: BMI has positive effects on the economic performance of SMEs in Pakistan, Malaysia, and China

H2b: BMI has positive effects on the environmental performance of SMEs in Pakistan, Malaysia, and China

H2c: BMI has positive effects on the social performance of SMEs in Pakistan, Malaysia, and China

2.4. Government Incentives and Sustainable Development Goals

As per the UN sustainability agenda 2030, economic, environmental, and social efficiencies are the prime motives of many organizations, agencies, and countries around the world to ensure energy conservation, the elimination of poverty and hunger, clean drinking water, employment, economy growth, health and protection, peace, and the reduction of pollution, and to improve the quality of life in societies to produce a sustainable world for all people. In this regard, government agencies can play a vital role in achieving the sustainable development goals, especially among SMEs, which most often face a lack of resources and guidance from the UN sustainability agenda [49]. State level initiatives and schemes, especially financial incentives, can play a significant role in monitoring pollution, providing direction, gaining growth, and achieving the sustainable development goals among firms [50]. However, this study focuses on the economic, environmental, and social efficiencies of SMEs in relation to government incentives.

In the view of the UN sustainability agenda, it is almost the key responsibility of every government to formulate a solid structure of SDG strategies for SMEs to ensure smooth operating functions [51], and to achieve economic, environmental, and social efficiencies, as well as the creation of values among societies. In the context of resource-based view theory, the firms with sufficient resources can efficiently transform their setup from the traditional production system to more the effective sustainable production system to achieve desirable performance [52]. Particularly, government incentives and technical assistance can play a major role in SMEs to launch more sustainable initiatives to ensure economic, environmental, and social efficiencies. Government financial schemes can lend support to control firms' crisis situations and help in survival, growth, achieving sustainable development, and overcoming imbalance of resources [53]. In this regard, the Chinese government has launched financial schemes to promote sustainability initiatives among firms and transition them from the traditional mode of business to a more sustainable and energy efficient production system [52].

Without financial incentives, it is often difficult to launch sustainability initiatives, environmental practices, and corporate social responsibilities among communities [54], while an efficient financial position can support the adoption of more desirable businesses practices among communities. In the same way, non-financial support is also essential for smooth functioning and performance to ensure firms' survival in a competitive environment [52]. In emerging economies, government incentives and favorable policies (such as low taxes, lower regulatory charges etc.) lead firms toward sustainability initiatives and green practices to achieve economic, environmental, and social efficiencies, especially among small firms. Therefore, we can hypothesize that:

H3a: Government incentives have positive effects on the economic performance of SMEs in Pakistan, Malaysia, and China

H3b: Government incentives have positive effects on the environmental performance of SMEs in Pakistan, Malaysia, and China

H3c: Government incentives have positive effects on the social performance of SMEs in Pakistan, Malaysia, and China

2.5. Circular Economy Innovation and Government Incentives

Circular economy innovation is the higher-level rethinking of manufacturing, industrial processes, services, product innovation, production process, consumption, and usage of raw material to achieve sustainable development goals [55], and is the "*transition from linear economic models based on take, make, use and waste towards circular models that minimize, recover, recycle, and reuse materials, water, and energy*." The transition process includes several steps and requires essential resources to implement a more efficient and sustainable business model. Circular economy innovation is the key driver for the elimination of environmental impacts to ensure sustainable economic growth in a competitive environment. In this regard, incentive schemes are the basic requirement among firms to bring a spirit of sustainable innovation, circular economy innovation, and business model innovation to achieve long lasting economic, environmental, and social efficiencies among communities. Accelerating the transition towards circular economy requires essential incentives to overcome the barriers to implementing circular economy innovation among firms [56]. Particularly, financial incentives play an instrumental role in empowering firms as well as consumers to adopt the habit of more sustainable choices. Incentives enable firms to initiate sustainability practices, create value for societies, and launch more desirable business activities [56]. Concisely, circular economy innovation attracts investor's interest in a competitive business environment and leads toward better financial outcomes [56]. Therefore, we can propose a hypothesis that:

H4: There is a positive relationship between circular economy innovation and government incentives among SMEs in Pakistan, Malaysia, and China

2.6. Business Model Innovation and Government Incentives

Firms, institutions, and government agencies are widely enforcing their workers to focus on discoveries, innovation, and R&D practices. In this regard, they offer a bundle of incentives to launch new projects, hire consultants, and train their workers for greater creativity and innovation purposes. Creation and discoveries are mysterious processes and require incentives for better innovation [57], and, without incentives, especially government funding, it is difficult for SMEs to create a higher level of innovation and show sustainable performance. Innovation and creativity increase the confidence of firms, lead to a competitive advantage, and grab investors' attention. Zhang and Guan [58] have observed that innovation performance affects government financial incentives in the Chinese context.

Furthermore, the innovation of business models can improve the economic, environmental, and social efficiencies among SMEs, growth, and value in societies, and gain government attention towards greater innovation and industrial development. Usually, the product, process, services, and model innovation among firms provide the direction and define the future for communities and government agencies to formulate more innovative policies and spare a good number of incentives for developmental and R&D purposes. Therefore, we propose in this study that:

H5: There is a positive relationship between BMI and the government incentives among SMEs in Pakistan, Malaysia, and China

2.7. Mediating Role of Government Incentives

In the interpretation of the UN sustainability outline, many organizations, agencies, and the EU have launched various developmental programs and schemes to achieve economic, environmental, and social efficiencies among communities and ensure the smooth survival of firms. Specifically, the EU and Chinese government have offered various schemes to transform from the linear production system to more efficient and sustainable business practices. Particularly, financial support can provide fresh motivation to SMEs which most often face a lack of financial resources, developmental fundings, and scarce knowledge of sustainable development goals. Financial support can play a significant role in achieving the economic, environmental, and social efficiencies among small enterprises [59]. In this regard, many countries have aggressively focused on boosting sustainability practices and accelerating the transition from linear to more energy efficient production systems and circular economy innovation [37,38]. Likewise, under the pressure of environmental degradation, many firms are trying to renovate their business models and adopting the practices of circular economy innovation to ensure the achievement of sustainable development goals.

Usually, government incentives and technical assistance put pressure on a firm's administration to implement sustainability practices by innovating their business model and adopting the activities of circular economy innovation. Government subsidies can support the controlling of environmental pollution, promote ecological innovation, green initiative, BMI, and circular economy innovation, and gain economic, environmental, and social efficiencies [60–62]. With a view towards a transition towards more sustainable business practices, many countries have initiated a program of circular economy and enforced firms to ensure its implementation to achieve economic, environmental, and social efficiencies. Especially, the EU has assumed the practices of circular economy as a tool to achieve the sustainable development goals [37]. In this regard, government incentives can play a significant role in achieving the sustainable development goals to transform the prominent business model from a high-pollution and energy-consumption process to more environment friendly business practices [63]. Environment-oriented policies can play a significant role in circular economy innovation, business model innovation [64,65], and achieving the sustainable development goals. Consequently, with the support of government incentives, SMEs can effectively prevent wastage and pollution by transforming the existing business model into a more innovative business model, promoting ecological practices and circular economy innovation to achieve the sustainable development goals [66]. Therefore, we can hypothesize that:

H6a: Government incentives mediate the relationship between circular economy innovation and economic performance among SMEs in Pakistan, Malaysia, and China

H6b: Government incentives mediate the relationship between circular economy innovation and environmental performance among SMEs in Pakistan, Malaysia, and China

H6c: Government incentives mediate the relationship between circular economy innovation and social performance among SMEs in Pakistan, Malaysia, and China

H7a: Government incentives mediate the relationship between BMI and economic performance among SMEs in Pakistan, Malaysia, and China

H7b: Government incentives mediate the relationship between BMI and environmental performance among SMEs in Pakistan, Malaysia, and China

H7c: Government incentives mediate the relationship between BMI and social performance among SMEs in Pakistan, Malaysia, and China

Based on the inclusive literature review and theoretical circumstantial, this study has developed a conceptual research framework that involves circular economy innovation, BMI, government incentives, and sustainable development goals as shown in (Figure 1).

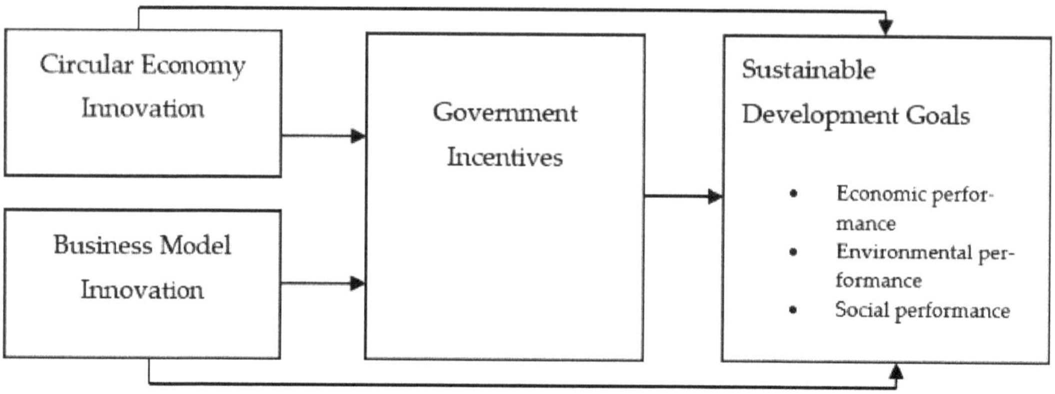

Figure 1. Conceptual model.

3. Methodology

3.1. Sample and Data Collection

The current study collected data through online questionnaires from the owners, CEO, and operational managers of registered SMEs in Pakistan, Malaysia, and China to analyze the practices of sustainable development goals in relation to circular economy innovation and BMI. The study derived the sample size with the support of G-Power which is perceived as an authentic procedure in PLS-SEM [67]. The outcome of the G-Power has shown that 129 is the minimum set of data to establish a relationship among the defined constructs for a single unit. However, we circulated 384 questionnaires among each country (Pakistan, Malaysia, and China) using the random sampling technique to achieve the minimum level of dataset from July 2021 to August 2022 among manufacturing and services firms. The profiles of firms are presented with their complete information in Table 1. Meanwhile, only 300 questionnaires were considered for each unit due to the minimum acceptable ratio of the respondent's participation. We clarified to the respondents that the data will be used only for research and publication purposes, and they happily cooperated in this regard.

Table 1. Profile of the firms.

Pakistan			Malaysia			China		
Description	**Frequency**	**Percentage**	**Description**	**Frequency**	**Percentage**	**Description**	**Frequency**	**Percentage**
Owners/Managers			**Owners/Managers**			**Owners/Managers**		
Owner	53	17.66%	Owner	61	20.33%	Owner	76	25.33%
CEO	64	21.33%	CEO	69	23.00%	CEO	57	19.00%
Managers	183	61.00%	Managers	170	56.66%	Managers	167	55.66%

Table 1. *Cont.*

Pakistan			Malaysia			China		
Description	Frequency	Percentage	Description	Frequency	Percentage	Description	Frequency	Percentage
	Industries			Industries			Industries	
Chemicals	41	13.66%	Chemicals	45	15.00%	Chemicals	39	13.00%
Software	57	19.00%	Software	64	21.33%	Software	63	21.00%
Pharmacy	47	15.66%	Pharmacy	43	14.33%	Pharmacy	52	17.33%
Cosmetics	62	20.66%	Cosmetics	54	18.00%	Cosmetics	59	19.66%
Engineering	37	12.33%	Engineering	42	14.00%	Engineering	50	16.66%
Electronics	56	18.66%	Electronics	52	17.33%	Electronics	37	12.33%
	Size of the Firm			Size of the Firm			Size of the Firm	
20–50 Employees	44	14.66%	20–50 Employees	51	17.00%	20–50 Employees	31	10.33%
51–100 Employees	72	24.00%	51–100 Employees	68	22.66%	51–100 Employees	57	19.00%
101–150 Employees	82	27.33%	101–150 Employees	74	24.66%	101–150 Employees	72	24.00%
151–200 Employees	56	18.66%	151–200 Employees	79	26.33%	151–200 Employees	81	27.00%
201–250 Employees	46	15.33%	201–250 Employees	28	9.33%	201–250 Employees	59	19.66%
	Age of the Firm			Age of the Firm			Age of the Firm	
10 Years or less	145	48.33%	10 Years or less	87	29.00%	10 Years or less	85	28.33%
11–20 Years	94	31.33%	11–20 Years	173	57.66%	11–20 Years	181	60.33%
21 and above Years	61	20.33%	21 and above Years	40	13.33%	21 and above Years	34	11.33%

3.2. Instruments

The instrument of the study was adopted and adapted for all the constructs as presented in Table 2. The questionnaire mentions the demographic information along with the important descriptions of defined constructs. The questionnaire was refined as per the study requirements. The study used structured questionnaires as most of the firms have no formal data for circular economy innovation, BMI, and SDGs. All the questions were closed-ended; however, the choice given to mark the most suitable option. The scale was defined in the range of strongly agree to strongly disagree on the five-point Likert scale. Highly expert researchers evaluated the accuracy of the scale to confirm the face validity. Some weak items were deleted as per the expert's advice. The scale was also evaluated with the strategy of pilot study to finalize the more reliable scale in this study (Appendix A).

Table 2. Instruments of the study.

Constructs		Number of Items	Authors
Circular Economy Innovation		8	Rodríguez-Espíndola et al., 2022 [3]
BMI		6	Anwar, 2018 [68]
Government Incentives	Financial Support	6	Anwar et al., 2020 [52]
	Non-Financial Support	6	Anwar et al., 2020 [52]
Sustainable Development Goals	Social Development	8	Rizwanullah et al., 2021; Anwar et al., 2020 [52,63]
	Environmental Development	12	Rizwanullah et al., 2021; Anwar et al., 2020 [52,63]
	Economics Development	6	Rehman& Anwar, 2019; Anwar, Khan& Shah, 2018 [69,70]

3.3. Technique

The current study evaluated the acquired data through Smart Partial Least Square Structural Equation Modeling (PLS-SEM) to test the developed hypothesis. PLS-SEM has the advantage of validating each step a in systematic way, displaying results only in one click, and having a better performance in predictive studies [71–73]. Therefore, we finalized the application of PLS-SEM in the proposed research model to obtain results due to its suitability in the defined theoretical framework.

4. Results

The current study has evaluated the collected information via PLS-SEM to display a layout of the results. In PLS-SEM analysis, measurement and structural models are the main steps to assess the validity of the model and test the developed hypothesis [74]. In the measurement model, factors loading, composite reliability, and average variance extracted (AVE) are the main factors of interest to ensure the convergent validity in the proposed conceptual model. The acceptable values of factors loading are 0.7 or higher, for composite reliability 0.7 or greater, for Cronbach's Alpha 0.7 or greater, and for AVE 0.5 or greater as indicated below (Table 3) for all three cases. The results revealed that all the concerned values are in the acceptable range. Further, the study applied the Fornell and Larcker [75] criteria to evaluate the accuracy of discriminant validity. The outcomes of the discriminant validity can be seen in Table 4 where the diagonal values are higher in the concerned rows and columns for each unit. The approach of HTMT was applied to verify the accuracy of discriminant validity in the assessment of the measurement model. The results of HTMT are presented in Table 4. In addition, the current study evaluated the issues of multicollinearity through variance inflation factor (VIF) among the available constructs and observed that there are no issues of multicollinearity in this study.

In the second step of PLS-SEM, in the assessment of the structural model, a bootstrapping procedure was applied for testing the developed hypothesis. The results indicated that the circular economy innovation, BMI, and government incentives have positive significant effects on economic, environmental, and social performance among SMEs in Pakistan and China. In the same way, circular economy innovation and BMI have positive significant effects on economic, environmental, and social performance among SMEs in Malaysia. Surprisingly, government incentives a have positive significant relationship with economic and social performance, but a non-significant relationship with environmental performance among SMEs in Malaysia, as shown in Table 5. In addition, the findings revealed that the government incentives mediate the relationship between circular economy innovation and the economic, environmental, and social performance among SMEs in Pakistan, Malaysia, and China, as shown in Table 6. In the same manner, the study observed that government incentives mediate the relationship between BMI and the economic, environmental, and social performance among SMEs in Pakistan, Malaysia, and China. Moreover, the findings revealed that the values of Q-squares are higher than zero which means that predictive relevance exists in this study.

Table 3. Factor Loading, Cronbach's Alpha, Composite Reliability, and AVE.

Variable	Items	Pakistan				Malaysia				China			
		F.L	C.A	C.R	AVE	F.L	C.A	C.R	AVE	F.L	C.A	C.R	AVE
Circular Economy Innovation	CEIN1	0.832	0.861	0.924	0.671	0.775	0.88	0.909	0.626	0.879	0.880	0.909	0.626
	CEIN 2	0.799				0.715				0.772			
	CEIN 3	0.749				0.786				0.719			
	CEIN 4	0.764				0.880				0.785			
	CEIN 5	0.682				0.801				0.803			
	CEIN 6	0.780				0.782				0.782			

Table 3. *Cont.*

Variable	Items	Pakistan				Malaysia				China			
		F.L	C.A	C.R	AVE	F.L	C.A	C.R	AVE	F.L	C.A	C.R	AVE
BMI	BMI1	0.898	0.901	0.896	0.591	0.828	0.851	0.889	0.573	0.653	0.886	0.914	0.643
	BMI2	0.783				0.701				0.899			
	BMI3	0.903				0.755				0.787			
	BMI4	0.777				0.700				0.900			
	BMI5	0.756				0.758				0.778			
	BMI6	0.783				0.790				0.768			
Government Incentives	GIN1	0.860	0.887	0.924	0.634	0.860	0.887	0.917	0.689	0.611	0.814	0.868	0.571
	GIN2	0.802				0.800				0.747			
	GIN3	0.889				0.890				0.750			
	GIN4	0.784				0.785				0.826			
	GIN5	0.812				0.812				0.822			
Economics Performance	EF1	0.816	0.903	0.928	0.636	0.816	0.904	0.924	0.636	0.855	0.907	0.926	0.642
	EF2	0.812				0.844				0.806			
	EF3	0.721				0.822				0.748			
	EF4	0.788				0.731				0.743			
	EF5	0.796				0.739				0.794			
	EF6	0.817				0.799				0.847			
	EF7	0.820				0.823				0.810			
Environmental Performance	ENF1	0.819	0.904	0.917	0.689	0.660	0.897	0.92	0.624	0.798	0.903	0.923	0.632
	ENF2	0.840				0.893				0.815			
	ENF3	0.825				0.779				0.798			
	ENF4	0.731				0.894				0.817			
	ENF5	0.738				0.765				0.745			
	ENF6	0.802				0.747				0.792			
	ENF7	0.818				0.767				0.797			
Social Performance	SF1	0.710	0.888	0.913	0.601	0.815	0.903	0.924	0.634	0.708	0.868	0.899	0.561
	SF2	0.785				0.814				0.774			
	SF3	0.702				0.721				0.691			
	SF4	0.790				0.791				0.771			
	SF5	0.888				0.791				0.878			
	SF6	0.743				0.816				0.743			
	SF7	0.792				0.821				0.656			

Table 4. Discriminant Validity and HTMT.

		Discriminant Validity						HTMT						
		CEIN	BMI	GIN	ED	END	SD		CEIN	BMI	GIN	ED	END	SD
Pakistan	CEIN	**0.919**						CEIN						
	BMI	0.507	**0.869**					BMI	0.407					
	GIN	0.651	0.494	**0.846**				GIN	0.535	0.497				
	ED	0.563	0.607	0.653	**0.897**			ED	0.549	0.513	0.440			
	END	0.393	0.429	0.423	0.414	**0.830**		END	0.335	0.401	0.357	0.463		
	SD	0.578	0.515	0.615	0.620	0.421	**0.775**	SD	0.438	0.582	0.282	0.388	0.464	
Malaysia	CEIN	**0.757**						CEIN						
	BMI	0.536	**0.891**					BMI	0.502					
	GIN	0.643	0.505	**0.897**				GIN	0.423	0.498				
	ED	0.622	0.428	0.568	**0.790**			ED	0.301	0.319	0.261			
	END	0.480	0.414	0.414	0.408	**0.830**		END	0.242	0.364	0.363	0.254		
	SD	0.641	0.286	0.353	0.456	0.423	**0.796**	SD	0.223	0.274	0.054	0.347	0.470	
China	CEIN	**0.802**						CEIN						
	BMI	0.509	**0.791**					BMI	0.453					
	GIN	0.424	0.492	**0.801**				GIN	0.318	0.482				
	ED	0.324	0.394	0.464	**0.795**			ED	0.219	0.304	0.066			
	END	0.516	0.446	0.496	0.348	**0.755**		END	0.386	0.411	0.354	0.530		
	SD	0.525	0.574	0.398	0.493	0.349	**0.749**	SD	0.492	0.428	0.261	0.303	0.418	

Table 5. Direct Relationship (Hypothesis Testing).

Country	Relationship	Estimate	SM	SD	T-Value	Decision	R-Square	F-Square	VIF	Q-Square
Pakistan	CEIN→EP	0.287	0.289	0.048	5.945	Supported	0.761	0.115	2.994	0.473
	BMI→EP	0.593	0.592	0.045	13.151	Supported		0.510	2.889	
	GIN→EP	0.067	0.066	0.031	2.193	Supported		0.015	1.235	
	CEIN→ENP	0.302	0.303	0.046	6.497	Supported	0.782	0.139	2.994	0.49
	BMI→ENP	0.600	0.599	0.04	15.191	Supported		0.571	2.889	
	GIN→ENP	0.049	0.048	0.029	2.665	Supported		0.009	1.235	
	CEIN→SP	0.068	0.068	0.081	2.839	Supported	0.38	0.003	2.987	0.22
	BMI→SP	0.436	0.436	0.076	5.760	Supported		0.106	2.853	
	GIN→SP	0.220	0.222	0.055	3.986	Supported		0.064	1.262	
	CEIN→GIN	0.321	0.326	0.096	3.340	Supported	0.19	0.044	2.867	0.128
	BMI→GIN	0.134	0.134	0.097	2.382	Supported		0.008	2.867	
Malaysia	CEIN→EP	0.645	0.643	0.035	18.367	Supported	0.711	0.978	1.470	0.446
	BMI→EP	0.295	0.297	0.04	7.335	Supported		0.190	1.582	
	GIN→EP	0.006	0.004	0.036	2.153	Supported		0.000	1.361	
	CEIN→ENP	0.693	0.693	0.03	23.069	Supported	0.729	1.208	1.470	0.445
	BMI→ENP	0.249	0.251	0.037	6.656	Supported		0.145	1.582	
	GIN→ENP	0.001	0.000	0.039	0.038	Not Supported		0.000	1.361	
	CEIN→SP	0.615	0.613	0.035	17.38	Supported	0.686	0.818	1.470	0.427
	BMI→SP	0.300	0.302	0.04	7.468	Supported		0.181	1.582	
	GIN→SP	0.025	0.023	0.04	2.012	Supported		0.001	1.361	
	CEIN→GIN	0.220	0.221	0.061	3.631	Supported	0.265	0.047	1.404	0.179
	BMI→GIN	0.362	0.361	0.06	5.990	Supported		0.127	1.404	
China	CEIN→EP	0.356	0.359	0.046	7.658	Supported	0.729	0.161	2.910	0.462
	BMI→EP	0.493	0.488	0.048	10.267	Supported		0.282	3.178	
	GIN→EP	0.083	0.088	0.04	2.068	Supported		0.019	1.367	
	CEIN→ENP	0.364	0.366	0.044	8.277	Supported	0.728	0.167	2.910	0.453
	BMI→ENP	0.498	0.493	0.047	10.671	Supported		0.287	3.178	
	GIN→ENP	0.061	0.066	0.04	2.534	Supported		0.010	1.367	
	CEIN→SP	0.172	0.169	0.07	2.460	Supported	0.47	0.019	2.750	0.243
	BMI→SP	0.330	0.33	0.063	5.223	Supported		0.065	3.168	
	GIN→SP	0.302	0.305	0.049	6.109	Supported		0.126	1.332	
	CEIN→GIN	0.081	0.086	0.086	1.988	Supported	0.269	0.003	2.900	0.142
	BMI→GIN	0.45	0.453	0.085	5.293	Supported		0.096	3.094	

Table 6. Indirect Effects (Hypothesis Testing).

Country	Relationship	Estimate	SM	SD	T-Value	CILL	CIUL	Decision
Pakistan	CEIN→GIN→EP	0.122	0.222	0.213	1.982	0.021	0.169	Supported
	CEIN→GIN→ENP	0.148	0.416	0.312	2.318	0.035	0.256	Supported
	CEIN→GIN→SP	0.061	0.173	0.203	2.341	0.071	0.289	Supported
	BMI→GIN→EP	0.239	0.309	0.108	2.105	0.043	0.160	Supported
	BMI→GIN→ENP	0.347	0.206	0.206	3.043	0.034	0.278	Supported
	BMI→GIN→SP	0.030	0.330	0.223	2.276	0.050	0.172	Supported
Malaysia	CEIN→GIN→EP	0.261	0.101	0.308	2.146	0.021	0.173	Supported
	CEIN→GIN→ENP	0.373	0.300	0.209	3.036	0.039	0.249	Supported
	CEIN→GIN→SP	0.145	0.405	0.209	2.575	0.045	0.266	Supported
	BMI→GIN→EP	0.182	0.202	0.114	2.147	0.039	0.161	Supported
	BMI→GIN→ENP	0.191	0.200	0.214	3.037	0.056	0.239	Supported
	BMI→GIN→SP	0.169	0.409	0.115	4.588	0.048	0.184	Supported
China	CEIN→GIN→EP	0.117	0.207	0.209	2.793	0.009	0.222	Supported
	CEIN→GIN→ENP	0.205	0.305	0.207	3.705	0.020	0.159	Supported
	CEIN→GIN→SP	0.125	0.226	0.126	2.944	0.017	0.222	Supported
	BMI→GIN→EP	0.238	0.204	0.121	3.813	0.003	0.132	Supported
	BMI→GIN→ENP	0.227	0.203	0.202	2.389	0.008	0.071	Supported
	BMI→GIN→SP	0.236	0.338	0.035	3.836	0.007	0.089	Supported

Notably, the results of PLS-SEM revealed that BMI has much greater effects on the economic performance of SMEs in Pakistan (Figure 2) as compared with circular economy innovation and government incentives. Likewise, the outcomes of PLS-SEM indicated that BMI has greater impacts on environmental performance among SMEs in Pakistan as compared to circular economy and government incentives. In the same way, the results have shown that BMI has much higher effects on social performance among SMEs in

Pakistan as compared to circular economy innovation and government incentives. In addition, we noted that circular economy innovation has a stronger relationship with government incentives as compared to BMI among SMEs in Pakistan. Further, the study observed that government incentives mediate the relationship between circular economy innovation and economic performance among SMEs in Pakistan. The study also noted that government incentives mediate the relationship between circular economy innovation and environmental performance among SMEs in Pakistan. Likewise, we also observed that government incentives mediate the relationship between circular economy innovation and social performance among SMEs in Pakistan.

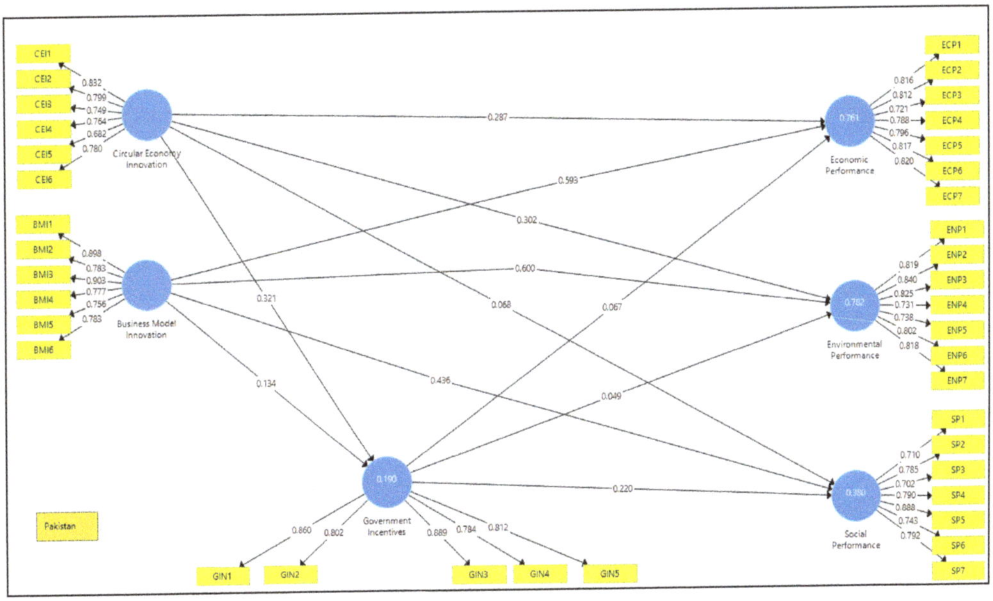

Figure 2. Measurement Model (Pakistan).

Additionally, the results of PLS-SEM have indicated that government incentives mediate the relationship between BMI and the economic performance among SMEs in Pakistan. Similarly, we noted that government incentives mediate the relationship between BMI and environmental performance among SMEs in Pakistan. Further, we also observed that government incentives mediate the relationship between BMI and social performance among SMEs in Pakistan. Therefore, based on these results, it has been noted that BMI has a much stronger role than the other factors in achieving economic, environmental, and social performance among SMEs in Pakistan (Figure 3).

Based on the results, it has been noted that BMI has a much stronger role in achieving economic and environmental performance among SMEs as compared to social performance. The study also observed that circular economy innovation has a stronger role in achieving environmental performance among SMEs in Pakistan as compared to economic and social performance. Therefore, on the basis of these results, it can be inferred that SMEs in Pakistan have a greater focus on BMI to achieve a higher level of economic and environmental performance. It can also be said that the government agencies may not be adequately supporting the achievement of social performance among SMEs due to their limited budget. It is also possible that the firms in Pakistan may not have enough internal sources to focus on social performance and, thus, prefer to focus widely on economic and environmental performance to ensure their firm's survival and environmental protection. It is also possible that the consumers in Pakistan are not strongly oriented to social activities and may not prefer to invest social wellbeing. It can also be said that the consumers in Pakistan are not strongly aware of the social circumstances, and, therefore, the SMEs are not widely focusing on the social issues. It is also possible that the Pakistani government is not adequately enforcing SMEs to improve social efficiencies as compared to economic and environmental efficiencies.

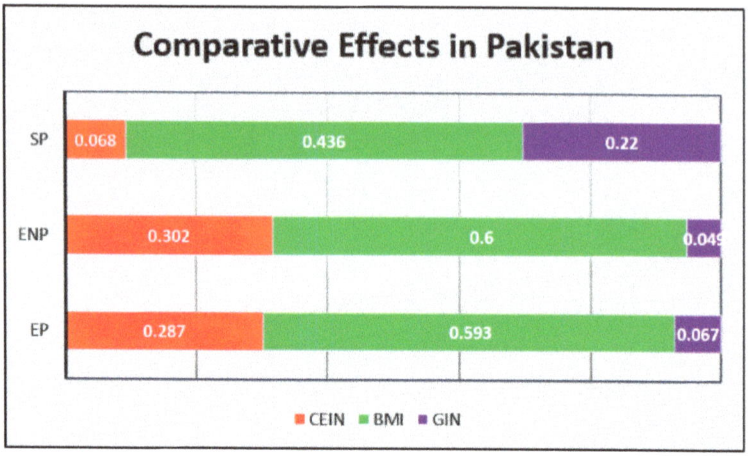

Figure 3. Comparative Effects in Pakistan.

Moreover, the results of PLS-SEM clarified that circular economy innovation has a greater role in achieving economic performance among SMEs in Malaysia as compared to BMI and government incentives. In the same way, the results have shown that circular economy innovation has a greater role in achieving environmental performance among SMEs in Malaysia as compared to BMI and government incentives. Likewise, the results have displayed that circular economy innovation has a greater role in achieving social performance among SMEs in Malaysia as compared to BMI and government incentives (Figure 4). Further, the study observed that government incentives mediate the relationship between circular economy innovation and economic performance among SMEs in Malaysia. The study also noted that government incentives mediate the relationship between circular economy innovation and environmental performance among SMEs in Malaysia. Likewise, we also observed that government incentives mediate the relationship between circular economy innovation and the social performance among SMEs in Malaysia.

Additionally, the results of PLS-SEM have indicated that government incentives mediate the relationship between BMI and economic performance among SMEs in Malaysia. Similarly, we also noted that government incentives mediate the relationship between BMI and environmental performance among SMEs in Malaysia. Further, we also observed that government incentives mediate the relationship between BMI and social performance among SMEs in Malaysia. Surprisingly, we noted that government incentives have weak positive and insignificant effects on environmental performance among SMEs in Malaysia. Perhaps the Malaysian government is not widely focusing on building science parks and SDG incubators in poor communities. It may be that the Malaysian government is not giving significant attention to the evaluation of suppliers who are not adequately involved in environmental practices. Therefore, based on these results, it can be noted that circular economy innovation has a much stronger role in achieving economic, environmental, and social performance among SMEs in Malaysia (Figure 5), as compared to BMI and government incentives. Therefore, on the basis of these results, it can be inferred that SMEs in Malaysia have a greater focus on circular economy innovation to achieve a higher level of economic, environmental, and social performance. It can also be said that the government agencies may not supporting too many SMEs in the context of BMI as compared to circular economy innovation.

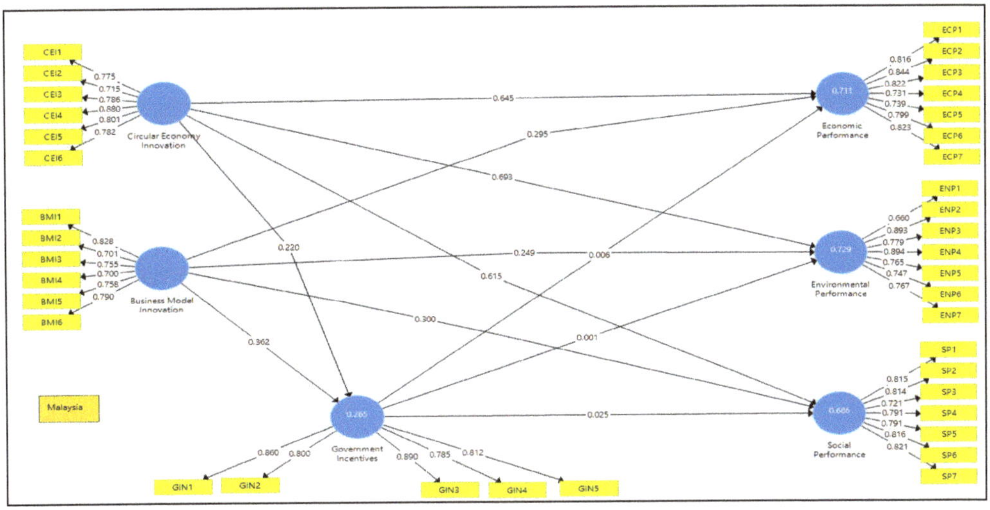

Figure 4. Measurement Model (Malaysia).

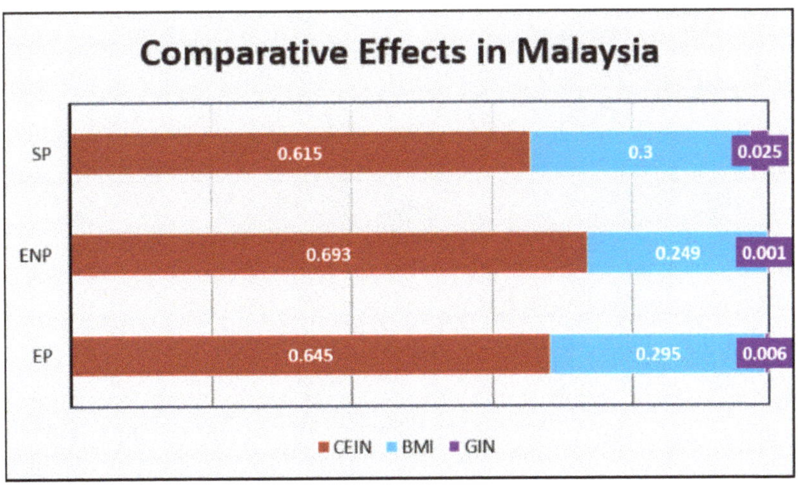

Figure 5. Comparative effects in Malaysia.

It is also possible that the firms in Malaysia may not have enough internal sources to focus on BMI and, thus, prefer to focus widely on circular economy innovation to achieve economic, environmental, and social efficiencies. It is also possible that the SMEs in Malaysia have already reinvented their business models and are now focusing widely on circular economy innovation. It is also imaginable that the consumers in Malaysia are more oriented towards circular economy innovation in the view of sustainable developmental goals. It can also be said that the consumers in Malaysia are too oriented towards circular economy innovation; therefore, the SMEs are not widely focusing on BMI. It is also possible that the Malaysian government is widely enforcing SMEs to improve their circular economy innovation practices in society as compared to BMI to achieve economic, environmental, and social efficiencies.

Likewise, the results of PLS-SEM have indicated that BMI has a greater role in achieving economic performance among SMEs in China as compared to circular economy innovation and government incentives (Figure 6). Likewise, the outcomes of PLS-SEM indicated that BMI has greater impacts on environmental performance among SMEs in China as compared to circular economy and government incentives. In the same way, the results have shown that BMI has much greater effects on the social performance of SMEs in China as compared to circular economy innovation and government incentives. In addition, we noted that circular economy innovation has a stronger relationship with government incentives as compared to BMI among SMEs in China. Further, the study observed that government incentives mediate the relationship between circular economy innovation and economic performance among SMEs in China. The study also found that government incentives mediate the relationship between circular economy innovation and environmental performance among SMEs in China. Likewise, we also observed that government incentives mediate the relationship between circular economy innovation and social performance among SMEs in China.

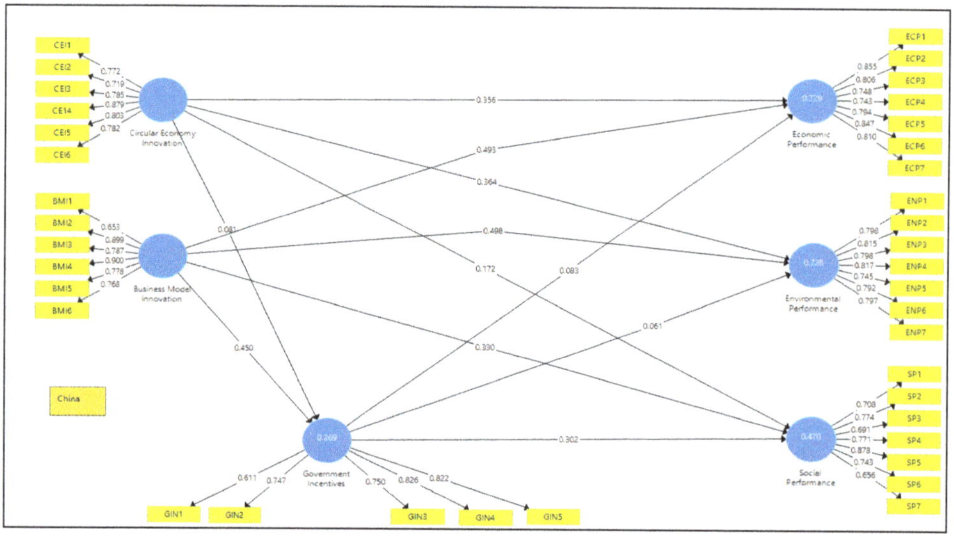

Figure 6. Measurement Model (China).

Additionally, the results of PLS-SEM have indicated that government incentives mediate the relationship between BMI and economic performance among SMEs in China. Similarly, we also noted that government incentives mediate the relationship between BMI and environmental performance among SMEs in China. Further, we also observed that government incentives mediate the relationship between BMI and social performance among SMEs in China. Therefore, based on these results, it can be noted that BMI has a much stronger role in achieving economic, environmental, and social performance among SMEs in China (Figure 7). Based on the results, it was found that BMI has a much stronger role in achieving economic and environmental performance among SMEs in China as compared to social performance. The study also observed that circular economy innovation has a stronger role in achieving environmental performance among SMEs in China as compared to economic and social performance.

Therefore, on the basis of these results, it can be inferred that SMEs in China have a greater focus on BMI to achieve a higher level of economic and environmental performance. It can also be said that the government agencies may not be adequately supporting the achievement of social performance among SMEs due to higher competition. It is also conceivable that the firms in China may not have enough internal sources to focus on social performance and thus prefer to focus widely on economic and environmental performance to ensure their firm's survival and environmental protection. It is also imaginable that the consumers in China are not strongly oriented towards social activities and may not prefer to invest too much in their social wellbeing. It can also be said that the consumers in China are not very interested in their social circumstances; therefore, the SMEs are not widely focusing on social issues. It is also possible that the Chinese government is not adequately enforcing the improvement of social efficiencies by SMEs as compared to economic and environmental efficiencies.

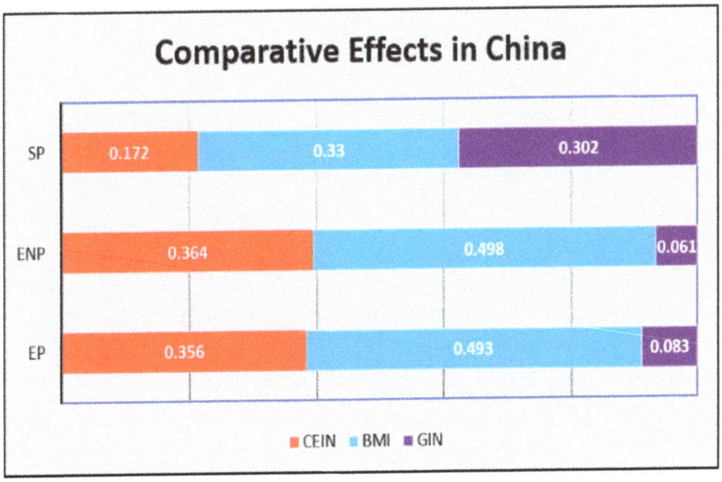

Figure 7. Comparative effects in China.

Based on the outcomes of this study, it can be said that SMEs have access to governmental financial schemes and invest a good amount of money in circular economy innovation practices, as well as BMI, to achieve the sustainable development goals in Pakistan, Malaysia, and China. It can also be inferred, based on the results, that there is a structure of stable policies for SMEs in Pakistan, Malaysia, and China, to follow the practices of circular economy innovation and reinvent their business models to achieve economic, environmental, and social efficiencies. Perhaps SMEs have defined some policies of circular economy innovation and strictly follow them while merging in new agreements. It can also be said, on the basis of the results, that SMEs engage their employees and other stakeholders in circular economy innovation strategies and practices to highlight the significance of sustainable development goals in all aspects of their organizations in Pakistan, China, and Malaysia. Similarly, based on the findings in this study, it can be inferred that SMEs are widely focusing on the innovation of their core products and services, as well as their production processes, on a continuing basis to increase their revenue using innovative business strategies. It can also be said that SMEs are aggressively involved in traditional sales practices as well as digital media sales landscapes to sustain their survival and economic, environmental, and social growth.

It can be said, based on the results, that the SMEs in Pakistan, China, and Malaysia are widely exchanging innovative information with their partners, such as pricing structures for raw material, products, and services, on a regular basis to improve their economic, environmental, and social performance in a competitive business environment. It is also possible that the firms are continuously evaluating their value propositions for their products and services by comparing with previous performance to validate their innovative strategies and strive for improvement. Likewise, it can be said, on the basis of the results, that the SMEs in Pakistan, Malaysia, and China have access to government funding opportunities for social, economic, and environmental initiatives and sufficient government financial subsidies are available to them for development purposes. It can also be imagined that the SMEs have access to interest-free government loans schemes, both for the short term and long term, that can improve their social, economic, and environmental performance in communities. It can also be said that there are an adequate number of public programs for social, economic, and environmental development where SMEs can reach out to government sources.

5. Discussion

The purpose of this study was to examine the effects of circular economy innovation and BMI on sustainable development goals (economic, environmental, and social development) along with the mediating role of government incentives in Pakistan, Malaysia, and China. The results indicated that circular economy innovation and BMI have positive significant effects on economic, environmental, and social performance in Pakistan, Malaysia, and China. Likewise, government incentives have positive significant effects on economic, environmental, and social performance among SMEs in Pakistan and China. Surprisingly, government incentives have positive significant effects on economic and social performance among SMEs in Malaysia, but insignificant effects on environmental performance. Therefore, the strategies of circular economy innovation, BMI, and the availability of government incentives have an advantage in achieving economic, environmental, and social performance objectives among SMEs, especially in Pakistan, Malaysia, and China. The application of circular economy innovation and innovating the business model as per the community requirements have the advantage of offering more fit services in economic, environmental, and social performance that can lead to achieving, for firms, long lasting objectives and the building of an image in the relevant communities. Specifically, government incentives play a major role in motivating firms to initiate circular economy innovation strategies and transform the firms from the traditional mode of production to a more energy efficient and pollution free production system to establish a more sustainable business environment.

The intent of this study was to examine the relationships between defined variables and provide answers to the raised research questions. The results of the study have clarified the relationships between defined constructs and provided answers to the raised research questions. Further, the results of this study are in line with Anwar [68] who examined the effects of business models on SMEs performance, along with the mediating role of competitive advantage; this study is in line in the sense of BMI and firm performance. This study is also in line with Rizwanullah et al. [63] who assessed the role of green innovation in achieving sustainable development goals along with the moderating role of government incentives; this study is in line in the sense of sustainable development goals and government incentives in Pakistani context. This study is also in line with Khan et al. [76] who investigated the role of sustainable development goals in firms' financial performance along with the moderating role of green innovation; this study is in line in the context of sustainable development goals. This study is also in line with Korsakiene and Raisiene [77], who highlighted the sustainability drivers in the context of SMEs. The study is also in line with Udeh and Akporien [78] in the context of evaluating the triple bottom line in the industrial aspects. This study is also in line with [79] in the context of circular economy and data collection online. However, the findings of this current study are unique as compared to in-line studies due to the analysis of circular economy innovation, BMI, and government incentives in achieving economic, environmental, and social efficiencies in a comparative context among SMEs in Pakistan, Malaysia, and China.

5.1. Implications of the Study

5.1.1. Practical Implications

These findings imply that firms can focus their efforts on circular economy innovation, BMI, and the combination of these variables. which would be an interesting strategy to achieve economic, environmental, and social performance among SMEs. The tactics of circular economy innovation and revising the business model as per the community requirement can inspire firms to initiate more efficient business practices that can result in better performance and gain a competitive advantage. Regular engagement in circular economy innovation activities and BMI practices can improve skills and confidence among employees and facilitate better economic, environmental, and social performance in a competitive business environment. By adopting the circular economy innovation strategies and innovating the business model as per the community requirement, firms can gain some unique production and business practices and skills which may lead them to be perceived as a market leader and difficult for competitors to beat in a competitive business environment. It can also lead firms to be perceived as a top brand among communities and help to capture market shares. The expertise of workers and skills can lead towards better innovation capabilities. The wise tactics of circular economy innovation and business model efficiencies can provide the opportunity for trust among communities and building a brand image which can lead to achieving economic, environmental, and social objectives. Market trust and confidence can lead firms to obtaining an advantage over competitors, enhancing commitment to innovation, and result in better business practices. Further, government assistance in terms of circular economy innovation and BMI could lead SMEs to achieve economic, environmental, and social efficiencies. In this scenario, private- and public-sector partnerships would be a wise strategy to boost the trend of circular innovation among SMEs.

Moreover, this study contributes to the prior literature by analyzing the relative importance of circular economy innovation, government incentives, and BMI in achieving economic, environmental, and social efficiencies, especially in Pakistan, Malaysia, and China. Therefore, managers can launch more effective business practices in developing business strategies in a competitive environment to ensure firm survival and gain sustainable objectives. The results of this study are also helpful for practitioners and policy makers to develop a more efficient business model per community requirements that can ensure more effective business services for those communities. The initiatives of circular economy innovation practices, BMI, and availability of government incentives allow practitioners to employ more accurate business techniques that can enhance the market attachment and customers' engagement with firms, as well as increasing economic, environmental, and social efficiencies. However, the study has noted that firms are widely focusing on BMI and circular economy innovation while developing strategies for sustainable development goals. Therefore, policy makers and practitioners need to widely focus on BMI and circular economy innovation while formulating strategies for economic, environmental and social efficiencies.

5.1.2. Theoretical Implications

This study merged the literature of circular economy innovation, BMI, government incentives, and sustainable development goals based on the findings from Pakistan, Malaysia, and China, and validated the proposed research model in a comparative context. The application of PLS-SEM in the defined comparative context is also a novel contribution in the emerging literature. This study has extended the resource-based view theory by developing a theoretical framework and validating that framework in a comparative way with data collection and analysis. The validated model adds to the relevant literature and can enhance practitioners' understanding of the strategies of circular economy innovation, BMI, government incentives, and the contribution of these factors to achieving economic, environmental, and social performance objectives. This study extends the theory of resource-based views in context to accommodate internal resources, capabilities, and efficiencies with the external market requirement to achieve long-lasting sustainable development goals. This study also extends the resource-based view theory in the defined context based on the findings from Pakistan, Malaysia, and China.

6. Conclusions and Future Directions

This study intends to examine the effects of circular economy innovation and BMI, in achieving economic, environmental, and social objectives along with the mediation of government incentives in the defined context. The results of the study have provided empirical evidence that the initiatives of circular economy innovation in relation with BMI and government incentives play a greater role in achieving economic, environmental, and social performance goals. The results of the study have also contributed to the relevant literature by testing the developed hypotheses. Eventually, it is believed that the findings of this study will contribute to the relevant theoretical literature and deliver significant information to policy-makers in formulating more effective economic, environmental, and social performance strategies. It will guide business practitioners to design more significant development strategies to achieve sustainable objectives. However, while interpreting results, the reader should know about the scope of the collected data and the analysis procedure. This study is limited to the resource-based view theory while the application of other theories (such as contingency theory and stakeholder theories) can interpret the results differently. To bring further perfection in results, future studies can examine the mediating role of the resources of management in the defined context and the application of stakeholder and contingency theories. Moreover, the study has only evaluated the defined theoretical framework by collecting data from firms in Pakistan, Malaysia, and China; conducting the study in other settings may show different results.

Author Contributions: Conceptualization: F.U.R., B.M.A.-G.; Methodology: F.U.R., B.M.A.-G.; Software: F.U.R., B.M.A.-G.; Validation: F.U.R., B.M.A.-G., M.R.M.F.; Investigation: F.U.R.; Resources: F.U.R., B.M.A.-G., M.R.M.F.; Writing—original draft preparation: F.U.R.; Writing—review and editing: F.U.R.; Visualization: F.U.R., M.R.M.F.; Supervision: F.U.R.; Project Administration: F.U.R.; Funding Acquisition: F.U.R., B.M.A.-G., M.R.M.F. All authors have read and agreed to the published version of the manuscript.

Funding: This research was funded by King Fahd University of Petroleum & Minerals, Dhahran 31261, Saudi Arabia.

Data Availability Statement: All the information available for the readers.

Acknowledgments: This publication is based upon work supported by King Fahd University of Petroleum & Minerals and the authors at KFUPM acknowledge the support.

Conflicts of Interest: The authors declare no conflict of interest.

Appendix A

Circular Economy Innovation

1. In our company we have replaced new recyclable system for raw materials with renewable, recyclable, or biodegradable inputs
2. In our company we have launched new processes to decrease the usage of non-recyclable raw materials in our processes
3. In our company we have launched new environmentally friendly packaging system
4. In our company we have launched new system to use the leftover raw material again in our processes
5. In our company we are using recycled materials as an input in our new processes
6. In our company we have launched new initiatives to collect leftover products from customers to recycle them
7. In our company we have introduced new alternative ways to use our products once they have served their initial purpose
8. In our company we have found new revenue streams for products/services after they have served their initial purpose

Business Model Innovation

1. How much of revenue from noncore products are generated through traditional revenue sources such as circulation, display advertising, and classified operations?
   Almost all 1 2 3 4 5 Almost none
2. How do you sell your noncore products?
   Existing salesforce sells both core and noncore products.
   1 2 3 4 5
   Noncore products are exclusively sold through digital media salesforce.
3. How many new formal or informal arrangements for information exchange with your partners have been created in the past 3 years?
   No new arrangements 1 2 3 4 5 Very many new arrangements
4. In the last 3 years, have you changed your pricing structure for raw material, product, and services?
   We have made no changes to our pricing structure.
   1 2 3 4 5
   We have completely changed our pricing structure.
5. Please compare the value propositions offered by your products/services now with those offered 3 years ago.
   They are pretty much the same 1 2 3 4 5 They are dramatically different
6. Please compare the cost structure of means employed to produce the noncore products with that employed to produce the core products.
   Cost structure for noncore product is much higher
   1 2 3 4 5
   Cost structure for noncore products is much lower

Government Incentives
Government Financial Support

1. We can easily access sufficient equity fundings provided by the government for SDGs
2. In our country, there are sufficient government financial subsidies available for SMEs, and we have easy access to it
3. We can easily access interest-free, and a low level of interest charged debt/loan fundings
4. We can easily access government short term and long-term financial services

Government Nonfinancial Support

5. Our government supports SMEs in building science parks and SDGs incubators in poor communities
6. We access a wide range of assistance provided by the government for SDGs activities and SDGs projects
7. Our government encourages SMEs to help in sustainable development by improving the corporate social responsibilities

Sustainable Development
Economics Performance

1. We have achieved return on asset
2. We have achieved return on equity
3. We have achieved return on investment
4. We have improved our profitability
5. We have improved the production cost
6. We have improved the sales growth
7. We have improved in work productivity

Environmental Performance

1. We raise awareness and/or train of the employees in water and/or energy conservation
2. We give priority to reusable, used or recycled materials.

3. We have established metrics that monitor (e.g., regarding risks, levels of pollution, of energy consumption, waste, etc.) the environmental initiatives
4. We consult stakeholders (e.g., employees, suppliers, clients, creditors, etc.) for environment-related decisions
5. We integrate environmental considerations in the conception and development of products and services in all phases of their life cycle (eco-conception and the analysis of the life cycle)
6. We integrate environmental considerations in purchase decisions and the evaluation of suppliers
7. We give priority to more water and energy-efficient equipment
8. We separate garbage and waste (recycling of materials: paper, plastic, glass and metal)
9. We encourage and support employees to use alternatives means of transportation to commute instead of single-occupancy cars (e.g., rideshare, public transport, bicycle, etc.)
10. We give priority to less polluting vehicles and modes of transportation and/or optimize distribution network
11. We communicate actions to internal stakeholders (e.g., meetings with staff, intranet, reports, etc.)
12. We communicate actions to external stakeholders (e.g., website, reports, etc.)

Social Performance

1. We have established metrics that monitor (e.g., amounts spent, allocated time, types of beneficiaries, etc.) to benefit the communities
2. We favor local suppliers in the regions
3. We favor job creation in the regions
4. We offer internships and contribute to student training in different communities
5. We contribute to community cultural, sporting or teaching activities (public organizations or associations with social, cultural, sporting or teaching activities)
6. We consult other stakeholders (employees, suppliers, clients, creditors, associations, firms, etc.) for decisions concerning local development
7. We communicate actions among internal stakeholders (e.g., meetings with staff, intranet, reports, etc.)
8. We communicate actions to external stakeholders (e.g., website, reports, etc.)

Pakistan Sample

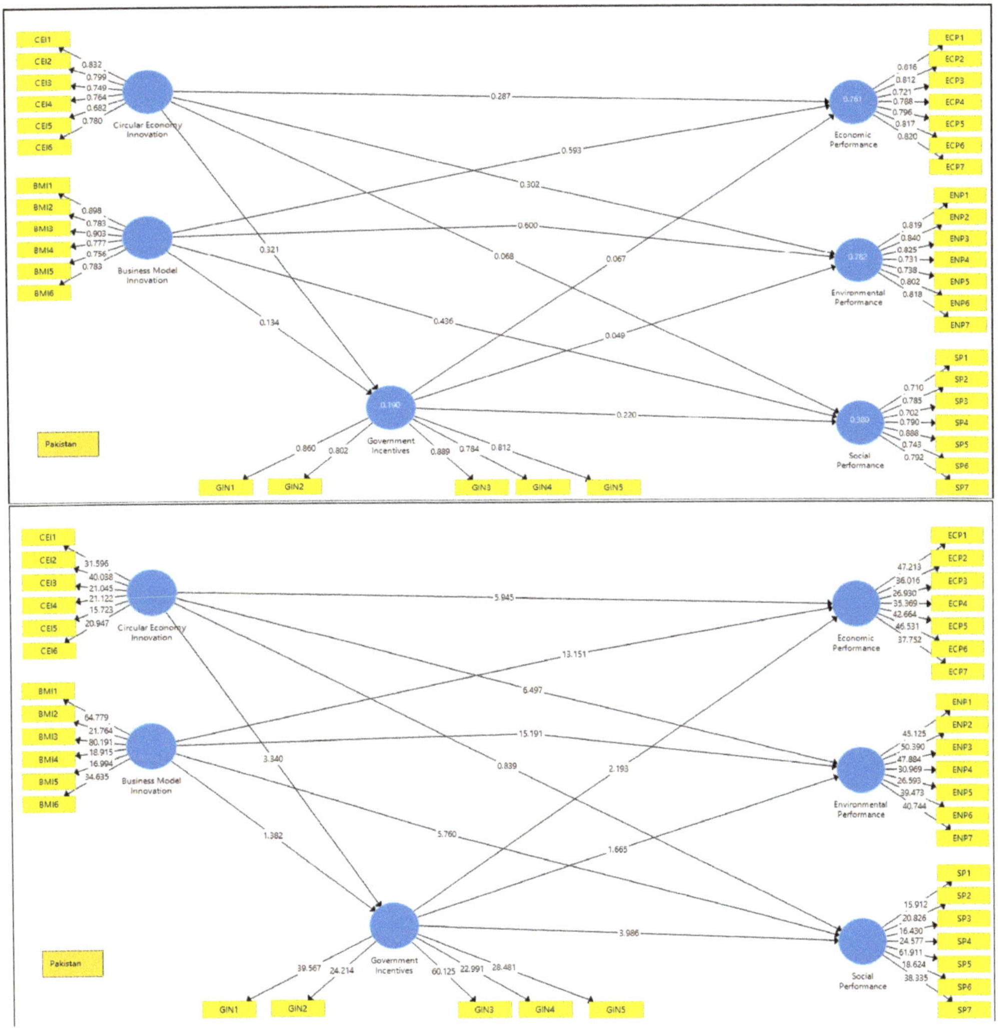

Outer Weight (Pakistan Sample)

	Business Model Innovation	Circular Economy Innovation	Economic Performance	Environmental Performance	Government Incentives	Social Performance
BMI1	0.219					
BMI2	0.19					
BMI3	0.219					
BMI4	0.183					
BMI5	0.177					
BMI6	0.232					
CEI1		0.243				
CEI2		0.227				

	Business Model Innovation	Circular Economy Innovation	Economic Performance	Environmental Performance	Government Incentives	Social Performance
CEI3		0.196				
CEI4		0.223				
CEI5		0.185				
CEI6		0.221				
ECP1			0.178			
ECP2			0.174			
ECP3			0.179			
ECP4			0.187			
ECP5			0.193			
ECP6			0.177			
ECP7			0.169			
ENP1				0.178		
ENP2				0.182		
ENP3				0.196		
ENP4				0.173		
ENP5				0.169		
ENP6				0.178		
ENP7				0.177		
GIN1					0.262	
GIN2					0.235	
GIN3					0.265	
GIN4					0.211	
GIN5					0.229	
SP1						0.177
SP2						0.177
SP3						0.149
SP4						0.19
SP5						0.199
SP6						0.168
SP7						0.225

Q-Square (Pakistan Sample)

	SSO	SSE	Q^2 (=1 − SSE/SSO)
Business Model Innovation	1824	1824	
Circular Economy Innovation	1824	1824	
Economic Performance	2128	1120.829	0.473
Environmental Performance	2128	1084.452	0.49
Government Incentives	1520	1325.282	0.128
Social Performance	2128	1660.263	0.22

Malaysia Sample

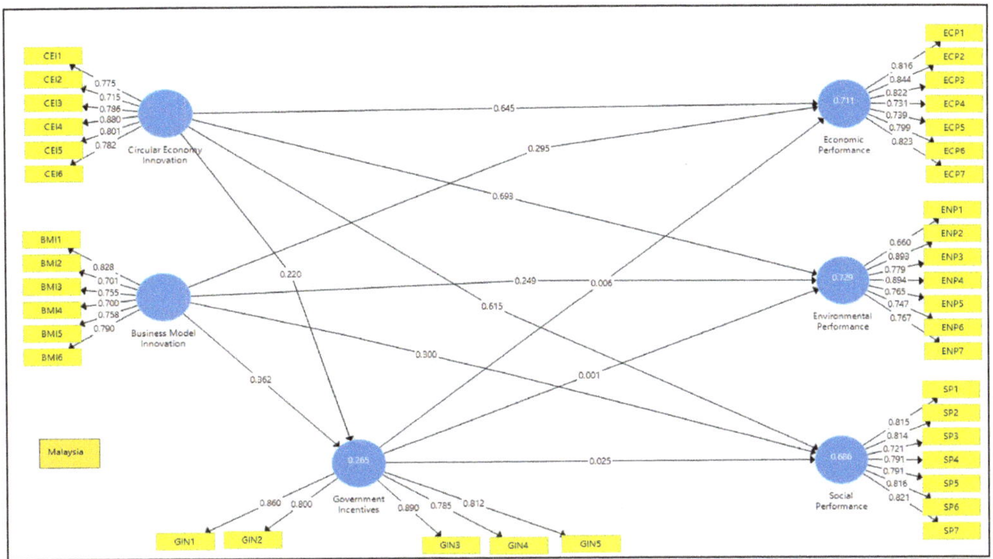

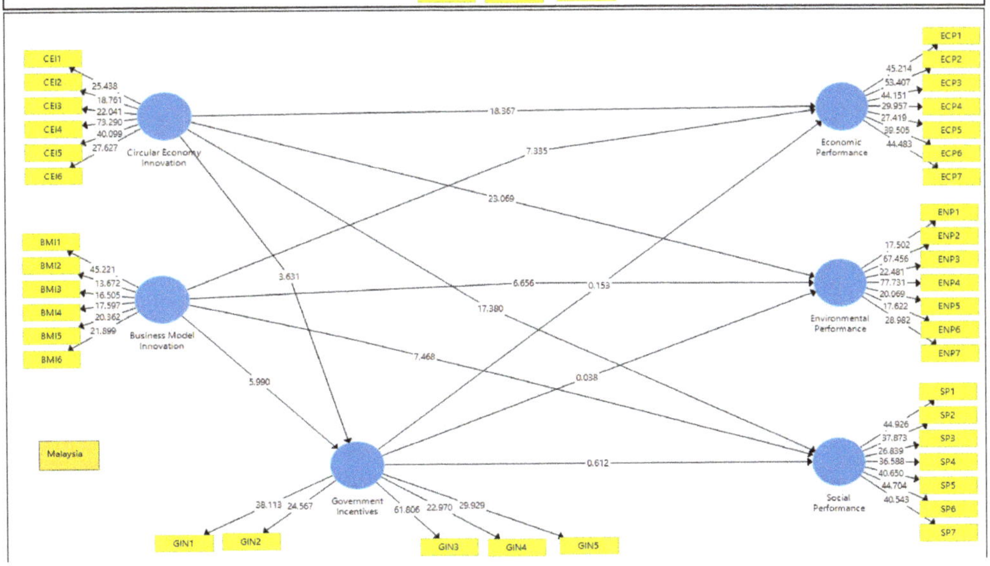

Outer Weight (Malaysia Sample)

	Business Model Innovation	Circular Economy Innovation	Economic Performance	Environmental Performance	Government Incentives	Social Performance
BMI1	0.258					
BMI2	0.179					
BMI3	0.209					
BMI4	0.25					
BMI5	0.208					
BMI6	0.216					
CEI1		0.217				
CEI2		0.175				
CEI3		0.211				
CEI4		0.254				
CEI5		0.205				
CEI6		0.196				
ECP1			0.175			
ECP2			0.189			
ECP3			0.187			
ECP4			0.171			
ECP5			0.171			
ECP6			0.176			
ECP7			0.184			
ENP1				0.179		
ENP2				0.206		
ENP3				0.168		
ENP4				0.202		
ENP5				0.16		
ENP6				0.161		
ENP7				0.188		
GIN1					0.262	
GIN2					0.231	
GIN3					0.267	
GIN4					0.211	
GIN5					0.229	
SP1						0.176
SP2						0.177
SP3						0.18
SP4						0.194
SP5						0.182
SP6						0.177
SP7						0.171

Q-Square (Malaysia Sample)

	SSO	SSE	Q^2 (=1 − SSE/SSO)
Business Model Innovation	1824	1824	
Circular Economy Innovation	1824	1824	
Economic Performance	2128	1178.145	0.446
Environmental Performance	2128	1181.213	0.445
Government Incentives	1520	1248.516	0.179
Social Performance	2128	1219.677	0.427

China Sample

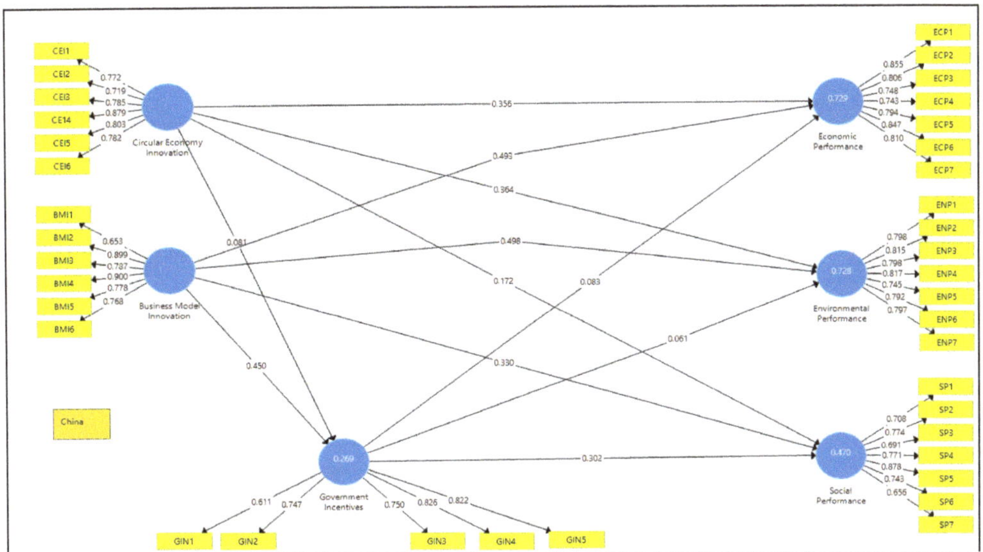

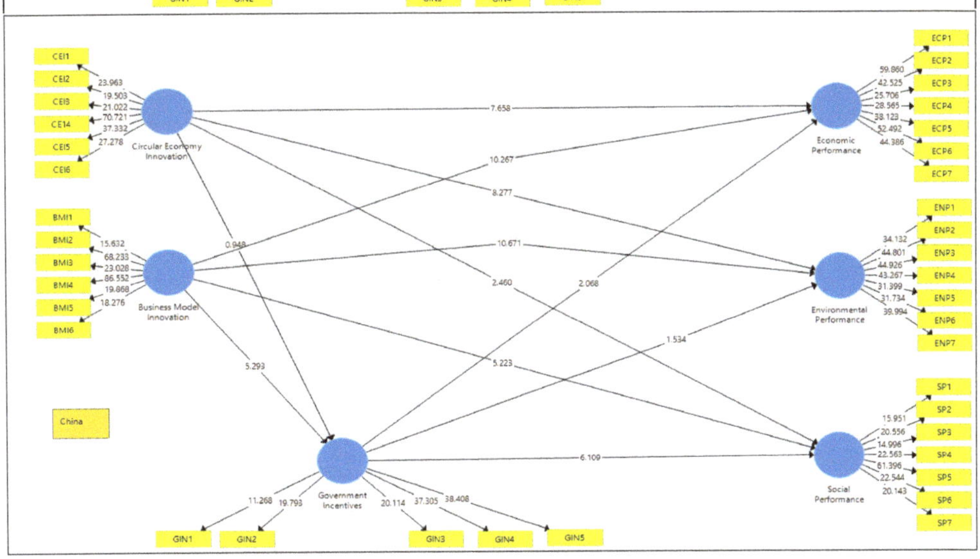

Outer Weight (China Sample)

	Business Model Innovation	Circular Economy Innovation	Economic Performance	Environmental Performance	Government Incentives	Social Performance
BMI1	0.182					
BMI2	0.236					
BMI3	0.202					
BMI4	0.234					
BMI5	0.197					

	Business Model Innovation	Circular Economy Innovation	Economic Performance	Environmental Performance	Government Incentives	Social Performance
BMI6	0.191					
CE14		0.253				
CEI1		0.21				
CEI2		0.18				
CEI3		0.21				
CEI5		0.209				
CEI6		0.197				
ECP1			0.183			
ECP2			0.178			
ECP3			0.17			
ECP4			0.178			
ECP5			0.187			
ECP6			0.177			
ECP7			0.176			
ENP1				0.167		
ENP2				0.184		
ENP3				0.19		
ENP4				0.187		
ENP5				0.177		
ENP6				0.171		
ENP7				0.183		
GIN1					0.177	
GIN2					0.224	
GIN3					0.251	
GIN4					0.287	
GIN5					0.363	
SP1						0.171
SP2						0.184
SP3						0.151
SP4						0.195
SP5						0.214
SP6						0.171
SP7						0.253

Q-Square (China Sample)

	SSO	SSE	Q^2 (=1 − SSE/SSO)
Business Model Innovation	1824	1824	
Circular Economy Innovation	1824	1824	
Economic Performance	2128	1145.61	0.462
Environmental Performance	2128	1163.374	0.453
Government Incentives	1520	1304.65	0.142
Social Performance	2128	1610.65	0.243

References

1. Mwangi, G.M.; Despoudi, S.; Espindola, O.R.; Spanaki, K.; Papadopoulos, T. A planetary boundaries perspective on the sustainability: Resilience relationship in the Kenyan tea supply chain. *Ann. Oper. Res.* **2021**, 1–35. [CrossRef]
2. Lii, P.; Kuo, F.I. Innovation-oriented supply chain integration for combined competitiveness and firm performance. *Int. J. Prod. Econ.* **2016**, *174*, 142–155. [CrossRef]
3. Rodríguez-Espíndola, O.; Cuevas-Romo, A.; Chowdhury, S.; Díaz-Acevedo, N.; Albores, P.; Despoudi, S.; Dey, P. The role of circular economy principles and sustainable-oriented innovation to enhance social, economic and environmental performance: Evidence from Mexican SMEs. *Int. J. Prod. Econ.* **2022**, *248*, 108495. [CrossRef]
4. Dey, P.K.; Malesios, C.; De, D.; Chowdhury, S.; Abdelaziz, F.B. The impact of lean management practices and sustainably oriented innovation on sustainability performance of small and medium-sized enterprises: Empirical evidence from the UK. *Br. J. Manag.* **2020**, *31*, 141–161. [CrossRef]
5. Korhonen, J.; Honkasalo, A.; Seppälä, J. Circular economy: The concept and its limitations. *Ecol. Econ.* **2018**, *143*, 37–46. [CrossRef]
6. Klewitz, J.; Hansen, E.G. Sustainability-oriented innovation of SMEs: A systematic review. *J. Clean. Prod.* **2014**, *65*, 57–75. [CrossRef]
7. Abad-Segura, E.; González-Zamar, M.D.; Infante-Moro, J.C.; Ruipérez García, G. Sustainable management of digital transformation in higher education: Global research trends. *Sustainability* **2020**, *12*, 2107. [CrossRef]
8. Cracolici, M.F.; Cuffaro, M.; Lacagnina, V. Assessment of sustainable well-being in the Italian regions: An activity analysis model. *Ecol. Econ.* **2018**, *143*, 105–110. [CrossRef]
9. Sehnem, S.; Jabbour, C.J.C.; Pereira, S.C.F.; de Sousa Jabbour, A.B.L. Improving sustainable supply chains performance through operational excellence: Circular economy approach. *Resour. Conserv. Recycl.* **2019**, *149*, 236–248. [CrossRef]
10. Ntsonde, J.; Aggeri, F. Stimulating innovation and creating new markets–the potential of circular public procurement. *J. Clean. Prod.* **2021**, *308*, 127303. [CrossRef]
11. Cristoni, N.; Tonelli, M. *Strategic Management and the Circular Economy*; Routledge: London, UK, 2018.
12. De los Ríos, C.; Charnley, F.J.S. Skills and capabilities for a sustainable and circular economy: The changing role of design. *J. Clean. Prod.* **2017**, *160*, 109–122. [CrossRef]
13. Rattalino, F. Circular advantage anyone? Sustainability-driven innovation and circularity at Patagonia, Inc. *Thunderbird Int. Bus. Rev.* **2018**, *60*, 747–755. [CrossRef]
14. Lüdeke-Freund, F.; Gold, S.; Bocken, N.M. A review and typology of circular economy business model patterns. *J. Ind. Ecol.* **2019**, *23*, 36–61. [CrossRef]
15. De PáduaPieroni, M.; McAloone, T.; Pigosso, D. Business model innovation for circular economy: Integrating literature and practice into a conceptual process model. In Proceedings of the Design Society: International Conference on Engineering Design; Cambridge University Press: Cambridge, UK, 2019; Volume 1, pp. 2517–2526.
16. Geissdoerfer, M.; Savaget, P.; Bocken, N.M.; Hultink, E.J. The Circular Economy—A new sustainability paradigm? *J. Clean. Prod.* **2017**, *143*, 757–768. [CrossRef]
17. Snihur, Y.; Bocken, N. A call for action: The impact of business model innovation on business ecosystems, society and planet. *Long Range Plan.* **2022**, *55*, 102182. [CrossRef]
18. Brown, P.; Bocken, N.; Balkenende, R. Why do companies pursue collaborative circular oriented innovation? *Sustainability* **2019**, *11*, 635. [CrossRef]
19. Aghelie, A. Exploring drivers and barriers to sustainability green business practices within small medium sized enterprises: Primary findings. *Int. J. Bus. Econ. Dev. (IJBED)* **2017**, *5*. Available online: https://ijbed.org/cdn/article_file/content_26562_17-03-27-10-42-33.pdf (accessed on 14 September 2022).
20. Ghisellini, P.; Cialani, C.; Ulgiati, S. A review on circular economy: The expected transition to a balanced interplay of environmental and economic systems. *J. Clean. Prod.* **2016**, *114*, 11–32. [CrossRef]
21. Ormazabal, M.; Prieto-Sandoval, V.; Puga-Leal, R.; Jaca, C. Circular economy in Spanish SMEs: Challenges and opportunities. *J. Clean. Prod.* **2018**, *185*, 157–167. [CrossRef]
22. Awan, U.; Sroufe, R.; Shahbaz, M. Industry 4.0 and the circular economy: A literature review and recommendations for future research. *Bus. Strat. Environ.* **2021**, *30*, 2038–2060. [CrossRef]
23. Groening, C.; Sarkis, J.; Zhu, Q. Green marketing consumer-level theory review: A compendium of applied theories and further research directions. *J. Clean. Prod.* **2018**, *172*, 1848–1866. [CrossRef]
24. Hudecheck, M.; Sire´n, C.; Grichnik, D.; Wincent, J. How Companies Can Respond to the Coronavirus'. MIT Sloan Management Review. Available online: https://sloanreview.mit.edu/article/how-companies-can-respondto-the-coronavirus/ (accessed on 31 May 2020).
25. Ranjbari, M.; Saidani, M.; Esfandabadi, Z.S.; Peng, W.; Lam, S.S.; Aghbashlo, M.; Tabatabaei, M. Two decades of research on waste management in the circular economy: Insights from bibliometric, text mining, and content analyses. *J. Clean. Prod.* **2021**, *314*, 128009. [CrossRef]
26. Wang, N.; Lee, J.C.K.; Zhang, J.; Chen, H.; Li, H. Evaluation of Urban circular economy development: An empirical research of 40 cities in China. *J. Clean. Prod.* **2018**, *180*, 876–887. [CrossRef]
27. White, K.; Habib, R.; Hardisty, D.J. How to SHIFT consumer behaviors to be more sustainable: A literature review and guiding framework. *J. Mark.* **2019**, *83*, 22–49. [CrossRef]

28. Han, D.; Konietzko, J.; Dijk, M.; Bocken, N. How do companies launch circular service business models in different countries? *Sustain. Prod. Consum.* **2022**, *31*, 591–602. [CrossRef]
29. Cronin, J.J.; Smith, J.S.; Gleim, M.R.; Ramirez, E.; Martinez, J.D. Green marketing strategies: An examination of stakeholders and the opportunities they present. *J. Acad. Mark. Sci.* **2011**, *39*, 158–174. [CrossRef]
30. Dangelico, R.M.; Vocalelli, D. Green Marketing: An analysis of definitions, strategy steps, and tools through a systematic review of the literature. *J. Clean. Prod.* **2017**, *165*, 1263–1279. [CrossRef]
31. Papadas, K.K.; Avlonitis, G.J.; Carrigan, M.; Piha, L. The interplay of strategic and internal green marketing orientation on competitive advantage. *J. Bus. Res.* **2019**, *104*, 632–643. [CrossRef]
32. Moraga, G.; Huysveld, S.; De Meester, S.; Dewulf, J. Development of circularity indicators based on the in-use occupation of materials. *J. Clean. Prod.* **2021**, *279*, 123889. [CrossRef]
33. Izzo, M.F.; Ciaburri, M.; Tiscini, R. The challenge of sustainable development goal reporting: The first evidence from Italian listed companies. *Sustainability* **2020**, *12*, 3494. [CrossRef]
34. Williams, A.; Whiteman, G.; Parker, J.N. Backstage interorganizational collaboration: Corporate endorsement of the sustainable development goals. *Acad. Manag. Discov.* **2019**, *5*, 367–395. [CrossRef]
35. Philip, J.; Graham, W. *A Framework for Codifying Business Models and Process Models in E-Business Design*; Butterworth-Heinemann: Oxford, UK, 2004; pp. 35–64.
36. Barney, J. Firm resources and sustained competitive advantage. *J. Manag.* **1991**, *17*, 99–120. [CrossRef]
37. Rodríguez-Antón, J.M.; Rubio-Andrada, L.; Celemín-Pedroche, M.S.; Ruíz-Peñalver, S.M. From the circular economy to the sustainable development goals in the European Union: An empirical comparison. *Int. Environ. Agreem. Politics Law Econ.* **2022**, *22*, 67–95. [CrossRef]
38. Rodríguez-Antón, J.M. La Economía circular comomodeloimpulsor de losObjetivos de Desarrollo Sostenible. In *Agenda 2030: Claves para la Transformaciónsostenible*; Alfaro, M., Arias, S., Gamba, A., Eds.; Los Libros de la Catarata: Madrid, Spain, 2019; pp. 296–311.
39. Rodríguez-Antón, J.M.; Alonso-Almeida, M.M. La estrategiaespañola de Economía circular y suadecuación al plan de acción de la Unión Europea para la Economía circular. In Proceedings of the 25th APDR Congress, Proceedings, Lisboa, Portugal, 5–6 July 2018; pp. 183–190.
40. Kirchherr, J.; Reike, D.; Hekkert, M. Conceptualizing the circular economy: An analysis of 114 definitions. *Resour. Conserv. Recycl.* **2017**, *127*, 221–232. [CrossRef]
41. Valverde, J.M.; Avilés-Palacios, C. Circular economy as a catalyst for progress towards the sustainable development goals: A positive relationship between two self-sufficient variables. *Sustainability* **2021**, *13*, 12652. [CrossRef]
42. DeCajas, A.U.; Seguros, B.Y. *La Economía Circular Enel Sector Agroalimentario*; Adicae: Brussels, Belgium, 2013.
43. Lewandowski, M. Designing the Business Models for Circular Economy—Towards the Conceptual Framework. *Sustainability* **2016**, *8*, 43. [CrossRef]
44. Khajuria, A.; Atienza, V.A.; Chavanich, S.; Henning, W.; Islam, I.; Kral, U.; Li, J. Accelerating circular economy solutions to achieve the 2030 agenda for sustainable development goals. *Circ. Econ.* **2022**, *1*, 100001. [CrossRef]
45. Khajuria, A. Integrated approach between DPSIR, planetary boundaries and sustainable development goals~towards 3Rs and resource efficiency. *World Environ.* **2020**, *10*, 52–56.
46. Suárez-Eiroa, B.; Fernández, E.; Méndez-Martínez, G.; Soto-Oñate, D. Operational principles of circular economy for sustainable development: Linking theory and practice. *J. Clean. Prod.* **2019**, *214*, 952–961. [CrossRef]
47. Mai, Y.; Yang, H.; Zhang, G. Does business model innovation enhance the sustainable development of new ventures? Understanding an inverted-U relationship. *Sustainability* **2020**, *13*, 262. [CrossRef]
48. Evans, S.; Vladimirova, D.; Holgado, M.; Van Fossen, K.; Yang, M.; Silva, E.A.; Barlow, C.Y. Business model innovation for sustainability: Towards a unified perspective for creation of sustainable business models. *Bus. Strategy Environ.* **2017**, *26*, 597–608. [CrossRef]
49. Madsen, H.L. Business model innovation and the global ecosystem for sustainable development. *J. Clean. Prod.* **2010**, *247*, 119102. [CrossRef]
50. Aryal, K.; Laudari, H.K.; Ojha, H.R. To what extent is Nepal's community forestry contributing to the sustainable development goals? An institutional interaction perspective. *Int. J. Sustain. Dev. World Ecol.* **2020**, *27*, 28–39. [CrossRef]
51. Boţa-Avram, C.; Groşanu, A.; Răchişan, P.R.; Gavriletea, M. The bidirectional causality between country-level governance, economic growth and sustainable development: A cross-country data analysis. *Sustainability* **2018**, *10*, 502. [CrossRef]
52. Aceleanu, M.I.; Şerban, A.C.; Ţîrcă, D.M.; Badea, L. The rural sustainable development through renewable energy. The case of Romania. *Technol. Econ. Dev. Econ.* **2018**, *24*, 1408–1434. [CrossRef]
53. Anwar, M.; Khattak, M.S.; Popp, J.; Meyer, D.F.; Máté, D. The nexus of government incentives and sustainable development goals: Is the management of resources the solution to non-profit organisations? *Technol. Econ. Dev. Econ.* **2020**, *26*, 1284–1310. [CrossRef]
54. Rodríguez Bolívar, M.P.; Navarro Galera, A.; Alcaide Muñoz, L.; López Subirés, M.D. Risk factors and drivers of financial sustainability in local government: An empirical study. *Local Gov. Stud.* **2016**, *42*, 29–51. [CrossRef]
55. Ayuso, S.; Navarrete-Báez, F.E. How does entrepreneurial and international orientation influence SMEs' commitment to sustainable development? Empirical evidence from Spain and Mexico. *Corp. Soc. Responsib. Environ. Manag.* **2018**, *25*, 80–94. [CrossRef]

56. Bimonte, G.; Romano, M.G.; Russolillo, M. Green innovation and competition: R&D incentives in a circular economy. *Games* **2021**, *12*, 68.
57. Katrakis, E.; Nacci, G.; Couder, N. *Incentives to Boost the Circular Economy: A Guide for Public Authorities*; European Commission: Minato City, Tokyo, 2021.
58. Scotchmer, S. *Innovation and Incentives*; MIT Press: Cambridge, MA, USA, 2004.
59. Zhang, J.; Guan, J. The time-varying impacts of government incentives on innovation. *Technol. Forecast. Soc. Change* **2018**, *135*, 132–144. [CrossRef]
60. Khattak, M.S. Does access to domestic finance and international finance contribute to sustainable development goals? Implications for policymakers. *J. Public Aff.* **2020**, *20*, e2024. [CrossRef]
61. Van Leeuwen, G.; Mohnen, P. Revisiting the Porter hypothesis: An empirical analysis of green innovation for The Netherlands. *Econ. Innov. New Technol.* **2017**, *26*, 63–77. [CrossRef]
62. Li, B.; Chen, W.; Xu, C.; Hou, P. Impacts of government subsidies for environmental-friendly products in a dual-channel supply chain. *J. Clean. Prod.* **2018**, *171*, 1558–1576. [CrossRef]
63. Owen, R.; Brennan, G.; Lyon, F. Enabling investment for the transition to a low carbon economy: Government policy to finance early stage green innovation. *Curr. Opin. Environ. Sustain.* **2018**, *31*, 137–145. [CrossRef]
64. Ullah, R.; Ahmad, H.; Rehman, F.U.; Fawad, A. Green innovation and Sustainable Development Goals in SMEs: The moderating role of government incentives. *J. Econ. Adm. Sci.* 2021, *in press*. [CrossRef]
65. Monasterolo, I.; Raberto, M. The EIRIN flow-of-funds behavioural model of green fiscal policies and green sovereign bonds. *Ecol. Econ.* **2018**, *144*, 228–243. [CrossRef]
66. Gerlach, H.; Zheng, X. Preferences for green electricity, investment and regulatory incentives. *Energy Econ.* **2018**, *69*, 430–441. [CrossRef]
67. Albort-Morant, G.; Leal-Millán, A.; Cepeda-Carrión, G. The antecedents of green innovation performance: A model of learning and capabilities. *J. Bus. Res.* **2016**, *69*, 4912–4917. [CrossRef]
68. Hair, J.F.; Hult, G.M.; Ringle, C.; Sarstedt, M. *A Primer on Partial Least Squares Structural Equation Modeling (PLS-SEM)*; Sage Publications: Thousand Oaks, CA, USA, 2014.
69. Anwar, M. Business model innovation and SMEs performance—Does competitive advantage mediate? *Int. J. Innov. Manag.* **2018**, *22*, 1850057. [CrossRef]
70. Rehman, A.U.; Anwar, M. Mediating role of enterprise risk management practices between business strategy and SME performance. *Small Enterp. Res.* **2019**, *26*, 207–227. [CrossRef]
71. Anwar, M.; Khan, S.Z.; Shah, S.Z.A. Big data capabilities and firm's performance: A mediating role of competitive advantage. *J. Inf. Knowl. Manag.* **2018**, *17*, 1850045. [CrossRef]
72. Rehman, F.U.; Yusoff, R.B.M.; Zabri, S.B.M.; Ismail, F.B. Determinants of personal factors in influencing the buying behavior of consumers in sales promotion: A case of fashion industry. *Young Consum.* **2017**, *18*, 408–424. [CrossRef]
73. Kabra, A.G.; Akhtar, R.P.; Dash, M.K. Understanding behavioural intention to use information technology: Insights from humanitarian practitioners. *Telemat. Inform.* **2017**, *34*, 1250–1261. [CrossRef]
74. Rehman, F.U.; Al-Ghazali, B.M. Evaluating the Influence of Social Advertising, Individual Factors, and Brand Image on the Buying Behavior toward Fashion Clothing Brands. *SAGE Open* **2022**, *12*, 21582440221088858. [CrossRef]
75. Ringle, C.M.; Wende, S.; Becker, J.M. Smart PLS 3. Smart PLS. 2015. Available online: http://www.smartpls.com (accessed on 14 September 2022).
76. Fornell, C.G.; Larcker, D.F. Evaluating structural equation models with unobservable variables and measurement error. *J. Mark. Res.* **1981**, *18*, 39–50. [CrossRef]
77. Khan, P.A.; Johl, S.K.; Akhtar, S. Vinculum of Sustainable Development Goal Practices and Firms' Financial Performance: A Moderation Role of Green Innovation. *J. Risk Financ. Manag.* **2022**, *15*, 96. [CrossRef]
78. Korsakienė, R.; Raišienė, A.G. Sustainability drivers of small and medium sized firms: A review and research agenda. *Sci. Pap. Univ. Pardubice. Ser. D Fac. Econ. Adm.* **2022**, *30*, 1–12. [CrossRef]
79. Udeh, F.N.; Akporien, F.O. Triple bottom line reporting practices and corporate relationship with host communities in Nigeria (a case study of Niger Delta region). *Sci. Pap. Univ. Pardubice. Ser. D Fac. Econ. Adm.* **2016**, *36*, 207–217.

MDPI
St. Alban-Anlage 66
4052 Basel
Switzerland
Tel. +41 61 683 77 34
Fax +41 61 302 89 18
www.mdpi.com

Sustainability Editorial Office
E-mail: sustainability@mdpi.com
www.mdpi.com/journal/sustainability

www.ingramcontent.com/pod-product-compliance
Lightning Source LLC
Chambersburg PA
CBHW061608220926
51850CB00005B/451
* 9 7 8 3 0 3 6 5 8 5 4 3 7 *